Neutrophil Methods and Protocols

METHODS IN MOLECULAR BIOLOGY™

John M. Walker, SERIES EDITOR

METHODS IN MOLECULAR BIOLOGY™

Neutrophil Methods and Protocols

Edited by

Mark T. Quinn

Deparment of Veterinary Molecular Biology
Montana State University, Bozeman, MT

Frank R. DeLeo

Laboratory of Human Bacterial Pathogenesis, Rocky Mountain
Laboratories, National Institute of Allergy and Infectious Diseases,
National Institutes of Health
Hamilton, MT

Gary M. Bokoch

Departments of Immunology and Cell Biology,
The Scripps Institute, La Jolla, CA

HUMANA PRESS ✳ TOTOWA, NEW JERSEY

Library of Congress Cataloging in Publication Data

Neutrophil methods and protocols / edited by Mark T. Quinn, Frank R. DeLeo, Gary M. Bokoch.
    p. ; cm. — (Methods in molecular biology ; 138)
 Includes bibliographical references and index.
 ISBN 978-1-58829-788-4 (alk. paper)
 1. Neutrophils—Laboratory manuals. I. Quinn, Mark T. II. Deleo, Frank. III. Bokoch, Gary M. IV. Series: Methods in molecular biology (Clifton, N.J.) ; 138.
 [DNLM: 1. Neutrophils—Laboratory Manuals. W1 ME9616J v.138 2007 / WH 25 N497 2007]
 QP95.8.N48 2007
 616.07'9—dc22

                                                                2006031694

Dedication

This volume is dedicated to Dr. John Badwey (1950–2004) and Dr. Bernard M. Babior (1935–2004) in recognition of their fundamental contributions to neutrophil biology. This volume is also dedicated to our families, who are exceedingly patient with all of the time we spend studying neutrophils.

Preface

Neutrophils (also known as polymorphonuclear leukocytes [PMNs] or granulocytes) are the most abundant white cell in humans. Granulocytes and/or granulocyte precursors normally comprise approx 60% of the nucleated cells in bone marrow and the bloodstream. Mature neutrophils have a typical circulating half-life of 6–8 h in the blood and then migrate through tissues for approx 2-3 d. Their relatively short life-span is devoted largely to surveillance for invading microorganisms. During infection, the neutrophil life-span is extended, granulopoiesis increases, and large numbers of neutrophils are rapidly recruited to the site(s) of infection. Following recognition (binding) and phagocytosis of microorganisms, neutrophils utilize an extraordinary array of oxygen-dependent and oxygen-independent microbicidal weapons to destroy infectious agents. Oxygen-dependent mechanisms involve the production of reactive oxygen species (ROS), while oxygen-independent mechanisms include degranulation and release of lytic enzymes and bactericidal peptides. Inasmuch as these processes are highly effective at killing most ingested microbes, neutrophils serve as the primary cellular defense against infection.

The aim of *Neutrophil Methods and Protocols* is to provide (1) a set of protocols to assess most basic neutrophil functions, (2) protocols for investigating specialized areas in neutrophil research, and (3) step-by-step diagnostic assays of common neutrophil disorders. A wide variety of methods have been developed to assess neutrophil function, and these methods have been instrumental in advancing our understanding of the role of neutrophils in host defense and inflammatory disease. For those researchers and clinicians interested in the study of neutrophils, the availability of a comprehensive source of protocols describing the most modern methodological advances in neutrophil biology is invaluable, as many publications do not provide information on the finer details critical to success of a given method. As such, we have compiled a series of protocols written by leading researchers in the field that provide detailed guidelines for establishing and performing the most common neutrophil function assays. Hints of the best way to perform these methods, as well as guidance in detecting associated problems are included so novice investigators will also be able to effectively utilize these assays. While the volume provides current protocols for evaluation of most basic neutrophil functions and certain specialized functions, a section is dedicated to diagnostic assays for common neutrophil disorders. Thus, this volume is designed for the basic researcher involved in the study of neutrophil function and clinical investigators interested in medical aspects of neutrophil function in health and disease.

Part I is an overview of neutrophils and their role in host defense and inflammation. Part II describes the most commonly used methods to isolate neutrophils from humans and other animal species and procedures for subcellular fractionation of human neutrophils. Part III encompasses protocols addressing neutrophil biochemistry, electrophysiology, signal transduction, and apoptosis. Part IV details methods for investigating adhesion and chemotaxis, and is followed by protocols to assess phagocytosis and bactericidal activity (Part V). Part VI provides an extensive set of assays for evaluating NADPH oxidase assembly and activation, production of reactive oxygen species, and methods for purification of flavocytochrome b. Part VII includes protocols to measure gene expression in neutrophils, and Part VIII provides assays for diagnosis of the most common neutrophil disorders. In addition to the step-by-step protocols, the Notes section of each chapter provides an outstanding repository of useful and interesting information not typically published in the Methods sections of standard journal articles.

We thank John M. Walker, Series Editor, and Humana Press, for the opportunity to assemble an outstanding collection of articles, and for help with the publication of the volume. We also thank the Montana State University COBRE Center for Immunotherapies to Zoonotic Diseases (NIH P20 RR020185) and the Intramural Research Program of the NIH, National Institute of Allergy and Infectious Diseases for sponsoring this volume. Finally, we thank the authors for taking time to write outstanding chapters.

Mark T. Quinn
Frank R. DeLeo
Gary M. Bokoch

Contents

PART I. NEUTROPHILS: AN OVERVIEW

PART II. NEUTROPHIL ISOLATION AND SUBCELLULAR FRACTIONATION

**PART III. BIOCHEMISTRY, BIOLOGY AND SIGNAL TRANSDUCTION
        OF NEUTROPHILS**

Contributors

LEE-ANN H. ALLEN • *Inflammation Program and Departments of Medicine and Microbiology, Roy J. and Lucille A. Carver College of Medicine, University of Iowa and the Vetrans Affairs Medical Center, Iowa City, IA*

ALISON BAMBERG • *Department of Immunology, The Scripps Research Institute, La Jolla, CA*

ROBERT BARGATZE • *LigoCyte Pharmaceuticals, Bozeman, MT*

FLAVIA BAZZONI • *Department of Pathology, Division of General Pathology, University of Verona, Verona, Italy*

JAMEL EL-BENNA • *INSERM U773, Centre de Recherche Biomédicale Bichat Beaujon, Faculté de Medecine, Paris, France*

YEVGENY BERDICHEVSKY • *The Julius Friedrich Cohnheim-Minerva Center for Phagocyte Research and the Ela Kodesz Institute of Host Defense against Infectious Diseases, Sackler School of Medicine, Tel Aviv University, Tel Aviv, Israel*

GARY M. BOKOCH • *Departments of Immunology and Cell Biology, The Scripps Research Institute, La Jolla, CA*

NIELS BORREGAARD • *The Granulocyte Research Laboratory, Department of Hematology, Rigshospitalet, Copenhagen University, Copenhagen, Denmark*

ERIC CAMPBELL • *Center for Experimental Therapeutics and Reperfusion Injury, Brigham and Women's Hospital and Harvard Medical School, Boston, MA*

MARCO A. CASSATELLA • *Department of Pathology, Division of General Pathology, University of Verona, Verona, Italy*

SEAN P. COLGAN • *Center for Experimental Therapeutics and Reperfusion Injury, Brigham and Women's Hospital and Harvard Medical School, Boston, MA*

IRIS DAHAN • *The Julius Friedrich Cohnheim-Minerva Center for Phagocyte Research and the Ela Kodesz Institute of Host Defense against Infectious Diseases, Sackler School of Medicine, Tel Aviv University, Tel Aviv, Israel*

CLAES DAHLGREN • *Department of Rheumatology and Inflammation Research, University of Göteborg, Göteborg, Sweden*

PHAM MY-CHAN DANG • *INSERM U773, Centre de Recherche Biomédicale Bichat Beaujon, Faculté de Medecine, Paris, France*

THOMAS E. DECOURSEY • *Department of Molecular Biophysics and Physiology, Rush University Medical Center, Chicago, IL*

FRANK R. DELEO • *Laboratory of Human Bacterial Pathogenesis, Rocky Mountain Laboratories, National Institute of Allergy and Infectious Diseases, National Institutes of Health, Hamilton, MT*

ISABELLE DE MENDEZ • *Pfizer Global Research and Development, Fresnes Laboratories, Fresnes, France*

SHARON DEWITT • *Neutrophil Signalling Group, School of Medicine, Cardiff University, Heath Park, United Kingdom*

MARY C. DINAUER • *Herman B Wells Center for Pediatric Research, Department of Pediatrics (Hematology/Oncology), Microbiology/ Immunology, and Medical and Molecular Genetics, Riley Hospital for Children, Indiana University School of Medicine, Indianapolis, IN*

IAN DRANSFIELD • *MRC/ University of Edinburgh Centre for Inflammation Research, Queen's Medical Research Institute, Edinburgh, United Kingdom*

HOUDA ZGHAL ELLOUMI • *Laboratory of Clinical Infectious Diseases, National Institute of Allergy and Infectious Diseases, National Institutes of Health, Bethesda, MD*

PER FOLLIN • *Department of Infectious Diseases, University Hospital, Linköping, Sweden*

MICHAEL GLOGAUER • *Canadian Institutes of Health Research Group in Matrix Dynamics, Faculty of Dentistry, University of Toronto, Toronto, Ontario, Canada*

JESSIE N. GREEN • *Free Radical Research Group, Department of Pathology, Christchurch School of Medicine, Christchurch, New Zealand*

SERGIO GRINSTEIN • *Cell Biology Program, Hospital for Sick Children, Toronto, Ontario, Canada*

MAURICE B. HALLETT • *Neutrophil Signalling Group, School of Medicine, Cardiff University, Heath Park, United Kingdom*

MARK B. HAMPTON • *Free Radical Research Group, Department of Pathology, Christchurch School of Medicine, Christchurch, New Zealand*

SIMON P. HART • *MRC/ University of Edinburgh Centre for Inflammation Research, Queen's Medical Research Institute, Edinburgh, United Kingdom*

ESTHER J. HILLSON • *Neutrophil Signalling Group, School of Medicine, Cardiff University, Heath Park, United Kingdom*

STEVEN M. HOLLAND • *Laboratory of Clinical Infectious Diseases, National Institute of Allergy and Infectious Diseases, National Institutes of Health, Bethesda, MD*

NEDA HOMAYOUNPOUR • *Laboratory of Host Defenses, National Institute of Allergy and Infectious Diseases, National Institutes of Health, Rockville, MD*

ANNA HUTTENLOCHER • *Departments of Pediatrics and Pharmacology, University of Wisconsin Medical School, Madison, WI*

ALGIRDAS J. JESAITIS • *Department of Microbiology, Montana State University, Bozeman, MT*

MARK A. JUTILA • *Department of Veterinary Molecular Biology, Montana State University, Bozeman, MT*

LILYA N. KIRPOTINA • *Department of Veterinary Molecular Biology, Montana State University, Bozeman, MT*

ULLA G. KNAUS • *Department of Immunology, The Scripps Research Institute, La Jolla, CA*

SCOTT D. KOBAYASHI • *Department of Microbiology, Molecular Biology, and Biochemistry, University of Idaho, Moscow, ID*

MARK C. LAVIGNE • *Cardiovascular and Metabolic Diseases, Wyeth Research, Cambridge, MA*

BENFANG LEI • *Department of Veterinary Molecular Biology, Montana State University, Bozeman, MT*

KRISTEN LEKSTROM • *Laboratory of Host Defenses, National Institute of Allergy and Infectious Diseases, National Institutes of Health, Rockville, MD*

THOMAS L. LETO • *Laboratory of Host Defenses, National Institute of Allergy and Infectious Diseases, National Institutes of Health, Rockville, MD*

GILDA LINTON • *Laboratory of Host Defenses, National Institute of Allergy and Infectious Diseases, National Institutes of Health, Bethesda, MD*

MARY A. LOKUTA • *Department of Pediatrics, University of Wisconsin Medical School, Madison, WI*

PER LÖNNBRO • *Section for Clinical and Experimental Infection Medicine, Department of Clinical Sciences, Biomedical Centre, Lund University, Lund, Sweden*

NANCY A. LOUIS • *Center for Experimental Therapeutics and Reperfusion Injury, Division of Newborn Medicine, Brigham and Women's Hospital and Harvard Medical School, Boston, MA*

HARRY L. MALECH • *Laboratory of Host Defenses, National Institute of Allergy and Infectious Diseases, National Institutes of Health, Bethesda, MD*

PATRICK P. MCDONALD • *Department of Pharmacology, University of Illinois, Chicago, IL*

LINDA C. MCPHAIL • *Department of Biochemistry and Program in Molecular Medicine, Wake Forest University School of Medicine, Winston-Salem, NC*

ARIEL MIZRAHI • *The Julius Friedrich Cohnheim-Minerva Center for Phagocyte Research and the Ela Kodesz Institute of Host Defense against Infectious Diseases, Sackler School of Medicine, Tel Aviv University, Tel Aviv, Israel*

DERI MORGAN • *Department of Molecular Biophysics and Physiology, Rush University Medical Center, Chicago, IL*

WILLIAM M. NAUSEEF • *Inflammation Program and Department of Medicine, Roy J. and Lucille A. Carver College of Medicine, University of Iowa and Veterans Affairs Medical Center, Coralville, Iowa City, IA*

PONTUS NORDENFELT • *Section for Clinical and Experimental Infection Medicine, Department of Clinical Sciences, Biomedical Centre, Lund University, Lund, Sweden*

PAUL A. NUZZI • *Molecular and Cellular Pharmacology, University of Wisconsin Medical School, Madison, WI*

AIYAPPA PALECANDA • *LigoCyte Pharmaceuticals, Bozeman, MT*

EDGAR PICK • *The Julius Friedrich Cohnheim-Minerva Center for Phagocyte Research and the Ela Kodesz Institute of Host Defense against Infectious Diseases, Sackler School of Medicine, Tel Aviv University, Tel Aviv, Israel*

MARK T. QUINN • *Department of Veterinary Molecular Biology, Montana State University, Bozeman, MT*

ADRIANO G. ROSSI • *MRC/ University of Edinburgh Centre for Inflammation Research, Queen's Medical Research Institute, Edinburgh, United Kingdom*

MELISSA R. SARANTOS • *Department of Biomedical Engineering, University of California, Davis, CA*

ULRICH Y. SCHAFF • *Department of Biomedical Engineering, University of California, Davis, CA*

IGOR A. SCHEPETKIN • *Department of Veterinary Molecular Biology, Montana State University, Bozeman, MT*

SUSAN SERGEANT • *Department of Biochemistry, Wake Forest University School of Medicine, Winston-Salem, NC*

DANIEL W. SIEMSEN • *Department of Veterinary Molecular Biology, Montana State University, Bozeman, MT*

SCOTT I. SIMON • *Department of Biomedical Engineering, University of California, Davis, CA*

BENJAMIN E. STEINBERG • *Cell Biology Program, Hospital for Sick Children, Toronto, Ontario, Canada*

DAN E. STURDEVANT • *Research Technologies Branch, Research Technologies Sections, Genomics Unit, Rocky Mountain Laboratories, National Institute of Allergy and Infectious Diseases, National Institutes of Health, Hamilton, MT*

NICOLA TAMASSIA • *Department of Pathology, Division of General Pathology, University of Verona, Verona, Italy*

HANS TAPPER • *Section for Clinical and Experimental Infection Medicine, Department of Clinical Sciences, Biomedical Centre, Lund University, Lund, Sweden*

EMMA L. TAYLOR • *MRC/ University of Edinburgh Centre for Inflammation Research, Queen's Medical Research Institute, Edinburgh, United Kingdom*

ROSS M. TAYLOR • *Department of Microbiology, Montana State University, Bozeman, MT*

HAROLD TING • *Department of Biomedical Engineering, University of California, Davis, CA*

MARTINE TORRES • *Saban Research Institute of Childrens Hospital Los Angeles and Department of Pediatrics, Keck School of Medicine, University of Southern California, Los Angeles, CA*

LENE UDBY • *The Granulocyte Research Laboratory, Department of Hematology, Rigshospitalet, Copenhagen University, Copenhagen, Denmark*

YELENA UGOLEV • *The Julius Friedrich Cohnheim-Minerva Center for Phagocyte Research and the Ela Kodesz Institute of Host Defense against Infectious Diseases, Sackler School of Medicine, Tel Aviv University, Tel Aviv, Israel*

BRUCE WALCHECK • *Department of Veterinary and Biomedical Sciences, University of Minnesota, St. Paul, MN*

CHRISTINE C. WINTERBOURN • *Free Radical Research Group, Department of Pathology, Christchurch School of Medicine, Christchurch, New Zealand*

RICHARD D. YE • *Department of Pharmacology, University of Illinois, Chicago, IL*

TIEMING ZHAO • *Department of Immunology, The Scripps Research Institute, La Jolla, CA*

COLOR PLATE

The following color illustrations are printed in the insert, appearing after page 300.

Chapter 4

Figure 5: Photograph of the three-layer Percoll density gradient before (middle) and after centrifugation (right). A two-layer gradient after centrifugation (left) is shown for comparison. The bands formed after centrifugation are indicated. (Reprinted from **ref. 8,** with permission from Elsevier.)

Chapter 14

Figure 1: Immunofluorescent staining for actin in a primary human neutrophil. Neutrophils were stimulated with 100 nM fMLP for 15 min at 37°C and 10% CO_2. Cells were fixed and stained with rhodamine-conjugated phalloidin and counterstained with DAPI, as described. Actin is shown in **A,** and the overlay is shown in **B.** Images were taken within a 100x oil-immersion objective. Bar, 10 μm.

Figure 2: Immunofluorescent staining for G_M3 in a primary human neutrophils were stimulated with 100 nM fMLP for 15 min at 37°C and 10% CO_2. Cells were fixed and stained with anti-G_M3 antibody, as described. A differential interference contrast image of the cells was also captured to show cell shape and position of immunofluorescent staining. G_M3 is shown in **A,** and the overlay is shown in **B.** Images were taken with a 60x oil-immersion objective. Bar, 10 μm.

Figure 3: Immunofluorescent staining for tubulin in a primary human neutrophil. Neutrophils were treated as described in **Subheading 3.2.** and stimulated with 100 nM fMLP for 15 min at 37°C and 10% CO_2. Cells were then fixed and stained with fluorescein isothiocyanate-conjugated anti-tubulin antibody and counterstained with DAPI as described in **Subheading 3.3.3.** Actin is shown in A and the overlay is shown in B. Images were taken with a 100x oil-immersion objective. Bar, 10 μm.

Chapter 15

Figure 2: Localizing free-actin barbed ends generated after fMLP stimulation in neutrophils. Wild-type and Rac2 null neutrophils were allowed to attach to bovine serum albumin-coated coverslips, and permeabilized as described above. Following the addition of fMLP, rhodamin-actin monomers were added to label the free barbed ends. After fixing the cells flourescently tagged phalloidin, which binds to actin filaments was added. We note there is a dramatic increase in free barbed ends at the leading edge (*) of the activated neutrophils. Also note that in the overlay (orange), the rhodamine tag is clearly at the advancing edge of the actin filaments (green). Neutrophils lacking the small GTPase Rac2, show no significant increase in free barbed ends or the development of an actin rich leading edge after fMLP stimulation.

Chapter 31

Figure 1: Nitroblue tetrazolium (NBT) assay. Following phorbol myristate acetate stimulation, reduction of yellow NBT to blue black formazan occurred in the healthy control cells (left), but is completely absent in cells from a patient with X-linked chronic granulonatous disease (center). The mother of the patient is a carrier of the mutation and has both NBT-positive and -negative cell populations (right).

Chapter 32

Figure 1: Comparison of polymorphonuclear neutrophils (PMN) from PMN from healthy subjects (Normal) and those with myeloperoxidase (MPO-deficient). Primary human PMN were isolated from peripheral blood using the protocol provided in Chapter 2.

Figure 2: Peroxidase activity in polymorphonuclear neutrophils (PMN) from a healthy subject (Normal PMN) or an individual with myeloperoxidase (MPO) deficiency (as indicated). Panel at far right indicates normal peroxidase activity in an eosinophil isolated from an individual with MPO deficiency. Histochemical staining was performed as described.

I

Neutrophils: An Overview

1

The Role of Neutrophils in the Immune System

An Overview

Harry L. Malech

1. Introduction

This valuable and unique book contains a compendium of methods and reviews that does much more than allow one to study the biology of neutrophils. What makes this collection of contributions so special is that it highlights and facilitates using the neutrophil as a simple, pure, single primary cell suspension model to study a remarkable array of generalized cellular functions (ameboid cellular motility, chemotaxis, adhesion, phagocytosis, degranulation, oxygen radical production, apoptosis), biochemical pathways (G protein-coupled receptor function and regulation, ion channel function, calcium transients, phosphorylation events, actin regulation, adhesion molecule regulation), as well as specialized functions and molecules important to host defense against infection, the mediation and resolution of inflammation, and the chemokine mediated modulation of acquired immunity (*see* **Fig. 1**). As I note the array of chapter topics, it evokes some of the past history of inquiry into how the neutrophil functions and how we evolved into the current widespread use of the neutrophil as a convenient model system for studying so many types of cellular processes and biochemical pathways.

2. Historical Overview

Only a few decades ago, in the 1970s and even the early 1980s, the biology and pathophysiology of the neutrophil was a boutique area of study involving a relatively small number of laboratories and investigators internationally. These investigators all tended to know each other and most of the active investigators in the field of neutrophil biology could easily meet together at the biannual

From: *Methods in Molecular Biology, vol. 412: Neutrophil Methods and Protocols*
Edited by: M. T. Quinn, F. R. DeLeo, and G. M. Bokoch © Humana Press Inc., Totowa, NJ

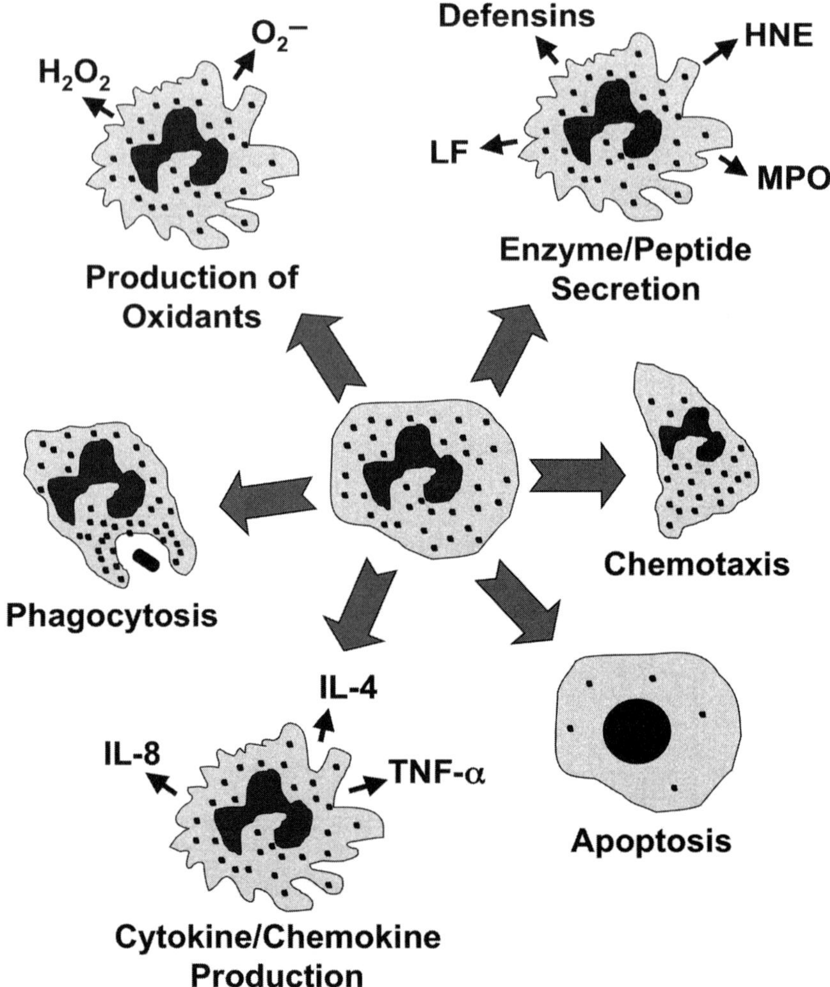

Fig. 1. Illustration of key neutrophil functions. Note that for production reactive oxygen species, secretion of granule components, and production of cytokines and chemokines, only a few representative molecules are shown. HNE, human neutrophil elastase; IL-8, interleukin-8; IL-4, interleukin-4; LF, lactoferrin; MPO, myeloperoxidase; TNF-α, tumor nectosis factor-α.

Gordon Conference on Phagocytes. Even as recently as the early 1980s, "real" immunologists were investigators who delineated the subtypes and life cycle of lymphocytes, and within this scheme the only phagocytes of significance for lymphocyte immunologists were the monocytes. This was because only monocytes, which were long-lived, and not neutrophils, which were short-lived, were thought to be capable of antigen presentation, differentiation into tissue macrophages and other fixed tissue cell types, or any significant protein synthesis, including production of potent immune modulating factors. The relatively recently coined phase, "innate immunity," encompasses in part the recent growing appreciation of the special role of neutrophils in host defense, immune regulation, and regulation of inflammation, reflecting a vast body of new knowledge about how the neutrophil functions and affects the classic lymphocyte-oriented area of immunity encompassed by the term "acquired immunity" (reviewed in **ref. *1***).

Although the rapid ameboid migration of neutrophils to sites of inflammation and its unique capacity to surround and engulf foreign bodies was known since the early 20th century, it was only in the 1960s that it was generally appreciated that neutrophils produced microbicidal activated oxygen products or contained other nonoxidative potently microbicidal substances (e.g., *see* **refs. *2–4***). It was only in the late 1960s and early 1970s that a more detailed understanding of the different types of granules was delineated (e.g., *see* **refs. *5,6***), and in the 1980s and 1990s that studies delineated the biochemistry of a large array of specialized cationic microbicidal proteins and a more complete understanding of the many proteolytic enzymes that were contained in those granules (reviewed in **ref. *7***). Only in the late 1980s and early 1990s were the biochemical details of the phagocyte oxidase delineated in fine detail (reviewed in **refs. *8–10***). Although investigators studying the biochemistry of nonmuscle actin in cell motility performed much of the critical early research in lower eukaryotic organisms, translation of this work to mammalian tissues was largely performed in the 1980s and 1990s in neutrophils and monocytes (e.g., *see* **refs. *11,12***).

Since the human tritium tracer studies of the 1960s, it has been appreciated that when neutrophils emerge from bone marrow to peripheral blood, the half-time in blood is only 6 to 10 h and even shorter in infected patients, and that the lifespan in tissues is 3 d or less (e.g., *see* **refs. *5,13,14***). This provided a basis for considering neutrophils as end-stage cells only minimally more capable of anabolic processes than erythrocytes. This impression was further engendered by the observation that, compared to other cell types, neutrophils produce the energy for survival primarily through anaerobic metabolism, reserving most use of oxygen for production of superoxide in the context of the stimulated microbicidal respiratory burst *(15)*. There is a paucity of mitochondria and ribosomes in neutrophils compared, for example, to monocytes, and most investigators in the 1970s assumed that mature neutrophils in blood or tissues were devoid of significant protein synthetic capacity, functioning entirely on the store of enzymes and other proteins that were contained within the granules, membranes, and cytoplasm of the neutrophils as these cells emerged from the bone marrow. It was not that investigators viewed the neutrophil as inactive, because after all, these are cells capable of remarkably rapid ameboid migration, rapid engulfment of microorganisms, a prodigious respiratory burst-induced production of superoxide, and extremely rapid degranulation into phagosomes (reviewed in **ref. *16***) (*see* **Fig. 1**). However, studies in the past decade have demonstrated unequivocally that neutrophils are capable of significant stimulated production of new proteins (e.g., *see* **refs. *17–19***); of note is production of a number of chemokines, in particular the production of large amounts of interleukin 8 *(20)*. This provides one important area of evidence for the important interface between the neutrophil component of innate immunity and the classic area of acquired immunity (e.g., *see* **ref. *1***).

One of the specialized motile properties of neutrophils is chemotaxis, and the delineation in the 1970s of bacteria-derived formyl peptides as chemotactic for neutrophils *(21)*, in particular the discovery of the simple formylated tripeptide formyl-methionyl-leucinyl-phenyalanine (fMLP) as a potent chemoattractant, began the process of making the neutrophil, in the 1980s, a model of choice for investigators interested in delineating a large array of biochemical signaling pathways whose diverse enzymes and regulatory proteins are still being worked out (e.g., *see* **refs. *22,23***). The formyl peptides induced chemotaxis, but they also induced degranulation, which was associated metabolically with an ionized calcium transient, changes in electric potential similar to neural signaling, phosphorylations, metabolism of GTP, and metabolism of certain membrane phospholipids (e.g., *see* **refs. *24–26***). This was the beginning of the use of the neutrophil as the model system of choice for an increasing number of investigators delineating many types of newly identified biochemical signaling pathways, including the G protein-coupled signaling pathway and the large array of small GTPases of the Ras and Rho family that regulate so many cell functions (e.g., *see* **refs. *27–30***).

Although it was appreciated that neutrophils have a relatively short life span following release from the bone marrow, the final fate of "old" neutrophils remained a sort of vague mystery until the emergence of the new paradigm of apoptosis defined the process of regulated cell death as a sort of final stage in the differentiation of cells and differentiated it from trauma and toxin- or immune-induced cell necrosis *(17,31–33)*. The recent application of the apoptosis paradigm to neutrophils has been used to explain how some processes lead to resolution of infection or inflammation without tissue damage by allowing neutrophil apoptosis to occur. The apoptotic process prevents release of the cytotoxic and proteolytic contents of the neutrophil by facilitating phagocytosis of apoptotic neutrophils by macrophages and dendritic cells through engagement of "death receptors" *(32,33)*. Not only does this facilitate "cleanup" and resolution of infection and inflammation without tissue damage, but probably comprises an important interface between innate and acquired immunity in that the engagement of death receptors modulates the function of antigen-presenting cells, and the contents of the phagocytosed apoptotic neutrophils include components of killed microorganisms, which are processed by antigen-presenting cells. Thus, it is appropriate that there is a chapter in this book on apoptosis of neutrophils (*see* Chapter 12). Apoptosis of neutrophils can be replaced prematurely by necrosis and release of the potent proteolytic enzymatic contents of neutrophils by microorganisms that contain toxins that lyse the neutrophil (reviewed in **ref. *34***), or by autoimmune processes such as antibodies to neutrophils or their contents that induce lysis or phagocytosis of neutrophils before normal apoptotic processes can occur.

Neutrophils are the primary mediators of the rapid innate host defense against most bacterial and fungal pathogens that occurs before the complex humoral and lymphocyte cellular processes of acquired immunity can be brought to bear on an infection. The importance of the neutrophil in this process is highlighted by the fact that iatrogenic neutropenia from cancer chemotherapy or reactions to cytotoxic drugs is the most common severe immune deficiency associated with significant morbidity encountered in medical practice *(35–38)*. Inherited forms of neutropenias are also associated with significant risk of infection from bacteria and fungi. There are a number of single-gene disorders that primarily affect specific neutrophil functions such as defects in the phagocyte oxidase responsible for the group of diseases with a common phenotype called chronic granulomatous disease (CGD), or the defect in the CD18 β integrin responsible for leukocyte adhesion deficiency (LAD) (*see* Chapter 30). The groups of bacterial and fungal organisms typically infecting patients with severe neutropenia, CGD, or LAD overlap a little, but in general are different and distinct, strongly suggesting that there are an array of microbicidal defense systems and molecules used by the neutrophil in host defense and that different systems have evolved to provide specific defense against different organisms. Also of note is that patients with CGD have excessive inflammation and a tendency to develop a variety of autoimmune disorders including inflammatory bowel disease and poor wound healing, suggesting that the superoxide or hydrogen peroxide products of the respiratory burst may play an important role in the resolution of inflammation and in wound healing *(39)*. Patients with LAD have large nonhealing ulcers, which also suggests a role for neutrophils in wound healing, as LAD neutrophils have trouble migrating into tissues *(40)*.

3. Future Prospects

Despite strong evidence in a variety of model systems of inflammation that superoxide dismutase, which catabolizes superoxide, catalase, which catabolizes hydrogen peroxide, or dimethylsulfoxide (DSMO), which scavenges hydroxyl radical protect against tissue injury from neutrophil-mediated tissue damage, there remains considerable controversy regarding the role of neutrophil products of oxidative metabolism in mediating any of the tissue-damage syndromes associated with autoimmune disease or septic shock (e.g., *see* **refs.** *41–44*). Stronger evidence exists for the accrual of actual clinical benefit from blocking the activity a number of the proteolytic enzymes derived from neutrophils in a variety of autoimmune or other disease processes (reviewed in **ref.** *45*). Because of this, the discovery and development of antiproteolytic small molecules active against specific proteases found in neutrophils are part of the anti-inflammatory drug development programs of a number of pharmaceutical companies (e.g., *see* **refs.** *46,47*). Finally, neutrophils are also a source of a number of small potent antimicrobial proteins and peptides that could be developed into new classes of antibiotics, and this is another area of interest to pharmaceutical development (reviewed in **ref.** *48*). Thus, the chapters in this book are likely to be of utility not only broadly to investigators interested in using the neutrophil as a model system, to investigators specifically interested in neutrophil function, or to investigators studying the interface between innate and acquired immunity, but will also be important to investigators developing new classes of natural biologicals as antibiotics or developing new classes of anti-inflammatory agents.

References

1. Hoebe, K., Janssen, E., and Beutler, B. (2004) The interface between innate and adaptive immunity. *Nat. Immunol.* **5,** 971–974.
2. Babior, B. M., Kipnes, R. S., and Curnutte, J. T. (1973) Biological defense mechanisms: production by leukocytes of superoxide,a potential bactericidal agent. *J. Clin. Invest.* **52,** 741–744.
3. Klebanoff, S. J. (1967) Iodination of bacteria: a bactericidal mechanism. *J. Exp. Med.* **126,** 1063–1078.
4. Lehrer, R. I., Hanifin, J., and Cline, M. J. (1969) Defective bactericidal activity in myeloperoxidase-deficient human neutrophils. *Nature* **223,** 78–79.
5. Bainton, D. F., Ullyot, J. L., and Farquhar, M. G. (1971) The development of neutrophilic polymorphonuclear leukocytes in human bone marrow. *J. Exp. Med.* **134,** 907–934.
6. Bainton, D. F. and Farquhar, M. G. (1968) Differences in enzyme content of azurophil and specific granules of polymorphonuclear leukocytes. I. Histochemical staining of bone marrow smears. *J. Cell Biol.* **39,** 286–298.
7. Borregaard, N. and Cowland, J. B. (1997) Granules of the human neutrophilic polymorphonuclear leukocyte. *Blood* **89,** 3503–3521.
8. Segal, A. W. and Abo, A. (1993) The biochemical basis of the NADPH oxidase of phagocytes. *Trends Biochem. Sci.* **18,** 43–47.
9. Babior, B. M. (1999) NADPH oxidase: an update. *Blood* **93,** 1464–1476.
10. Clark, R. A. (1990) The human neutrophil respiratory burst oxidase. *J. Infect. Dis.* **161,** 1140–1147.

11. Zigmond, S. H. (1978) Chemotaxis by polymorphonuclear leukocytes. *J. Cell. Biol.* **77**, 269–287.
12. Southwick, F. S. and Stossel, T. P. (1983) Contractile proteins in leukocyte function. *Semin. Hematol.* **20**, 305–321.
13. Fliedner, T. M., Cronkite, E. P., and Robertson, J. S. (1964) Granulocytopoiesis. I. Senescence and random loss of neutrophilic granulocytes in human beings. *Blood* **24**, 402–414.
14. Athens, J. W., Haab, O. P., Raab, S. O., et al. (1961) Leukokinetic studies. IV. The total blood, circulating and marginal granulocyte pools and the granulocyte turnover rate in normal subjects. *J. Clin. Invest.* **40**, 989–995.
15. Rossi, F. and Zatti, M. (1964) Changes in the metabolic pattern of polymorphonuclear leukocytes during phagocytosis. *Br. J. Exp. Pathol.* **45**, 548–559.
16. Nauseef, W. M. and Clark, R. A. (2000) Granulocytic phagocytes, in *Basic Principles in the Diagnosis and Management of Infectious Diseases, 5th Ed.* (Mandell, G. L., Bennett, J. P., and Dolin, R., eds.), Churchill Livingstone: New York, pp. 89–112.
17. Kobayashi, S. D., Voyich, J. M., Buhl, C. L., Stahl, R. M., and DeLeo, F. R. (2002) Global changes in gene expression by human polymorphonuclear leukocytes during receptor-mediated phagocytosis: cell fate is regulated at the level of gene expression. *Proc. Natl. Acad. Sci. USA* **99**, 6901–6906.
18. Theilgaard-Mönch, K., Knudsen, S., Follin, P., and Borregaard, N. (2004) The transcriptional activation program of human neutrophils in skin lesions supports their important role in wound healing. *J. Immunol.* **172**, 7684–7693.
19. Zhang, X. Q., Kluger, Y., Nakayama, Y., et al. (2004) Gene expression in mature neutrophils: early responses to inflammatory stimuli. *J. Leukoc. Biol.* **75**, 358–372.
20. Strieter, R. M., Kasahara, K., Allen, R. M., et al. (1992) Cytokine-induced neutrophil-derived interleukin-8. *Am. J. Pathol.* **141**, 397–407.
21. Schiffmann, E., Corcoran, B. A., and Wahl, S. M. (1975) *N*-formylmethionyl peptides as chemoattractants for leucocytes. *Proc. Natl. Acad. Sci. USA* **72**, 1059–1062.
22. Snyderman, R. and Goetzl, E. J. (1981) Molecular and cellular mechanisms of leukocyte chemotaxis. *Science* **213**, 830–835.
23. Allen, R. A., Jesaitis, A. J., and Cochrane, C. G. (1990) The N-formyl peptide receptor, in *Cellular and Molecular Mechanisms of Inflammation: Receptors of Inflammatory Cells, Structure-Function Relationships* (Cochrane, C. G. and Gimbrone, M. A., eds.), Academic: San Diego, pp. 83–112.
24. O'Flaherty, J. T., Showell, H. J., and Ward, P. A. (1977) Influence of extracellular Ca^{2+} and Mg^{2+} on chemotactic factor-induced neutrophil aggregation. *Inflammation* **2**, 265–276.
25. Serhan, C. N., Broekman, M. J., Korchak, H. M., Smolen, J. E., Marcus, A. J., and Weissmann, G. (1983) Changes in phosphatidylinositol and phosphatidic acid in stimulated human neutrophils. Relationship to calcium mobilization, aggregation and superoxide radical generation. *Biochim. Biophys. Acta* **762**, 420–428.

26. McPhail, L. C., Clayton, C. C., and Snyderman, R. (1984) A potential second messenger role for unsaturated fatty acids: Activation of Ca^{++}-dependant protein kinase. *Science* **224,** 622–625.

27. Aharoni, I. and Pick, E. (1990) Activation of the superoxide-generating NADPH oxidase of macrophages by sodium dodecyl sulfate in a soluble cell-free system: evidence for involvement of a G protein. *J. Leukoc. Biol.* **48,** 107–115.

28. Quinn, M. T., Parkos, C. A., Walker, L., Orkin, S. H., Dinauer, M. C., and Jesaitis, A. J. (1989) Association of a ras-related protein with cytochrome b of human neutrophils. *Nature* **342,** 198–200.

29. Abo, A., Pick, E., Hall, A., Totty, N., Teahan, C. G., and Segal, A. W. (1991) Activation of the NADPH oxidase involves the small GTP-binding Protein p21[rac1]. *Nature* **353,** 668–670.

30. Knaus, U. G., Heyworth, P. G., Evans, T., Curnutte, J. T., and Bokoch, G. M. (1991) Regulation of phagocyte oxygen radical production by the GTP-binding protein Rac2. *Science* **254,** 1512–1515.

31. Serhan, C. N. and Savill, J. (2005) Resolution of inflammation: the beginning programs the end. *Nat. Immunol.* **6,** 1191–1197.

32. Savill, J. S., Wyllie, A. H., Henson, J. E., Walport, M. J., Henson, P. M., and Haslett, C. (1989) Macrophage phagocytosis of aging neutrophils in inflammation. Programmed cell death in the neutrophil leads to its recognition by macrophages. *J. Clin. Invest* **83,** 865–875.

33. Whyte, M. K., Meagher, L. C., MacDermot, J., and Haslett, C. (1993) Impairment of function in aging neutrophils is associated with apoptosis. *J. Immunol.* **150,** 5124–5134.

34. DeLeo, F. R. (2004) Modulation of phagocyte apoptosis by bacterial pathogens. *Apoptosis* **9,** 399–413.

35. Tobias, J. D. and Schleien, C. (1991) Granulocyte transfusions—a review for the intensive care physician. *Anaesth. Intensive Care* **19,** 512–520.

36. Froland, S. S. (1984) Bacterial infections in the compromised host. *Scand. J. Infect. Dis. Suppl* **43,** 7–16.

37. Bodey, G. P., Buckley, M., Sathe, Y. S., and Freireich, E. J. (1966) Quantitative relationships between circulating leukocytes and infection in patients with acute leukemia. *Ann. Intern. Med.* **64,** 328–340.

38. Dale, D. C., Guerry, D., Wewerka, J. R., Bull, J. M., and Chusid, M. J. (1979) Chronic neutropenia. *Medicine (Baltimore)* **58,** 128–144.

39. Kobayashi, S. D., Voyich, J. M., Braughton, K. R., et al. (2004) Gene expression profiling provides insight into the pathophysiology of chronic granulomatous disease. *J. Immunol.* **172,** 636–643.

40. Bunting, M., Harris, E. S., McIntyre, T. M., Prescott, S. M., and Zimmerman, G. A. (2002) Leukocyte adhesion deficiency syndromes: adhesion and tethering defects involving beta 2 integrins and selectin ligands. *Curr. Opin. Hematol.* **9,** 30–35.

41. Finkel, T. and Holbrook, N. J. (2000) Oxidants, oxidative stress and the biology of ageing. *Nature* **408,** 239–247.

42. Rahman, I., Biswas, S. K., and Kode, A. (2006) Oxidant and antioxidant balance in the airways and airway diseases. *Eur. J. Pharmacol.* **533,** 222–239.
43. Temple, M. D., Perrone, G. G., and Dawes, I. W. (2005) Complex cellular responses to reactive oxygen species. *Trends Cell Biol.* **15,** 319–326.
44. Weiss, S. J. (1989) Tissue destruction by neutrophils. *N. Engl. J. Med.* **320,** 365–376.
45. Altieri, D. C. (1995) Proteases and protease receptors in modulation of leukocyte effector functions. *J. Leukoc. Biol.* **58,** 120–127.
46. Zaidi, S. H., You, X. M., Ciura, S., Husain, M., and Rabinovitch, M. (2002) Overexpression of the serine elastase inhibitor elafin protects transgenic mice from hypoxic pulmonary hypertension. *Circulation* **105,** 516–521.
47. Zeiher, B. G., Matsuoka, S., Kawabata, K., and Repine, J. E. (2002) Neutrophil elastase and acute lung injury: prospects for sivelestat and other neutrophil elastase inhibitors as therapeutics. *Crit. Care Med.* **30,** S281–S287.
48. Ganz, T. (2004) Antimicrobial polypeptides. *J. Leukoc. Biol.* **75,** 34–38.

II

Neutrophil Isolation and Subcellular Fractionation

2

Isolation of Human Neutrophils From Venous Blood

William M. Nauseef

Summary

Venous blood provides a ready source of large numbers of unstimulated granulocytes and mononuclear cells. Exploiting the differences in the relative densities of the leukocytes circulating in venous blood, one can separate leukocytes from erythrocytes as well as isolate the individual leukocyte populations in high purity for use in ex vivo studies.

Key Words: Granulocytes; mononuclear cells; Hypaque-Ficoll; dextran sedimentation.

1. Introduction

Under normal conditions, there are approx 7400/mm^3 (range of 4500–11,000) white blood cells in circulation, approx 59% neutrophils (polymorphonuclear neutrophils [PMN]), 2.7% eosinophils, and 4% monocytes. The different densities of the circulating hematopoietic cells are exploited in order to separate erythrocytes from leukocytes and then to isolate the individual leukocyte populations. Although a variety of methods for rapid, one-step isolation of PMN have been developed recently and used successfully, we routinely use one of two variants of a method first described by Böyum in 1968 *(1)*. PMN isolated in this way do not consume oxygen and maintain their secretory vesicles intracellularly, two features that suggest that the PMN are *bona fide* resting cells and thus not activated by the isolation procedure.

Two major steps are involved, each using specific conditions to separate cells based on their intrinsic density: sedimentation in dextran at 1 g and differential sedimentation in a discontinuous density gradient of Hypaque-Ficoll. In dextran, erythrocytes form rouleaux and thus sediment more rapidly than do granulocytes in suspension. In the Hypaque-Ficoll, granulocytes and erythrocytes pellet to the bottom, whereas mononuclear cells (i.e., lymphocytes and monocytes), basophils, and platelets remain at the interface between plasma/buffer and the

From: *Methods in Molecular Biology, vol. 412: Neutrophil Methods and Protocols*
Edited by: M. T. Quinn, F. R. DeLeo, and G. M. Bokoch © Humana Press Inc., Totowa, NJ

Hypaque-Ficoll. Because we routinely process relatively large quantities of blood (e.g., ≥200 mL) and are primarily interested in recovering PMN (not monocytes), we generally perform dextran sedimentation first, followed by Hypaque-Ficoll sedimentation. When smaller volumes of blood are processed or when monocytes are the targeted cell of interest, the sedimentation in Hypaque-Ficoll can be performed first. Both approaches are described below.

2. Materials

1. The preferred anticoagulant is preservative-free sodium heparin (1000 U/mL), although lithium heparin, ethylenediamine tetraacetic acid (EDTA), or sodium citrate (37.9 mg/mL in sodium phosphate buffer) is acceptable.
2. 3% (w/v) Dextran solution: dissolve 30 g/L of Dextran-500 (average molecular weight 200,000–500,000) in endotoxin-free, sterile 0.9% NaCl (*see* **Note 1**). Heat the solution if necessary to promote dissolution of the dextran and handle the solution using sterile technique.
3. Hypaque-Ficoll solution: originally prepared by mixing isopaque and Ficoll 400 to create a solution with a density of 1.077 g/mL *(1)*. Currently premixed, sterile Ficoll-Paque plus can be purchased from GE Healthcare (formerly Amersham-Pharmacia Biotech). Each 100 mL contains 5.7 g Ficoll 400 and 9.0 g diatrizoate solution with 0.0231 g of disodium calcium EDTA in endotoxin-free water. Protected from light, the solution is stable for 3 yr when stored at 4–30°C. The appearance of yellow color or particulate material indicates deterioration.
4. Sterile, endotoxin-free H_2O and 1.8% (w/v) NaCl.
5. Sterile, endotoxin-free Hank's balanced salt solution (HBSS) without Ca^{2+} or Mg^{2+}.

3. Methods (see Note 2)

3.1. Dextran Sedimentation Followed by Hypaque-Ficoll Density Centrifugation

1. Draw blood into a syringe containing sufficient preservative-free heparin to have a final concentration of 20 U/mL in the blood sample.
2. Add equal volumes of 3% dextran solution and whole blood to 50-mL conical tubes. Mix by repeated inversion (10 times), and set tubes upright for 18–20 min at room temperature (*see* **Note 3**).
3. Aspirate the straw-colored, leukocyte-rich, erythrocyte-poor upper layer with a sterile plastic pipet or a 10- to 15-mL sterile syringe and transfer the aspirate to a sterile 50-mL conical tube (*see* **Note 4**).
4. Pellet leukocytes by centrifugation at 500*g* for 10 min at 4°C; aspirate and discard the fluid.
5. Resuspend pellets in 10 mL; mix well but **do not** vortex. Three resuspended pellets can be pooled into a single conical tube to achieve a final volume of 35 mL. Top off with sterile saline if needed to reach 35 mL.
6. Carefully underlay the leukocyte suspension with 10 mL of Hypaque-Ficoll, using a sterile, plastic 10-mL pipet (**Fig. 1**).

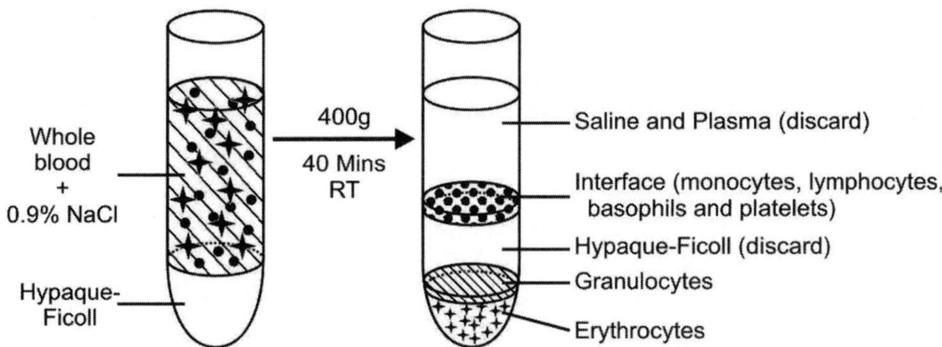

Fig. 1. Hypaque-Ficoll gradient separation of granulocytes and peripheral blood mononuclear cells. Hypaque-Ficoll gradient before (left side) and after (right side) centrifugation.

7. Centrifuge gradients at 400g for 40 min at room temperature (*see* **Note 5**).
8. After centrifugation, two bands should be apparent in the conical tube (**Fig. 1**). The lighter band contains mononuclear cells, whereas the denser band has both granulocytes and erythrocytes. Aspirate and discard supernatant above the PMN–erythrocyte layer, taking care not to lose any of the PMN-rich pellet (*see* **Note 6**).
9. Resuspend each pellet in sterile H_2O and mix well (but **do not** vortex) for 28 s.
10. Promptly restore tonicity by adding an equal volume of 1.8% saline and mixing (*see* **Note 7**).
11. Centrifuge at 500g for 5 min at 4°C. Discard supernatant, and repeat **step 10** *if* more lysis is needed, but repeat **step 10** only once (*see* **Note 7**).
12. Resuspend cells in Ca^{2+} and Mg^{2+}-free HBSS at ≤30 × 10^6 cells/mL. Determine the cell concentration by manually counting with a hemacytometer. The differential of leukocytes can be assessed by examining a stained slide microscopically (*see* **Note 8**).

3.2. Using Hypaque-Ficoll Density Centrifugation First-Mononuclear Cell Recovery

The use of Hypaque-Ficoll gradients as the first step in cell isolation allows for the isolation of both mononuclear cells and neutrophils.

1. Draw blood into a syringe as above and dilute with 1 volume of 0.9% NaCl at room temperature to a total volume of 40 mL in a sterile 50-mL conical plastic tube (*see* **Note 9**).
2. Carefully underlay the diluted blood with 10 mL of Hypaque-Ficoll.
3. Centrifuge at 400g for 40 min at room temperature.
4. After centrifugation, remove the HBSS and plasma above the mononuclear band at the interface and discard.
5. Recover the fraction containing mononuclear cells, combining bands from no more than two gradients into a separate 50-mL conical tube.

6. Bring each of the pooled fractions to 50 mL with sterile HBSS without Ca^{2+} and Mg^{2+} to dilute the Hypaque-Ficoll (*see* **Note 10**).
7. Pellet mononuclear cells at 600*g* for 10 min at 4°C.
8. Discard supernatant. Pool and wash pellets twice more with HBSS without Ca^{2+} and Mg^{2+}.
9. Resuspend final pellet in 5 mL of HBSS without Ca^{2+} and Mg^{2+} and count.

3.3. Using Hypaque-Ficoll Density Centrifugation First-PMN Recovery

1. After removal of the supernatant and the mononuclear band described above, aspirate and discard the remaining gradient above the PMN–erythrocyte pellet. Be careful not to lose any of the PMN-rich pellet (*see* **Note 6**).
2. Resuspend the pellet to 25 mL using HBSS without Ca^{2+} and Mg^{2+}.
3. Add 25 mL of 3% dextran and mix by inverting the tube 10 times.
4. Set tubes upright for 18–20 min at room temperature (*see* **Note 3**).
5. Aspirate the straw-colored, leukocyte-rich upper layer with a sterile plastic pipet or a 10- to 15-mL sterile syringe and transfer the aspirate to a sterile 50-mL conical tube (*see* **Note 4**).
6. Pellet leukocytes by centrifugation at 500*g* for 10 min at 4°C; aspirate and discard the fluid.
7. Resuspend each pellet in sterile H_2O and mix well (but **do not** vortex) for 28 s.
8. Promptly restore tonicity by adding an equal volume of 1.8% saline and mixing (*see* **Note 7**).
9. Centrifuge at 500*g* for 5 min at 4°C. Discard supernatant, and repeat **step 10** *if* more lysis is needed, but repeat **step 10** only once (*see* **Note 7**).
10. Resuspend cells in Ca^{2+} and Mg^{2+}-free HBSS at ≤30 × 10^6/mL.
11. Determine the cell concentration by manually counting using a hemacytometer. The differential of leukocytes can be assessed by examining a stained slide microscopically.

4. Notes

1. The concentration and the molecular weight range of the dextran is a critical determinant in the speed with which cells sediment *(2)*. The preparation of 3% in the 200,000–500,000 weight range provides excellent sedimentation of erythrocytes within 18–20 min, with rates slower using either the lower molecular weight dextrans (<20,000) or the high molecular weight (>7,000,000) preparations. Given the range in size of the components in the dextran being used, one can anticipate corresponding variation in the rate of sedimentation (reflected as differences in the number of cells recovered). For example, we currently allow sedimentation to occur for 20 min at room temperature to obtain maximal number of granulocytes. Be prepared to modify the time allotted for sedimentation accordingly. The volume of the dextran–blood mixture has no effect on the efficiency of sedimentation. In the event that sedimentation has proceeded for longer than the intended time, the sample can be mixed again and the sedimentation repeated with no adverse effect on yield or purity.

2. In general, always use polypropylene tubes, as glass and polyethylene may activate the PMN during isolation. Pipets should be sterile and buffers should be prepared with sterile, endotoxin-free reagents, as endotoxin will prime PMN. Cell suspensions should never be vortexed.

3. Adequate mixing of the sample is essential for reproducible sedimentation, presumably to distribute the erythrocytes throughout the suspension and provide maximal opportunities for rouleaux formation. When directly examined, 10 mixing inversions were sufficient to guarantee reproducible and maximum sedimentation; fewer inversions compromised the reproducibility in yield.

4. Once the leukocyte-rich supernatant from the dextran sedimentation has been centrifuged, proceed as quickly as possible with the isolation procedure. Do not leave cell pellets at intermediate steps. Aspirates and cell pellets are biohazard waste and should be handled accordingly.

5. The centrifugation of the Hypaque-Ficoll gradient should be at 20°C and with no brake on the centrifuge. When performed at 4°C, there is more PMN contamination of the mononuclear band than seen when centrifugation is at 20°C (2.3% vs 0.1%, respectively), and more lymphocyte and mononuclear contamination of the PMN pellet (6.9 and 0.6% vs 1.4 and 0.1%). It is best to have a dedicated room temperature centrifuge for this use; alternatively, have the refrigeration in the centrifuge turned off overnight before running the gradient the following morning.

6. If monocytes are desired, aspirate the less-dense band using a sterile plastic pipet and mix with an equal volume of cold sterile saline in a separate conical tube.

7. The hypotonic lysis of erythrocytes exploits the relative resistance of leukocytes to osmotic stress. This difference is relative and prolonged or repeated hypotonic lysis will damage PMN. Attention to limiting the time of exposure to hypotonicity to 18–28 s should be strictly maintained. More than two cycles of hypotonic lysis must be avoided, as they will not lyse additional erythrocytes but will begin to damage PMN.

8. When counting the PMN suspension, one can directly count the PMN by diluting the cell suspension 1:20 in 3% acetic acid and observe under ×40 objective. In the acetic acid solution, the nuclear morphology of PMN is clearly identified, allowing direct counting of the PMN in suspension. Alternatively, PMN suspension can be diluted 1:20 in HBSS and counted to determine the leukocyte concentration. A separate sample (e.g., 10 μL) can be diluted 1:5 in HBSS or saline and placed on a slide (either using a cytospin or simply maneuver the slide to obtain a thin layer) and allowed to air-dry. The slide can then be stained with Wright Stain (or an equivalent) and a differential performed. With the total number of leukocytes and the percent of PMN on differential staining, the number of PMN isolated can be calculated. In general, the yield should be 2–4 million PMN per mL blood drawn and the suspension should be approx 98% PMN with a few contaminating eosinophils.

9. Efficient separation is achieved by 1:1 dilution of blood with saline prior to centrifugation in the Hypaque-Ficoll gradient. The tendency of lymphocytes to be trapped in erythrocyte–granulocyte aggregates decreases the yield of lymphocytes while simultaneously contaminating the granulocyte pellet with lymphocytes. Dilution of whole blood at the start of the isolation will decrease both problems.

10. Neither mononuclear cells nor granulocytes tolerate long incubation in Hypaque-Ficoll, so it is best to dilute the gradient matrix in the isolated cell fractions as soon as feasible.

References

1. Böyum, A. (1968) Isolation of mononuclear cells and granulocytes from human blood. *Scan. J. Clin. Lab. Invest.* **21,** 77–89.
2. Skoog, W. A. and Beck, W. S. (1956) Studies on the fibrinogen, dextran, and phytohemagglutinin methods of isolating leukocytes. *Blood* **11,** 436–454.

3

Neutrophil Isolation From Nonhuman Species

Daniel W. Siemsen, Igor A. Schepetkin,
Liliya N. Kirpotina, Benfang Lei, and Mark T. Quinn

Summary

Advances in the understanding of neutrophil biochemistry require the development of effective procedures for isolating purified neutrophil populations. Although methods for human neutrophil isolation are now standard, similar procedures for isolating neutrophils from many of the nonhuman species used to model human diseases are not as well developed. Because neutrophils are reactive cells, the method of isolation is extremely important to avoid isolation technique-induced alterations in cell function. We present methods here for reproducibly isolating highly-purified neutrophils from large (cow, horse, sheep) and small (mouse, rabbit) animal models and describe optimized details for obtaining the highest cell purity, yield, and viability. We also describe methods to verify phagocytic capacity in the purified cell populations using a flow cytometry-based phagocytosis assay.

Key Words: Inflammation; phagocytosis; large animal model; granulocyte; polymorphonuclear leukocyte; cell isolation; flow cytometry; blood, bone marrow.

1. Introduction

Over the years, various animal models have been developed for investigation of the pathogenesis of human inflammation and infectious disease (reviewed in **ref. 1**). Although small animal models, such as rodents, are easier to handle, breed easily, require much less in the way of housing facilities, and are generally less expensive, they often do not provide an accurate reflection of human physiology *(2)*. Thus, large animal models, such as sheep, are often desirable as models for human disease pathogenesis *(2)*. In these models, it is important to characterize neutrophil function, which requires efficient methods for purification of these phagocytic cells.

Currently, much of our understanding of neutrophil biology is based on studies using human cells, whereas much less is known regarding the biology of these

From: *Methods in Molecular Biology, vol. 412: Neutrophil Methods and Protocols*
Edited by: M. T. Quinn, F. R. DeLeo, and G. M. Bokoch © Humana Press Inc., Totowa, NJ

cells in nonhuman species. It is clear, however, that neutrophils from other species differ from their human counterpart in a number of important functional characteristics *(3)*, and one cannot assume that the basic features of human neutrophil biochemistry and function are representative of neutrophils in all other species. Many species are used as models to understand human pathophysiology; thus, it is essential that we gain a thorough understanding of neutrophil functions in these models. It is also important to characterize neutrophil biology in various nonhuman species in order to determine immune status and host defense mechanisms in these organisms. Additionally, comparison of conserved features among neutrophils from different species can contribute to our understanding of neutrophil biochemistry in general.

One of the major challenges associated with neutrophil studies is the isolation of highly purified cell preparations that are morphologically and functionally similar to cells found in the blood in vivo. Neutrophils are temperamental cells that can be easily altered by improper handling *(4)*. Thus, the method of cell isolation is extremely important to avoid isolation technique-induced alterations in neutrophil function *(4)*. For instance, neutrophils can be primed during isolation, resulting in altered neutrophil responses to subsequent stimuli and changes in surface antigen expression *(5–7)*. Furthermore, some neutrophil functions, such as chemotaxis, can actually be inhibited by isolation procedures *(4)*. To completely eliminate any isolation artifacts, methods have been developed for analysis of neutrophils in whole blood (e.g., *see* **refs. *5,8***). Conversely, the use of whole blood can be impractical for many biochemical studies, which require purified cells in the absence of other contaminating cells and serum proteins.

Although effective methods for isolation of human neutrophils have been extensively developed (*see* Chapter 2), these methods do not work in many nonhuman species. We present methods here for neutrophil isolation from a range of large (cow, horse, sheep) and small (mouse, rabbit) animal models and describe optimized details for obtaining the highest cell purity, yield, and viability. For reference, we compare these results to those obtained from a standard human neutrophil purification protocol. We also verify functional phagocytic capacity in the purified cell populations using a flow cytometry-based phagocytosis assay. Importantly, all of these isolation methods are relatively inexpensive, utilize commonly available reagents, and do not require the acquisition of specialized equipment. Thus, they can be easily implemented in any lab.

2. Materials

1. Fifteen milliliter Vacutainer tubes (Becton Dickinson) containing 150 μL of 500 m*M* disodium ethylenediamine tetraacetic acid (EDTA), pH 7.4, prepared in sterile H_2O and filtered (*see* **Note 1**).

2. Sterile endotoxin-free disposable plastic pipets, polypropylene centrifuge tubes, and disposable polyethylene transfer pipets (Fisher Scientific) (*see* **Note 2**).

2.1. Buffers

1. Sterile injection-grade H_2O and 0.9% NaCl solution (Baxter Healthcare Corporation) (*see* **Note 3**).
2. Dulbecco's modified Eagle's medium (DMEM), Dulbecco's phosphate-buffered saline (DPBS), 10X Hank's balanced salt solution (HBSS) (without Ca^{2+}, Mg^{2+}, and phenol red) (Gibco/Invitrogen).
3. HBSS: Dilute 10X HBSS in sterile H_2O, adjust pH to 7.4, and sterile-filter. Store at 4°C (*see* **Notes 3** and **4**).
4. Hetastarch solution containing 6% hetastarch in 0.9% NaCl (Abbott Laboratories).
5. Solutions of 9% (w/v) and 10% (w/v) NaCl in sterile H_2O. Prepare fresh and sterile-filter (*see* **Notes 3** and **4**).
6. Acid citrate dextrose (ACD): 65 mM citric acid, 85 mM sodium citrate, 2% dextrose dissolved in sterile H_2O and sterile-filtered. Store at 4°C (*see* **Notes 3** and **4**).
7. Murine neutrophil buffer: HBSS containing 0.1% (w/v) bovine serum albumin, 1% (w/v) glucose. Prepare fresh and sterile-filter (*see* **Notes 3** and **4**).
8. Rabbit neutrophil buffer: 138 mM NaCl, 27 mM KCl, 8.1 mM Na_2HPO_4, 1.5 mM KH_2PO_4, and 5.5 mM glucose dissolved in sterile H_2O and sterile-filtered. Store at 4°C (*see* **Notes 3** and **4**).
9. Dextran solution: 6% (w/v) nonpyrogenic Dextran 500 (Amersham Biosciences) dissolved in sterile 0.9% NaCl solution and sterile-filtered. Store at 4°C (*see* **Notes 3–5**).
10. 10X PBS/EDTA buffer: Dissolve 100 mM KH_2PO_4, 9% (w/v) NaCl, 2 mg/L EDTA in sterile H_2O, adjust pH to 7.4, and sterile-filter (*see* **Notes 3** and **4**).
11. PBS/EDTA buffer: Dilute 10X PBS/EDTA 1:10 in sterile H_2O, readjust pH to 7.4 if needed, and sterile-filter (*see* **Notes 3** and **4**).

2.2. Density Gradient Solutions

1. Percoll, Histopaque 1077, and Histopaque 1119 (Sigma-Aldrich).
2. Histopaque 1077/1119 solution: mix equal volumes of Histopaque 1077 and Histopaque 1119 (*see* **Note 6**).
3. Percoll stock solution (100% Percoll): Percoll and 10X HBSS, pH 7.4 mixed at a ratio of 9:1 (v/v) (*see* **Note 6**).
4. Percoll solutions: mix 100% Percoll stock with the appropriate volumes of 1X HBSS to obtain 85, 81, 70, 65, 62, 55, 50, and 45% (v/v) Percoll solutions (*see* **Note 6**).
5. Percoll/EDTA stock solution (100% Percoll/EDTA): Percoll and 10X PBS/EDTA, pH 7.4 mixed at a ratio of 9:1 (v/v) (*see* **Note 6**).
6. Percoll/EDTA solution: mix 100% Percoll/EDTA stock with the appropriate volume of 1X PBS/EDTA to obtain 65% (v/v) Percoll/EDTA solution (*see* **Note 6**).

2.3. Cell Analysis Reagents

1. Acetic acid (2%) prepared in sterile H_2O.
2. Trypan Blue Solution (0.4%) (Sigma-Aldrich)
3. Vybrant Phagocytosis Assay Kit (Molecular Probes).

3. Methods

In the methods described below, we outline the steps to obtain highly purified and functionally active neutrophils from bovine, equine, ovine, and rabbit blood, as well as murine bone marrow. Note that some of the methods detailed here have been adapted with modifications from previously published methods for isolation of equine *(9)*, murine *(10)*, ovine *(11)*, and rabbit neutrophils *(12, 13)*. In addition to the basic neutrophil isolation procedures, we also provide details for quantifying cell yield, evaluating cell purity and viability, and measuring phagocytic function of the purified cells. These parameters are compared with human neutrophils purified by a standard method. Although neutrophil isolation from all species is based on density gradient separation techniques, the gradient media and composition vary widely because of slight differences in neutrophil density between species. In addition, differences in red blood cell reactivity to aggregating reagents between species are reflected in the methods described below. Overall, these methods are efficient, easy to perform, and reproducibly generate high-quality neutrophil populations for biochemical and functional studies.

3.1. Bovine Neutrophil Isolation

1. Collect bovine blood into Vacutainer tubes containing EDTA. For the method outlined here, we collected 50 mL of blood. If different volumes of blood are required, adjust the indicated volumes proportionally.
2. Pool 50 mL of blood into a conical 50-mL polypropylene centrifuge tube and centrifuge at 740g for 10 min at room temperature with low brake.
3. Remove the upper plasma layer and buffy coat found at the plasma–red blood cell interface with a plastic transfer pipet. Transfer the remaining red blood cell layer into a conical 250-mL polypropylene tube.
4. Lyse red blood cells by adding 50 mL of sterile H_2O. Mix by gently inverting the tube for 20 s at room temperature (*see* **Note 7**).
5. Immediately add 5 mL of 10% NaCl solution, and mix well by gently inverting the tube.
6. Centrifuge at 585g for 10 min at room temperature with low brake.
7. Remove the supernatant using a plastic transfer pipet, and resuspend the cell pellet in 50 mL of HBSS.
8. Lyse any remaining red blood cells by repeating **steps 4–6**.
9. Resuspend the leukocyte pellet in 10 mL of HBSS.

10. Prepare Histopaque gradients by first pipetting 15 mL of Histopaque 1077 into the bottom of a conical 50-mL centrifuge tube. Place a borosilicate glass Pasteur pipet into the tube so that the pipet tip rests on the bottom of the tube. Use this pipet as a funnel to carefully underlay 15 mL of Histopaque 1077/1119 solution.
11. Layer the 10-mL leukocyte suspension on top of the Histopaque gradient using a plastic transfer pipet. This must be done carefully to avoid mixing the cell suspension with the Histopaque.
12. Centrifuge the gradient at 440g for 25 min at room temperature with no brake.
13. Remove the supernatant with a plastic transfer pipet and discard (*see* **Note 8**).
14. Wash the neutrophil pellet by resuspending the cells in 50 mL of HBSS and centrifuging at 585g for 10 min at room temperature.
15. Resuspend purified cells in the desired assay buffer.

3.2. Equine Neutrophil Isolation

1. Collect equine blood into Vacutainer tubes containing EDTA. For the method outlined here, we collected 24 mL of blood. If different volumes of blood are required, adjust the indicated volumes proportionally.
2. Prepare Percoll gradients by underlaying 2.5 mL of 85% Percoll solution below 2.5 mL of 70% Percoll solution in a conical 15-mL polypropylene centrifuge tube. Use a borosilicate glass Pasteur pipet as a funnel to underlay the Percoll solution (*see* **Subheading 3.1., step 11**).
3. Carefully layer 3 mL of blood on top of each gradient using a plastic transfer pipet.
4. Centrifuge the gradients for 20 min at 400g with no brake at room temperature.
5. The neutrophil band sediments at the interface between 70% and 85% Percoll solutions. Carefully remove all of the supernatant above the neutrophil band with a plastic transfer pipet and discard (*see* **Note 8**).
6. Collect the neutrophil band with a clean plastic transfer pipet.
7. Wash the cells by resuspending them in 50 mL of HBSS and centrifuging at 200g for 10 min at room temperature.
8. Wash the cells twice more by repeating **step 7** above.
9. Resuspend purified cells in the desired assay buffer.

3.3. Human Neutrophil Isolation (For Comparison)

1. Collect human blood into Vacutainer tubes containing EDTA. For the method outlined here, we collected 30 mL of blood. If different volumes of blood are required, adjust the indicated volumes proportionally.
2. Combine 6.7 mL of dextran solution with 30 mL of blood in a conical 50-mL polypropylene centrifuge tube. Mix by gently inverting the tube.
3. Allow the blood–dextran mixture to sediment for 45 min at room temperature. Dextran causes the red blood cells to form aggregates, which sediment to the bottom of the tube. This leaves a clear, red blood cell-depleted layer above the red blood cell-rich lower layer.
4. Transfer the upper cell layer to a clean conical 50-mL polypropylene centrifuge tube with a plastic transfer pipet.

5. Centrifuge the tube at 740*g* for 10 min with low brake at room temperature.
6. Remove the supernatant using a plastic transfer pipet and discard.
7. Resuspend the white blood cell pellet in 7 mL of sterile 0.9% NaCl solution.
8. Place 7 mL of Histopaque 1077 into a conical 50-mL polypropylene centrifuge, and carefully layer the white blood cell suspension on top of the Histopaque. This must be done carefully to avoid mixing the cell suspension with the Histopaque.
9. Centrifuge at 700*g* for 15 min with no brake at room temperature.
10. Remove the supernatant using a plastic transfer pipet and discard (*see* **Note 8**).
11. Resuspend the neutrophil pellet in 6 mL of sterile 0.9% NaCl solution.
12. Lyse contaminating red blood cells by adding 20 mL of sterile H_2O. Mix by gently inverting tubes for 20 s at room temperature (*see* **Note 7**).
13. Immediately add 1.8 mL of 10% NaCl solution, and mix well by gently inverting tubes.
14. Centrifuge at 740*g* for 10 min at room temperature with low brake.
15. Remove the supernatant using a plastic transfer pipet and discard.
16. Resuspend the neutrophil pellet in 6 mL of sterile 0.9% NaCl solution.
17. Lyse any remaining red blood cells by repeating **steps 12–15** above.
18. Remove the supernatant with a plastic transfer pipet.
19. Wash the neutrophil pellet by resuspending the cells in 50 mL of 0.9% NaCl solution and centrifuging at 740*g* for 10 min at room temperature.
20. Resuspend purified cells in the desired assay buffer.

3.4. Murine Neutrophil Isolation

1. Dissect femurs and tibias from 8- to 12-wk-old male mice. BALB/c mice were used here, but this procedure should also work for other strains of mice.
2. Clip the ends of each tibia and femur with dissecting scissors to expose the marrow.
3. Flush bone marrow cells from the tibias and femurs with murine neutrophil buffer using a syringe with 27-G needle. Use two 1-mL volumes of buffer for tibias and three 1-mL volumes of buffer for femurs.
4. Resuspend the pooled bone marrow eluates by gentle pipetting, followed by filtration through a 70-µm nylon cell strainer (Becton Dickinson) to remove cell clumps and bone particles.
5. Centrifuge pooled bone marrow cells at 600*g* for 10 min at 4°C with low brake.
6. Remove the supernatant with a plastic transfer pipet and discard.
7. Resuspend the cell pellet in 3 mL of 45% Percoll solution.
8. Prepare Percoll gradients by layering 2 mL each of the 62, 55, and 50% Percoll solutions successively on top of 3 mL of 81% Percoll solution in a conical 15-mL polypropylene tube.
9. Carefully layer the bone marrow cell suspension on top of the gradient.
10. Centrifuge at 1600*g* for 30 min with no brake at 10°C.
11. Remove the supernatant down to the 62% Percoll layer using a plastic transfer pipet and discard (*see* **Note 8**).
12. Collect the cell band located between the 81 and 62% Percoll layer.

13. Wash the collected cells by resuspending them in 10 mL of murine neutrophil buffer and centrifuging at 600g for 10 min at 10°C.
14. Wash the cells again by repeating **step 13** above, and resuspend the final pellet in 3 mL of murine neutrophil buffer.
15. Carefully layer the cell suspension on top of 3 mL of Histopaque 1119 in conical 15-mL polypropylene tubes.
16. Centrifuge the gradients at 1600g for 30 min at 10°C and no brake to remove contaminating red blood cells.
17. Remove the supernatant using a plastic transfer pipet and discard (*see* **Note 8**).
18. Collect the cell layer between the Histopaque and buffer layers with a plastic transfer pipet.
19. Wash the cells by resuspending them in 10 mL of murine neutrophil buffer and centrifuging at 600g for 10 min at 10°C.
20. Wash the cells again by repeating **step 19** above.
21. Resuspend purified cells in the desired assay buffer.

3.5. Ovine Neutrophil Isolation

1. Collect ovine blood into Vacutainer tubes containing EDTA. For the method outlined here, we collected 50 mL of blood. If different volumes of blood are required, adjust the indicated volumes proportionally.
2. Transfer 50 mL of blood into a conical 50-mL polypropylene tube and centrifuge at 400g for 20 min with low brake at room temperature.
3. Remove the upper plasma layer and buffy coat found at the plasma–red blood cell interface with a plastic transfer pipet.
4. Dilute the red blood cell layer up to the starting blood volume (50 mL in this case) with PBS/EDTA buffer.
5. Pipet 25 mL of the diluted cells into each of two conical 250-mL polypropylene tubes.
6. Lyse red blood cells by adding 150 mL of sterile H_2O into each tube. Mix by gently inverting tubes for 20 s at room temperature (*see* **Note 7**).
7. Immediately add 15 mL of 9% NaCl solution, and mix well by gently inverting tubes.
8. Centrifuge at 250g for 5 min at room temperature with low brake.
9. Remove the supernatant using a plastic transfer pipet, and resuspend the cell pellet in 50 mL of PBS/EDTA buffer.
10. Centrifuge at 250g for 5 min at room temperature.
11. Resuspend the leukocyte pellet in 9 mL of PBS/EDTA buffer.
12. Carefully layer 3 mL of the white blood cell suspension on top of 5 mL of 65% Percoll/EDTA solution using a plastic transfer pipet.
13. Centrifuge the gradients at 400g for 20 min at room temperature with no brake.
14. Remove supernatant using a plastic transfer pipet and discard (*see* **Note 8**).
15. Wash the cells by resuspending them in 50 mL of PBS/EDTA buffer and centrifuging at 400g for 10 min at room temperature.
16. Wash the cells again by repeating **step 15**.
17. Resuspend purified cells in the desired assay buffer.

3.6. Rabbit Neutrophil Isolation

1. Collect rabbit blood into a conical 50-mL polypropylene tube containing ACD so that a 4:1 (v/v) ratio of blood:ACD is achieved. For the method outlined here, we collected 24 mL of blood into a tube containing 6 mL of ACD. If different volumes of blood are required, adjust the indicated volumes proportionally.
2. Transfer 30 mL of blood into a 250-mL conical centrifuge tube and add 5 volumes (150 mL in this case) of Hetastarch to each tube. Mix by gently inverting the tube.
3. Allow the blood–hetastarch mixture to sediment for 40 min at room temperature. Hetastarch causes the red blood cells to form aggregates, which sediment to the bottom of the tube. This leaves a clear, red blood cell-depleted layer above the red blood cell-rich lower layer (*see* **Note 9**).
4. Transfer the upper red blood cell-depleted layer to a clean conical 250-mL polypropylene tube with a plastic transfer pipet.
5. Centrifuge the solutions at 585g for 10 min with low brake at room temperature.
6. Remove the supernatant using a plastic transfer pipet and discard.
7. Resuspend the white blood cell pellet in 10 mL of rabbit neutrophil buffer.
8. Lyse red blood cells by adding 100 mL of sterile H_2O. Mix by gently inverting tubes for 20 s at room temperature (*see* **Notes 7** and **10**).
9. Immediately add 10 mL of 10% NaCl solution, and mix well by gently inverting tubes.
10. Centrifuge at 585g for 10 min with low brake at room temperature.
11. Remove the supernatant using a plastic transfer pipet and discard.
12. Resuspend the cell pellet in 10 mL of rabbit neutrophil buffer.
13. Lyse any remaining red blood cells by repeating **steps 8–10** above.
14. Resuspend the leukocyte pellet in 5 mL of rabbit neutrophil buffer.
15. Carefully layer cell suspension on top of 7 mL of Histopaque 1077 in a conical 50-mL polypropylene tube.
16. Centrifuge the gradients at 475g for 25 min with no brake at room temperature.
17. Remove the supernatant using a plastic transfer pipet and discard (*see* **Note 8**).
18. Wash the neutrophil pellet by resuspending the cells in 50 mL of rabbit neutrophil buffer and centrifuging at 585g for 10 min at room temperature.
19. Resuspend purified cells in the desired assay buffer.

3.7. Quantifying Cell Number and Viability

1. Resuspend the final neutrophil pellet into the desired volume of assay buffer to achieve the appropriate cell concentration (usually 2 to 5 mL) and remove an aliquot for counting.
2. To quantify cell number, dilute 10 μL of the final cell suspension in 190 μL of 2% acetic acid. Pipet a few microliters onto a hemacytometer, and count the cells contained in the 25 squares inside the central double lines. Count only neutrophils, which are easily identified by their characteristic multilobed nuclei. Divide the neutrophil count by 25 to obtain the average per square. Multiply the average per square by 5×10^6 and then by the volume (in milliliters) of the final cell suspen-

Table 1
Average Neutrophil Purity, Yield,
and Cell Viability Using the Described Methods

Species	Total cells	Total neutrophils	Neutrophil purity (%)	Yield (per mL blood)[a]	Viability (%)
Bovine	1.04×10^8	9.79×10^7	93.67	6.94×10^5	>99
Equine	3.98×10^7	3.9×10^7	97.76	1.63×10^6	>99
Human	2.55×10^7	2.53×10^7	99.10	8.42×10^5	>99
Murine	6.06×10^6	5.12×10^6	85.88	1.71×10^6	>99
Ovine	1.93×10^7	1.8×10^7	93.57	4.89×10^5	>99
Rabbit	2.02×10^6	1.83×10^6	90.69	7.62×10^4	>99

[a]Murine neutrophil yield is presented as cells per mouse. The data represent the average from at least three separate neutrophil preparations per species.

    sion to determine the total number of isolated neutrophils. A summary of the neutrophil recovery data determined for all species is shown in **Table 1**.

3. Cell viability is determined by mixing equal aliquots of neutrophil suspension and trypan blue, pipetting the mixture onto microscope slides, and viewing the cells under a microscope. Cells that exclude the trypan blue and appear transparent are counted as viable, whereas cells that turn blue are counted as dead cells. A summary of the cell viability data determined for all species is shown in **Table 1**.

3.8. Analysis of Cell Purity

1. Purity can be evaluated with the hemacytometer (*see* **Subheading 3.8., step 2**) by differential counting of neutrophils vs nonneutrophils.
2. Analysis of cell purity can also be performed by flow cytometric analysis, which provides an effective approach to evaluate the cells present and their level of activation.
3. Collect 10,000 events for each sample using a flow cytometer with linear amplification of forward and side scatter channels.
4. Create a forward-scatter versus side-scatter dot plot, and gate out any cellular debris. Set a gate around the neutrophil population to obtain gate statistics, such as percent of total events (a measure of purity) and relative size and granularity. **Figure 1A** shows a representative dot plot, where the neutrophils form a relatively uniform profile (*see* **Note 12**).
5. A summary of the neutrophil purity data obtained for all species is shown in **Table 1**.

3.8. Phagocytosis Assay

1. Phagocytosis assays were preformed using a Vybrant Phagocytosis Assay Kit with modifications for use with flow cytometry.
2. Thaw one vial each of fluorescent *Escherichia coli* K-12 bioparticles and concentrated HBSS solution.

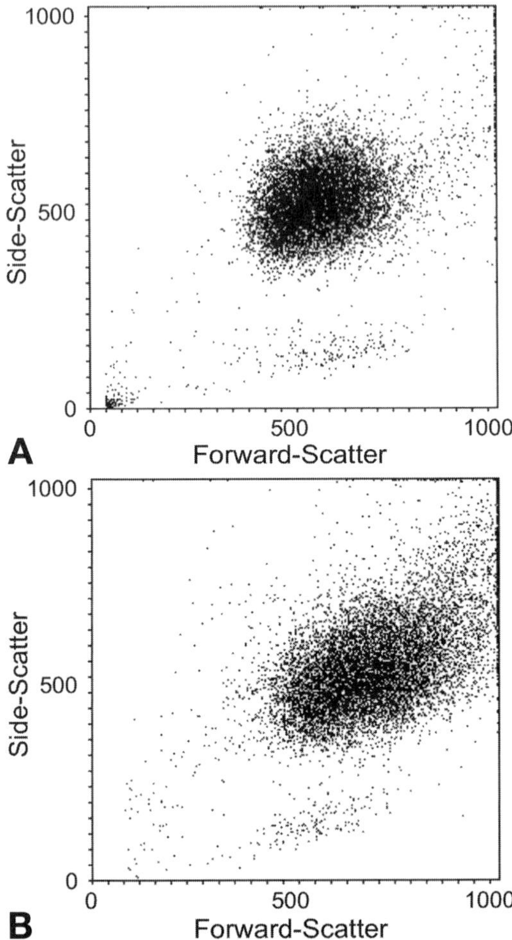

Fig. 1. Analysis of neutrophil purity by flow cytometry. A forward-scatter vs side-scatter dot plot was used to evaluate neutrophil purity (equine cells shown here as an example). Resting (**A**) and mildly activated neutrophils (**B**) are shown (*see* **Note 12**). Neutrophils from all species showed similar forward-scatter vs side-scatter profiles.

3. To prepare stock bioparticles, pipet concentrated HBSS solution into the vial containing the bioparticles and sonicate. Transfer the solution into a clean glass test tube, add 4.5 mL of sterile H_2O, and sonicate again until the beads are completely dispersed.
4. Thaw one vial of trypan blue solution, transfer the trypan blue solution to a polycarbonate test tube, dilute with 4 mL of sterile H_2O, and sonicate.
5. Dilute isolated neutrophils from any species to a final concentration of 1×10^6 cells/ mL in DMEM.

6. Dilute stock bioparticles by mixing 1 mL of bioparticles and 0.5 mL of DMEM.
7. Aliquot 150 μL DMEM, 100 μL neutrophils, and 10 μL of diluted bioparticles into 1.5-mL microcentrifuge tubes. For control samples, substitute 10 μL DMEM for the bioparticles. This dilution gives a final bioparticle to neutrophil ratio of 20:1 (*see* **Note 13**).
8. Incubate triplicate samples for 2, 5, 10, and 15 min at 37°C. We normally incubate control samples of neutrophils alone and neutrophils with trypan blue quench control for 15 min to evaluate any effects resulting from the extended incubation time.
9. After each incubation time, the neutrophils are pelleted by centrifuging at 3000g for 30 s in a microfuge at room temperature.
10. The pellet is very small and easy to lose, so carefully aspirate the supernatant.
11. To quench free bioparticles and neutrophil surface-associated bioparticles, add 100 μL of the trypan blue solution and mix well.
12. Incubate for 1 min at room temperature, and centrifuge at 3000g for 30 s in a microfuge at room temperature.
13. Carefully aspirate the supernatant. Again, use care because the neutrophil pellet is very small and easily lost.
14. Resuspend each pellet in 250 μL DPBS, and transfer the samples to flow cytometer tubes.
15. Collect 10,000 events for each sample using a flow cytometer with linear amplification of forward-scatter and side-scatter channels and logarithmic amplification for the FL1 channel. Analyze the data using flow cytometry software (e.g., CellQuest software) to determine the percent of neutrophils containing fluorescent bioparticles. A summary of the neutrophil phagocytosis data obtained for all species is shown in **Fig. 2** (*see* **Note 14**).

4. Notes

1. Add EDTA to Vacutainer tubes by injection with a 1-mL syringe and 27-G needle.
2. It is essential that the blood and subsequently isolated neutrophils do not ever come into contact with glass, which leads to cell activation. Thus, plasticware should be used throughout all procedures, with exception of the Vacutainer tubes, which are silicone-coated.
3. Neutrophils are highly susceptible to priming and/or activation by endotoxin or lipopolysaccharide (LPS) (e.g., *see* **ref.** *14*), which is often a contaminant in biological reagents. Thus, all plasticware must be endotoxin-free. In addition, all buffers and reagents are prepared in sterile H_2O or saline and sterile-filtered to avoid endotoxin contamination.
4. All buffers were sterile-filtered through 0.2-μm filter units (Fisher Scientific).
5. To avoid possible contamination, which is a common problem with dextran, weigh 30 g of Dextran 500 directly into a sterile plastic 500 mL Nalgene container, and dissolve in sterile 0.9% NaCl solution. Sterile-filter the solution.
6. Use extreme care and accuracy when preparing Percoll mixtures, as small variations in the final density of Percoll mixtures affects the purity and yield of neutrophil preparation.

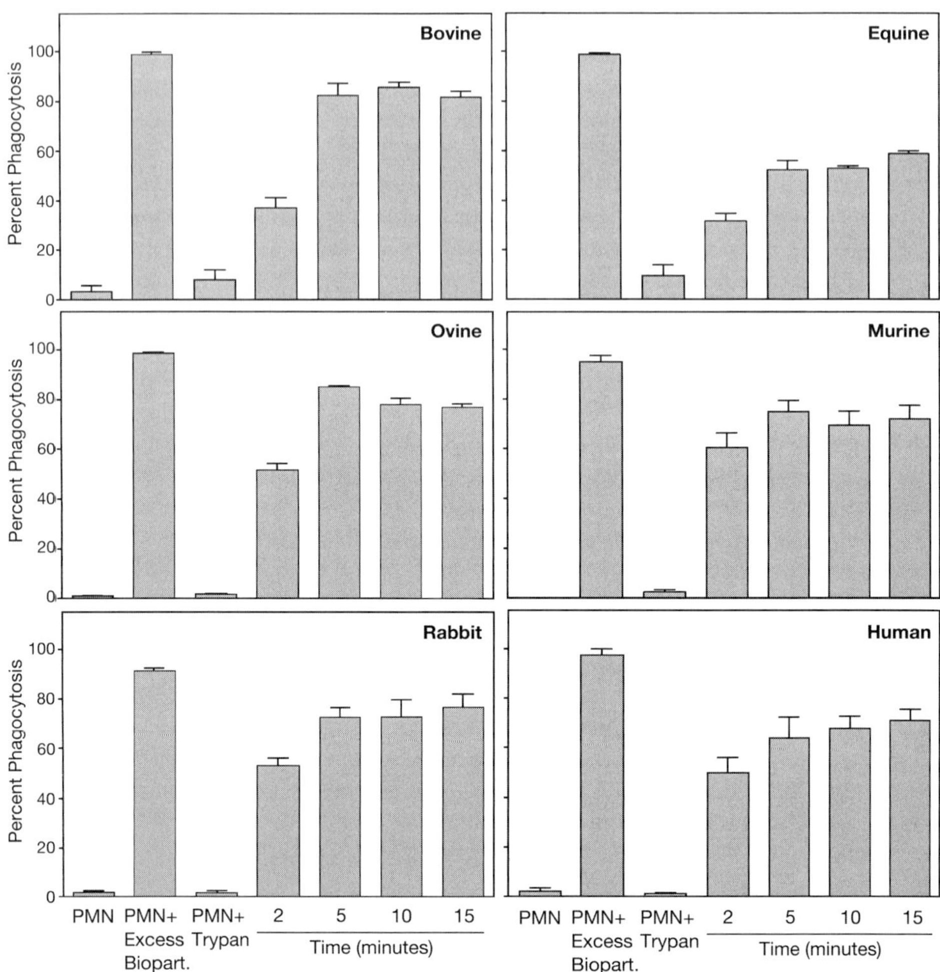

Fig. 2. Functional analysis of purified neutrophils. Neutrophils purified from the indicated species were analyzed for their ability to phagocytose fluorescent bioparticles, as described. Cells were incubated with bioparticles (1:20 ratio) for 2, 5, 10, and 15 min at 37°C. Control samples include neutrophils alone (negative control), neutrophils incubated with trypan blue (trypan control), and neutrophils incubated with excess (1:200 ratio) bioparticles (positive control). In each panel, the results are presented as the mean ± SD of triplicate samples. Representative of at least three experiments for each panel.

7. Do not extend this incubation longer than 20 s, as longer incubation in hypotonic solution can alter and/or damage the neutrophils.
8. After the supernatant has been removed from the gradients, cotton applicators may be used to wipe the walls of the centrifuge tube to remove any adherent debris,

which may contaminate the preparation. Be sure to avoid touching the neutrophil band or pellet with the applicator.

9. As an alternative, rabbit red blood cells can also be aggregated with 6% dextran (100,000–200,000 molecular weight) for 30–40 min *(13)*; however, hetastarch seems to be more efficient.

10. For some reason, rabbit red blood cells do not lyse as readily as those from other species. Even after two rounds of H_2O lysis, some red blood cells may still be present. If this is the case, remaining red blood cells may be removed by very gently washing the surface of the neutrophil pellet.

11. Forward-scatter vs side-scatter plots yield different profiles as a result of varying lots of trypan blue used to quench external fluorescence.

12. Note that neutrophil priming or activation causes an increase in cell size and granularity, which can also be evaluated with these dot plots (*see* **Fig. 1B**).

13. The relative amount of bioparticles and DMEM can be adjusted up or down to achieve different neutrophil:bioparticle ratios. However, we found that increased bioparticle concentrations resulted in close to 100% phagocytosis at the 2-min time point, making it difficult to distinguish differences in rates of phagocytosis between samples.

14. Most phagocytosis experiments showed a decrease in the percent of positive cells after 5 min, which has been reported to be a result of acidification of the phagosomal compartments over time, resulting in quenching of the bioparticle fluorescence *(12)*.

Acknowledgments

This work was supported in part by National Institutes of Health grants R01 AR42426 and P01 RR020185, US Department of Agriculture National Research Inititative/Competitive Grants Programs grants 2006-35204-16563 (M. T. Q.) and 2004-35204-14637 (B. L.), and the Montana State University Agricultural Experimental Station.

References

1. Wiles, S., Hanage, W. P., Frankel, G., and Robertson, B. (2006) Modelling infectious disease—time to think outside the box? *Nat. Rev. Microbiol.* **4,** 307–312.

2. Casal, M. and Haskins, M. (2006) Large animal models and gene therapy. *Eur. J. Hum. Genet.* **14,** 266–272.

3. Styrt, B. (1989) Species variation in neutrophil biochemistry and function. *J. Leukoc. Biol.* **46,** 63–74.

4. Glasser, L. and Fiederlein, R. L. (1990) The effect of various cell separation procedures on assays of neutrophil function. A critical appraisal. *Am. J. Clin. Pathol.* **93,** 662–669.

5. Watson, F., Robinson, J. J., and Edwards, S. W. (1992) Neutrophil function in whole blood and after purification—changes in receptor expression, oxidase activity and responsiveness to cytokines. *Biosci. Rep.* **12,** 123–133.

6. Forsyth, K. D. and Levinsky, R. J. (1990) Preparative procedures of cooling and re-warming increase leukocyte integrin expression and function on neutrophils. *J. Immunol. Methods* **128,** 159–163.

7. Macey, M. G., Jiang, X. P., Veys, P., McCarthy, D., and Newland, A. C. (1992) Expression of functional antigens on neutrophils. Effects of preparation. *J. Immunol. Methods* **149,** 37–42.

8. Alvarez-Larrán, A., Toll, T., Rives, S., and Estella, J. (2005) Assessment of neutrophil activation in whole blood by flow cytometry. *Clin. Lab. Haematol.* **27,** 41–46.

9. Pycock, J. F., Allen, W. E., and Morris, T. H. (1987) Rapid, single-step isolation of equine neutrophils on a discontinuous Percoll density gradient. *Res. Vet. Sci.* **42,** 411–412.

10. Lowell, C. A., Fumagalli, L., and Berton, G. (1996) Deficiency of Src family kinases p59/61hck and p58c-fgr results in defective adhesion-dependent neutrophil functions. *J. Cell Biol.* **133,** 895–910.

11. Woldehiwet, Z., Scaife, H., Hart, C. A., and Edwards, S. W. (2003) Purification of ovine neutrophils and eosinophils: Anaplasma phagocytophilum affects neutrophil density. *J. Comp. Pathol.* **128,** 277–282.

12. White-Owen, C., Alexander, J. W., Sramkoski, R. M., and Babcock, G. F. (1992) Rapid whole-blood microassay using flow cytometry for measuring neutrophil phagocytosis. *J. Clin. Microbiol.* **30,** 2071–2076.

13. Doerschuk, C. M., Allard, M. F., Martin, B. A., MacKenzie, A., Autor, A. P., and Hogg, J. C. (1987) Marginated pool of neutrophils in rabbit lungs. *J. Appl. Physiol.* **63,** 1806–1815.

14. DeLeo, F. R., Renee, J., Mccormick, S., et al. (1998) Neutrophils exposed to bacterial lipopolysaccharide upregulate NADPH oxidase assembly. *J. Clin. Invest.* **101,** 455–463.

4

Subcellular Fractionation of Human Neutrophils and Analysis of Subcellular Markers

Lene Udby and Niels Borregaard

Summary

The neutrophil has long been recognized for its impressive number of cytoplasmic granules that harbor proteins indispensable for innate immunity. Analysis of isolated granules has provided important information on how the neutrophil grades its response to match the challenges it meets on its passage from blood to tissues. Nitrogen cavitation was developed as a method for disruption of cells on the assumption that sudden reduction of the partial pressure of nitrogen would lead to aeration of nitrogen dissolved in the lipid bilayer of plasma membranes. We find that cells are broken by the shear stress that is associated with passage through the outlet valve under high pressure, and that this results in disruption of neutrophils while leaving granules intact. The unique properties of Percoll as a sedimentable density medium with no inherent tonicity or viscosity are exploited for the creation of continuous density gradients with shoulders in the density profile created to optimize the physical separation of granule subsets and light membranes. Immunological methods (sandwich enzyme-linked immunosorbent assays) are used for quantitation of proteins that are characteristic constituents of the granule subsets of neutrophils.

Key Words: Nitrogen cavitation; Percoll density gradient centrifugation; subcellular fractionation; granules; secretory vesicles; enzyme-linked immunosorbent assay; myeloperoxidase; lactoferrin; gelatinase; latent alkaline phosphatase.

1. Introduction

Human neutrophils undergo remarkable changes of activity to adjust to the different environments that they encounter and to meet the demands of fulfilling their role as the primary mobile innate immune defense system. Because most of the neutrophil's anti-microbial defense mechanisms can damage host tissues, inappropriate activation of neutrophils may cause severe tissue destruction and can be life-threatening, as exemplified by Wegener's Granulomatosis and adult respiratory distress syndrome (ARDS). To minimize the risk of inappropriate

From: *Methods in Molecular Biology, vol. 412: Neutrophil Methods and Protocols*
Edited by: M. T. Quinn, F. R. DeLeo, and G. M. Bokoch © Humana Press Inc., Totowa, NJ

activation while circulating and yet to secure a rapid and effective alertness for activation when needed, the circulating neutrophil has stored most of its antimicrobial armory in granules, and most of the receptors that are needed for activation of granule exocytosis and activation of the respiratory burst, inside vesicles. Thus, the circulating neutrophil is a dormant cell that does not readily respond to activation by the receptors designed for sensing microbial invasion and the system's response to such *(1,2)*. The primary key to activation is the endothelium in the postcapillary venules, which responds to microbial invasion or tissue damage by exposing selectins on the surface in order to capture bypassing dormant neutrophils and to cause their awakening by signals that mobilize secretory vesicles as result of selectin and P-selectin glycoprotein ligand (PSGL)-1-mediated interaction. Secretory vesicles are the most readily mobilizable of the neutrophil's storage organelles and are created by endocytosis during terminal neutrophil maturation in the bone marrow. Secretory vesicles contain plasma proteins and thus do not release toxic substances when they fuse with the plasma membrane and empty their content. Their importance lies in their membrane, which is the main store of receptors that, when translocated to the cell surface, allow the neutrophil to interact with other ligands presented by the epithelium and to respond to soluble mediators released from microorganisms and other cells in response to the microbial threat. Thus, with the exception of a minor amount of the membrane components of the NADPH oxidase, secretory vesicles are not the store of neutrophil antibacterial proteins but are better characterized as the key to neutrophil responsiveness.

Not all stored effector mechanisms are needed at the same time. Proteases that disrupt the collagenous network of the basement membrane are needed before the complement receptors that mediate phagocytosis of microorganisms opsonized by soluble innate bacterial recognition systems. Also, pro-antibacterial proteins are stored away from their activating proteases as a means of preventing the unleashing of the potential toxic effector mechanisms before creation of a phagocytic vacuole sets the stage for full orchestration of the antibacterial systems. Therefore, several different subsets of neutrophil granules exist, each with its characteristic profile of proteins and with its specific set-point for degranulation. In fact, neutrophil granules are best described as a continuum from the earliest-appearing azurophil granules to the latest-appearing gelatinase granules, each filled with proteins that are synthesized when the granules are formed. The differences and continuum are determined by the highly individual profiles for biosynthesis of granule proteins during neutrophil maturation in the bone marrow from promyelocyte to segmented neutrophil *(3)*. At the same time, the set-point of the trigger that determines the propensity for degranulation is most likely determined by the density of fusion proteins on the surface of the granules, which increases during myelopoiesis such that azurophil granules contain the

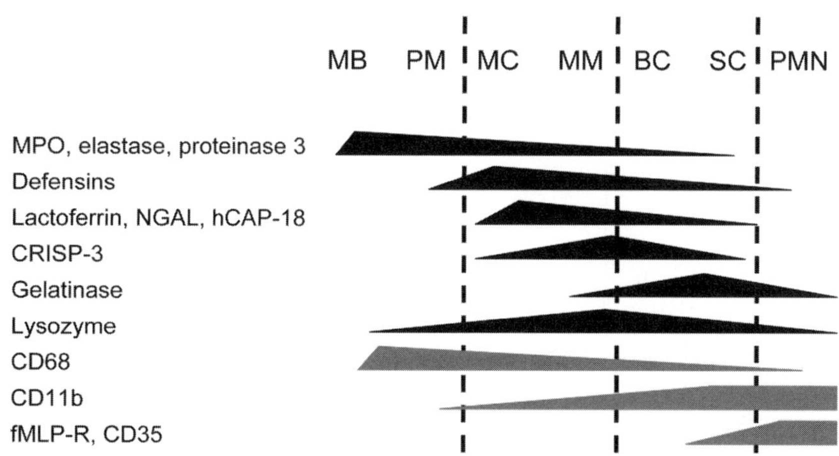

Fig. 1. Targeting by timing hypothesis. Hypothetical mRNA transcription profiles of neutrophil granule proteins during myelopoiesis (adapted from **ref. 3**). The distribution of transcripts for selected proteins is shown in black (matrix proteins) or grey (membrane proteins). MB, myeloblasts; PM, promyelocytes; MC, myelocytes; MM, metamyelocytes; BC, band cells; SC, segmented cells; PMN, polymorphonuclear neutrophils.

lowest amount and secretory vesicles the highest amount of fusion proteins. Thus, there is heterogeneity among both peroxidase-positive (azurophil) granules and peroxidase-negative (specific and gelatinase) granules, but this is only associated with a heterogeneity of the set-points for exocytosis with regard to the peroxidase-negative granules, whereas no difference in extent and direction of degranulation of the subsets of azurophil granules, defined by their presence or absence of defensins, has been documented *(4)*.

The heterogeneity of neutrophil granules can be understood from the targeting-by-timing hypothesis (*see* **Fig. 1**). Although this hypothesis explains the fundamentals of neutrophil granule heterogeneity, it does not completely explain the presence or absence of proteins in all granules, because another as yet not well described mechanism(s) determines whether proteins are sorted to granules or to the constitutive secretory pathway. This is particularly clear for proteins such as defensins and bactericidal permeability-increasing protein (BPI) that are only routed to granules during the promyelocyte stage and largely routed to the constitutive secretory pathway during the myelocyte stage, at which most of the proteins are synthesized.

Another fortuitous consequence of the heterogeneity of granules is that they differ by density. In general, the earliest-formed granules are the densest, with the marked exception being granules that are high in defensins. Because defensins are present in such high concentration in the defensin-positive azurophil granules, these are the granules of highest isopycnic density of the neutrophil.

1.1. Subcellular Fractionation on Percoll Density Gradients

The method for subcellular fractionation that we have developed takes advantage of the excellent properties of Percoll *(5)*. In contrast to the alternative, sucrose, density and tonicity are completely dissociated in Percoll, because Percoll owes its density to the concentration of silica macromolecules. The tonicity is thus determined by the medium in which Percoll is dissolved or diluted. Another fundamental advantage is that Percoll has close to no viscosity, again in contrast to the sucrose solutions. This means that in a Percoll gradient, subcellular organelles reach their isopycnic density in a matter of minutes by high-speed centrifugation in contrast to the hours of high-speed centrifugation needed in a sucrose gradient. Finally, and perhaps most importantly, Percoll gradients are dynamic even during high-speed centrifugation. Because the large polyvinylpyrrolidone (PVP)-covered silica molecules sediment by ultracentrifugation, Percoll was introduced as a self-generating density profile medium with the understanding that if broken cells were mixed with a uniform solution of Percoll, this would then form a self-generated density profile. Although this is certainly true, there increase in density is shallow from the top of the gradient almost to the bottom where the density then rises almost exponentially, making Percoll less well suited for separation of organelles denser than plasma membranes and other light membranes, as these are separated by only a few microliters of density medium although they may differ significantly in isopycnic density (*see* **Fig. 2**).

We took advantage of the fact that Percoll does modulate its preset density profile even during short-term, high-speed centrifugation, and thus initially constructed a two-layer Percoll density gradient that would create a major physical separation between granules with the density of peroxidase-negative granules and the density of azurophil (peroxidase-positive) granules *(6)*. Thus, although constructed as a discontinuous density gradient, the two-layer Percoll density gradient works as a continuous density gradient with two main separating shoulders, one that separates light membranes from granules and one that separates peroxidase-negative from peroxidase-positive granules (*see* **Fig. 2**). When we subsequently discovered that gelatinase granules are distinctly lighter than specific granules (by analyzing the shoulder that separates azurophil from specific granules), we put in yet another major shoulder that gives a large physical separation of gelatinase-rich granules from lactoferrin-rich granules *(7,8)*. This three-layer gradient gives the same separation of light membranes as the two-layer gradient, and thus allows the identification of secretory vesicles as distinct from plasma membranes but does not allow a major physical separation (*see* **Fig. 3**). Such has been achieved by high-voltage free-flow electrophoresis *(9)*, and also by floatation on density gradients *(10)*.

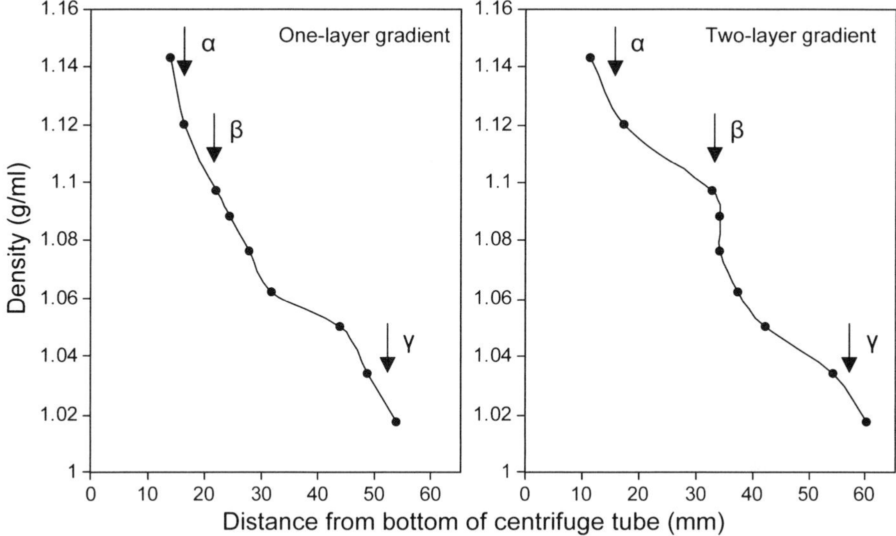

Fig. 2. Density profile of Percoll gradients after centrifugation. One-layer gradient: initial density 1.080 g/mL. Two-layer gradient: initial density 1.050/1.120 g/mL. Arrows indicate positions of the three visible bands (α, β, and γ bands) representing azurophil (peroxidase-positive) granules, peroxidase-negative (specific and gelatinase) granules, and light membranes, respectively.

1.2. Disruption of Neutrophils by Nitrogen Cavitation

Nitrogen cavitation is our method of choice for the disruption of neutrophils. Neutrophils are pressurized in a Parr Nitrogen bomb in cavitation buffer that mimics the intracellular environment to "welcome" organelles when these are released from the plasma membrane and relaxes the cytoskeleton to minimize trapping of granules to nuclei *(6,11)*. We do not believe that the disruption of cells is a matter of aeration of nitrogen that has been dissolved in the lipid membranes when the pressure suddenly drops from the initial 375 psi to atmospheric pressure, but that it is a matter of breaking by shear stress when the neutrophils are forced through the valve at the outlet nozzle of the bomb. Our arguments are that breaking is ineffective if the passage through the valve is too rapid—the cavitate should be let out dropwise, not as a continuous flow. Second, the same cavitation efficiently is achieved whether neutrophils have been pressurized for 2 min or 20 min, which clearly has consequences for the amount of nitrogen that is dissolved in the lipid membranes. Finally, as the bomb wears (i.e., the valve wears) the nitrogen pressure needed to obtain efficient cavitation must be increased. We are currently operating at 550 psi. Thus, we believe that the breaking is achieved by shear force and not by cavitation. This also explains

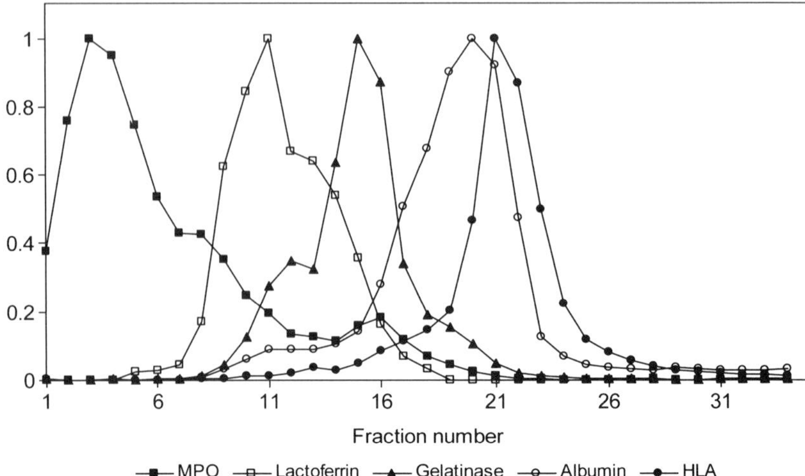

Fig. 3. Subcellular fractionation of human neutrophils on a three-layer Percoll density gradient. The distribution profiles of marker proteins for the different granules and light membranes were determined by enzyme-linked immunosorbent assay of individual fractions collected from the bottom of the tube. Markers are: myeloperoxidase (MPO) for azurophil granules, lactoferrin for specific granules, gelatinase for gelatinase granules, albumin for secretory vesicles, and human leukocyte antigen (HLA) for plasma membrane. Concentrations for each protein are given as measured concentration in the fraction relative to the maximal concentration.

the exquisite, gentle nature of this method for breaking of cells: only the surface membrane is broken, not the granule membrane.

After breaking of cells, the plasma membrane vesicles reseal, but this is not by random orientation. Based on the study of alkaline phosphatase, it has been documented that the plasma membrane vesicles are oriented right-side-out *(12)*. Previous studies arguing to the contrary most likely were unaware of the presence of secretory vesicles in the plasma membrane preparations.

Although the method is very gentle, it is recommended that the netrophils be pretreated by an effective inhibitor of serine proteases before disruption. We recommend treatment with diisopropylfluorophosphate (DFP).

1.3. Analysis of Subcellular Markers

Methods for detection and quantification of granule proteins with a known distribution in subcellular organelles and hence localization in the gradient are needed to evaluate the quality of the subcellular fractionation and to define the localization of a given protein. In general, immunological methods are preferred, because they may be less sensitive to proteases than activity-based assays. Also, the serine proteases in azurophil granules are irreversibly inhibited by the treatment of cells with DFP and the metalloproteases in peroxidase-negative granules are present in latent forms, which requires activation before measuring activity. We have developed an array of immunological assays (enzyme-linked immunosorbent assays [ELISAs]) that are suitable for quantification of proteins from each of the subcellular compartments, which are separated in the three-layer Percoll gradient (**Table 1**). One advantage of the assays is that removal of Percoll is not necessary, as we have not observed interference of this substance with any of the ELISAs in the sample dilutions used. Also, the same buffers and incubation times are used for all the ELISAs, and the sample dilutions can be reused for more than one assay. This means that several marker proteins can be measured simultaneously in a few hours. Unfortunately, not all of the antibodies used in our assays are commercially available, and some require modification (affinity purification and/or biotinylation) before use. It should, however, be possible to develop similar assays using commercially available antibodies.

The granules and vesicles are characterized by their content of both matrix proteins and membrane proteins *(1,2)*, and both protein types are useful as subcellular markers *(13)*. Azurophil (primary) granules are traditionally characterized by their content of myeloperoxidase, which also is our preferred marker for this type of granule. The other major azurophil granule protein, defensin, is absent from azurophil granules with the lowest density and its precursor (pro-defensin) is present in specific granules, which makes this protein less suitable as an azurophil granule marker *(4,14)*. The two subsets of peroxidase-negative granules, specific (secondary) and gelatinase (tertiary) granules, are separated in the three-layer Percoll gradient. Although lactoferrin is present in both subtypes, the majority of lactoferrin is present in specific granules, and lactoferrin is often used as a marker for this subset. Other proteins with the same localization are neutrophil gelatinase-associated lipocalin (NGAL) and human cationic antimicrobial protein of 18 kDa (hCAP-18), and these may be preferred as specific granule markers. The gelatinase granules are best characterized by their high content of gelatinase (matrix metalloproteinase 9), which defines this subtype *(7)*. The secretory vesicles are characterized by their endocytic origin, which makes the plasma protein, albumin, a suitable marker *(15)*. Although we prefer matrix proteins for markers of subcellular organelles, a membrane protein is obviously necessary for detection of the plasma membrane. We use a mixed enzyme-linked immunosorbent assay for human leukocyte antigen (HLA) class 1, which catches the histocompatibility complexes with an antibody against β_2-microglobulin and detects the complexes with an antibody against the HLA class 1 heavy chain *(16)*. Uncomplexed β_2-microglobulin, which is a constituent of peroxidase-negative granules matrix, is not detected in this assay.

Alkaline phosphatase is present in unperturbed cells both in the plasma membrane and in the membrane of secretory vesicles with a distribution of 30% and

Table 1
Enzyme-Linked Immunosorbent Assay (ELISA) Reagents

ELISA *(Marker for)*	Capture Ab *(Supplier, product)* (Dilution)	Detecting Ab *(Supplier, product)* (Dilution)	HRP-conjugated reagent *(Supplier, product)* (Dilution)	Standard *(Supplier, product)* (Range)
Myeloperoxidase (MPO) *(18)* *(Azurofil gr.)*	Affinity-purified rabbit- anti-MPO *(Dako, A0398)*[a] *(Determined after purification)*[a]	Biotinylated rabbit-anti-MPO *(Dako, A0398)*[b] *(Determined after biotinyl)*[b]	HRP-Avidin *(Dako, P0347)* (1:3000)	MPO from neutrophils[c] (0–100 ng/mL)
Lactoferrin (Lfr) *(19,20)* *(Specific gr.)*	Goat-anti-lactoferrin *(NordicLabs, GAHu/Lfr)* (1:50,000)	Rabbit-anti-Lactoferrin *(Dako, A0186)* (1:2000)	HRP-goat-anti rabbit-Ig *(Dako, P0448)* (1:1000)	Lactoferrin from milk[c] *(Gift from H. Birgens)* (0–400 ng/mL)
NGAL *(21)* *(Specific gr.)*	Rabbit-anti-recombinant NGAL *(Not commercially available)*	Biotinylated capture Ab *(Not commercially available)*	HRP-Avidin *(Dako, P0347)* (1:2000)	NGAL monomer from neutrophils[c] (0–2 ng/mL)
Gelatinase *(19,22)* *(Gelatinase gr.)*	Rabbit-anti-gelatinase (-NGAL) *(Not commercially available)*	Biotinylated, affinity purified rabbit-anti gelatinase[a,b] *(Not commercially available)*	HRP-Avidin *(Dako, P0347)* (1:3500)	Gelatinase from neutrophils[c] (0–5 ng/mL)
Albumin (HSA) *(15)* *(Secretory ves.)*	Rabbit-anti-albumin *(Dako, A0001)* (1:2500)	Biotinylated, affinity purified rabbit-anti HSA *(A0001)*[a,b] *(Determined after biotinyl)*[b]	HRP-Avidin *(Dako, P0347)* (1:2000)	Human serum albumin *(Sigma, A1887)* (0–250 ng/mL)
HLA Class I *(16)* *(Plasma membr.)*	Rabbit-anti-β-2-microglobulin *(Dako, A0072)* (1:1000)	Mouse-anti-HLA-ABC *(Dako, M0736)* (1:750)	HRP-rabbit anti-mouse Ig *(Dako, P0260)* (1:1500)	Dilution buffer for sub- traction of background (Arb. Units)

[a]*See* **Note 22.**
[b]*See* **Note 23.**
[c]*See* **Note 24.**
The dilutions of antibodies are only intended as a guide and may need to be optimized in each laboratory (also batch-dependable). Ab, antibody; HRP, horseradish peroxidase; NGAL, neutrophil gelatinase-associated lipocalin; HLA, human leukocyte antigen.

70%, respectively *(12)*. Because plasma membrane fragments reseal right-side-out following disruption by nitrogen cavitation, the enzyme activity is present on the outside of plasma membrane vesicles. In contrast to this, the enzyme activity is present inside secretory vesicles, from where it is translocated to the outside of the plasma membrane after activation of the neutrophil. Using a substrate that does not penetrate membranes, it is possible to distinguish these two activities. Only activity in the plasma membrane is measured in the absence of detergent, whereas both plasma membrane and secretory vesicle-associated activity is measured in the presence of a detergent that solubilizes membranes. The activity hidden inside the secretory vesicles can thus be calculated as the difference between the two measures and is termed latent alkaline phosphatase activity.

One should bear in mind that co-localization of proteins in the gradient is not necessarily a result of localization in the same organelles. It may be that the proteins are localized in different organelles of equal density. Serglycin is for example found in the light membrane fractions as a result of its localization in Golgi vesicles *(17)*. The argument for co-localization is strengthened if the proteins also follow the same pattern of mobilization in response to different stimuli. The degree of exocytosis can be evaluated using intact cells, but the mobilization is more elegantly visualized by comparing the distribution in subcellular fractionations of unperturbed and stimulated cells, respectively *(8)*. In this way, it is possible to demonstrate the disappearance of proteins from individual fractions in the gradient and the appearance of the same proteins in the supernatant or in the plasma membrane fractions, depending on their matrix or membrane localization.

2. Materials

2.1. Subcellular Fractionation of Neutrophils

A total of $4–5 \times 10^8$ neutrophils is recommended for one Percoll gradient. This can usually be obtained from 400–500 mL of venous blood.

2.1.1. Inactivation of Neutrophil Serine Proteases

1. Freshly isolated neutrophils suspended in isotonic saline or Krebs-Ringer phosphate solution supplemented with glucose (KRG; 130 mM NaCl, 5 mM KCl, 1.27 mM MgSO$_4$, 0.95 mM CaCl$_2$, 10 mM NaH$_2$PO$_4$/Na$_2$HPO$_4$, pH 7.4, 5 mM glucose; *see* **Note 1**).
2. DFP (Sigma-Aldrich). Store at 2–8°C. (**Caution:** DFP is very toxic; *see* **Note 2**).
3. Hamilton microliter syringe with blunt needle (Hamilton Company).
4. 2% NaOH in water (*see* **Note 3**).

2.1.2. Disruption of Neutrophils by Nitrogen Cavitation

1. Disruption buffer (1X): 100 mM KCl, 3 mM NaCl, 3.5 mM MgCl$_2$, 10 mM 1,4-piperazinediethanesulfonic acid (PIPES) (*see* **Note 4**), pH 7.2. Stable for 1 mo at 2–8°C.

2. Adenosine 5'-triphosphate disodium salt (ATP[Na]$_2$) stock solution: 50 mM in water. Titrate to pH 7.0 with 1 N KOH. Store in aliquots at –20°C.
3. Phenylmethylsulfonylfluoride (PMSF) stock solution: 100 mM in anhydrous ethanol. Stable for months at 2–8°C. (**Caution:** PMSF is toxic; *see* **Note 5**).
4. EGTA stock solution: 100 mM in water. Titrate to pH 7.0 (*see* **Note 4**). Store at 2–8°C.
5. Parr Cell Disruption Bomb model 4635 including an 1831 nitrogen filling connection (Parr Instrument Company) and a nitrogen cylinder.

2.1.3. Density Centrifugation on Percoll Gradient

1. Percoll™ (Amersham Biosciences). Density 1.130 ± 0.005 g/mL. Store at 2–8°C.
2. Disruption buffer (10X) stock: 1 M KCl, 30 mM NaCl, 35 mM MgCl$_2$, 15 mM EGTA, 100 mM PIPES (*see* **Note 4**). Titrate pH to 6.8 and adjust volume to 80% of the final volume. Stable for 1 mo at 2–8°C. (10 mM ATP[Na]$_2$ is added from a 50 mM stock solution just before use [*see* **Subheadings 2.1.2** and **3.1.3**] and makes up the final volume.)

2.1.4. Fractionation of Percoll Gradient

1. Automated fraction collector (e.g., FRAC-200 from Amersham Biosciences)
2. Peristaltic pump (e.g., Peristaltic Pump P-1 from Amersham Biosciences).

2.2. Analysis of Subcellular Markers

2.2.1. ELISA Assays

1. Nunc-Immuno™ Plates MaxiSorp (96-well flat-bottom ELISA plates, cat. no. 439454).
2. Capture antibodies, protein standards, detecting antibodies, and horseradish peroxidase (HRP)-conjugated detection reagents (*see* **Table 1**).
3. Carbonate buffer: 50 mM Na$_2$CO$_3$/NaHCO$_3$, pH 9.6. Buffer is colored with 20 mg/L phenol red (optional). Stable for months at 2–8°C.
4. Dilution buffer: 0.5 M NaCl, 3 mM KCl, 8 mM Na$_2$HPO$_4$/KH$_2$PO$_4$, 1% (w/v) bovine serum albumin, 1% (v/v) Triton X-100, pH 7.2. Buffer is colored with 20 mg/L phenol red (optional). Stable for 2 wk at 2–8°C.
5. Wash buffer: 0.5 M NaCl, 3 mM KCl, 8 mM Na$_2$HPO$_4$/KH$_2$PO$_4$, 1% (v/v) Triton X-100, pH 7.2 (*see* **Note 6**).
6. Color buffer: 0.1 M Na$_2$HPO$_4$, 0.1 M citric acid monohydrate, pH 5.0. Store at 2–8°C.
7. Color reagents: ortho-phenylenediamine (OPD) tablets, 2 mg (Kem-En-Tec Diagnostics) and 30% H$_2$O$_2$.
8. Stop solution: 2 N H$_2$SO$_4$. Store at room temperature.
9. Automated ELISA plate washer (e.g., Skanwasher from Skatron Instruments).
10. ELISA plate reader (e.g., Multiskan Ascent microplate photometer from Thermo Electron Corporation).

2.2.2. Assay for Latent Alkaline Phosphatase

1. Nunc 96 MicroWell™ Plates (96-well flat-bottom plates, cat. no. 269620).
2. Barbital buffer: 50 mM sodium diethylbarbiturate, 0.6 mM MgCl$_2$, pH 10.5.

3. Phosphatase substrate: *p*-Nitrophenyl phosphate disodium hexahydrate (Sigma) 2 mg/mL in water. Prepare fresh.
4. Triton X-100 stock: 10% (v/v) in water.
5. ELISA plate reader (e.g. Multiskan Ascent microplate photometer from Thermo Electron Corporation).

2.2.3. Removal of Percoll

An ultracentrifuge is needed for the removal of Percoll.

3. Methods

3.1. Subcellular Fractionation of Neutrophils

All steps (*see* **Note 7**) are performed at 4°C with precooled buffers and equipment. Cell suspensions are kept on ice (*see* **Note 8**).

3.1.1. Inactivation of Neutrophil Serine Proteases

1. Suspend the isolated neutrophils in isotonic saline or KRG at 3×10^7 cells/mL (*see* **Note 9**). Use a microliter syringe to aspirate the desired volume of DFP directly from the container (*see* **Notes 2** and **3**). Add 5 mM DFP to the cell suspension (approx 5 µL DFP per 10 mL). Carefully mix the cell suspension and leave it on ice for 5 min.
2. Centrifuge for 6 min at 200g and decant the supernatant into an excess of 2% NaOH (*see* **Note 3**).

3.1.2. Disruption of Neutrophils by Nitrogen Cavitation

1. Add 1 mM ATP(Na)$_2$ and 0.5 mM PMSF to the desired volume of disruption buffer (1X) just before use (for 10 mL of disruption buffer [1X], add 0.2 mL from a 50 mM ATP[Na]$_2$ stock and 0.05 mL from a 100 mM PMSF stock).
2. Resuspend the DFP-treated cells in the initial volume of this buffer in a 50-mL conical Falcon™ centrifuge tube. If the volume of the cell suspension exceeds 30–35 mL, it should be divided into two or more tubes.
3. Place the Falcon tube inside the precooled Parr bomb so that the bottom of the tube is as close as possible to the dip tube leading to the release valve (to ensure complete sample recovery). The Falcon tube should be placed in a plastic beaker cushioned with paper towel or similar (*see* **Fig. 4**).
4. Set the head of the bomb (with the dip tube extending into the cell suspension) on the bottom part and attach the split ring cover clamp. Secure the parts by turning the thumbscrew finger tight. Close the two valves (nitrogen inlet valve and release valve) on the head of the bomb.
5. Attach the filling connection to a nitrogen cylinder and connect the filling hose to the Parr bomb inlet valve. Open the outlet valve on the nitrogen cylinder. Carefully open the nitrogen inlet valve on the Parr bomb and slowly raise the bomb pressure. Close the inlet valve when the pressure inside the bomb reaches 375 psi (or more; *see* **Note 10**). Close the nitrogen cylinder outlet valve, depressurize the filling hose, and disconnect from the Parr bomb inlet valve. Leave the Parr bomb for approx 5 min.

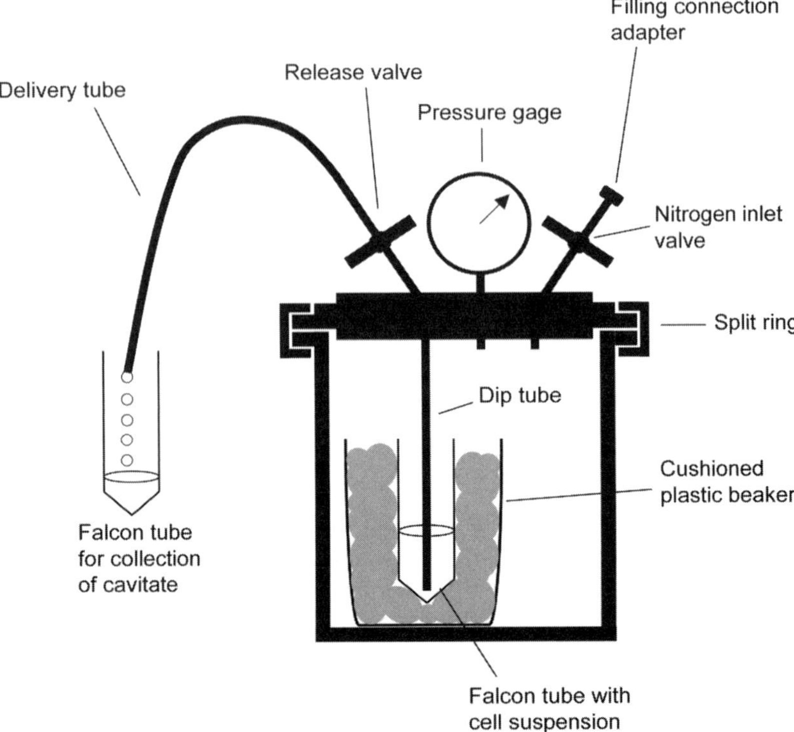

Fig. 4. Schematic illustration of the Parr bomb arrangement during nitrogen cavitation.

6. Prepare a 50-mL Falcon tube for the disrupted cells (i.e., the cavitate): add an amount of 100 m*M* EGTA that will result in 1.5 m*M* after collection of the cavitate (0.15 mL EGTA stock per 10 mL of cell suspension).
7. Place the EGTA-containing Falcon tube under the Parr bomb delivery tube and carefully open the release valve. Hold the Falcon tube in one hand and operate the release valve with the other hand. By adjusting the valve opening, the cavitate should be released quickly dropwise and collected in the Falcon tube. Close the release valve when the majority of the cell suspension has been collected (*see* **Note 11**).
8. Centrifuge the cavitate 400g for 15 min. Carefully decant the postnuclear supernatant (S_1) into a new tube. Resuspend the pellet of nuclei and unbroken cells (P_1) in 1 mL of disruption buffer and save it for determination of disruption efficiency, if desired (*see* **Note 10**).
9. Depressurize the bomb by opening the nitrogen inlet valve. Dismantle and clean the bomb later.

3.1.3. Density Centrifugation on a Three-Layer Percoll Gradient
1. Add 10 m*M* (final concentration) ATP(Na)$_2$ to the 10-fold concentrated disruption buffer stock (*see* **Subheading 2.1.3., item 2**).

Table 2
Preparation of Percoll Solutions of Different Densities

Final density (g/mL)	Final volume (mL)	Percoll stock (1.130 g/mL)[a] (mL)	Disruption buffer (10X) (including 10 mM ATP[Na]$_2$) (mL)	H$_2$O (mL)
1.050	30	10.25	3.0	16.75
1.090	30	19.48	3.0	7.52
1.120	30	26.40	3.0	0.60

[a]If a Percoll stock solution with a different density is used (e.g., 1.129 g/mL) or if one desires to prepare a solution with a different final density, the volume of Percoll stock (V_P) can be calculated using the following formula: $V_P = V_0(D_0 - 1.0056)/(D_P - 1)$, where V_0 is the final volume and D_0 the desired density of the Percoll solution, D_P is the density of the Percoll stock and 1.056 is the density of the 10-fold concentrated disruption buffer. The volume of H$_2$O (V_{H_2O}) should be adjusted to make up the desired final volume [$V_{H_2O} = (0.9 \times V_0) - V_P$]. The volume of 10-fold concentrated disruption buffer (including 10 mM ATP[Na]$_2$) is always one-tenth the final volume.

2. Prepare three Percoll solutions with different densities: 1.05, 1.09, and 1.12 g/mL as indicated in **Table 2** (*see* **Note 12**). For one gradient, 10 mL of each is needed, but preparation of a larger volume is recommended. Add PMSF to a final concentration of 0.5 mM (0.15 mL PMSF stock for 30 mL of Percoll solution) and titrate the solutions to pH 7.0 by addition of a small volume of 1 N HCl.
3. Aspirate 10 mL of the solution with the lowest density (1.05 g/mL) using a 10-mL syringe and a pleurascenthesis needle (14 G × 3.25 inches). Place 9 mL of the solution in a 50-mL round bottom polycarbonate centrifuge tube (Sorvall).
4. Aspirate 10 mL of the medium density solution (1.09 g/mL) with a new syringe and needle. Carefully place 9 mL of this solution in the tube underneath the 1.05 g/mL solution. Make sure that a sharp demarcation between the two densities is obtained.
5. Likewise, aspirate 10 mL of the densest solution (1.12 g/mL) and place 9 mL of this underneath the medium dense Percoll solution.
6. Cautiously, place 10 mL of the postnuclear supernatant (S_1) on top of the gradient without disturbing the Percoll solution. Use a 10-mL syringe or a pipet for this procedure. Save a sample of the S_1 for determination of disruption efficiency or recovery, if desired (*see* **Notes 10** and **20**).
7. Centrifuge the gradient 37,000g for 30 min in an SS-34 fixed angle rotor in a Sorvall RC-5B Superspeed centrifuge. After centrifugation, four distinct bands containing granules and light membranes should be clearly visible: from the bottom, the α-band containing azurophil granules, the β_1-band containing specific granules, the β_2-band containing gelatinase granules, and the γ-band containing secretory vesicles, plasma membranes, and other light membranes (ER and Golgi vesicles). The clear cytosol (S_2) is present on top of the γ-band (*see* **Fig. 5**).

3.1.4. Fractionation of Percoll Gradient

1. Cover the opening of the centrifuge tube with Parafilm and place the tube in a rack to secure the vertical position. Be careful not to disturb the gradient.

Fig. 5. Photograph of the three-layer Percoll density gradient before (middle) and after centrifugation (right). A two-layer gradient after centrifugation (left) is shown for comparison. The bands formed after centrifugation are indicated. (Reprinted from **ref. 8**, with permission from Elsevier.) (color insert will appear after page 300)

2. Stick a glass capillary tube through the center of the Parafilm to the bottom of the tube.
3. Attach a polyethylene tube to the glass capillary tube and connect it to a peristaltic pump and a fraction collector.
4. Collect 37 fractions of 1 mL each. Use clear tubes, which reveal the different granule content (*see* **Note 13**).

3.2. Analysis of Subcellular Markers

3.2.1. ELISA Assays

All steps are performed at room temperature. The necessary volume of buffers, except carbonate buffer, should be brought to room temperature before use. It is recommended to use an eight-channel pipet for application of buffer and antibody dilutions.

3.2.1.1. Day 1

1. Dilute capture antibodies in carbonate buffer according to **Table 1** and immediately coat ELISA plates with 100 µL/well. Incubate plates overnight at room temperature (*see* **Note 14**).
2. Dilute samples (i.e. subcellular fractions and $S_1 + P_1$) in dilution buffer according to **Table 3**. Store samples at 2–8°C (*see* **Note 15**).

Table 3
Dilution of Subcellular Fractions for Marker Assays

ELISA[a]	Fractions[b]	Peak fractions	S_1	P_1
Myeloperoxidase	All fractions 1:2000	Fractions 2–5 1:5000	1:2000 and 1:5000	1:5000 and 1:10000
Lactoferrin and NGAL	All fractions 1:2000	Fractions 9–12 1:5000	1:2000 and 1:5000	1:5000 and 1:10000
Gelatinase	Fractions 1–11 and 18–36 1:2000	Fractions 12–14 1:5000 / Fractions 15–17 1:10000	1:5000 1:10000	1:10000 and 1:20000
Albumin	All fractions 1:15	Fractions 19–23 1:50	1:15	1:15 and 1:50
HLA Class I	All fractions 1:15	Fractions 19–23 1:50	1:20 and 1:100	1:100 and 1:200

[a]Refers to the enzyme-linked immunosorbent assays (ELISAs) described in **Table 1** and under **Subheading 3.2.1.**

[b]In the subcellular fractions containing the cytosol, it is sufficient to measure granule markers in every second fraction (i.e., fractions 30, 32, 34, and 36).

The sample dilutions are adjusted to a subcellular fractionation of a three-layer Percoll gradient with 3×10^8 cells fractionated in 37 fractions of 1 mL each, as described under **Subheading 3.1.** The dilutions are only intended as a guide and may need to be optimized in each laboratory, especially if other antibodies or ELISA kits are used in marker assays.

NGAL, neutrophil gelatinase-associated lipocalin; HLA, human leukocyte antigen.

3.2.1.2. Day 2

1. Wash ELISA plates in wash buffer (*see* **Note 16**). Quench the plates by applying dilution buffer 200 µL/well and incubate for 1 h.
2. Bring samples to room temperature and prepare standard solutions according to **Table 1** (*see* **Note 17**).
3. Wash ELISA plates in wash buffer. Apply standards and samples 100 µL/well. Standards are applied in duplicate in a twofold dilution in lanes 1–2. Samples are applied in duplicate in lanes 3–12 (*see* **Table 4**). Incubate for 1 h.
4. Wash ELISA plates in wash buffer. Apply detecting antibody 100 µL/well and incubate for 1 h.
5. Wash ELISA plates in wash buffer. Apply HRP-conjugated reagent 100 µL/well and incubate for 1 h.
6. Prepare color solution: for one plate, dissolve three OPD tablets in 15 mL color buffer. Protect the solution from light (*see* **Note 18**; **Caution:** OPD is toxic).
7. Wash ELISA plates in wash buffer. Equilibrate plates in color buffer by pouring 50–100 mL buffer over each plate. Remove buffer after 30 s to 5 min.
8. Add H_2O_2 to the color solution just before use: add 1 µL 30% H_2O_2/5 ml color solution. Apply 100 µL/well of this solution and incubate ELISA plates for 10–30 min protected from light (*see* **Notes 18** and **19**).

Table 4
**Distribution of Standards and Samples
on the Enzyme-Linked Immunosorbent Assay (ELISA) plate for Myeloperoxidase ELISA**

	1	2	3	4	5	6	7	8	9	10	11	12
A	0^a	12.5 ng/mL	fr. 1 1/2000	fr. 1 1/2000	fr. 9 1/2000	fr. 9 1/2000	fr. 17 1/2000	fr. 17 1/2000	fr. 25 1/2000	fr. 25 1/2000	fr. 36 1/2000	fr. 36 1/2000
B	0^a	12.5 ng/mL	fr. 2 1/2000	fr. 2 1/2000	fr. 10 1/2000	fr. 10 1/2000	fr. 18 1/2000	fr. 18 1/2000	fr. 26 1/2000	fr. 26 1/2000	fr. 2 1/5000	fr. 2 1/5000
C	1.56 ng/mL	25 ng/mL	fr. 3 1/2000	fr. 3 1/2000	fr. 11 1/2000	fr. 11 1/2000	fr. 19 1/2000	fr. 19 1/2000	fr. 27 1/2000	fr. 27 1/2000	fr. 3 1/5000	fr. 3 1/5000
D	1.56 ng/mL	25 ng/mL	fr. 4 1/2000	fr. 4 1/2000	fr. 12 1/2000	fr. 12 1/2000	fr. 20 1/2000	fr. 20 1/2000	fr. 28 1/2000	fr. 28 1/2000	fr. 4 1/5000	fr. 4 1/5000
E	3.13 ng/mL	50 ng/mL	fr. 5 1/2000	fr. 5 1/2000	fr. 13 1/2000	fr. 13 1/2000	fr. 21 1/2000	fr. 21 1/2000	fr. 29 1/2000	fr. 29 1/2000	fr. 5 1/5000	fr. 5 1/5000
F	3.13 ng/mL	50 ng/mL	fr. 6 1/2000	fr. 6 1/2000	fr. 14 1/2000	fr. 14 1/2000	fr. 22 1/2000	fr. 22 1/2000	fr. 30 1/2000	fr. 30 1/2000	S_1 1/2000	S_1 1/2000
G	6.25 ng/mL	100 ng/mL	fr. 7 1/2000	fr. 7 1/2000	fr. 15 1/2000	fr. 15 1/2000	fr. 23 1/2000	fr. 23 1/2000	fr. 32 1/2000	fr. 32 1/2000	S_1 1/5000	S_1 1/5000
H	6.25 ng/mL	100 ng/mL	fr. 8 1/2000	fr. 8 1/2000	fr. 16 1/2000	fr. 16 1/2000	fr. 24 1/2000	fr. 24 1/2000	fr. 34 1/2000	fr. 34 1/2000	P_1 1/2000	P_1 1/2000

^{a}Dilution buffer is applied for determination of background.

50

9. Terminate color reaction by addition of 2 N H$_2$SO$_4$ 100 µL/well and read the optical density at 492 nm in an ELISA plate reader.
10. Calculate the actual concentration of marker proteins in the undiluted samples (*see* **Note 20**).

3.2.2. Assay for Latent Alkaline Phosphatase

Alkaline phosphatase activity is simultaneously measured in the presence and absence of detergent.

1. Prepare two 96-well plates identically with material from each fraction (5 µL/well). Apply all fractions in duplicate on both plates.
2. Mix 15 mL of barbital buffer with 15 mL of phosphatase substrate. Divide the solution in two equal volumes. Add 300 µL of Triton X-100 stock to one and 300 µL of water to the other assay solution.
3. Apply assay solution 100 µL/well. Use assay solution with detergent (Triton X-100) on one plate and assay solution without detergent on the other.
4. Wrap plates in tin foil and incubate at 37°C for 40 min.
5. Read the optical density at 405 nm in an ELISA plate reader.
6. Calculate the latent alkaline phoshatase activity: for each fraction, subtract the mean activity in the absence of detergent from the mean activity in the presence of detergent (*see* **Note 20**).

3.2.3. Removal of Percoll

We have not experienced any interference of Percoll with the assays described above (**Subheading 3.2.1.** and **3.2.2.**), which is why Percoll is not removed from fractions prior to measurements. In other assays, e.g., Western blotting or functional assays, Percoll may pose a problem, especially in the densest fractions with the highest content of Percoll. It is possible to remove Percoll by ultracentrifugation at 100,000g for 45 min for light membranes or for 90 min for granules (*see* **Note 21**). After centrifugation, the Percoll will form a hard pellet and the membranous material will form a disc or lump above the Percoll. The membranous material can easily be collected using a glass Pasteur pipet and subsequently resuspended in the desired buffer.

4. Notes

1. Unless stated otherwise, all solutions are prepared in water of MilliQ quality (i.e., deionized, depleted from organic content, and filtered). This is referred to as "water" in the text.

   To avoid precipitation, KRG should be prepared fresh from stock solutions: (a) 154 mM NaCl, (b) 150 mM KCl, (c) 150 mM Mg$_2$SO$_4$, (d) 110 mM CaCl$_2$, (e) 100 mM NaH$_2$PO$_4$/Na$_2$HPO$_4$, pH 7.4. Add 100 mL a, 4 mL b, 1 mL c, 1 mL d, and 12 mL e. Adjust to pH 7.4 if necessary. Supplement with 5 mM glucose.
2. DFP is an organophosphorus compound, which is a potent inhibitor of serine proteases including neutrophil elastase, proteinase 3, and cathepsin G. DFP is also an

irreversible inhibitor of acetylcholinesterase, which makes it especially toxic by all routes of administration (inhalation, skin contact, injection, etc.). All work with DFP should be carried out in a well ventilated hood while wearing suitable protective gloves and eye/face protection. Use of a blunt needle for aspiration of DFP from the container and delivery to the cell suspension is recommended to minimize the risk of skin penetration. DFP should be handled only by persons familiar with its properties, and a competent person (e.g., a physician) should be present and alerted of the plan to work with DFP. The antidote atropine sulfate as well as syringes and needles should be present if their use is required.

3. DFP is readily hydrolyzed by alkali. A beaker of 2% aqueous NaOH should be ready in the hood. When aspirating DFP from the container, the needle should be wiped off using a soft tissue wetted in the NaOH solution. All contaminated equipment should be immersed in a large excess of the NaOH solution and left in the fume hood for at least 48 h to ensure complete hydrolysis. The solution can hereafter be neutralized and flushed with a large amount of water into an appropriate disposal system.

4. PIPES is added to the disruption buffer (1X) from a 500 mM stock solution titrated to pH 7.0. For disruption buffer (10X), both PIPES and EGTA are added from stock solutions. The solubility of PIPES and EGTA in water is low, but increases with the addition of NaOH (dissolve PIPES/EGTA directly in a small volume of 1 N NaOH or add 10 N NaOH or NaOH pellets to an aqueous solution).

5. PMSF is an inhibitor of serine proteases and cholinesterase, but is less potent and toxic than DFP (*see* **Note 2**). For preparation of the stock solution, PMSF should be weighed in a fume hood while wearing protective gloves. PMSF is unstable in aqueous solution, with a half-life of <1 h at neutral pH and 25°C, and should be added to buffers just before use.

6. A (5X) wash buffer stock can be prepared and stored for months at 2–8°C: dissolve 584.4 g NaCl, 4.4 g KCl, 4.0 g KH_2PO_4, and 23.0 g Na_2HPO_4: 2 H_2O in 4 L water. Do not adjust pH. Before use, the stock is diluted 5X, titrated to pH 7.2, and supplemented with 1% Triton X-100. This working solution can be stored for at least 1 wk at room temperature.

7. Isolation of neutrophils, except for dextran sedimentation of erythrocytes, is carried out at 4°C to minimize unintended activation with exocytosis of, especially, the secretory vesicles. Nitrogen cavitation and collection of fractions are preferably carried out in a cold room.

8. The described method is also applicable to stimulated neutrophils: cells are preincubated 5 min at 37°C in KRG before addition of stimulus (e.g., fMLP, PMA, or Ionomycin). After stimulation for 15 min at 37°C, cells are pelleted by centrifugation (200g for 6 min) and the supernatant, termed S_0, is aspirated. The cells are resuspended in the original volume of cold buffer before treatment with DFP. Exocytosis of a granule protein can be calculated as the amount in the supernatant divided by the total amount in the cells (i.e., in the supernatant after stimulation and in the postnuclear supernatant and unbroken cells and nuclei after cavitation and centrifugation, see later): $S_0 / (S_0 + S_1 + P_1)$.

9. It is possible to use cell concentrations between 0.5×10^7 and 1.5×10^8 cells/mL throughout the fractionation procedure, but 3×10^7 cells/mL gives an excellent resolution of organelles and the described marker assays (**Subheadings 3.2.1** and **3.2.2.**) are optimized for this cell concentration.

10. As the bomb (i.e., valves) wears, it may be necessary to increase the pressure. The disruption efficiency should be approx 90%, and can be calculated if the amount of a granule marker protein (**Subheadings 3.2.1** and **3.2.2.**) is measured in the postnuclear supernatant (S_1) and the pellet (P_1) of nuclei and unbroken cells following centrifugation of the cavitate {disruption efficiency = $[S_1 / (S_1 + P_1)] \times 100\%$}. Granules should be intact, i.e., less than 0.5% of the granule proteins should be found in the cytosolic fractions after gradient centrifugation *(8)*.

11. Prevent jet, bubbles, or foam when releasing the cell suspension. The last few drops of the disrupted cells will be ejected with high speed, if collection is not terminated. This may cause heavy splattering in the collection tube and loss of sample, because this is practically blown out of the tube. The opening of the Falcon tube may be covered with Parafilm with a slit for the delivery tube and release of excess nitrogen. It is, however, recommended that a face shield and gloves be used for personal protection during the procedure.

12. If separation of gelatinase granules from specific granules is not desired, a two-layer Percoll gradient can be used instead. Percoll solutions of 1.05 and 1.12 g/mL are used, and 14 mL of each are placed in the centrifuge tube. The applied volume of S_1 and density centrifugation is unaltered. Following centrifugation, it is sufficient to collect 25 fractions of 1.5 mL each.

13. It is possible to pool fractions corresponding to one or more types of granule/vesicle, preferably after marker assays have been performed (**Subheading 3.2.**). Usually, fractions 1–7 correspond to azurophil granules, fractions 8–15 to specific granules, fractions 16–20 to gelatinase granules, and fractions 21–25 to secretory vesicles and plasma membranes. Alternatively, the distinct bands can be harvested manually using a plastic Pasteur pipet.

14. Prepare 12 mL antibody solution for each ELISA plate. It is possible to analyze 40 samples on each plate, which is enough for one three-layer Percoll gradient.

15. Dilution of samples can be postponed to d 2, but it is often convenient to do this in advance. Diluted samples can be stored in a refrigerator for 1 wk before ELISA measurements.

16. Between each step, the ELISA plates are washed three times in wash buffer (3×300 μL) and left with 150 μL in each well to prevent drying of the surface. Pour out the buffer over a sink and tap the plate against a towel to remove the last drops just prior to application of the next layer.

17. Standard solutions, except for albumin and HLA ELISA, are made from concentrated stock solutions stored in aliquots at $-20°C$. For preparation of standard stocks, pure standard protein is diluted in 50 mM Na-phosphate, pH 7.4 supplemented with 0.5% (w/v) bovine serum albumin and 0.01% (w/v) Na-azide. Aliquots should not be refrozen after use. Standard solution for the albumin ELISA should be prepared

fresh from lyophilized albumin. No standard is used for HLA ELISA, but dilution buffer is used for subtraction of background.

18. Color solution should be prepared fresh, but dissolution of OPD tablets requires mixing (rotation end-over-end) for 10–15 min. If color solution is prepared for more than one plate, just make 10 mL/plate and 5 mL extra. Color solution will develop color if exposed to light, and this will cause high background in the ELISA. OPD is toxic and possibly carcinogenic; it is also harmful to the aquatic environment. Color solution should be used in a fume hood while wearing protective gloves. Dispose excess color solution and developed ELISA plates according to local regulations.

19. Incubation time may differ between the ELISAs and from day to day partly depending on temperature. After termination of color reaction, the most concentrated standard should result in an optical density (at 492 nm) about 1.5 (1.2–1.8), and background should be below 0.1.

20. It is possible to calculate the recovery as the total content of a marker protein in subcellular fractions expressed in percentage of the content in the postnuclear supernatant applied to subcellular fractionation. Recovery is usually 90–100%.

21. For ultracentrifugation of small volumes, i.e., if Percoll should be removed from individual fractions, it is convenient to use a Beckman Airfuge CLS ultracentrifuge with an A-95 fixed angle rotor and Ultra-Clear (8×20 mm) centrifuge tubes. Centrifuge 300–450 µL of each fraction at 28 psi for 10–20 min (the densest fractions require the longest centrifugation time). To assure rotor balance, fill opposing tubes with fractions of similar density. After centrifugation, the membranous material can be collected above the Percoll pellet with a small pipet and resuspended in the desired volume of an appropriate buffer. A disadvantage of this method may be that cooling is not an option and samples are subjected to temperatures a little above room temperature.

22. Antibodies are affinity-purified using columns containing the relevant marker protein (= ELISA standard) coupled to CNBr-activated Sepharose. The eluted antibodies are dialyzed against phosphate-buffered saline (PBS) and kept at 4°C with the addition of 0.1% (w/v) Na-azide. The proper dilution of antibody in the ELISA depends, among other things, on the IgG concentration and is determined by titration.

23. Biotinylation of antibodies is performed as described by Bayer and Wilchek *(23)*. In short, dialyze antibodies (2–5 mg/mL) against 0.1 *M* Na-carbonate, pH 8.0. For each milliliter, add 50–125 µL biotin-*N*-hydroxysuccinimid (Sigma H1759) from a 2 mg/mL stock in dimethylformamide. Incubate for 1 h at room temperature. Dialyze against PBS, with several changes of buffer to remove excess of free biotin. Keep at 4°C with the addition of 0.1% (w/v) Na-azide, protected from light. The proper dilution of antibody in the ELISA depends on the actual IgG concentration and efficiency of biotinylation and should be determined by titration.

24. Myeloperoxidase is purified from azurophil granules harvested manually from a two-layer Percoll gradient followed by removal of Percoll by ultracentrifugation (*see* **Notes 12** and **13** and **Subheading 3.2.3**). Lactoferrin is purified from human milk. NGAL and Gelatinase are purified from exocytosed material from PMA-stimulated neutrophils.

References

1. Borregaard, N. and Cowland, J. B. (1997) Granules of the human neutrophilic polymorphonuclear leukocyte. *Blood* **89**, 3503–3521.
2. Faurschou, M. and Borregaard N. (2003) Neutrophil granules and secretory vesicles in inflammation. *Microbes Infect.* **5**, 1317–1327.
3. Cowland, J. B. and Borregaard, N. (1999) The individual regulation of granule protein mRNA levels during neutrophil maturation explains the heterogeneity of neutrophil granules. *J. Leukoc. Biol.* **66**, 989–995.
4. Faurschou, M., Sørensen, O., Johnsen, A., Askaa, J., and Borregaard, N. (2002) Defensin-rich granules of human neutrophils: characterization of secretory properties. *Biochim. Biophys. Acta* **1591**, 29–35.
5. Pertoft, H. Laurent, T. C., Låås, T., and Kågedal, L. (1978) Density gradients prepared from colloidal silica particles coated by polyvinylpyrrolidone (Percoll). *Anal. Biochem.* **88**, 271–282.
6. Borregaard, N., Heiple, J. M., Simons, E. R., and Clark, R. A. (1983) Subcellular localization of the b-cytochrome component of the human neutrophil microbicidal oxidase: translocation during activation. *J. Cell Biol.* **97**, 52–61.
7. Kjeldsen, L., Sengeløv, H., Lollike, K., Nielsen, M. H., and Borregaard, N. (1994) Isolation and characterization of gelatinase granules from human neutrophils. *Blood* **83**, 1640–1649.
8. Kjeldsen, L., Sengeløv, H., and Borregaard, N. (1999) Subcellular fractionation of human neutrophils on Percoll density gradients. *J. Immunol. Methods* **232**, 131–143.
9. Sengeløv, H. and Borregaard, N. (1999) Free-flow electrophoresis in subcellular fractionation of human neutrophils. *J. Immunol. Methods* **232**, 142–152.
10. Dahlgren, C., Carlsson, S. R., Karlsson, A., Lundqvist, H., and Sjölin, C. (1995) The lysosomal membrane glycoproteins Lamp-1 and Lamp-2 are present in mobilizable organelles, but are absent from the azurophil granules of human neutrophils. *Biochem J.* **311**, 667–674.
11. Klempner, M. S., Mikkelsen, R. B., Corfman, D. H., and Andre-Schwartz, J. (1980) Neutrophil plasma membranes. I. High-yield purification of human neutrophil plasma membrane vesicles by nitrogen cavitation and differential centrifugation. *J. Cell Biol.* **86**, 21–28.
12. Borregaard, N., Miller, L. J., and Springer, T. A. (1987) Chemoattractant-regulated mobilization of a novel intracellular compartment in human neutrophils. *Science* **237**, 1204–1206.
13. Sørensen, O. and Borregaard, N. (1999) Methods for quantitation of human neutrophil proteins, a survey. *J. Immunol. Methods* **232**, 179–190.
14. Faurschou, M., Kamp, S., Cowland, J. B., Udby, L., Johnsen, A. H., Winther, H., and Borregaard, N. (2005) Prodefensins are matrix proteins of specific granules in human neutrophils. *J. Leukoc. Biol.* **78**, 785–793.
15. Borregaard, N., Kjeldsen, L., Rygaard, K., Bastholm, L., Nielsen, M. H., Sengeløv, H., Bjerrum, O. W., and Johnsen, A. H. (1992) Stimulus-dependent secretion of plasma proteins from human neutrophils. *J. Clin. Invest.* **90**, 86–96.

16. Bjerrum, O. W. and Borregaard, N. (1990) Mixed enzyme-linked immunosorbent assay (MELISA) for HLA class I antigen: a plasma membrane marker. *Scand. J. Immunol.* **31**, 305–313.
17. Niemann, C. U., Cowland, J. B., Klausen, P., Askaa, J., Calafat, J., and Borrregaard, N. (2004) Localization of serglycin in human neutrophil granulocytes and their precursors. *J. Leukoc. Biol.* **76**, 406–415.
18. Borregaard, N., Kjeldsen, L., Sengeløv, H., et al. (1994) Changes in subcellular localization and surface expression of L-selectin, alkaline phosphatase, and Mac-1 in human neutrophils during stimulation with inflammatory mediators. *J. Leukoc. Biol.* **56**, 80–87.
19. Kjeldsen, L., Bjerrum, O. W., Askaa, J., and Borregaard, N. (1992) Subcellular localization and release of human neutrophil gelatinase, confirming the existence of separate gelatinase-containing granules. *Biochem. J.* **287**, 603–610.
20. Birgens, H. S. (1985) Lactoferrin in plasma measured by an ELISA technique: evidence that plasma lactoferin is an indicator of neutrophil turnover and bone marrow activity in acute leukaemia. *Scand. J. Haematol.* **34**, 326–331.
21. Kjeldsen, L., Koch, C., Arnljots, K., and Borregaard, N. (1996) Characterization of two ELISAs for NGAL, a newly described lipocalin in human neutrophils. *J. Immunol. Methods* **198**, 155–164.
22. Kjeldsen, L., Bjerrum, O. W., Hovgaard, D., Johnsen, A. H., Sehested, M., and Borregaard, N. (1992) Human neutrophil gelatinase: a marker for circulating blood neutrophils. Purification and quantitation by enzyme linked immunosorbent assay. *Eur. J. Haematol.* **49**, 180–191.
23. Bayer, E. A. and Wilchek, M. (1990) Protein biotinylation. *Methods Enzymol.* **184**, 138–160.

III

BIOCHEMISTRY, BIOLOGY, AND SIGNAL TRANSDUCTION OF NEUTROPHILS

5

Rac and Rap GTPase Activation Assays

Ulla G. Knaus, Alison Bamberg, and Gary M. Bokoch

Summary

The detection of Ras superfamily GTPase activity in neutrophils is important when studying signaling events elicited by various ligands and cellular processes. Substantial progress in monitoring GTPase activation has been made in recent years by the development of high-affinity probes for small GTPases. These probes are created by fusing a high-affinity GTPase-binding domain derived from a specific downstream effector protein to glutathione-*S*-transferase (GST). Such domains bind preferentially to the GTP-bound form of specific Rho or Ras GTPases. Coupling these probes to beads enables extraction of the complex and subsequent quantification of active GTP-binding protein by immunoblotting. Although effector domains that discriminate efficiently between GDP- and GTP-bound states and highly specific antibodies are not yet available for many small GTPases, analysis of certain members of the Rho and Ras GTPase family are now routinely performed. Here, we describe affinity-based pull-down assays for detection of Rac/Cdc42 and Rap activity in stimulated neutrophils.

Key Words: Rho GTPases; Rac; Rap; affinity-based activity probe; PBD assay; RBD assay; pull-down assay.

1. Introduction

Leukocytes are key cellular components of innate immunity. Almost all of their biological functions are under the control of small GTPases of the Ras superfamily. Specifically, the GTPases of the Rho family have been recognized for their crucial role in neutrophil phagocytic responses including chemotaxis, phagocytosis, and bacterial killing *(1,2)*. The ability to directly measure Rho GTPase activity in cell samples has become an important biochemical tool and has greatly aided our understanding of leukocyte responses to many different stimuli or matrix-mediated changes. GTPases cycle from inactive (GDP-bound)

From: *Methods in Molecular Biology, vol. 412: Neutrophil Methods and Protocols*
Edited by: M. T. Quinn, F. R. DeLeo, and G. M. Bokoch © Humana Press Inc., Totowa, NJ

forms to active (GTP-bound) forms that interact with and regulate components of intracellular signaling pathways. The identification of binding domains in these downstream targets that specifically recognize the active, GTP-bound form of the upstream GTPase has provided the basis for affinity-based GTPase activation assays.

To date, several different probes for monitoring small GTPase activity have been developed; all are based on the use of high-affinity domains of various effectors. For analysis of leukocyte function, two examples of these probes are described here: a Rho family member (Rac/Cdc42)-specific probe and a Ras family member (Rap)-specific probe. The Rac/Cdc42-specific probe is based on a domain within the N-terminal regulatory region of p21-activated kinase (Pak)1. This region, referred to as the Cdc42/Rac interactive binding (CRIB) domain, has a minimal sequence (amino acids 74–89) required for interaction with the active GTP forms of these two GTPases *(3,4)*. For enhanced affinity, a larger region of Pak1 is typically used as a Rac/Cdc42 activity probe, termed the p21-binding domain (PBD) (amino acids 67–150). This PBD domain will bind to the active, GTP-bound form of several Rho GTPases expressed in leukocytes, namely Rac1, Rac2, and Cdc42 *(5–8)*. For the Ras family member Rap, a specific binding site is contained within residues 1–97 of the RalGDS protein, and this domain provides the basis for a Rap-GTP specific probe (RBD) *(9,10)*. The R-ras binding domain (RBD) provides a tool to study the activation of members of the Rap family in leukocytes, which primarily express Rap1A, Rap1B, Rap2B, and Rap2C *(11)*. Rap1 is activated by numerous stimuli in leukocytes *(12)* and may mediate signaling events controlling the oxidative burst.

Using such probes, it is possible to selectively bind GTP-GTPases as this active form is generated during cell stimulation (**Fig. 1**). The isolated GTP-GTPases are then detected through the use of specific antibodies for the particular GTPase being assayed. We describe here in detail the methods for preparing these two probes (PBD, RBD) and for performing Rac/Cdc42 activation assays based on the use of the Pak1-binding domain (PBD probe) *(13)*, and Rap activation assays based on the use of the RalGDS-binding domain (RBD probe) *(14,15)*.

2. Materials

1. The Rac/Cdc42-binding domain of Pak1 (glutathione-*S*-transferase [GST]-PBD, amino acids 67–150) *(3,4)* and the Rap1/Rap2-binding domain of RalGDS *(9,10)* (GST-RBD, amino acids 1–97) cloned into pGEX bacterial expression vectors (Amersham Biosciences) (*see* **Notes 1** and **2**).
2. Luria-Bertani (LB) broth.
3. Ampicillin (final concentration 100 µg/mL).
4. Isopropyl-β-D-thio-galactopyranoside (IPTG).

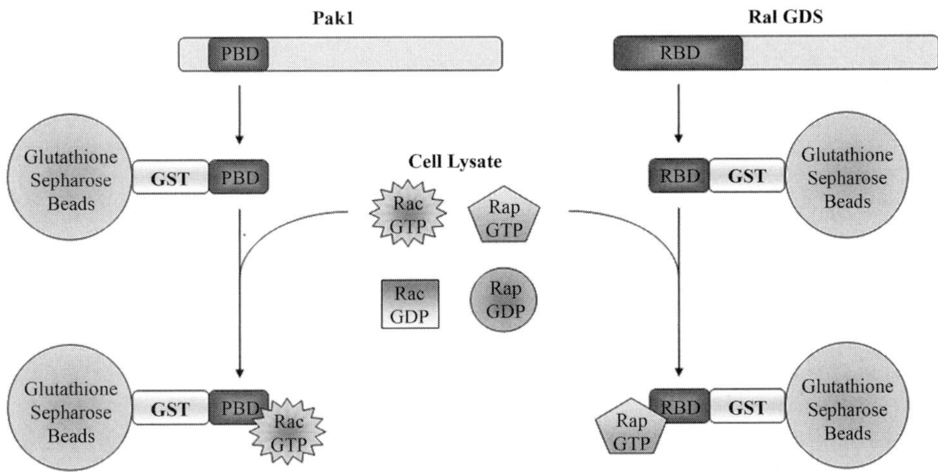

Fig. 1. Principle of the affinity precipitation assay for detection of active Rac and active Rap in cell lysates using the GST-PBD/RBD.

5. Bacterial lysis buffer: 50 mM Tris-HCl, pH 7.5, 150 mM NaCl, 5 mM MgCl$_2$, 1 mM ethylenediamine tetraacetic acid (EDTA), 1mM dithiothreitol (DTT), 1 mM phenylmethylsulfonylfluoride (PMSF), 1 µg/mL aprotinin. Add DTT, PMSF, and aprotinin to the freshly made, ice cold buffer prior to use.

6. Bacterial wash buffer: 50 mM Tris-HCl, pH 8.0, 150 mM NaCl, 5 mM MgCl$_2$, 1 mM DTT, 1 mM PMSF, 1 µg/mL aprotinin. Add DTT, PMSF, and aprotinin to the freshly made, ice cold buffer prior to use.

7. Glutathione Sepharose 4B beads (Amersham Biosciences).

8. Cell lysis buffer (Rac, Rap): 50 mM Tris-HCl, pH 7.5, 200 mM NaCl, 5 mM MgCl$_2$, 1% NP-40, 10% glycerol, 1 mM DTT, 1 mM PMSF, 5 µg/mL aprotinin, 1 µg/mL leupeptin, 2 mM sodium orthovanadate. Add DTT, protease inhibitors, and phosphatase inhibitors to the freshly made, ice-cold buffer just prior to use (*see* **Note 3**).

9. PBD binding buffer (Rac): 25 mM Tris-HCl, pH 7.5, 40 mM NaCl, 30 mM MgCl$_2$, 1% NP-40, 1 mM DTT, 1 mM PMSF, 5 µg/mL aprotinin, 1 µg/mL leupeptin. Add DTT and protease inhibitors to the freshly made, ice-cold buffer prior to use.

10. Phosphate-buffered saline (PBS).

11. GTPγS tetralithium salt (Roche), GDP Tris salt (Sigma).

12. Laemmli sample buffer: 2% sodium dodecylsulfate (SDS), 10% glycerol, 2% β-mercaptoethanol, 60 mM Tris-HCl, pH 6.8, trace of bromophenol blue.

13. Ponceau S solution: 0.2% Ponceau S in 3% trichloroacetic acid.

14. Antibodies. Multiple commercial sources are available. Suggested antibodies are anti-Rac1 monoclonal antibody (23A8) (Upstate), anti-Rac2 polyclonal antibody (Upstate), anti-Rap1 monoclonal antibody (BD Biosciences), or anti-Rap1 polyclonal antibody (Santa Cruz Biotech), anti-Rap2 monoclonal antibody (BD Biosciences).

3. Methods

3.1. Expression of GST-PBD and GST-RBD Fusion Proteins in Bacteria

1. Transform the competent *Escherichia coli* strain BL21(DE3)LysE with pGEX-PBD or pGEX-RBD plasmid by standard procedures and grow bacteria on LB/agar plates with 100 µg/mL ampicillin.
2. Prepare a bacterial stock from a log phase LB/ampicillin culture after inoculating the culture with an individual bacterial colony. Dilute the stock culture 1:1 with sterile 30% v/v glycerol, snap-freeze in liquid nitrogen, and store at −80°C (*see* **Note 4**).
3. From the frozen stock or a streaked plate, inoculate a 50-mL LB/ampicillin culture and grow the culture overnight with shaking at 37°C. The next morning, transfer 25 mL of this culture to 1 L of LB/ampicillin broth. Continue bacterial growth until the optical density $(OD)_{600nm}$ reaches 0.6–0.8 (usually 2 h). Save 50 µL of bacteria culture for final SDS-polyacrylamide gel electrophoresis (PAGE) analysis (this is the uninduced sample).
4. Induce transcription by addition of 0.8 m*M* IPTG (GST-PBD) or 0.1 m*M* IPTG (GST-RBD), and grow bacteria for 3 h at 30°C. Save about 50 µL of this culture for final SDS-PAGE analysis (this is the induced sample).

3.2. Preparation of GST-PBD and GST-RBD Probes

1. Prepare the *E. coli* lysate by spinning down the culture for 10 min at 2300*g* and 4°C.
2. Wash the pellet once with cold PBS, followed by centrifugation, and resuspend the pellet in 10 mL cold bacterial lysis buffer.
3. Incubate on ice for 10 min, and sonicate three times for 30 s on ice without foaming. If necessary, the *E. coli* pellet can be stored after the PBS wash at −80°C. In this case, addition of 1 mg/mL lysozyme and 20 µg/mL DNAse I is recommended for efficient bacterial lysis and DNA shredding.
4. Clear the lysate by centrifugation (15 min, 9300*g*, 4°C). Collect the supernatant and save 50 µL of supernatant sample for final SDS-PAGE analysis (this is the GST-PBD/RBD total sample).
5. Prepare glutathione Sepharose 4B beads by centrifuging the equivalent of 1 mL of dry beads (for 1 L *E. coli* culture) for 5 min at 400*g* and 4°C in a table-top microcentrifuge. Wash beads twice with 10 mL H_2O and twice with 5 mL bacterial lysis buffer.
6. Combine washed glutathione Sepharose 4B beads and *E. coli* supernatant from **step 2** and incubate for 2 h (recommended) or overnight at 4°C while mixing by inverting.
7. Centrifuge the sample for 5 min at 400*g* and at 4°C and save 50 µL of supernatant as GST-PBD/GST-RBD unbound protein sample for final SDS-PAGE analysis.
8. Wash beads five times with 10 mL of ice-cold bacterial wash buffer.
9. Aliquot beads in washing buffer with 10% v/v glycerol (1:1 slurry) and store at −80°C. Determine the protein concentration of GST-PBD or GST-RBD bound to beads with commercially available protein assays.
10. Analyze approx 5 µL of samples from uninduced, induced, GST–PBD/RBD total, GST-PBD/RBD unbound, and final purified GST-PBD/RBD bound to beads ali-

quots (use 5 and 10 µL of 1:1 bead slurry) by SDS-PAGE and Coomassie blue staining to determine protein induction, amounts of soluble and bound protein, and purity of the final GST-PBD/RBD probe (*see* **Note 5**).

3.3. Identification of GTP-Bound Rac and Rap With Activity-Specific Probes in Neutrophils

1. Prepare stimulated neutrophil extracts for PBD/RBD assay by resuspending freshly isolated neutrophils in the appropriate amount of warm stimulation media and preincubate for 5 min at 37°C in a water bath with gentle shaking.
2. Remove an aliquot of the cell suspension, which represents the unstimulated sample.
3. Add desired stimulus, and remove aliquots of the cell suspension at different time points.
4. Immediately place the aliquots on ice, and stop the stimulation by adding an equal volume of ice-cold, 2X-concentrated cell lysis buffer. For adherent neutrophils, the stimulation media is removed at different time points, cells are washed rapidly with ice-cold PBS before cold, 1X cell lysis buffer is added, and cells are scraped from the plates while on ice (*see* **Notes 6** and **7**).
5. Keep the resulting neutrophil lysates on ice for several minutes for efficient extraction, then vortex and clarify by centrifugation for 10 min at 9300g and 4°C. The supernatants can be aliquoted and quick-frozen in liquid nitrogen or placed on ice for immediate use.
6. Determine the protein concentration of each sample and normalize total cell protein for each sample.
7. The maximal final volume is 500 µL, and if required, the samples are diluted in PBD binding buffer (Rac) or cell lysis buffer (Rap). Keep one-tenth to one-twentieth of the original sample for immunoblot determination of total Rac/Rap GTPase content in the lysate (*see* **Note 8**).
8. Add approx 10 µg of purified GST-PBD/RBD beads to samples, and incubate for 1 h at 4°C while mixing by inverting (*see* **Note 9**).
9. Centrifuge the samples at 400g for 2 min, aspirate the supernatant, and wash three times in PBD binding buffer (Rac) or cell lysis buffer (Rap).
10. Add Laemmli sample buffer, heat at 65°C for 5 min, and perform SDS-PAGE on a 12–15% SDS-PAGE gel.
11. Transfer the proteins to nitrocellulose or polyvinylidene difluoride (PVDF) membrane and stain the membrane with Ponceau S to insure equal distribution of the GST-PBD/RBD in different samples.
12. Block the membrane with blocking buffer, and immunoblot with appropriate Rac or Rap antibodies (*see* **Notes 10–12**). A representative example of Rac1 activation in Toll receptor 2-stimulated murine neutrophils, including the recommended control reactions, is depicted in **Fig. 2**.

3.4. Control Reactions

Positive and negative controls for the assay should be performed with unstimulated neutrophil lysates in which the endogenous GTPases are loaded with

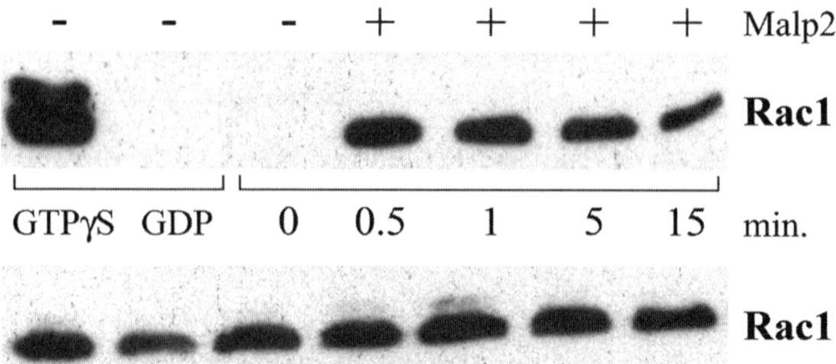

Fig. 2. Rac1 activation in murine neutrophils. Bone marrow-derived neutrophils were stimulated with 1 ng/mL of Malp2 for the indicated time points. Cells were lysed, and Rac1-GTP was bound to the GST-PBD probe using 1000 μg of protein per time point. Bound Rac1 was then detected by immunoblot analysis. The total amount of active Rac1 was determined after loading of the neutrophil lysate with GTPγS. A GDP-loaded lysate served as additional control. The lower panel shows a Rac1 immunoblot using one-twentieth of the cell lysate for each sample.

either GDP (negative control) or with GTPγS (positive control). In addition, lysates derived from cell lines (293T, HeLa, etc.) overexpressing constitutively active Rac1 or Rap1 mutants can be used as positive binding controls, whereas dominant negative Rac1 or Rap1 mutants serve as negative binding controls. Alternatively, a GTPase-regulatory protein such as a constitutively active guanine nucleotide exchange factor (GEF) can be expressed if Rac or Rap represent an in vivo target of a particular GEF (*see* **Note 13**).

1. As a positive control (GTP-bound form), simultaneously add GTPγS at a final concentration of 100 μM and EDTA at a final concentration of 10 mM to cell lysates prepared in cell lysis buffer.
2. Incubate 15 min at 30°C and place sample on ice. Add $MgCl_2$ to a final concentration of 5 mM above initial $MgCl_2$ concentration of buffer to stop nucleotide exchange.
3. As a negative control (GDP-bound form), simultaneously add GDP at a final concentration of 1 mM and EDTA at a final concentration of 10 mM to cell lysates prepared in cell lysis buffer (*see* **Note 14**).
4. Incubate and stop of the reaction is as described above (**step 2**).
5. The GTPγS- or GDP-loaded lysates should be kept on ice and used within 15–30 min by following the procedure for the binding assays as described in **Subheading 3.3.**, **steps 3** and **4**.

4. Notes

1. The described GST-PBD probe will bind the active, GTP-bound form of all three Rac GTPases (Rac1, Rac2, Rac3), as well as the related GTPases Cdc42 and TC10.

The GST-RBD probe described herein will bind to GTP-Rap1 isoforms, GTP-Rap2 isoforms and GTP-Ras isoforms. The RalGDS-RBD should not be confused with the Rhotekin-RBD, a Rho-specific probe. Because other small GTPases may also have the ability to bind to GST-PBD/RBD, identification of an activated GTPase is dependent on the specificity of the antibody used for immunoblotting.

2. Additional high-affinity probes for small GTPases have been developed, including the CRIB domain of WASP (Cdc42), the RBD domain of Rhotekin (Rho proteins), the RBD domain of Raf-1 (Ras proteins), and the RBD domain of RalBP1/RLIP76 (Ral), to name a few. The experimental protocols for preparing and using these probes must be adjusted to their specific requirements. Commercial kits are available for some of the GTPase affinity probes (Upstate, Cytoskeleton, Pierce, etc.).

3. Sodium orthovanadate should be activated for maximal inhibition of protein phosphotyrosyl-phosphatases *(16)*. This procedure depolymerizes the vanadate, converting it into a more potent inhibitor of protein tyrosine phosphatases. Please note that adding DTT rapidly inactivates sodium orthovanadate. Store the activated sodium orthovanadate as aliquots at $-20°C$.

4. Various strains of *E. coli* can be tried to maximize protein expression and purity of the final product. Freshly inoculated and induced cultures give higher protein yields.

5. Breakdown products of the isolated GST-PBD/RBD are often seen. These are usually not detrimental to the assay as long as the intact GST-PBD/RBD is the major product (>75% of total GST fusion protein).

6. Certain ligands may require priming or starvation of neutrophils before stimulation of the cells. These procedures can alter the GTPase activation state.

7. The short wash step is only recommended for adherent cells because it can be done very rapidly, whereas centrifugation of suspension cells would take too long for maintaining the accuracy of short time points.

8. The choice of the cell lysis buffer and the wash conditions for the beads after GTPase binding can be adjusted as long as GTP hydrolysis is minimized. Positive controls should give a strong clear signal, whereas negative controls and untreated control lysates should give little or no signal. Immediate use of cell lysates is recommended for better results; avoid freeze–thawing cycles.

9. Binding of GTP-GTPase to the PBD reaches 75% by 30 min and is maximal by 1 h at 4°C. This time may need to be shortened if GTP hydrolysis in the cell lysate is high, although the binding of GTPase to the PBD or RBD inhibits hydrolysis. It is preferable to add the beads to the sample just after the lysis step to avoid any rapid hydrolysis of GTP to GDP.

10. This method only determines the relative increase of the GTP-bound form of a specific GTPase. One can assume that only about 5–10% of the total GTPase pool will be activated by a given stimulus. This amount is further reduced by the intrinsic GTP hydrolysis rate of the GTPase and the activity of GAP proteins in the cell lysates. These losses can be minimized by shorter incubations and strict adherence to ice-cold conditions. Release of granule contents and high protease activation pose

specific problems in neutrophil activation assays. This can be ameliorated if intact neutrophils are treated before stimulation with the irreversible protease inhibitor diisopropylfluorophosphate (DFP) for 15 min on ice (final concentration 2.7 mM). As DFP is a neurotoxin, use appropriate precautions and perform in a chemical hood. Note that DFP treatment affects some receptor-dependent responses.

11. Unspecific binding of the GTPase to the affinity probe in unstimulated lysates may be due to stress-dependent activation of the GTPase during the isolation or handling of the cells. Careful handling of the cells as well as increasing the NaCl and/or detergent concentration in the wash buffer may reduce background effects.

12. Sensitivity of the assay will be determined to a large extent by the antibody used for detection. Certain antibodies are recommended here. However, before using the affinity probe, the sensitivity and specificity of the antibody must be tested. The abundance of the GTPase in the cell together with the sensitivity of the antibody determines the amount of lysate necessary for efficient detection of the activated GTPase.

13. It is important to determine the maximal signal obtainable from GTPγS-bound GTPase in cell lysates. This allows adjusting the number of cells per sample in order to obtain a signal in the detectable range, assuming that the level of endogenously activated GTPase will usually be on the order of 5–10% of the total Rac or Rap pool. These controls also assure the quality of different preparations of the GST-PBD/RBD probe.

14. The concentration of EDTA has to be adjusted to the magnesium concentration in the cell lysis buffer (5 mM in this protocol). Note that an excess of EDTA is critical for guanine nucleotide loading (a molar excess of two- to fivefold is recommended), whereas an excess of MgCl$_2$ is required to stop the guanine nucleotide exchange.

Acknowledgments

The authors thank Ms. Katrina Schreiber for excellent editorial assistance. This work was supported by grants from the National Institutes of Health.

References

1. Bokoch, G. M. (2005) Regulation of innate immunity by Rho GTPases. *Trends Cell Biol.* **15,** 163–171.

2. Dinauer, M. C. (2003) Regulation of neutrophil function by Rac GTPases. *Curr. Opin. Hematol.* **10,** 8–15.

3. Burbelo, P. D., Drechsel, D., and Hall, A. (1995) A conserved binding motif defines numerous candidate target proteins for both Cdc42 and Rac GTPases. *J. Biol. Chem.* **270,** 29,071–29,074.

4. Thompson, G., Owen, D., Chalk, P. A., and Lowe, P. N. (1998) Delineation of the Cdc42/Rac-binding domain of p21-activated kinase. *Biochemistry* **37,** 7885–7891.

5. Benard, V., Bohl, B. P., and Bokoch, G. M. (1999) Characterization of rac and cdc42 activation in chemoattractant-stimulated human neutrophils using a novel assay for active GTPases. *J. Biol. Chem.* **274,** 13,198–13,204.

6. Geijsen, N., van Delft, S., Raaijmakers, J. A., et al. (1999) Regulation of p21rac activation in human neutrophils. *Blood* **94,** 1121–1130.

7. Akasaki, T., Koga, H., and Sumimoto, H. (1999) Phosphoinositide 3-kinase-dependent and -independent activation of the small GTPase Rac2 in human neutrophils. *J. Biol. Chem.* **274,** 18,055–18,059.

8. Arbibe, L., Mira, J. P., Teusch, N., et al. (2000) Toll-like receptor 2-mediated NF-kappa B activation requires a Rac1-dependent pathway. *Nat. Immunol.* **1,** 533–540.

9. Herrmann, C., Horn, G., Spaargaren, M., and Wittinghofer, A. (1996) Differential interaction of the Ras family GTP-binding proteins H-Ras, Rap1A, and R-Ras with the putative effector molecules Raf kinase and Ral-guanine nucleotide exchange factor. *J. Biol. Chem.* **271,** 6794–6800.

10. Franke, B., Akkerman, J. W. N., and Bos, J. L. (1997) Rapid Ca^{2+}-mediated activation of Rap1 in human platelets. *EMBO J.* **16,** 252–259.

11. Paganini, S., Guidetti, G. F., Catricala, S., et al. (2005) Identification and biochemical characterization of Rap2C, a new member of the Rap family of small GTP-binding proteins. *Biochimie* **88,** 285–295.

12. M'Rabet, L., Coffer, P., Zwartkruis, F., et al. (1998) Activation of the small GTPase Rap1 in human neutrophils. *Blood* **92,** 2133–2140.

13. Benard, V. and Bokoch, G. M. (2002) Assay of Cdc42, Rac, and Rho GTPase activation by affinity methods. *Methods Enzymol.* **345,** 349–359.

14. Van Triest, M. and Bos, J. L. (2004) Pull-down assays for guanoside 5'-triphosphate-bound Ras-like guanosine 5'-triphosphatases. *Methods Mol Biol.* **250,** 97–102.

15. Castro, A. F., Rebhun, J. F., and Quilliam, L. A. (2005) Measuring Ras-family GTP levels in vivo-running hot and cold. *Methods* **37,** 190–196.

16. Gordon, J. (1991) Use of vanadate as protein-phosphotyrosine phosphatase inhibitor *Methods Enzymol.* **201,** 477–482.

6

Measurement of Phospholipid Metabolism in Intact Neutrophils

Susan Sergeant and Linda C. McPhail

Summary

Phospholipid metabolizing enzymes are important participants in neutrophil signal transduction pathways. The methods discussed herein describe assays for assessing the activities of phospholipase (PL)A_2, PLC, PLD, and phosphoinositide 3-OH-kinase (PI3-K) in intact neutrophils. PLA_2 activity is measured as the release of radiolabed arachidonic acid. PLC activity is measured as the accumulation of inositol 1,4,5-trisphosphate (IP_3), a water-soluble product, using a commercially available radioreceptor assay kit. PLD activity is measured as the appearance of its radiolabeled products, phosphatidic acid and phosphatidylethanol. PI3-K activity is measured as the appearance of its radiolabeled product, phosphatidylinositol-3,4,5-trisphosphate.

Key Words: Neutrophil; phospholipase; lipid kinase; phospholipid; signal transduction; lipid second messengers.

1. Introduction

Numerous phospholipids-metabolizing enzymes are expressed in neutrophils, and many appear to play important roles in intracellular signaling pathways that regulate the functional responses of these cells. Thus, it is important to be able to accurately assess the activity of these enzymes in order to better define their role(s) in signaling pathways. Two approaches are generally used to assess the activity of these enzymes. The first measures the appearance of the radiolabeled product(s) and the second measures the increase in the mass of a product. Neither approach alone gives a complete picture of the activity of a particular enzyme because the products themselves are efficiently metabolized, interconverted, and recycled. Moreover, each approach has its limitations, which must be considered when evaluating the results. The ability to measure radiolabeled products requires a reasonably short (30 min to 2 h) incubation of neutrophils with radiolabeled precursors that will sufficiently label phospholipids

From: *Methods in Molecular Biology, vol. 412: Neutrophil Methods and Protocols*
Edited by: M. T. Quinn, F. R. DeLeo, and G. M. Bokoch © Humana Press Inc., Totowa, NJ

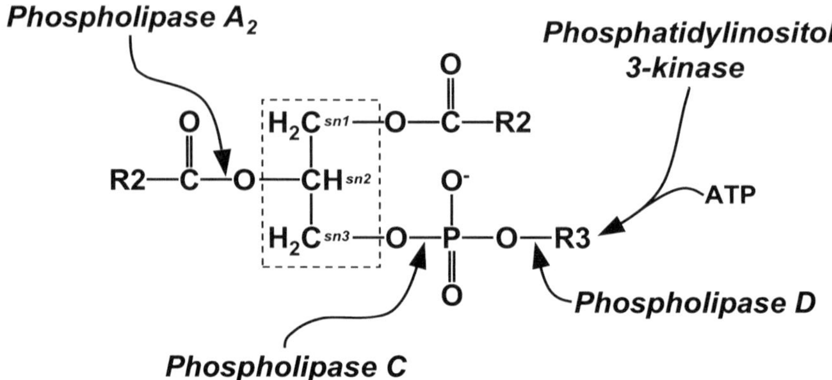

Fig. 1. Sites of phospholipid cleavage by the phospholipases A_2, C, and D and site of phosphorylation by phosphoinositide 3-OH-kinase. The glycerol backbone (dashed box) is esterified to fatty acids (R1 and R2) through the *sn*-1 (top) and *sn*-2 (middle) carbons. The head group (R3; e.g., choline, inositol) is located on the *sn*-3 carbon (bottom) through a phosphodiester bond.

without compromising cell viability and cell function. Although this scheme does not allow for equilibrium labeling *(1)*, it can allow for the detection of product and its metabolites. The determination of product mass requires only detection methods with sufficient sensitivity to measure the small amounts of the transiently produced product(s). Here, we present the most widely used techniques for the activity measurements of phospholipase (PL)A_2, PLC, PLD, and phosphoinositide 3-OH-kinase (PI3-K) in intact neutrophils. However, we have included references to detailed protocols for alternative activity measurements for each enzyme because of the limitations inherent in any single method. No attempt has been made to catalog or reference assays used to assess activity for these enzymes in cell-free systems using exogenous substrates.

PLA_2 cleaves the fatty acid chain from the *sn*-2 position of phospholipids (**Fig. 1**), generating a free fatty acid and a lyso-phospholipid. In neutrophils, the *sn*-2 position is enriched in arachidonic acid (AA), especially in phosphatidylcholine (PC), the most abundant (41%) phospholipid *(2)*. The method for measuring PLA_2 activity described here assesses the release of [³H]-AA from labeled cellular phospholipids into the medium. This method compares favorably to [³H]-AA release assessed by total lipid analysis of cells using thin-layer chromatography (TLC) *(3)*. A more involved method designed to examine the release of [³H]-AA (extra- and intracellularly) and its metabolites can be found, described in detail, elsewhere *(4)*.

PLC primarily utilizes the low abundance (~1% *[2]*) phosphatidylinositol (PI) as a substrate, generating diacylglycerol (DAG) and a water-soluble inositol phosphate as products (**Fig. 1**). The inositol head group of PI undergoes multiple phosphorylations by PI lipid kinases. A major substrate for PLC is phosphatidyl-inositol *(4,5)* bisphosphate (PIP$_2$). When metabolized by PLC, PIP$_2$ yields inositol 1,4,5-trisphosphate (IP$_3$), and its appearance is used as a measure of PLC activation. The method for evaluating PLC activation in neutrophils has been simplified in recent years by the appearance of commercially available kits for the measurement of IP$_3$ mass. This method eliminates a prolonged incubation to label neutrophils with tritiated *myo*-inositol, needed for its incorporation into the phosphatidylinositide pool. However, this *myo*-inositol labeling method *(5–7)* is still suitable when it is necessary to examine all inositol phosphate products, which are fractionated by liquid chromatography *(8)*.

PLD primarily utilizes PC as a substrate, generating phosphatidic acid (PA) and choline as products (**Fig. 1**). PA production can be monitored as mass accumulation *(9)* or as generation of a radiolabeled product *(10,11)*. The latter will be described here. A unique feature of PLD is its transphosphatidylation activity, which is often exploited to assay the activity of the enzyme. In the presence of a primary alcohol (e.g., ethanol, 1-butanol), a phosphatidylalcohol is generated at the expense of PA *(12)*. Metabolically more stable than PA, the phosphatidylalcohol is not subject to the normal degradative pathways of PA. This approach also provides unequivocal evidence of PLD-derived PA, because PA can also be generated by the phosphorylation of DAG (derived from PLC activation) by DAG kinase. However, care must be taken with the use of alcohols because high concentrations can compromise cell viability, and 1-butanol has recently been observed to inhibit one PLD isoform *(13)*. Although not discussed here, a detailed description of a chemiluminescent method for the detection of choline mass as an alternative means of assessing PLD activation has been published *(14)*.

Phospholipids can also be metabolized by lipid kinases, which add phosphate groups to the headgroup (**Fig. 1**). PI, with its inositol ring, is the principal substrate of these enzymes. Of particular interest is PI3-K, because its activation is highly regulated and it appears to be involved in intracellular signaling *(15)*. Three methods are in use to assess PI3-K activation. One assesses the activity of PI3-K in immunoprecipitates by monitoring the generation of a radiolabeled lipid product from an exogenous substrate *(16)*. This method utilizes isoform-specific PI3-K or phosphotyrosine antibodies and it has been used to assess PI3-K activation in neutrophils *(17,18)*. The other two methods detect total product formed in intact cells either as product mass or as a radiolabeled product. The mass method is comparable to the radioreceptor assay used to detect IP$_3$ for PLC activation, except that phosphatidylinositol-3,4,5-trisphosphate (PIP$_3$) in lipid extracts of cells must first be hydrolyzed to a water-soluble ligand (IP$_4$) before use in the binding assay *(19)*. Unfortunately, commercially available kits for this assay are not yet available. However, a detailed description can be found in an earlier volume in this series *(20)*, and this type of assay has been used to assess PI3-K activation in neutrophils *(21)*. The method described here detects total radiolabeled PIP$_3$ by thin layer chromatography after cells are incubated with [^{32}P] inorganic phosphate to label the cellular ATP pool *(22)*.

2. Materials

2.1. Common Materials (Used in Two or More Methods)

1. Buffer A (for labeling incubations): 138 mM NaCl, 5.3 mM KCl, 10 mM HEPES, 0.34 mM Na$_2$HPO$_4$, 0.44 mM KH$_2$PO$_4$, 4.2 mM NaHCO$_3$, 0.49 mM MgCl$_2$, 0.4 mM MgSO$_4$, 5.5 mM glucose, pH 7.4 (*see* **Note 1**).
2. Buffer B (for stimulation incubations): Buffer A plus 1.2 mM CaCl$_2$.
3. Fatty acid-free bovine serum albumin (BSA).
4. Chromatography grade solvents: chloroform, methanol, ethyl acetate, iso-octane, acetone (*see* **Note 2**).
5. Acetic acid (glacial).
6. Silica Gel-60 (20 × 20 cm and 5 × 20 cm) TLC plates (EM Science).
7. Silica Gel-G TLC plates (20 × 20 cm, scored for breakage into half and quarter plates) (Analtech).
8. Solvent System I (for phospholipid analysis): chloroform/methanol/acetic acid/water (100:50:16:6, v/v/v/v).
9. TLC plate guide.
10. TLC tank(s) with lid.
11. Oven (to 110°C).
12. Screw-top glass tubes (13 × 100 mm) with Teflon-lined caps (*see* **Note 3**).
13. Nitrogen (N$_2$) tank and manifold (for drying lipid extracts).
14. Dry (glass bead) bath.
15. Pasteur pipets and bulbs.
16. Glass (capillary) micropipets (5, 10, 25, and 50 μL; Drummond) and pipettor (Drummond) (*see* **Note 4**).
17. Iodine (solid) and a dedicated TLC tank with lid (*see* **Note 5**).
18. Weigh paper (10 × 10 cm).
19. Single-edge razor blades.
20. Scintillation vials.
21. Liquid scintillation counting fluid.

2.2. PLA$_2$ Activity

1. [5,5,8,9,11,12,14,15-^{3}H]-arachidonic acid (1 mCi/mL, 150–230 Ci/mmol, in ethanol; Amersham/GE Healthcare). Store at −20°C. Flush with N$_2$ after each opening.
2. 0.1 N NaOH in H$_2$O.

2.3. PLC Activity

1. Ice-cold 20% perchloric acid (PCA; v/v).
2. 5 M KOH.

3. IP$_3$ radioreceptor assay kit (Amersham, TRK1000 or PerkinElmer, NEK064). Kits contain [^{3}H]-IP$_3$ (*see* **Note 6**).

2.4. PLD Activity

1. 1-O-[^{3}H]-octadecyl lyso platelet activating factor (lysoPC; 1 mCi/mL, 110–210 Ci/mmol in toluene:ethanol; Amersham/GE Healthcare). Store at −20°C (*see* **Note 7**).
2. PLD label buffer: Buffer A containing 2.5 mg/mL BSA.
3. Unlabeled phospholipids for use as standards: PC, phosphatidylethanolamine (PE), phosphatidylserine (PS), PI, PA, lyso-platelet activating factor (lysoPAF) from naturally occurring sources (brain, egg, or liver, monoglyceride [18:1], diglyceride [18:1], triglyceride, and phosphatidylethanol [18:1]; Avanti Polar Lipids).
4. PLD Quench mixture: chloroform/methanol/2%formic acid (1:2:0.2, v/v/v) (*see* **Note 8**).
5. Solvent system II (for PA/phosphatidylethanol): ethyl acetate/iso-octane/acetic acid/water (110:50:20:100, v/v/v/v). Combine components in a 1-L separatory funnel. Mix thoroughly three times, venting between mixings. Let stand (1 h to overnight) in a fume hood. Discard bottom (aqueous) phase. Pour the upper phase in the TLC tank and let stand (covered, ~1 h) before use.

2.5. PI3-K Activity

1. [^{32}P]-orthophosphate (HCl-free; PerkinElmer) (*see* **Note 9**).
2. Buffer C (for labeling incubation): 30 mM HEPES, pH 7.4, 110 mM NaCl, 10 mM KCl, 1 mM MgCl$_2$, and 5.5 mM glucose.
3. Buffer D (for stimulation incubation): Buffer C containing 1.2 mM CaCl$_2$.
4. PI3-K Quench mixture: chloroform/methanol (1:2, v/v).
5. Unlabeled PI, phosphatidylinositol-4-phosphate (PIP$_4$), PIP$_2$, and PIP$_3$ (Avanti Polar Lipids).
6. Solvent system III (for plate preparation): 1.2% potassium oxalate in methanol/water (2:3, v/v).
7. Solvent system IV (for PIP$_n$ separation): chloroform/acetone/methanol/acetic acid/water (80:30:26:24:14, v/v/v/v/v).
8. Lift-Away cleaner (Research Products International) or comparable cleaner for radioactive decontamination.
9. Plexiglass shields and boxes (to transport tubes), neoprene apron.

3. Methods

3.1. Common TLC Methods and Assessment of Labeling

3.1.1. Preparation of Plates

1. Silica Gel-60 plates (general use hard plates): activate plates (100°C, 30 min) and let cool before using (*see* **Note 10**). Mark the origin 1 cm above bottom of plate with a light pencil line using the TLC plate guide as a straightedge. Mark lanes (maximum of six on a 20 × 20 cm plate) of equal width in a similar manner.

2. Silica Gel-G plates (soft plates with no binder): activate plates (100°C, 30min) and let cool before using. Score lines for lanes with a pencil to remove silica gel between lanes using the TLC plate guide. Place the plate guide approx 1 cm above bottom of plate to indicate the origin, but do not mark the origin with any pencil or scoring line (*see* **Note 11**). Lanes can be labeled in the silica gel-free top of the plate with a wax-pencil (*not* a Sharpie-type marker).

3.1.2. Test Plates

1. Use a 5 × 20 cm plate for test plates.
2. Use microcapillary pipets to spot plates at the origin.
3. Let spots dry between application of standard (or sample). Keeping spots as small as possible, apply each lipid standard individually and to a spot called "Mix," which contains all of the standards (*see* **Note 12**). No lane markings are needed on test plates.
4. Develop the test plate in the appropriate solvent system to ensure that lipids will separate well before running radioactive samples. Let the plate air-dry in a fume hood before visualization of lipids.

3.1.3. Visualization of Separated Lipids

Iodine vapor is used to reversibly visualize separated lipids on TLC plates by virtue of the ability of iodine to interact with double bonds (*see* **Note 5**).

1. Sprinkle a small amount of the solid in the TLC tank and allow it to sublimate.
2. Place an air-dried, developed plate in the iodine tank and observe until yellowish-brown spots appear.
3. Remove plate from tank and lightly mark (with pencil) the lipid standard spots or the fractions to be scraped and counted. Allow iodine to evaporate in hood before scraping or analyzing plate.

3.1.4. Scraping TLC Plates and Quantification

After TLC plates have been developed and lipids visualized, scrape the separated lipid fractions and count the samples.

1. Prepare the required number (based on number of samples and fractions for each) of scintillation vials.
2. Carefully scrape the silica gel fractions from the plate (lane by lane) onto a piece of weigh paper with a clean razor blade (*see* **Note 13**).
3. Pour each fraction into a separate vial.
4. Add 0.5 mL of methanol (to help elute lipids/counts from silica gel) and a suitable volume of liquid scintillation counting fluid to each vial.
5. Cap tightly and shake well by hand. Place in liquid scintillation counter, and count for 1 min with a program for [^{3}H]-disintegrations per minute (DPM).

3.1.5. Estimation of Tritiated-Lipid Incorporation into Cells

In initial experiments, it is necessary to determine the extent of tritiated-lipid incorporation into cells.

1. Collect small (known) volumes of samples after the labeling incubation and count by liquid scintillation counting to calculate the percent of total radiolabel incorporated into cells.
2. The samples are: unincorporated (cell-free supernatant and washes) and incorporated (washed cells after labeling incubation).
3. Count samples, and calculate DPM in the final sample volume.
4. Percent incorporation = [total DPM in cells/(total DPM in cells + total DPM in supernatant + washes)] × 100.

3.1.6. Evaluation of Tritiated-Lipid Incorporation into Phospholipids (for PLA_2 and PLD Analysis)

The radiolabeled precursors [^{3}H]-AA and [^{3}H]-lysoPC will predominantly label PC. Still, in initial experiments it is necessary to evaluate which phospholipid(s) the tritiated-lipid has labeled.

1. When labeled, washed cells are aliquoted for stimulation, reserve duplicate aliquots of cell suspension on ice for analysis of phospholipids.
2. Extract lipids, add approx 50 µg lysoPAF standard per tube (only for PLD), and dry the chloroform extract.
3. Prepare a TLC tank for Solvent system I (use a filter paper wick [*see* **Note 14**]), and run a test plate (Silica Gel-G, ~1.5 h development time) with the following lipid standards: PC, PS, lysoPAF (only for PLD), PE, and Mix.
4. Visualize separated lipids in iodine vapor (*see* **Note 5**), and carefully encircle separated lipids with etch lines. In this solvent system, lipids will separate from the origin as follows: lysoPAF, PC, PS, PE, and neutral lipids (near solvent front).
5. Run samples on a prepared G plate, visualize lipids, and scrape fractions. Scrape the entire sample lane, even if no lipid is visualized (*see* **Note 15**).
6. Count fractions by liquid scintillation counting. Average the DPM in lipid fractions of duplicates. Add the average DPM in all fractions for a sample to yield total DPM.
7. Calculate the amount of radiolabel in each fraction as the % of total DPM.

3.2. PLA_2 Analysis

To determine PLA_2 activity, neutrophil cellular phospholipids are labeled with [^{3}H]-AA, and the release of [^{3}H]-AA into the medium is measured after cell stimulation. AA release is then normalized to the total cell-associated radioactivity in unstimulated cells *(23,24)*.

1. Incubate isolated neutrophils (2×10^7 cells/mL in Buffer A) with [^{3}H]-AA (1 µCi/mL) for 30 min at 37°C in a shaking water bath.
2. Pellet cells, and wash cells three times in cold Buffer A containing 1 mg/mL BSA. Take an aliquot of labeling supernatant and of washes to evaluate percent incorporation (*see* **Note 16**).
3. Resuspend cells to 1×10^7 cells/mL in Buffer B containing 1 mg/mL BSA.
4. For each stimulation condition, aliquot cell suspension (2×10^6 cells/condition) to duplicate tubes on ice. Duplicate aliquots should also be added directly to liquid

   scintillation vials to determine total radioactivity per condition (termed as Total samples; *see* **Note 17**).

5. Warm tubes to 37°C, and add desired stimulus or solvent (final vol = 0.4 mL).
6. Add 0.4 mL of ice-cold Buffer B to stop the reactions (except Total samples).
7. Pellet cells (600*g*, 10 min, 4°C), and remove an aliquot of the supernatant to a scintillation vial (AA released).
8. Discard any remaining supernatant (to radioactive waste), and solubilize cells in 0.4 mL of 0.1 N NaOH.
9. Transfer an aliquot of the lysate to a scintillation vial.
10. Add scintillation fluid, cap vials, and count vials for 1 min in a liquid scintillation counter with a program for an output of [^{3}H]-DPM.
11. For a sample, average the supernatant (AA released) DPM and the cell-associated DPM from duplicates. The sum of these two fractions should be constant for all conditions and should be similar to the counts in the Total samples.
12. Express [^{3}H]-AA released as the percent of the total DPM in the unstimulated cell suspension (Total samples).

3.3. PLC Analysis

IP$_3$ is the primary substrate utilized by PLC, resulting in the formation of DAG and inositol phosphate. Thus, PLC activity can be determined by measuring IP$_3$ mass in stimulated neutrophils. In this assay, IP$_3$ in neutrophil samples competes with radiolabeled IP$_3$ for binding to IP$_3$ binding protein. The IP$_3$ concentration in samples is then calculated from a standard curve *(25–27)*.

1. Warm isolated neutrophils (5×10^6 cells/mL in Buffer B), and stimulate cells with desired agonist (final volume, 1–2 mL) (*see* **Note 18**).
2. Add 0.4 volume of ice-cold 20% PCA to stop the reactions (*see* **Note 19**).
3. Vortex samples, and let tubes sit on ice for 20 min.
4. Pellet precipitated proteins (1000*g*, 10 min, 4°C), and transfer extracts to clean tubes.
5. Neutralize the extracts with 5 *M* KOH, and pellet (1000*g*, 10 min, 4°C) the resultant precipitate.
6. Transfer neutralized extracts to a clean tube and use as samples for the radioreceptor assay, as described in the manufacturer's instructions (*see* **Note 20**).

3.4. PLD Analysis

To determine PLD activity, neutrophil phospholipids are labeled with [^{3}H]-lysoPC, and the labeled cells are stimulated in the presence and absence of ethanol. Cleavage by PLD results in the formation of radioactive PA and/or phosphatidylethanol, which is normalized to total lipid-associated radioactivity *(10)*.

1. Resuspend cells at 3.5×10^7 cells/mL in Buffer A and keep on ice.
2. Dry an appropriate volume of stock [^{3}H]-lysoPC (1.6 µCi/mL of cell suspension) under N$_2$ in a microcentrifuge tube.

3. Add 100 µL of PLD label buffer, and vigorously vortex for 5 min to reconstitute the label.
4. Count a small volume (0.5–1 µL) of the reconstituted label to estimate recovery (*see* **Note 21**).
5. Add reconstituted label to cell suspension, and incubate (30 min, 37°C) in a shaking water bath.
6. Pellet cells (225g, 10 min, 4°C), and save the supernatant (to estimate incorporation of label).
7. Wash cells twice in 10 volumes of cold Buffer A. Save the supernatants from washes (to estimate incorporation of label).
8. Resuspend cells in cold Buffer B at 1×10^7cells/mL. Take a small volume (50–100 µL) of this cell suspension to estimate incorporation of label.
9. For each condition, aliquot 0.5 mL of cell suspension into duplicate tubes (e.g., capped 15-mL polypropylene tubes) for stimulation and into tubes for analysis of phospholipids. Keep tubes on ice.
10. To stimulation tubes, add 0.5 mL of Buffer B ± ethanol (0.8% final concentration).
11. Warm tubes for 5 min at 37°C, and add vehicle or stimulus for the desired time.
12. Add 3 mL of PLD quench mixture to stop the reaction, and vortex tubes (*see* **Note 22**).
13. Add 1 mL each of chloroform and Buffer A to the tubes, vortex, and centrifuge (225g, 5 min, 25°C) to split phases.
14. Transfer bottom phase (chloroform) with a glass Pasteur pipet to screw-top glass tubes (13 × 100 mm). Work in a fume hood.
15. Wash the aqueous (upper) phase twice with chloroform (1 mL). Centrifuge as in **step 13**.
16. Pool bottom phases with the original extract in the glass tube (*see* **Note 23**).
17. Add 5–10 µg of unlabeled PA (egg), monoglyceride (18:1), and phosphatidylethanol (di-18:1) to each tube (*see* **Note 24**).
18. Dry samples under N_2. Wash sides of tube two to three times during the drying process with chloroform/methanol (9:1, v/v) to recover sample on the walls of the tube.
19. Dissolve dried samples in 50 µL of chloroform/methanol (9:1, v/v) (see **Note 23**).
20. Prepare solvent system II, and run a test plate (Silica Gel-60 plate, 5 × 20 cm) with lipid standards (PC, PE, PA, phosphatidylethanol, mono-, di-, triglycerides) individually and as a mixture to ensure good separation, as judged by visualization with iodine.
21. Develop plates for approx 2 h (or ~1 cm from top of plate). In this system, the lipids separate from the origin as follows: PC/PE at the origin, PA, phosphatidylethanol, monoglyceride, diglyceride, triglyceride (near solvent front).
22. Prepare Silica Gel-60 plate(s) for samples by spotting samples at origin (*small* spots along the origin line within a lane). Add approx 30 µL of chloroform/methanol (9:1, v/v) to tube as a rinse and spot the rinse in the appropriate lane. Allow lanes to dry before developing the plate in solvent system II.
23. Air-dry developed plate (horizontal orientation) in fume hood, and then visualize separated lipid in iodine vapors.

24. Lightly draw pencil lines to mark five fractions to scrape (#1, PC/PE at origin; #2, PA; #3, phosphatidylethanol; #4, monoglycerides, #5, di-and triglycerides).
25. Scrape TLC plates and quantify scraping as described under **Subheading 3.1.4**.
26. For each sample, average the DPM in lipid fractions of duplicates, and add the average DPM in all fractions for a sample to yield total DPM.
27. Calculate the data as the percentage of total DPM in PA or phosphatidylethanol (if ethanol was included in the stimulation conditions).

3.5. PI3-K Analysis

To measure PI3-K activity, the cellular ATP pool is labeled with $[^{32}P]$-orthophosphate. Cell stimulation induces PI3-K-mediated phosphorylation of PIP_3, resulting in ^{32}P-labeled PIP_3. Radioactivity in the PIP_3 fraction is then normalized to total lipid-associated radioactivity *(22)*.

1. Resuspend cells at 5×10^7 cells/mL in Buffer C.
2. Add $[^{32}P]$-orthophosphate (0.5 mCi/mL), and tightly cap the cell suspension tube (*see* **Note 25**).
3. Incubate the suspension at 37°C for 60 min (or 25°C for 90 min) in a shaking water bath, suitably shielded.
4. Pellet the labeled cells by centrifugation ($225g$, 10 min, 4°C), and wash cells three times in cold Buffer C.
5. Transfer the initial supernatant and washes to a well shielded radioactive liquid waste container. Because this is the bulk of the initial radioactivity, dispose of it as soon as possible.
6. Resuspend the washed cells in cold Buffer D at 2×10^7 cells/mL.
7. For each condition, aliquot cell suspension (0.5 mL) to duplicate tubes (e.g., capped 15 mL polypropylene tube) on ice.
8. Warm tubes for 5 min at 37°C, and add vehicle or stimulus for the desired time.
9. Add 3 mL of PI3-K Quench mixture to stop the reaction, and vortex tubes (*see* **Note 22**).
10. Add 2.1 mL of chloroform, 2.1 mL of 2.4 M HCl, and unlabeled phosphoinositides (10 µg each; *see* **Note 24**) to each tube. Vortex tubes and centrifuge ($225g$, 5 min, 25°C) to separate phases.
11. Transfer bottom phase (chloroform) with a glass Pasteur pipet to a screw-top glass tube (13 × 100 mm). Work in fume hood.
12. Wash remaining aqueous phase four times with 1 mL of chloroform, and centrifuge as in **step 5**. Pool bottom phases with original extract in glass tube.
13. Wash each pooled bottom phase once with methanol/1 M HCl (1:1, v/v), and transfer the washed bottom phase to a clean glass tube (*see* **Note 23**).
14. Dry samples under N_2. Wash sides of tubes two to three times during the drying process with chloroform/methanol (9:1, v/v) to recover sample on tube walls of tube. Dissolve dried samples in 50 µL of chloroform/methanol (9:1, v/v) (*see* **Note 23**).

15. Impregnate Silica Gel-60 plates with potassium oxalate by developing in Solvent system III overnight (*see* **Note 26**). Allow plate to air-dry, and activate plate by heating at 110°C for 15 min. Let plate cool before use.

16. Run a prepared test plate with lipid standards (PIP, PIP_2, PIP_3, PA, individually and as a mixture) to ensure good separation in solvent system IV, as judged by visualization with iodine. Develop plates to the top of the silica gel, and then develop for an additional 15 min. In this system, the lipids separate from the origin as follows: PIP_3, then PIP_2 and PIP (as a doublet in mid-plate) and PA (near solvent front). When lipid separation is satisfactory, proceed with radiolabeled samples.

17. On prepared Silica Gel-60 plates (**step 15**), spot samples at origin (*small* spots along the origin line within a lane). Add approx 30 µL of chloroform/methanol (9:1, v/v) to tube as a rinse, and spot the rinse in the appropriate lane. Allow lanes to dry before developing plate in solvent system IV (*see* **Note 27**).

18. Let developed plate air-dry (horizontal orientation) in fume hood.

19. Visualize and quantify separated, radiolabeled lipids with a Phosphorimager or Typhon scanner (Molecular Dynamics) or comparable instrument. These instruments typically yield radioactivity in terms of density of user-selected areas of the image. This technology eliminates the need to scrape plates.

20. For each sample, average the density output of the duplicates in each radiolabeled lipid fraction. Add the average density in all fractions for a sample to yield total density.

21. Calculate the data as the % of total density in PIP_3.

4. Notes

1. Comparable buffered solutions that maintain cell viability and function can be used. Ca^{2+} is excluded from the labeling incubations to minimize cell activation. Note that Buffer A, which is phosphate buffered, is not suitable for labeling cells with inorganic phosphate for the measurement of PI3-K activity.

2. Use solvents in fume hood and collect solvent and radioactive wastes for disposal as prescribed by your institutional environmental safety department.

3. Only Teflon-lined caps should be used with chloroform-containing lipid extracts. Chloroform will dissolve other plastic-lined caps and Parafilm, resulting in contaminated samples. Do not use polystyrene plastics, because they are unstable to chloroform. However, polyprolyene tubes, which are commonly used for cell incubations, are resistant to chloroform for the short periods needed for the lipid extraction described here.

4. Use the pipettor designed for the glass capillary micropipets to dispense stock radiolabeled compounds and for spotting TLC plates. Do not mouth-pipet solvents or radioactive solutions. Minimize the use of plastic pipet tips for lipid-handling, especially when chloroform is the solvent.

5. Iodine vapor is poisonous. This tank must remain in a fume hood, and always wear gloves to handle any TLC plate.

6. These kits only differ in the protein used as the IP_3 receptor. A complete instruction manual is supplied with these kits. However, additional detailed information concerning the use of these kits can be found in an earlier volume of this series *(26)*.

7. Neutrophils are efficiently labeled with this lysoPC, which is acylated in the *sn-2* position to form PC *(10)*. The use of this lysoPC to label neutrophils is advantageous because labeling requires relatively short incubations (30–75 min) rather than overnight incubations with a tritiated fatty acid, which is a common practice for cultured cells *(28)*. A disadvantage of the lysoPC labeling method is that only the PC pool is labeled.

8. This is a modification (acidified) of the Bligh and Dyer *(29)* extraction mixture.

9. ^{32}P is a strong β-emitting radionuclide with a short half-life, and mCi amounts are required to adequately label cells. From a safety standpoint, it is recommended that fresh radionuclide be obtained for each experiment and be disposed of quickly after use (through a Radiation Safety department) rather than storing the material in a laboratory setting. A thorough understanding of this radionuclide, its safe handling, and required shielding (plexiglass) is imperative before use.

10. Silica gel absorbs humidity. Heating the plates evaporates the water, which might otherwise compromise the separation of lipids. Let plates cool vertically or in a plate rack. Laying a hot plate on a cooler bench top will crack the plate.

11. Removal of the silica gel at the origin will prevent solvent migration up the plate. Avoid using side edges (~1 cm) of G plates. Lacking a binder, the silica gel on the edges of the plate can be easily scraped off if plates are not handled carefully.

12. Lipids can separate slightly differently when chromatographed in a mixture compared to when they are alone.

13. Scraping of plates should be performed on the bench (not in a fume hood) on a clean piece of bench paper. Blades will become dull; use a fresh blade as necessary. Weigh papers can be reused for several samples.

14. For solvent system I, add a wick (~50 × 20 cm piece of chromatography paper) to the inside perimeter of the tank when it is prepared. The wick aids in plate development.

15. Iodine interacts with unsaturated lipids. Thus, not all lipids can be visualized with iodine. Analysis of the entire plate will ensure that the extent of tritiated-label incorporation is not overestimated.

16. After counting, collect all supernatant and washes in a tritium waste container for disposal.

17. In initial experiments, reserve additional sets of duplicate aliquots for assessment of ^{3}H incorporation into cells and phospholipid analysis.

18. The cell concentration and final volume can and should be adjusted for experimental dictates. However, an appropriate number of cells per sample should be between 10^6 and 10^7 cells.

19. Water-soluble inositol phosphates can also be extracted by adding an equal volume of ice-cold 15% trichloroacetic acid. The extract is then washed three times with water-saturated diethyl ether and neutralized to pH 7.5 before assay. Diethyl ether is flammable; work in a fume hood and use suitable precautions.

20. The basal IP_3 content of neutrophils is reported to be 1–2 pmol/million cells *(27,30)*.
21. Alkyl-[^{3}H]-lysoPC is a sticky lipid, and recovery of ≥80% of the theoretical amount of radioactivity added is desirable using the conversion factor of 2.2×10^6 DPM/ μCi. If recovery is low, vortex label again.
22. If two phases are apparent after vortexing, add a few drops of methanol.
23. Extracts can by stored overnight at –20°C after flushing the tubes with N_2 and capping tightly. Tubes should be warmed to room temperature before proceeding.
24. A mixture of these lipids can be prepared for use as unlabeled carrier lipids. Unlabeled PC and di- and triglycerides need not be added because they are of suitable abundance in the neutrophil sample to be visualized with iodine.
25. Because of the safety issues involved in using these quantities of ^{32}P, it is important to carefully plan all aspects of the experiment, especially the cell labeling procedure, during which the risks of exposure and contamination are the greatest. An initial dry run experiment (without radioactivity) is recommended to help develop a procedure that will minimize exposure and maximize containment of all radioactive material and waste (solid and liquid). It is often necessary to dedicate certain equipment (pipettors, centrifuge buckets, vacuum flask) to ^{32}P-labeling protocols as a means of reducing contamination.
26. The potassium oxalate chelates cations that would otherwise interact with the polyphosphoinositides and thereby interfere with their ability to be separated by TLC.
27. Ensure that plates are run in the same direction in solvent system IV as for the pretreatment in solvent system III.

Acknowledgments

Development of the phospholipase D assay was partially supported by National Institutes of Health grant R01 AI-22564 to L.C.M.

References

1. Vadakekalam, J. and Metz, S. (1998) Isotopic efflux studies as indices of phospholipase activation, in *Phospholipid Signaling Protocols* (Bird, I. M., ed.), Humana, Totowa, NJ, pp. 175–185.
2. Mueller, H. W., O'Flaherty, J. T., Greene, D. G., Samuel, M. P., and Wykle, R. L. (1984) 1-*O*-alkyl-linked glycerophospholipids in human neutrophils: distribution of arachidonate and other acyl residues in the ether-linked and diacyl species. *J. Lipid Res.* **25**, 383–388.
3. Cockcroft, S. and Stutchfield, J. (1989) The receptors of ATP and fMetLeuPhe are independently coupled to phospholipases C and A_2 via G-protein(s): relationship between phospholipases C and A_2 and exocytosis in HL60 cells and human neutrophils. *Biochem. J.* **263**, 715–723.
4. Bauldry, S. A., Wykle, R. L., and Bass, D. A. (1998) Phospholipase A_2 activation in human neutrophils. Differential actions of diacylglycerols and alkylacylglycerols in priming cells for stimulation by N-formyl-Met-Leu-Phe. *J. Biol. Chem.* **263**, 16,787–16,795.

5. Di Virgilio, F., Vicentini, L. M., Treves, S., Riz, G., and Pozzan, T. (1985) Inositol phosphate formation in fMet-Leu-Phe-stimulated human neutrophils does not require an increase in the cytosolic free Ca^{2+} concentration. *Biochem. J.* **229,** 361–367.

6. Ferretti, M. E., Nalli, M., Biondi, C., et al. (2001) Modulation of neutrophil phospholipase C activity and cyclic AMP levels by fMLP-OMe analogues. *Cell. Signal.* **13,** 233–240.

7. Skippen, A., Swigart, P., and Cockcroft, S. (2006) Measurement of phospholipase C by monitoring inositol phosphates using [^{3}H]inositol-labeling protocols in permeabilized cells, in *Calcium Signaling Protocols, Second Edition* (Lambert, D. G., ed.), Humana, Totowa, NJ, pp. 183–193.

8. Berridge, M. J., Dawson, R. M. C., Downes, C. P., Heslop, J. P., and Irvine, R. F. (1983) Changes in the levels of inositol phosphates after agonist-dependent hydrolysis of membrane phosphoinositides. *Biochem. J.* **212,** 473–482.

9. Agwu, D. E., McPhail, L. C., Wykle, R. L., and McCall, C. E. (1989) Mass determination of receptor-mediated accumulation of phosphatidate and diglycerides in human neutrophils by Coomassie blue staining and densitometry. *Biochem. Biophys. Res. Com.* **159,** 79–86.

10. Agwu, D. E., McPhail, L. C., Chabot, M. C., Daniel, L. W., Wykle, R. L., and McCall, C. E. (1989) Choline-linked phosphoglycerides, a source of phosphatidic acid and diglycerides in stimulated neutrophils. *J. Biol. Chem.* **264,** 1405–1413.

11. Bauldry, S. A., Elsey, K. L., and Bass, D. A. (1992) Activation of NADPH oxidase and phospholipase D in permeabilized human neutrophils. Correlation between oxidase activation and phosphatidic acid production. *J. Biol. Chem.* **267,** 25,141–25,152.

12. Pai, J.-K., Liebl, E. C., Tettenborn, C. S., Ikegwuonu, F. I., and Mueller, G. C. (1987) 12-*O*-tetradecanylphorbol-13-acetate activates the synthesis of phosphatidylethanol in animal cells exposed to ethanol. *Carcinogenesis* **8,** 173–178.

13. Hu, T. and Exton, J. H. (2005) 1-Butanol interferes with phospholipase D1 and protein kinase Cα association and inhibits phospholipase D1 basal activity. *Biochem. Biophys. Res. Com.* **327,** 1047–1051.

14. Pedruzzi, E., Hakim, J., Giroud, J. P., and Perianin, A. (1998) Analysis of choline and phosphorylcholine content in human neutrophils stimulated by f-Met-Leu-Phe and phorbol myristate acetate: contribution of phospholipase D and C. *Cell. Signal.* **10,** 481–489.

15. Wymann, M. P., Sozzani, S., Altruda, F., Mantovani, A., and Hirsch, E. (2000) Lipids on the move: phosphoinositide 3-kinases in leukocyte function. *Immunol. Today* **21,** 260–264.

16. Endemann, G., Yonezawa, K., and Roth, R. A. (1990) Phosphatidylinositol kinase or an associated protein is a substrate for the insulin receptor tyrosine kinase. *J. Biol. Chem.* **265,** 396–400.

17. Ding, J., Vlahos, C. J., Liu, R., Brown, R. F., and Badwey, J. A. (1995) Antagonists of phosphatidylinositol 3-kinase block activation of several novel protein kinases in neutrophils. *J. Biol. Chem.* **270,** 11,684–11,691.

18. Naccache, P. H., Levasseur, S., Lachance, G., Chakravarti, S., Bourgoin, S. G., and McColl, S. R. (2000) Stimulation of human neutrophils by chemotactic factors is associated with the activation of phosphatidylinositol 3-kinase γ. *J. Biol. Chem.* **275,** 23,636–23,641.

19. van der Kaay, J., Batty, I. H., Cross, D. A. E., Watt, P. W., and Downes, C. P. (1997) A novel, rapid, and highly sensitive mass assay for phosphatidylinositol 3,4,5-trisphosphate (PtdIns(3,4,5)P$_3$) and its application to measure insulin-stimulated PtIns(3,4,5) production in rat skeletal muscle in vivo. *J. Biol. Chem.* **272,** 5477–5481.

20. van der Kaay, J., Cullen, P. J., and Downes, C. P. (1998) Phosphatidylinositol(3,4,5)-trisphosphate (Ptdins(3,4,5)P3) mass measurements using a radioligand displacement assay, in *Phospholipid Signaling Protocols* (Bird, I. M., ed.), Humana, Totowa, NJ, pp. 109–125.

21. Cadwallader, K. A., Condliffe, A. M., McGregor, A., et al. (2002) Regulation of phosphatidylinositol 3-kinase activity and phosphatidylinositol 3,4,5-trisphosphate accumulation by neutrophil priming agents. *J. Immunol.* **169,** 3336–3344.

22. Traynor-Kaplan, A. E., Thompson, B. L., Harris, A. L., Taylor, P., Omann, G. M., and Sklar, L. A. (1989) Transient increase in phosphatidylinositol 3,4-bisphosphate and phosphatidylinositol trisphosphate during activation of human neutrophils. *J. Biol. Chem.* **264,** 15,668–15,673.

23. Cockcroft, S. (1991) Relationship between arachionate release and exocytosis in permeabilized human neutrophils stimulated with formylmethionyl-leucylphenylalanine fMetLeuPhe, guanosine 5'-[γ-thio]triphosphate (GTP[S]) and Ca^{2+}. *Biochem. J.* **275,** 127–131.

24. Briand, S. I., Bernier, S. G., and Guillemente, G. (1998) Monitoring of phospholipase A$_2$ activation in cultured cells using tritiated arachidonic acid. In: *Meth. Mol. Biol.* (Bird, I. M., ed.), Humana Press, Totowa, NJ, **105,** 161–166.

25. Anderson, R., Steel, H. C., and Tintinger, G. R. (2005) Inositol 1,4,5-triphosphate-mediated shuttling between intracellular stores and the cytosol contributes to the sustained elevation in cytosolic calcium in fMLP-activated human neutrophils. *Biochem. Pharmacol.* **69,** 1567–1575.

26. Zhang, L. (1998) Inositol 1,4,5-trisphosphate mass assay, in *Phospholipid Signaling Protocols* (Bird, I. M., ed.), Humana, Totowa, NJ, pp. 77–87.

27. Fruman, D. A., Gamache, D. A., and Earnst, M. J. (1991) Changes in inositol 1,4,5-trisphosphate mass in agonist-stimulated human neutrophils. *Agents Actions* **34,** 16–19.

28. Bollag, W. B. (1998) Measurement of phospholipase D activity, in *Phospholipid Signaling Protocols* (Bird, I. M., ed.), Humana, Totowa, NJ, pp. 151–160.

29. Bligh, E. G. and Dyer, W. J. (1959) A rapid method of total lipid extraction and purification. *Can. J. Biochem. Physiol.* **37,** 911–917.

30. Thompson, N. T., Bonser, R. W., Tateson, J. E., et al. (1991) A quantitative investigation into the dependence of Ca^{2+} mobilisation on changes in inositol 1,4,5-trisphosphate levels in the stimulated neutrophil. *Br. J. Pharmacol.* **103,** 1592–1596.

7

Analysis of Protein Phosphorylation in Human Neutrophils

Jamel El-Benna and Pham My-Chan Dang

Summary

Polymorphonuclear neutrophils play a key role in host defense and inflammation. Neutrophils can be activated by a variety of soluble and particulate factors, leading to an increase in the phosphorylation of numerous proteins on tyrosines, serines and threonines. Upon covalent binding of phosphates to these amino acids, the charge and conformation of the corresponding proteins are modified, generally leading to changes in protein functions such as protein–protein interactions and enzymatic activity. Protein phosphorylation in neutrophils can be studied by following protein incorporation of radiolabeled inorganic phosphate. This technique is based on labeling the intracellular ATP pool with ^{32}P prior to applying a stimulus that induces changes in protein phosphorylation status. The proteins are extracted from the cells, immunoprecipitated or not, resolved on polyacrylamide gel electrophoresis, then transferred onto nitrocellulose membranes and visualized by means of autoradiography. Phosphorylated sites of a multisite-phosphorylated protein can be analyzed by using two-dimensional tryptic phosphopeptide mapping. This chapter describes a phosphorylation protocol and the analysis of the phosphorylation of a neutrophil protein, p47phox, by two-dimensional tryptic phosphopeptide mapping.

Key Words: Neutrophils; protein phosphorylation; respiratory burst; NADPH oxidase; p47phox; two-dimensional phosphopeptide mapping.

1. Introduction

Neutrophils play a crucial role in host defense against invading pathogens. Neutrophils circulate in a "dormant" state. They migrate to sites of infection, where they engulf the pathogen and release toxic substances such as reactive oxygen species (ROS), proteolytic enzymes, and bactericidal proteins contained in granules (*1*). Engulfment of bacteria triggers the activation of enzyme systems such as NADPH oxidase, which is responsible for superoxide anion production and ROS generation (*2*). This multicomponent enzyme system is composed

From: *Methods in Molecular Biology, vol. 412: Neutrophil Methods and Protocols*
Edited by: M. T. Quinn, F. R. DeLeo, and G. M. Bokoch © Humana Press Inc., Totowa, NJ

of cytosolic proteins (p47phox, p67phox, p40phox, and Rac1/2) and membrane-bound proteins (p22phox and gp91phox, which form flavocytochrome b_{558}) that assemble together at membrane sites upon cell activation *(3–5)*.

Neutrophils can be activated by a variety of soluble and particulate factors, such as opsonized bacteria, opsonized zymosan, latex particles, complement fragment C5a, formylated peptides such as formyl-Met-Leu-Phe (fMLP), leukotriene B_4 (LTB4), platelet activating factor (PAF), diacylglycerol (DAG), calcium ionophores (ionomycin, A23187), and proteoin kinase C (PKC) activators (phorbol myristate acetate [PMA]). Neutrophil activation by these agonists increases the phosphorylation status of proteins such as p47phox *(6)*, p67phox *(7)*, p40phox *(8)*, p22phox *(9)*, and several other proteins *(10,11)*.

Protein phosphorylation on tyrosines, serines, and threonines is a common intracellular signaling mechanism found in almost all cell types. Upon covalent binding of phosphate groups to these amino acids, the charge and conformation of the corresponding proteins are modified, generally leading to changes in protein functions such as protein–protein interaction, enzymatic activity, and translocation *(12)*. Historically, protein phosphorylation was studied by following the incorporation of radiolabeled inorganic phosphate. When available, anti-phospho amino acid or anti-phospho-site antibodies were subsequently used to analyze protein phosphorylation. The first technique is based on labeling the intracellular ATP pool with ^{32}P prior to applying a stimulus that induces changes in protein phosphorylation status. The proteins are then extracted from the cells, immunoprecipitated or not, resolved on sodium dodecyl sulfate (SDS)-polyacrylamide gel electrophoresis (PAGE), transferred to membranes and visualized by means of autoradiography. If the protein is phosphorylated on several sites, then these can be analyzed by using two-dimensional tryptic phosphopeptide mapping. The phosphorylated protein is first digested with trypsin, then the peptides are separated by thin-layer electrophoresis (TLE) in one dimension and by ascending thin-layer chromatography (TLC) in the second dimension, and the phosphorylated peptides are detected by autoradiography. This chapter describes the protocol used to study protein phosphorylation in human neutrophils, based on the example of p47phox immunoprecipitation and analysis by two-dimensional phosphopeptide mapping.

2. Materials

2.1. Reagents

1. Diisopropylfluorophosphate (DFP), purchased as a 5.4 *M* solution (Sigma).
2. PMA (Sigma): dissolve in sterile dimethylsulfoxide (DMSO) at 1 mg/mL, aliquot, and store at −70°C.
3. *N*-formylmethionyl-leucyl-phenylalanine (fMLP; Sigma): dissolve in sterile DMSO at 10 m*M*, aliquot, and store at −20°C.

4. Leupeptin, aprotonin, and pepstatin: dissolve at 10 mg/mL in phosphate buffer and DMSO, respectively, aliquot, and store at −20°C.
5. Phenylmethylsulfonylfluoride (PMSF): prepare fresh at 0.5 M in DMSO.
6. ^{32}P-orthophosphoric acid ($H_3{}^{32}PO_4$) in water, HCl-free and carrier free (10 mCi/ mL) (Perkin Elmer-NEN or Amersham).
7. Trypsin, sequencing grade (Roche-Bohringer): prepare fresh at 50 µg/mL in ammonium bi-carbonate buffer (50 mM).
8. Tosylphenylalanylchloromethane (TPCK)-treated trypsin (Sigma): prepare fresh at 1 mg/mL in carbonate buffer.
9. p-Nitroblue tetrazolium (NBT) chloride: dissolve 30 mg of NBT in 1 mL of 70% dimethylformamide.
10. 5-bromo-4-chloro-3-indolylphosphate p-toluidine salt (BCIP): dissolve 15 mg of BCIP in 1 mL of dimethylformamide.
11. BCIP/NBT: combine 10 mL of carbonate buffer (100 mM, pH 9.8), 100 µL of NBT solution, and 100 µL of BCIP solution. Make fresh before use.
12. Protein G sepharose beads (Gamma-bind, Amersham-Pharmacia)

2.2. Buffers and Solutions

1. 2% Dextran T500: dissolve 10 g of Dextran T500 (Amersham-Pharmacia) in 500 mL of 0.9% NaCl and filter-sterilize. Store the solution at 4°C for up to 4 wk (*see* **Note 1**).
2. Phosphate-free loading buffer: 20 mM HEPES, pH 7.4 (make from a 1 M sterile HEPES solution, pH 7.4) (Invitrogen), 150 mM NaCl, 5.4 mM KCl, 5.6 mM D-glucose, 0.8 mM MgCl$_2$, and 0.025% bovine serum albumin (BSA). Filter-sterilize and store for up to 2 wk at 4°C.
3. Immunoprecipitation buffer: 20 mM Tris-HCl, pH 7.5, 150 mM NaCl, 0.25 M sucrose, 0.5% (w/v) Triton-X-100, 5 mM EDTA, 5 mM EGTA, 10 µg/mL leupeptin, 10 µg/mL pepstatin, 10 µg/mL aprotonin, 25 mM NaF, 2.5 mM sodium pyrophosphate, 5 mM β-glycerophosphate, 5 mM NaVO$_3$, 1 mM p-nitrophenyl phosphate, and 1 mM levamisole. This buffer can be stored at −20°C for up to 6 mo; add 0.5 mM of PMSF just before use (*see* **Note 2**).
4. Lysis buffer: immunoprecipitation buffer containing 1 mM DFP and 1 mg/mL DNAse I. Prepare the required volume of lysis buffer fresh before use (*see* **Note 3**). For DFP use, *see* precautions under **Subheading 3.2.**
5. 2X Laemmli sample buffer for electrophoresis of phosphorylated proteins: 125 mM Tris-HCl, pH 6.8, 6% SDS, 8% β-mercaptoethanol, 18% glycerol, 5 mM EDTA, 5 mM EGTA, 10 µg/mL leupeptin, 10 µg/mL pepstatin, 10 µg/mL aprotonin, 10 mM NaF, 5 mM NaVO$_3$, and 2 mM p-nitrophenyl phosphate. This buffer is stored at −20°C for up to 6 mo.
6. Transfer buffer: 50 mM Tris base, 95 mM glycine, 0.08% SDS, and 20% methanol.
7. Blotto buffer: 25 mM Na$_2$B$_4$O$_7$-10H$_2$O, 100 mM H$_3$BO$_3$, 75 mM NaCl, 1% nonfat milk, 0.02% NaN$_3$, and 0.002% anti-Foam.
8. Carbonate buffer (pH 9.8): 100 mM NaHCO$_3$, 1 mM MgCl$_2$-6H$_2$O, adjust pH to 9.8 with NaOH.

9. Ammonium bicarbonate buffer: 50 mM in H_2O (make fresh, do not adjust the pH).
10. Polyvinylpyrrolidone (PVP)-360 solution: 0.5% of PVP in 100 mM acetic acid.
11. TLE marker dye: 0.5 mg/mL phenol red in H_2O (can be stored at room temperature for several months).
12. TLE buffer pH 1.9: 88% formic acid/water (3v/17v). This buffer is stable for several months at room temperature.
13. TLC buffer: isobutyric acid/butanol/pyridine/acetic acid/water (25/1/2.5/1.8/10, v/v). This buffer is stable for several months at room temperature; it is stored in the chromatography tank under the hood.

3. Methods

3.1. Isolation of Human Neutrophils

Isolation methods are described in detail elsewhere in this volume (*see* Chapters 2 and 3). Here, we describe the technique used by our group to study protein phosphorylation with phosphate-free reagents.

1. Collect venous blood from healthy adult volunteers with citrate as anticoagulant.
2. To 1 volume of whole blood (up to 400 mL is used in some experiments) add 1 volume of 2% Dextran T500 solution. Gently mix by inverting the tubes several times and allow to sediment at room temperature for 20–30 min. Erythrocytes, which are permeable to dextran, sediment more rapidly than leukocytes.
3. Collect the upper layer in centrifuge tubes and discard the red cell pellet. Centrifuge at 400g for 8 min at room temperature (*see* **Note 4**).
4. After centrifugation, the pellets contain leukocytes and contaminating erythrocytes. Recover and resuspend each pellet in 5 mL of phosphate free loading buffer, pool several pellets, and gently layer on Ficoll (2v cells/1v Ficoll) (*see* **Note 5**).
5. Centrifuge tubes at 400g for 30 min at 22°C.
6. Discard the upper layers and mononuclear cells, which are at the interface of the buffer and Ficoll.
7. Disperse the pellet containing neutrophils and contaminating erythrocytes. Remove contaminating erythrocytes by hypotonic lysis: add 12 mL of ice-cold sterile distilled water for 30 s, mix gently, and restore isotonicity by adding 4 mL of 0.6 M KCl (*see* **Note 6**).
8. Centrifuge the tubes at 400g for 8 min at 4°C. Decant the red supernatant, and resuspend the neutrophil pellet in 5 to 10 mL of phosphate-free loading buffer.
9. Count the cells by diluting in trypan blue (1/100) and using a hemocytometer.

3.2. DFP Treatment

DFP, a serine protease inhibitor, protects proteins from degradation by neutrophil serine proteases but is extremely toxic. Appropriate safety precaution must be used, including handling only in a fume hood, and wearing gloves and lab coats. All tips and other materials contaminated with DFP must be discarded in a 1 N NaOH-containing waste container placed in the hood (*see* **Note 7**).

1. Resuspend the cells at $50–100 \times 10^6$/mL in phosphate-free loading buffer.
2. Add the serine protease inhibitor DFP (2.7 m*M*) to dilute the stock solution (5.4 *M*) to a final concentration of 2.7 m*M* (1/2000 dilution), and incubate 15 min at 4°C.
3. Add 10 volumes of phosphate-free loading buffer and centrifuge at 400*g* and 4°C.
4. Discard the supernatant in 1 *N* NaOH, and resuspend the neutrophil pellet in phosphate-free loading buffer at 10^8 cells/mL.

3.3. ^{32}P-labeling and Stimulation of Neutrophils

As large amounts of [^{32}P]-orthophosphoric acid are used during cell labeling, good shielding with 1.5 cm-thick Plexiglas is necessary at all times. Waste containers and bags must be kept in shielded boxes. The investigator should wear protective eyewear, a laboratory coat, and two pairs of gloves. To avoid laboratory contamination, it is important to check the gloves with a Geiger counter after each manipulation of ^{32}P.

1. Dilute the cells at 100×10^6/mL in phosphate-free loading buffer in a 50-mL conical tube.
2. Add [^{32}P]-orthophosphoric acid at 0.5 mCi/10^8 cells, and incubate for 60 min at 30°C with gentle mixing every 10–15 min.
3. Centrifuge the cells at 400*g* for 8 min, discard the supernatant in a shielded waste container, and resuspend the cells in loading buffer containing 1 m*M* CaCl$_2$ and 0.7 m*M* MgCl$_2$ (*see* **Note 8**).
4. Dilute the cells to 20×10^6 cells/mL, and preincubate at 37°C for 10 min.
5. Add 200 ng/mL PMA for 6 min or 1 µ*M* fMLP for 2 min to stimulate the cells.
6. Stop the reaction by adding ice-cold loading buffer, mixing gently and centrifuging at 400*g* for 8 min at 4°C.
7. Discard the supernatants, and remove residual buffer with a pipet.

3.4. Neutrophil Lysis

It is difficult to preserve proteins and their phosphorylation status in neutrophil lysates because of the presence of active proteases and phosphatases. Here, we describe a lysis technique that we found useful to preserve phosphorylated proteins.

1. Keep the tubes on ice and resuspend the cell pellets (to 50×10^6 cells/mL) with a pipet in ice-cold lysis buffer containing DFP and DNAse I. For DFP treatment of cells (*see* **Subheading 3.2.**), all manipulations with DFP must be done under the hood following safety rules.
2. Sonicate the cells (3×10 s) at 30% power with a microtip sonicator.
3. Incubate the sonicate on ice for 15 min and centrifuge at 100,000*g* for 30 min at 4°C in a TL100 ultracentrifuge (Beckman Coulter).
4. Collect the cleared cell lysate for immunoprecipitation. If required, mix a sample of this lysate (e.g., 50 µL) with 2X Laemmli sample buffer, denature by boiling, and analyze phosphorylated proteins by electrophoresis, blotting, and autoradiography (**Fig. 1**) (*see* **Note 9**).

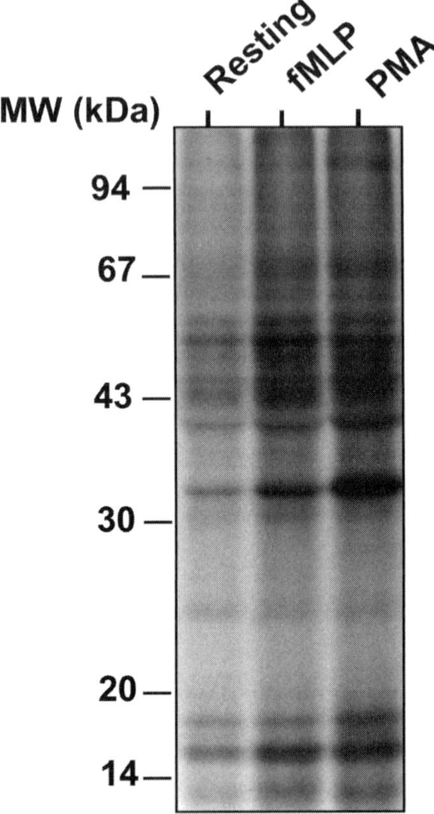

Fig. 1. A representative autoradiogram showing protein phosphorylation in human neutrophils. Human neutrophils were isolated and labeled with $[^{32}P]H_3PO_4$ then stimulated with fMLP (1 μ*M* for 2 min) or PMA (200 ng/mL for 6 min). Resting, fMLP- and PMA-activated neutrophils were lysed, centrifuged and proteins (2.5×10^6 cell equivalents) were analyzed by SDS-PAGE, transferred to nitrocellulose and autoradiographed.

3.5. p47phox Immunoprecipitation

1. Wash protein G sepharose beads (use 50 μL of beads/tube) in 10 volumes of phosphate-buffered saline (PBS). Mix by inverting, and centrifuge at 200*g* for 3 min. To minimize contamination, use screw-cap Eppendorf tubes for immunoprecipitation and replace the caps with new ones when they become very radioactive.
2. Pellet beads, and discard the supernatant.
3. Add 10 volumes of immunoprecipitation buffer containing 1 mg/mL BSA, equilibrate on a rotating wheel for at least 30 min, centrifuge at 200*g* for 3 min, and discard the supernatant.
4. Add the cleared cell lysate from **Subheading 3.4., step 3** to the beads and add anti-p47phox antibody (dilution 1/200). We routinely use a rabbit polyclonal antibody against p47phox that was generously provided by Dr. B.M. Babior (The Scripps

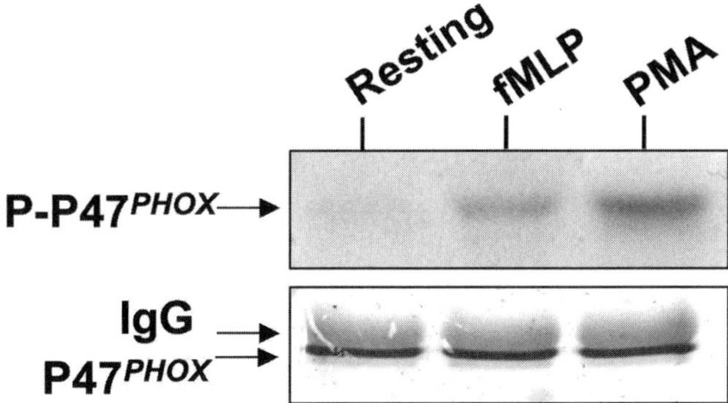

Fig. 2. Representative autoradiogram and Western blot showing phosphorylated and total p47phox in human neutrophils. Human neutrophils were isolated and labeled with [^{32}P] then stimulated with fMLP (1 μM for 2 min) or PMA (200 ng/mL for 6 min). Resting and fMLP- or PMA-activated neutrophils were lysed and p47phox was immunoprecipitated, then analyzed by SDS-PAGE. After transfer on membrane, phosphorylated-p47phox (p-p47phox) was detected by autoradiography (upper) and the whole protein (p47phox) was detected with a specific antibody (lower). IgG represents the heavy chain of immunoglobulin G.

Research Institute, La Jolla, CA), but others can be substituted after verification of their efficacy to immunoprecipitate.

5. Incubate overnight at 4°C on a rotating wheel with end-over-end rotation.
6. Centrifuge the tubes at 200g for 3 min, and transfer the supernatant to new tubes. Store these tubes at −70°C in a shielded box for further experiments or controls.
7. Wash the beads four times with 1 mL of immunoprecipitation buffer by gently inverting the tubes and centrifuging at 200g for 3 min to recover the bead pellet.
8. Elute p47phox by adding 100 μL of Laemmli sample buffer, vortexing, and boiling for 5 min.
9. Centrifuge the samples at 15,000g for 30 s, and use the supernatant for electrophoresis, blotting, and autoradiography (Fig. 2).

3.6. Electrophoresis and Western Blotting

The samples are subjected to SDS-PAGE in 10% polyacrylamide gels, using standard techniques *(13)*. The separated proteins are transferred to nitrocellulose *(14)* in the transfer buffer described above. When two-dimensional tryptic phosphopeptide mapping analysis is performed, the samples are divided in half and run on two different gels. After transfer, one nitrocellulose membrane is wrapped in plastic film, placed in a cassette with two screens, and incubated for autoradiography at −70°C for 24–72 h (**Fig. 2**). The second nitrocellulose membrane is used to detect p47phox with a specific antibody in order to check that the same amount of p47phox is immunoprecipitated from each sample:

1. Incubate the membrane in blotto-buffer for 30 min at room temperature on a shaker to block any nonspecific protein binding sites.
2. Add anti-p47phox antibody (1/5000 dilution, or as appropriate) in blotto buffer, and incubate for 1 h at room temperature.
3. Wash three times with blotto, incubating 10 min for each wash.
4. Add alkaline phosphatase-conjugated goat anti-rabbit antibody (1/5000 dilution), and incubate for 1 h at room temperature.
5. Wash three times with blotto incubating 10 min for each wash.
6. Develop with 10 mL of BCIP/NBT solution. Incubate the membrane for 15 to 30 min at room temperature, and stop the reaction by washing with H_2O (**Fig. 2**).

3.7. Two-Dimensional Tryptic Phosphopeptide Mapping of p47[phox]

This technique is adapted from Boyle et al. *(15)*. Phosphorylated protein is first digested with trypsin, and the peptides are then separated by TLE in one dimension and by ascending TLC in the second dimension. Phosphorylated peptides are detected by autoradiography.

3.7.1. Digestion of p47phox With Trypsin

1. After autoradiography, cut the nitrocellulose area containing the band of interest (^{32}P-labeled p47phox) into pieces about 2 mm × 2 mm and place in an Eppendorf tube (*see* **Note 10**).
2. Add 100 µL PVP solution, and incubate for 30 min at 37°C for saturation. Do not vortex at this step.
3. Wash three times with 1 mL of H_2O and twice with ammonium bicarbonate buffer, simply by aspirating the liquid with a transfer pipet and adding a new one.
4. Add 100 µL of sequencing-grade trypsin and incubate overnight at 37°C with vortexing from time to time.
5. Add 10 µL of fresh TPCK-treated trypsin, vortex, and incubate for another 2 h.
6. Stop the reaction by centrifugation at 16,000g for 2.5 min, transfer the supernatant containing the released peptides to a new tube, and dry in a Speedvac concentrator.
7. Wash the peptides by adding 100 µL of H_2O, vortexing, and drying in a Speedvac concentrator. Repeat this step four times.

3.7.2. Thin-Layer Electrophoresis

Although some groups use the HTLE-7000 thin-layer electrophoresis apparatus, we routinely use the horizontal gel apparatus (Multiphor II electrophoresis system, Pharmacia) cooled with a circulating water bath.

1. Redissolve the peptides in 10 µL of TLE buffer using a pipet tip, and vortex.
2. Centrifuge the sample briefly to ensure that any residual particulate material is applied to the cellulose plate.
3. Before applying the sample, choose a cellulose plate (20 × 20 cm, VWR-Merck) with no irregularities, and mark the sample and dye marker origins on the back of the plate with a permanent marker.

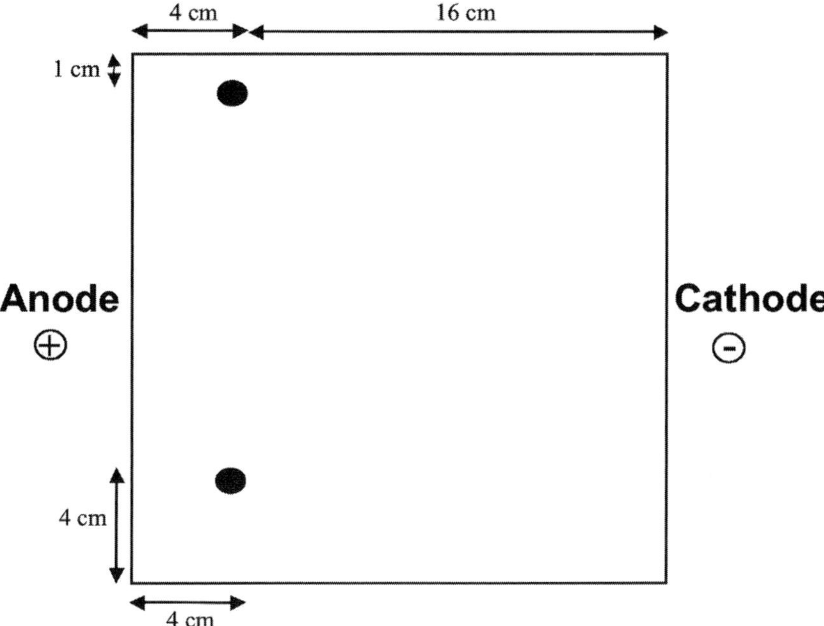

Fig. 3. Diagram of a cellulose plate showing marks used for p47phox phosphopeptide analysis. The sample is spotted 4 cm from the left side (anode side) and 4 cm from the bottom of the plate. Thin-layer electrophoresis marker dye is spotted above the sample origin, 1 cm from the top of the plate.

4. Spot the sample 4 cm from the left side (anode side) and 4 cm from the bottom of the plate.

5. Spot TLE-marker dye above the sample origin at 1 cm from the top of the plate. Apply 2-µL drops of the sample, and air-dry between applications (**Fig. 3**).

6. Fill the buffer tanks with TLE buffer, and place double-layer 3-mm Whatman electrophoresis wicks in the tank.

7. Take a new double-layer 3-mm Whatman paper with the same dimensions as the cellulose plate (20 cm × 20 cm) and make two holes corresponding to the positions of the sample and dye marker. Wet the blotter paper by soaking it in TLE buffer and let the excess buffer drain off.

8. Wet the plate by placing the blotter paper on it with the holes aligned over the sample origin and the red marker origin. Press gently on the paper until the plate is evenly wetted, then remove the blotter. This step allows the sample and marker to concentrate into one spot.

9. Place the wet plate on the horizontal gel apparatus, apply the electrophoresis wicks to the anode and cathode side of the plate by gently pressing so that they cover 1.5 cm of the plate, place the cover on the apparatus, connect the electrodes to the power supply, and start the run towards the cathode at 1100 V at 6°C for 25 min. Run until the phenol red marker runs 6 cm.

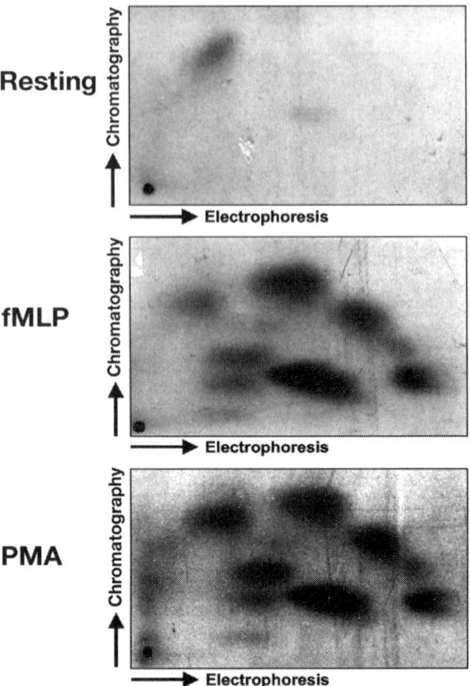

Fig. 4. A representative autoradiogram showing two-dimensional phosphopeptide mapping of p47phox. ^{32}P-labeled neutrophils were stimulated with fMLP (1 μM for 2 min) or PMA (200 ng/mL for 6 min) and lysed. p47phox was immunoprecipitated and analyzed by SDS-PAGE, then transferred to membranes before being subjected to two-dimensional tryptic phosphopeptide mapping and autoradiography.

10. Stop the electrophoresis, disconnect the apparatus, and air-dry the plate under the hood.

3.7.3 Thin-Layer Chromatography

Once the plate is dry, TLC can be performed up to several days later. We routinely perform TLC the day after TLE.

1. Place the chromatography tank under the hood, and pour TLC buffer to a depth of 2 cm.
2. Place the dried plate in the tank with the phenol red marker (used for TLE) at the top, and put the lid on. Ascending chromatography occurs perpendicular to the direction of electrophoresis.
3. Let the buffer run until it reaches 2 cm from the top (this usually takes 6 to 8 h).
4. Dry the plate under the hood and expose it to X-ray film or use an Instantimager (**Fig. 4**).

4. Notes

1. All solutions used for cell preparation and cell manipulation, such as Dextran solution and phosphate-free loading buffer, are prepared with sterile water or sterile 0.9% NaCl (injection-grade).
2. Solutions used for cell lysis and protein analysis are prepared in very pure water (resistivity: 18.2 MΩ). Because PMSF is unstable in aqueous solution, it is added to lysis buffer just before use.
3. Addition of DFP to this buffer prevents protein degradation during the lysis and immunoprecipitation steps. DNAse I dissolves the cytoskeleton and releases all proteins bound to particles.
4. This centrifugation step is used to concentrate the cells and thus to use less Ficoll. If a small volume of blood is used, the supernatant can be layered directly on Ficoll.
5. It is important to use phosphate-free buffer for all procedures in order to starve the cells. When layering the cells on Ficoll, take care to avoid mixing the two media.
6. Never vortex cells, and never lyse erythrocytes without resuspending the pellet, otherwise the cells will aggregate.
7. Other serine proteases inhibitors have been used, but DFP yields the best results, despite its toxicity.
8. To avoid radioactive contamination, radioactive supernatants are best discarded by pipetting with plastic disposable transfer pipets rather than by decanting.
9. For whole-cell lysates, the gels can be dried and exposed to a film. However, we have found that protein transfer to a membrane and exposure of this membrane gives a cleaner signal.
10. To locate the phosphorylated band of interest, before autoradiography, the nitrocellulose membrane is Sellotaped to a used film, and the new film is Sellotaped to the old one.

References

1. Segal, A. S. (2005) How neutrophils kill microbes. *Annu. Rev. Immunol.* **23,** 197–223.
2. Roos, D., van Bruggen, R., and Meischl, C. (2003) Oxidative killing of microbes by neutrophils. *Microbes Infect.* **14,** 1307–1315.
3. El-Benna, J., Dang, P. M., Gougerot-Pocidalo, M. A., and Elbim, C. (2005) Phagocyte NADPH oxidase: a multicomponent enzyme essential for host defenses. *Arch. Immunol. Ther. Exp. (Warsz)* **3,** 199–206.
4. Babior, B. M. (1999) NADPH Oxidase: an update. *Blood* **93,** 1464–1476.
5. Quinn, M. T. and Gauss, K. A. (2004) Structure and regulation of the neutrophil respiratory burst oxidase: comparison with nonphagocyte oxidases. *J. Leukoc. Biol.* **76,** 760–781.
6. El-Benna, J., Faust, L. P., and Babior, B. M. (1994) The phosphorylation of the respiratory burst oxidase component p47phox during neutrophil activation. Phosphorylation of sites recognized by protein kinase C and by proline-directed kinases. *J. Biol. Chem.* **269,** 23,431–23,436.

7. El Benna, J., Dang, M. C. P., Gaudry, M., et al. (1997) Phosphorylation of the respiratory burst oxidase subunit p67phox during human neutrophil activation. Regulation by protein kinase C-dependent and independent pathways. *J. Biol. Chem.* **272,** 17,204–17,208.

8. Bouin, A. P., Grandvaux, N., Vignais, P. V., and Fuchs, A. (1998) p40(phox) is phosphorylated on threonine 154 and serine 315 during activation of the phagocyte NADPH oxidase. Implication of a protein kinase C-type kinase in the phosphorylation process. *J. Biol. Chem.* **273,** 30,097–30,103.

9. Regier, D. S., Waite, K. A., Wallin, R., and McPhail, L. C. (1999) A phosphatidic acid-activated protein kinase and conventional protein kinase C isoforms phosphorylate p22(phox), an NADPH oxidase component. *J. Biol. Chem.* **274,** 36,601–36,608.

10. Bokoch, G. M. (1995) Chemoattractant signalling and leukocyte activation. *Blood* **86,** 1649–1660.

11. Ohira, T., Zhan, Q., Ge, Q., VanDyke, T., and Badwey, J. A. (2003) Protein phosphorylation in neutrophils monitored with phosphospecific antibodies. *J. Immunol. Meth.* **281,** 79–94.

12. Cohen, P. (2000) The regulation of protein function by multisite phosphorylation—a 25 year update. *Trends Biochem. Sci.* **25,** 596–601.

13. Laemmli, U. K. (1970) Cleavage of structural proteins during the assembly of the head of bacteriophage T4. *Nature* **227,** 680–685.

14. Towbin, H., Staehlin, T., and Gordon, J. (1979) Electrophoretic transfer of proteins from polyacrylamide gels to nitrocellulose sheets: procedure and some applications. *Proc. Natl. Acad. Sci. USA* **76,** 4350–4354.

15. Boyle, W. J., Van Der Geer, P., and Hunter, T. (1991) Phosphopeptide mapping and phosphoamino acid analysis by two-dimensional separation on thin-layer cellulose plates (Hunter, T. and Sefton, B. M. eds.), Academic, San Diego, pp. 110–149.

8

Mitogen-Activated Protein Kinase Assays

Martine Torres

Summary

Polymorphonuclear neutrophils (PMN) play an essential role in host defense against bacteria and fungi through coordinated responses such as adhesion, migration, phagocytosis, secretion, and activation of the NADPH oxidase. The mitogen-activated protein kinases (MAPKs) and their activation kinase cascades, which transduce signals from the plasma membrane to the cytosol and nucleus, are an integral part of signaling pathways involved in many cellular responses. PMN express several members of the MAPK family that have been shown, mainly through the use of pharmacological inhibitors, to mediate the cellular activities triggered by a variety of extracellular agonists. Methods to determine MAPK activation have been greatly simplified with the availability of antibodies raised to active MAPKs. The recent development of novel inhibitors for the MAPK pathways may further our understanding of their role in neutrophil function.

Key Words: MAP kinases; extracellular signal-regulated kinase; c-Jun N-terminus kinase; p38MAPK; immunoblotting; pharmacological inhibitors; peptide-based inhibitors.

1. Introduction

The mitogen-activated protein kinases (MAPKs), namely the extracellular signal-regulated kinase (ERK), c-Jun N-terminus kinase (JNK), and p38MAPK, are a large family of proline-directed serine/threonine kinases that are ubiquitously expressed and are activated through a common mechanism involving a three-tiered kinase cascade, also referred to as a module. MAPKs are grouped into subfamilies, mostly defined by their TXY activation motif and by the dual-specificity MAPK kinases (MKKs) that exclusively phosphorylate this motif. Discovered in mammalian cells in the late seventies, the MAPKs are highly conserved throughout evolution and have been shown to regulate many cellular activities, ranging from gene expression and cell survival to metabolism and movement. Furthermore, the study of MAPK regulation has greatly expanded our general understanding of the organization of cellular signaling pathways (*1–3*).

From: *Methods in Molecular Biology, vol. 412: Neutrophil Methods and Protocols*
Edited by: M. T. Quinn, F. R. DeLeo, and G. M. Bokoch © Humana Press Inc., Totowa, NJ

The MAPK and their activation pathways have been extensively investigated in polymorphonuclear neutrophils (PMN) after the initial demonstration of the expression of the ERKs and their activation by *N*-formylmethionyl-leucyl-phenylalanine (fMLP) and granulocyte/macrophage colony-stimulating factor (GM-CSF) *(4–6)*. It is now well known that ERK1/2 and p38MAPK are activated in PMN stimulated with many pro-inflammatory stimuli, including fMLP and cytokines *(1,7)*. Furthermore, p38MAPK and JNK are activated by bacterial lipopolysaccharide *(8,9)*, and JNK may also be activated by tumor necrosis factor (TNF)-α in adherent PMN *(10)*. The involvement of the MAPKs in a particular response has been more difficult to ascertain as PMN studies have primarily relied on pharmacological inhibitors; nevertheless, keeping in mind the pitfalls associated with inhibitors (discussed later), ERK1/2 and p38MAPK have been implicated in the respiratory burst, adherence, exocytosis, priming, motility, and apoptosis *(11–14)*. Pharmacological inhibitors have also been used in animal studies that have confirmed the role of p38MAPK as a critical component of the inflammatory response *(7,15)*.

The methodology for assessing MAPK activation has evolved over the years from the rather complex to the relatively simple. Separation by ion exchange chromatography, fast protein liquid chromatography (FPLC), or immunoprecipitation, followed by measurement of kinase activity with an appropriate substrate, was initially used to measure ERK activation *(4–6)*. Currently, the method of choice is the analysis of whole-cell lysates by sodium dodecyl sulfate (SDS)-polyacrylamide gel electrophoresis (PAGE) and Western blotting with antibodies that specifically recognize the phosphorylated and active form of the MAPK (and their upstream activators). These antibodies are commercially available through many companies, and some can also be used for immunofluorescence studies. In this chapter, we will concentrate on the methodology for Western blotting and immunoprecipitation and on the use of inhibitors.

2. Materials

2.1. Polymorphonuclear Neutrophils

1. Krebs Ringer Phosphate Glucose (KRPG) buffer: 10 mM sodium phosphate, 125 mM NaCl, 5 mM KCl, 1 mM MgSO$_4$, 1.3 mM CaCl$_2$, 10 mM HEPES, pH 7.4.
2. Prepare human PMN under endotoxin-free conditions from whole blood obtained from healthy volunteers according to the protocols approved by the institutional review board. A detail of the methodology is available in Chapter 2.
3. Resuspend purified PMN (>95% pure and >95% viable) in KRPG, and keep on ice until use.
4. Dissolve agonists in the appropriate solvent, and store in single-use aliquots at –70°C. Prepare working solutions by dilution in KRPG just before starting the experiment.

2.2. Diisopropylfluorophosphate Treatment

1. Diisopropylfluorophosphate (DFP; Sigma): available as a 1 g/mLsolution (approx 5.73 M). Treat the cells with a 5-mM final concentration.
2. DFP is a highly toxic cholinesterase inhibitor that should be handled in a chemical hood. Spills and materials should be neutralized by immersion into a 5-M NaOH solution and disposed as toxic waste.

2.3. Lysis Buffers

1. Suspension Buffer A: KRPG containing 1 mM phenylmethylsulfonylfluoride (PMSF), 1 mM DFP, 10 mM ethylenediamine tetraacetic acid (EDTA), 2 mM sodium ortho-vanadate (VAN), 10 mM sodium fluoride, and 10 μg/mL each of leupeptin and aprotinin. Prepare fresh from stock solutions, and keep at 4°C. Add PMSF and VAN just prior to use (*see* **Note 1**).
2. Sample Buffer B (2X): 125 mM Tris-HCl pH 6.8, 4% SDS, 10% glycerol, 80 mM dithiothreitol (DTT), 0.003% Pyronin-Y (in place of the usual bromophenol blue; *see* **Note 2**). Store at room temperature.
3. Lysis buffer C (immunoprecipitation): 50 mM Tris-HCl, pH 7.5, 150 mM NaCl, 0.5% Nonidet P-40, 1% Triton X-100, 0.25% sodium deoxycholate, 10 mM sodium pyrophosphate, 25 mM β-glycerophosphate, 5 mM EDTA, 5 mM EGTA, 1 mM PMSF, 1 mM VAN, 100 mM sodium fluoride, and 10 μg/mL each of aprotinin, leupeptin, and pepstatin A. Prepare fresh from stock solutions, and keep at 4°C. Add PMSF and VAN just prior to use (*see* **Note 1**).

2.4. Sodium Dodecyl Sulfate-Polyacrylamide Gel Electrophoresis

1. Precast 10% Tris-glycine gels (Novex/Invitrogen) (*see* **Note 3**).
2. Running buffer 10X: 250 mM Tris-HCl, 1.92 M glycine, 1% (w/v) sodium dodecylsulfate. Store the stock solution at RT, and dilute 1:10 prior to running the gel.

2.5. Immunoprecipitation

1. Protein G Agarose Fast Flow (UBI).
2. Protein A Sepharose CL-4B (Amersham Pharmacia) (*see* **Note 4**).
3. Primary antibodies to ERK1 (C16; SC-93), ERK2 (C14; SC-154), JNK (SC-571), and p38MAPK (C-20; SC-535) (all from Santa Cruz Biotechnology).
4. Sample buffer (2X): 125 mM Tris-HCl pH 6.8, 4% SDS, 10% glycerol, 80 mM DTT, 0.003% Pyronin-Y. Store at room temperature.
5. Mini-Rotator (2–80 rpm; Glas-Col, Terre Haute, IN).

2.6. Kinase Assay

1. Kinase Wash Buffer A: 25 mM HEPES, pH 7.5, 25 mM MgCl$_2$, 25 mM sodium β-glycerophosphate, 2 mM DTT, 1 mM VAN, 250 μM ATP, and 25 mM sodium fluoride. Prepare from stock solutions, and keep at 4°C, except for VAN, which is added freshly (*see* **Note 1**).
2. Kinase substrates: glutathione-S-transferase (GST)-ATF-2$_{19-96}$ for p38 MAPK (Cell Signaling Technology #9224), GST-c-Jun$_{1-79}$ for JNK (Biomol; SE-151), and GST-

Elk1$_{307-428}$ (Cell Signaling Technology #9184) or T-669 (H-KRELVEPL**TP**SGE APNQALLR-OH, Calbiochem #454857) for ERK.
3. [γ-^{32}P]-ATP (10 mCi/mL) with specific activity no less than 3000 Ci/mmol (GE-Healthcare/Amersham).
4. Kinase Buffer: buffer A containing the appropriate substrate (1 µg or 400 µ*M* of T-669) and 10 µCi [γ-^{32}P]-ATP in 40 µL per sample. Prepare a working solution fresh in a sufficient amount for the number of samples in the experiment.
5. The general rules for handling radioactive materials are to be followed according to the Safety Manual of the Institution. Plexiglas shielding, storage, and waste containers are required.

2.7. Western Blotting

1. Polyvinylidene difluoride (PVDF) membranes (Hybond-P; Amersham/GE Healthcare).
2. Transfer Buffer (10X): 250 m*M* Tris-HCl, 1.92 *M* glycine (do not adjust pH), stored at 4°C. Prepare a 1:10 dilution containing 20% methanol, and store on ice until used.
3. Wash buffer (TBS-TW) (10X): 200 m*M* Tris-HCl, 1.37 *M* NaCl, pH 7.6. Prepare a 1:10 dilution containing 0.1% Tween-20 (v/v).
4. Blocking/dilution buffer: Tris-buffered saline-Tween 20 containing 5% nonfat dry milk (TBS-TW/NFDM).
5. Rabbit polyclonal antibodies to dual-phosphorylated ERK1/2 (1:5000; Promega #V8031), JNK (1:1000; Cell Signaling #9251), and p38MAPK (1:1000; Cell Signaling #9211). Prepare dilutions in TBS-TW/NFDM (*see* **Note 5**).
6. Rabbit polyclonal antibodies to ERK1/2 (1:1000; Cell Signaling #9102), p38 MAPK (1:1000; Santa Cruz Biotechnology #SC-535), and JNK (1:1000; Cell Signaling #9252) to detect total MAPK protein. Prepare dilutions in TBS-TW/NFDM (*see* **Note 5**).
7. Secondary antibody: peroxidase-labeled, affinity-purified goat anti-rabbit (1 mg/mL; Kirkegaard & Perry Laboratories). Prepare a 1:30,000 dilution in TBS-TW/NFDM for use in conjunction with the enhanced chemiluminescence (ECL) detection system.
8. ECL detection system (ECL™ products, Amersham/GE Healthcare).

2.8. Blot Stripping

Blot stripping buffer: 0.1 *M* glycine, pH 2.9, stored at room temperature. This buffer gives efficient stripping in a short time without using high incubation temperatures or noxious β-mercaptoethanol.

3. Methods

As mentioned above, the first assays developed to measure the activation of ERK1/2 were rather cumbersome. It was then noted that phosphorylation led

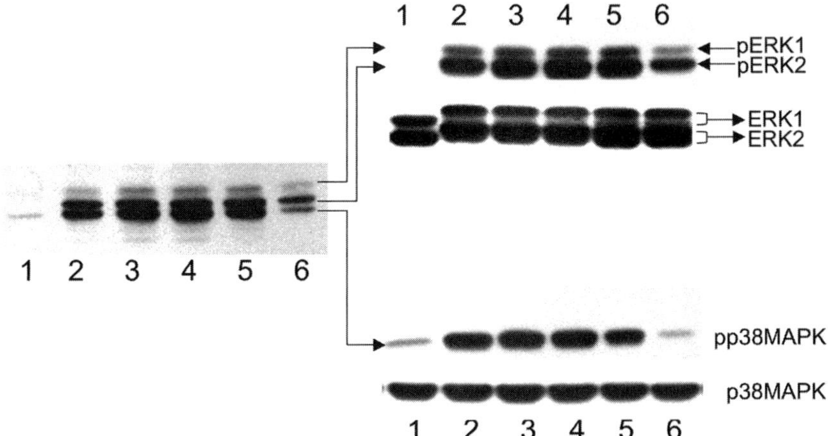

Fig. 1. Time course of extracellular signal-regulated kinase (ERK) and p38 mitogen-activated protein kinase (MAPK) activation. Polymorphonuclear neutrophils were stimulated with $10^{-7} M$ fMLP for 15', 30', 1', 2', and 5' (lanes 2–6) or with buffer (lane 1) and lysed in sample buffer B. Whole-cell lysates were analyzed by SDS-PAGE and Western blotting using an antibody to phosphotyrosine (PY, left), and antibodies to either dually-phosphorylated (top) or total (bottom) ERK1/2 and p38MAPK. Note the migration pattern of the three MAPKs when analyzed with the PY antibody and the changes in migration induced by stimulation when blotting with total ERK1/2 antibody.

to decreased electrophoretic mobility of ERK1/2 on SDS-PAGE, which could be analyzed in whole-cell lysates by immunoblotting (**Fig. 1**). Activation of ERK1/2 was also studied on whole-cell lysates using phosphotyrosine antibodies; however, this method may have led to errors as other MAPKs migrating similarly or closely to ERK1/2 may have been activated by the same agonist. In fact, p38MAPK migrates very closely to ERK2 and may have been mistaken for ERK2 (and ERK2 for ERK1) under conditions where ERK1 activation was minimal (**Fig. 1**); however, when the proper gel system is used, ERK1 is well separated from ERK2 (*see* **Note 3**). Furthermore, tyrosine phosphorylation may occur before phosphorylation of the threonine on the TXY motif and may not represent the fully activated ERK1/2. Nevertheless, the development of phosphospecific antibodies that recognize the dual-phosphorylated and active form of MAPK has made the analysis of whole-cell lysates by Western blotting the method of choice for assessing the activation of all MAPKs. Immunoprecipitation and kinase assay may also be used, and better antibodies are now available that efficiently immunoprecipitate MAPKs. In both cases, the data will be only as good as the preparation of the samples. Thus, it is important to include inhibitors of proteases (EDTA/EGTA, PMSF, 4-[2-aminoethyl]-benzenesulfonyl-fluoride [AEBSF], leupeptin, aprotinin, and others) and phosphatases (VAN, sodium fluoride, sodium pyrophosphate) to minimize the action of these enzymes. Proteases being quite abundant in the granules of PMN, it is recommended that PMN be additionally pretreated with the protease inhibitor, DFP, to ensure maximum protection.

3.1. Preparation of Samples for Western Blotting

1. As with any other functional PMN study, the experiment must proceed shortly after isolation of the cells, and all reagents must be gathered during cell isolation for timely use.
2. To ensure that all samples are treated within the shortest time frame and that cell lysis occurs as closely as possible to the intended time of treatment, establish a time sheet that includes a 5-min incubation to bring the samples up to 37°C and leaves sufficient time for the lysis step of each sample. In addition, organize the benchtop so that the microcentrifuge, several preset pipets, vacuum flask, ice bucket, and water bath are all in close proximity.
3. Resuspend PMN in KRPG at 1×10^7 cells/mL, and treat with DFP (0.92 µL of stock solution per mL of cells) for 15 min on ice.
4. Centrifuge ($400g$), and transfer the supernatant into an appropriate waste disposal container.
5. Wash PMN twice in KRPG and resuspended at 1×10^7 cells/mL.
6. Aliquot DFP-treated cells (200 µL, i.e., 2×10^6 PMN per sample) into labeled, 1.5-mL flat-bottom microcentrifuge tubes and place on ice in the order in which they will be processed, according to the pre-established time sheet.
7. Start the reaction by addition of the stimulus (or the inhibitor followed by the stimulus) after 5 min pre-incubation at 37°C.
8. At the end of the predetermined incubation, quickly centrifuge each tube (10 s) in a microcentrifuge to pellet the cells.
9. Carefully remove the supernatant, and immediately resuspend the cells in 0.5 volume suspension buffer A to which 0.5 volume of hot 2X lysis buffer B is added while vortexing the sample.
10. Place all samples in a microtube holder, which is then transferred to boiling water for 15 min.
11. After rapid cooling on ice, centrifuge the samples for 15 min at 15,000g.
12. Aliquot the supernatants and store at −70°C for later processing. Use one aliquot to determine the protein concentration (*see* **Note 2**).

3.2. Preparation of Samples for immunoprecipitation

1. Prepare the setup as above, except that the lysis buffer (C) is maintained on ice.
2. Stimulate DFP-treated cells with the desired agonist at 37°C and quick-spin to pellet the cells.
3. Lyse PMN (1×10^7 cells/mL) in an original volume of cold lysis buffer C, and place on ice for 25 min with occasional vortexing.

4. Centrifuge the whole-cell lysates at 15,000*g* for 20 min at 4°C, and transfer the supernatants to clean tubes. Save an aliquot to measure the protein concentration (*see* **Note 2**).

3.3. Immunoprecipitation

1. Add anti-MAPK antibody (10 µL) to each tube containing protein-adjusted whole-cell lysates (500 µg protein).
2. Incubate mixture overnight at 4°C on a rotating wheel for thorough mixing.
3. Add an aliquot (40 µL) of protein G Agarose slurry to the lysates, and incubate for 90 min on the rotating wheel at 4°C.
4. At the end of the incubation, recover the beads by quick centrifugation and wash three times with lysis buffer C (*see* **Note 6**).
5. Add 40 µL 2X sample buffer to the bead pellets, and boil for 5 min to release the immunocomplexes.
6. Centrifuge the samples, and carefully collect the supernatants. Keep samples frozen at −70°C until analysis by SDS-PAGE and Western blotting. Alternatively, the beads are used for a kinase activity assay (see below).

3.4. Kinase Assay

1. Pellet the immune complexes by brief centrifugation and washing as above.
2. Wash the beads twice in kinase wash buffer before addition of 40 µL of kinase buffer containing the substrate relevant to the immunoprecipitated MAPKs and 10 µCi [γ-^{32}P]-ATP.
3. Incubate samples at 30°C for 10–30 min (*see* **Note 7**).
4. When using a GST-fusion protein substrate, the reaction can be stopped and analyzed either of two methods:
5. Assay method 1:
   a. Add 10 µL of 5X hot sample buffer, boil for 15 min, and centrifuge.
   b. Analyze the supernatant by SDS-PAGE and Western blotting (*see* **Note 8**).
   c. Either expose the membranes to X-ray films or quantify the incorporated ^{32}P with a Phosphorimager (*see* **Note 9**).
   d. After exposure, re-hydrate the membranes and use for immunoblotting with an antibody to the immunoprecipitated MAPKs to determine efficiency of the immunoprecipitation.
6. Assay method 2:
   a. Use quick centrifugation to pellet the beads, and spot the entire supernatant onto prelabeled 1-inch square phosphocellulose papers (P81, Whatman).
   b. Immerse the papers in 180 m*M* phosphoric acid, and wash five times to remove ATP.
   c. Use a final rinse in acetone to remove the phosphoric acid, and let the papers air-dry.
   d. Insert the papers into small flasks, and estimate incorporated ^{32}P radioactivity by Cerenkov or liquid scintillation counting.

    e. Perform control incubations by replacing the immunoprecipitated MAPK with kinase buffer and spot onto P81 papers as above.

    f. Subtract control values from the values obtained with the immune complexes.

7. If using the T-669 peptide, stop the reaction by addition of 5 µL of perchloric acid to precipitate most proteins.

8. Centrifuge for 5 min, spot the supernatant on phosphocellulose papers, and proceed as described above (Method 2).

3.5. Immunoblotting

1. Use standard methods for electrophoretic transfer of proteins separated by SDS-PAGE.
2. Block the membranes in TBS-TW/NFDM for 1 h at room temperature with shaking.
3. Wash the membranes twice in TBS-TW.
4. Place the membranes in a dish with the appropriate dilution of primary antibody, and incubate overnight at 4°C with shaking on a rotating shaker.
5. Wash the membranes in TBS-TW with vigorous shaking, once for 15 min and three times for 5 min each (*see* **Note 10**).
6. Incubate the membranes with peroxidase-labeled anti-species secondary antibody for 90 min at room temperature, followed by a minimum of six washes for 10 min each.
7. After washing, remove excess liquid, and treat the membranes with chemiluminescent detection reagent according to the manufacturer's instructions.
8. Terminate the incubation by shaking off and blotting excess reagent.
9. Mount the membranes on a filter paper fitting an X-ray cassette and cover with plastic wrap.
10. Expose the membranes to films for various times, and develop the films (*see* **Note 11**).

3.6. Stripping Blots

It is critical that the protein loading and recovery be verified on the same membrane as the one used for immunoblotting with the phosphospecific antibodies.

1. Carefully remove the plastic wrap, and re-hydrate the membranes.
2. Wash membranes twice in TBS-TW, and incubate with a large volume of stripping buffer on a rotating shaker for 20 min at room temperature.
3. Wash the membranes twice with TBS-TW, and incubate in blocking buffer as described previously.
4. Re-probe the membrane with the antibody to total MAPK.

3.7. Old and New Inhibitors of the MAPK Pathways

1. The role of the MAPKs in neutrophils has been studied mostly via the use of pharmacological inhibitors. Fortunately, the critical role of the MAPKs in proliferation, differentiation, and apoptosis and the potential for pharmacological intervention

in many diseases has raised the interest of the pharmaceutical industry and sped up the development of many inhibitors for each of the MAPKs and their pathways, providing the research community with several alternatives. A summary of the most commonly used pharmacological inhibitors can be found in **Table 1** (*see* **Notes 12–16**).

2. Pharmacological kinase inhibitors are typically ATP-competitive small molecules. Despite the fact that the ATP-binding pockets of kinases have great similarities, it has been possible to design inhibitors with greater specificity through the use of crystallography and computer modeling, as studies have shown that only a few residues regulate specificity. Nevertheless, inhibitors often target more than one kinase, albeit with different sensitivity, which makes the assessment of their effects on the targeted kinase in the cell of choice even more critical. Moreover, inhibitor specificity often comes into question with time. For example, PD98059 and U0126, previously considered to have good specificity toward MEK1/2, have recently been shown to also inhibit MEK5 and thereby ERK5 *(16,17)*, raising doubts about the conclusions obtained with this inhibitor. Studies with MAPK inhibitors in PMN have sometimes given contradictory results. For example, the MEK/ERK pathway inhibitor PD98059 has been shown to inhibit the fMLP-stimulated phosphorylation of p47^{phox} and respiratory burst by some (50 µM, 45 min *[12]*) whereas others found it ineffective (30–50 µM, 30 min *[18,19]*). Novel inhibitors may then be helpful to resolve such discrepancies.

3. It has become clear that the MAPK pathways are organized and regulated through protein–protein interactions and that substrate specificity involves more than the interaction with the phosphoacceptor site (XS/TP). Several docking domains have now been identified in the MAPKs or the MAPK-interacting proteins (substrate, phosphatase, scaffold, upstream kinases) and shown to be required for information flow or substrate targeting. These include the D domains of MAPK-interacting proteins that are comprised of at least basic residues followed by an *LxL* motif, the common docking (CD) domain found in MAPKs that features a cluster of negatively charged residues at the opposite side of the active site, the domain for versatile docking (DVD) in MKK, and the docking site for ERK, FXFP (DEF) domain (*FxFP*) in substrates *(20)*. Disrupting these interactions with a peptide derived from these domains would be predicted to interrupt the signaling pathway, if the peptide had access to the components of the module. The discovery of short sequences from HIV TAT or the *Drosophila* transcription factor *Antennapedia* that exhibit cell membrane translocation properties has led to the design of cell-penetrating peptides (CPPs) that represent an efficient cell delivery strategy once the protein transduction domains are fused to various biological molecules ranging from liposomes to oligonucleotides and peptides *(21)*. Combining CPP and MAPK docking domains, novel peptide-based inhibitors have been developed that show promising inhibitory activity *(22,23)*, and some are now commercially available (**Table 2**). Although CPP-derived inhibitors have been tested in a variety of cells, including alveolar macrophages *(24)*, such studies have yet to be published in neutrophils. Nevertheless, CPPs coupled to other peptides such as the inhibitor domain of p21-

Table 1
Pharmacological Inhibitors of Mitogen-Activated Protein Kinases (MAPKs) and Their Activation Pathways

Inhibitors	Target(s)	Mechanism	Comments
PD98059	MEK1,2,5	Non-comp. Binds MEKs	Inhibits activation of MEK (NOT activity), thereby inhibits ERK1, 2, 5 activation.
PD184352	MEK1,2,5	Non-comp. Binds MEKs	Derived from PD98059. *Not commercially available.* Used in clinical trials.
U0126	MEK1,2,5	Non-comp. Binds MEKs	Inhibits activation of MEK (NOT activity), thereby inhibits ERK1, 2, 5 activation.
U0125 U0124	MEK1,2 Negative Control	Non-comp. Binds MEKs	Inhibits activation of MEK (NOT activity).
SL327	MEK1,2	Non-comp. Binds MEKs	Structural analog of U0126. Used for in vivo studies in CNS.
ERK inhibitor $C_{14}H_{16}N_2O_3S$-HCl	ERK	ATP competitor	Inhibits ERK kinase activity.
ERK CD inhibitor	ERK	Compound disrupting ERK interaction with substrate via CD domain	Recently published novel approach. Could selectively inhibit substrates. *Not commercially available (28).*
SB203580	p38 MAPK	ATP competitor	Inhibits α and β isoforms only. Can inhibit Raf and some JNK isoforms at higher doses. Suppresses Cyt. P450 1A1 transcription and recently found to inhibit RIP2 kinase.
SB202190	p38MAPK	ATP competitor	Structurally similar to SB203580. Inhibits alpha and beta isoforms only. Suppresses Cyt. P450 1A1 transcription and recently found to inhibit RIP2 kinase.
SB239063	p38MAPK	ATP competitor	Inhibits α and β isoforms only. Greater selectivity than SB203580. Given orally to mice.
SB220025	p38MAPK	ATP competitor	Recently found to inhibit RIP2 kinase.

SB202474	Negative Control		
ML3163	p38MAPK	ATP competitor	Structurally related to SB compounds.
BIRB796	p38MAPK	Allosteric competition binding site	Structurally different from SB compounds. Inhibits all isoforms in vivo and in vitro. *Not commercially available.*
PD169316	p38MAPK	ATP competitor	Structurally related to SB compounds. Recently found to inhibit RIP2 kinase.
SC68376	p38MAPK		Seems to be best suited for in vivo studies.
SP600125	JNK	ATP competitor	NOT specific. Inhibited 13 out of 28 kinases with similar or greater potency.
AS601245	JNK	ATP competitor	Has been used in vivo, known to cross the blood brain barrier and to inhibit brain-specific JNK3. May inhibit p38 and ERK at higher doses.
CEP-1347	MLK1-3	ATP competitor	Derived from K-252a. Used in clinical trials. Inhibits JNK and p38 MAPK pathways.

MEK, MAPK or extracelular signal-regulated kinase kinase; ERK, extracellular signal-regulated kinase; CNS, central nervous system; JNK, c-Jun N-terminus kinase; CD, common docking; MLK, mixed lineage kinase.

Table 2
Peptide-Based Inhibitors of Mitogen-Activated Protein Kinases (MAPKs)

Inhibitors	Target(s)	Mechanism	Comments
Ste-MPKKKPTPIQLNP-NH$_2$	ERK	N-terminus of MEK1, binds to ERK2 and prevent sits activation.	Stearated at the C terminus, cell permeable
MTP$_{TAT}$-G-MEK1$_{13}$ *H-GYGRKKRRQRRR-G-* *MPKKKPTPIQLNP-NH$_2$*	ERK	N-terminus of MEK1, binds to ERK2 and prevents its activation.	Fused to TAT PTD for better delivery
HIV-TAT$_{48-57}$-PP-JBD$_{20}$ *H-GRKKRRQRRR-* *PPRPKRPTTLNLFPQVPRSQDT-NH$_2$*	JNK	Sequence derived from the JBD of scaffold JIP. Inhibits JNK activation.	Fused to TAT PTD
HIV-TAT$_{47-57}$-gaba-c-Jun δ$_{33-57}$ *Ac-YGRKKRRQRRR-gaba-* *ILKQSMTLNLADPVGSLKPHLRAKN-* *NH$_2$*	JNK	Sequence from the JBD of c-Jun (δ domain), disrupts interaction of JNK-c-Jun and inhibits c-Jun phosphorylation.	Fused to TAT PTD with a GABA spacer

The inhibitors listed above are commercially available (Calbiochem). JIP is a family of scaffold proteins that bind the components of the JNK module. JBD, JNK-binding domain; TPD, transduction protein domain. *See* **Table 1** for other definitions.

activated kinase (Pak) (aa 83–149) *(25)* or the 11-residue peptide for the NEMO-binding domain that interferes with IKK/NFκB activation *(26)* have successfully been used in neutrophils. Thus, these newly developed inhibitors are likely to be useful to confirm data obtained with pharmacological inhibitors (*see* **Notes 12–16**).

4. Notes

1. VAN should be activated for maximal inhibition of protein phosphotyrosyl-phosphatases *(27)*. This procedure depolymerizes the vanadate, converting it into a more potent inhibitor of protein tyrosine phosphatases. Please note that adding DTT rapidly inactivates VAN. Store the activated VAN as aliquots at −20°C.

2. When lysing PMN in sample buffer, high concentrations of SDS and DTT interfere with Lowry-based assay for protein determination. We have used a modified Lowry assay in which proteins are precipitated for 1 h on ice in 1 mL of precipitating solution (10% phosphotungstic acid in 1% perchloric acid). The Pyronin-Y used in the sample buffer instead of the usual bromophenol blue stays pink under these conditions, providing greater display of the precipitates, which are re-suspended in 1 mL of Lowry reagent for 10 min followed by addition of 0.1 mL of Folin Ciocalteu reagent (Sigma) for 10 min. Optimal densities are read at 650 nm. A DC-Biorad kit (500-0112), which we found compatible with NP-40, TX-100, and low doses of SDS, is used to determine protein concentration in the immunoprecipitation lysates.

3. Precast mini gels are recommended for gain-of-time and consistency, ensuring better separation. A 10% acrylamide concentration provides best migratory resolution of ERK1 from ERK2 and is adequate for the other MAPKs.

4. We recently started using Protein G Agarose Fast Flow from UBI, which is supplied as a 1:1 slurry and is used straight out of the bottle. We had previously used Protein A Sepharose, which is adequate for recognition of rabbit polyclonal immunoglobulin (Ig)G. The later requires washing of the beads and swelling in phosphate saline buffer prior to make the 1:1 slurry.

5. Although it has been suggested that dilutions of the antibodies can be re-used many times, we noted that re-using them more than twice led to significant decrease in signal:noise ratio. The dilutions are kept at 4°C no longer than 30 d without addition of sodium azide.

6. After each washing of the beads, the supernatant must be removed carefully without loss of any beads. We have used gentle suction through a vacuum line and a fine and elongated gel pipet tip. If analysis by Western blotting reveals that the background is too high, an additional wash step with lysis buffer containing 1 *M* NaCl can be performed.

7. The time of incubation can be adjusted from 10 to 30 min, depending on the cell concentration/protein content, extent of activation, etc. We have routinely used a 20-min incubation time for p38MAPK and ERK and a 30-min incubation for JNK.

8. Instead of exposing a dry gel to X-ray films, we have found that transferring the proteins to membranes and exposing the membranes to the X-ray films leads to a sharper image with reduced background (*see* **Fig. 2**).

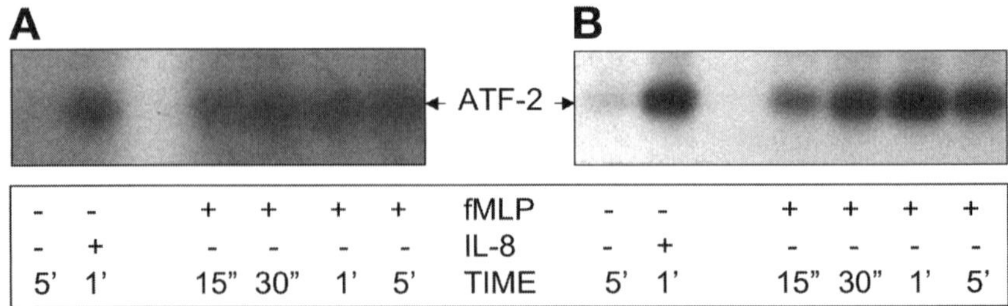

Fig. 2. p38 mitogen-activated protein kinase (MAPK) immunoprecipitation and kinase assay. Polymorphonuclear neutrophils were stimulated with 25 nM IL-8 and 10^{-7} M fMLP for the times indicated. Cells were lysed in lysis buffer C and p38 MAPK was immunoprecipitated as described in the text. Kinase activity was measured using GST-ATF-2 as substrate. The reaction was stopped with 5X hot sample buffer and the supernatants were divided into two aliquots to analyze ^{32}P incorporated in ATF-2 either by SDS-PAGE and drying of the gel (**A**) or by SDS-PAGE and Western blotting (**B**). Gel and membrane were exposed to films overnight. Note the greater sharpness of the film in **B**.

9. If a phosphoimager is not available and quantitation of incorporated ^{32}P is desirable, excise the band of interest after staining the gel (Pierce; 24590) and drop the slice in scintillation liquid for counting.

10. The protocols for immunoblotting can be adapted for convenience. We have used with similar results an overnight incubation at 4°C with TBS-TW/NFDM to block the membranes, followed by a 2-h incubation at room temperature with the primary antibody.

11. When densitometric analysis of the immunoblot is needed, it is important to do the analysis on films that have not been overexposed and to determine the range of linearity using standards, especially if a chemiluminescent detection system such as ECL is used. We recently started using LumiGlo (KPL; 54-12-50) as the detection system, and it seems to exhibit greater linear sensitivity. It is recommended that the films be exposed for multiple lengths of time and that the densitometric analysis be performed on more than one film.

12. It is important to be aware of the mechanism of the inhibitor of choice. For example, SB203580 and related compounds inhibit the kinase activity of p38MAPK and do not affect the activation and phosphorylation of p38MAPK. As most of the inhibitors are ATP-reversible, the activity of the inhibited kinase cannot be measured in an immunoprecipitation/kinase assay because the inhibitor may be partly or entirely lost during the process and the washes. The effects of the inhibitor must then be assessed on the phosphorylation of endogenous downstream substrates such as MK2 through a kinase assay or ATF-2 using a phosphospecific antibody.

13. Do not believe everything in the catalog or even the literature; verify the information.

14. What is true for NIH3T3 cells may not be true for neutrophils. Initial studies with PD98059 reported the need for long incubation time (1 h) with high doses (50–

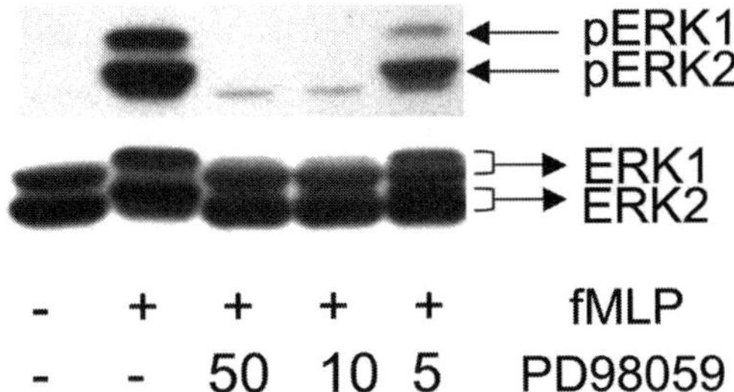

Fig. 3. Effect of PD98059 on fMLP-stimulated activation of extracellular signal-regulated kinase (ERK)1/2. Polymorphonuclear neutrophils were treated with 5, 10, or 50 μ*M* PD98059 for 20 min prior to addition of 10^{-7} *M* fMLP for 1 min. Cells were lysed in sample buffer B and the whole-cell lysates were analyzed by SDS-PAGE and Western blotting with antibodies to either dual-phosphorylated (top) or total (bottom) ERK1/2.

100 μ*M*) of the inhibitors; however, we have obtained 85–90% ERK inhibition in fMLP-stimulated PMN after 20 min incubation with 10 μ*M* PD98059 (*see* **Fig. 3**).

15. *Always* assess the effects of the inhibitor on the target kinase and use the minimal dose and incubation time, which will have a greater effect on the target and allow one to study the pathway's role in a particular response.
16. Use more than one inhibitor for any individual kinase in any study.

References

1. Torres, M. (1996) Mitogen-activated protein kinases and signal transduction in myeloid cells. *Int. J. Pediat. Hematol. Oncol.* **3,** 387–405.
2. Torres, M. (2003) Mitogen-activated protein kinase pathways in redox signaling. *Front. Biosci.* **8,** D369–D391.
3. Morrison, D. K. and Davis, R. J. (2003) Regulation of MAP kinase signaling modules by scaffold proteins in mammals. *Annu. Rev. Cell Dev. Biol.* **19,** 91–118.
4. Grinstein, S. and Furuya, W. (1992) Chemoattractant-induced tyrosine phosphorylation and activation of microtubule-associated protein kinase in human neutrophils. *J. Biol. Chem.* **267,** 18,122–18,125.
5. Gomez-Cambronero, J., Huang, C.-K., Gomez-Cambronero, T. M., Waterman, W. H., Becker, E. L., and Sha'Afi, R. I. (1992) Granulocyte-macrophage stimulating factor induced protein tyrosine phosphorylation of microtubule-associated protein kinase in human neutrophils. *Proc. Natl. Acad. Sci. USA* **89,** 7551–7555.
6. Torres, M., Hall, F. L., and O'Neill, K. A. (1993) Stimulation of human neutrophils with formyl-methionyl-leucyl-phenylalanine induces tyrosine phosphorylation and activation of two distinct mitogen-activated protein kinases. *J. Immunol.* **150,** 1563–1576.
7. Powell, D. W., Pierce, W. M., and McLeish, K. R. (2005) Defining mitogen-activated protein kinase pathways with mass spectrometry-based approaches. *Mass Spectrom. Rev.* **24,** 847–864.

8. Nick, J. A., Avdi, N. J., Gerwins, P., Johnson, G. L., and Worthen, G. S. (1996) Activation of a p38 mitogen-activated protein kinase in human neutrophils by lipopolysaccharide. *J. Immunol.* **156,** 4867–4875.

9. Arndt, P. G., Suzuki, N., Avdi, N. J., Malcolm, K. C., and Worthen, G. S. (2004) Lipopolysaccharide-induced c-Jun NH2-terminal kinase activation in human neutrophils: role of phosphatidylinositol 3-Kinase and Syk-mediated pathways. *J. Biol. Chem.* **279,** 10,883–10,891.

10. Avdi, N. J., Nick, J. A., Whitlock, B. B., et al. (2001) Tumor necrosis factor-alpha activation of the c-Jun N-terminal kinase pathway in human neutrophils. Integrin involvement in a pathway leading from cytoplasmic tyrosine kinases apoptosis. *J. Biol. Chem.* **276,** 2189–2199.

11. Brown, G. E., Stewart, M. Q., Bissonnette, S. A., Elia, A. E., Wilker, E., and Yaffe, M. B. (2004) Distinct ligand-dependent roles for p38 MAPK in priming and activation of the neutrophil NADPH oxidase. *J. Biol. Chem.* **279,** 27,059–27,068.

12. Dewas, C., Fay, M., Gougerot-Pocidalo, M. A., and El Benna, J. (2000) The mitogen-activated protein kinase extracellular signal-regulated kinase 1/2 pathway is involved in formyl-methionyl-leucyl-phenylalanine-induced p47phox phosphorylation in human neutrophils. *J. Immunol.* **165,** 5238–5244.

13. El Benna, J., Han, J. H., Park, J. W., Schmid, E., Ulevitch, R. J., and Babior, B. M. (1996) Activation of p38 in stimulated human neutrophils: Phosphorylation of the oxidase component p47phox by p38 and ERK but not by JNK. *Arch. Biochem. Biophys.* **334,** 395–400.

14. Benna, J. E., Dang, P. M., Gaudry, M., et al. (1997) Phosphorylation of the respiratory burst oxidase subunit p67phox during human neutrophil activation-regulation by protein kinase C-dependent and independent pathways. *J. Biol. Chem.* **272,** 17,204–17,217.

15. Kaminska, B. (2005) MAPK signalling pathways as molecular targets for anti-inflammatory therapy—from molecular mechanisms to therapeutic benefits. *Biochim. Biophys. Acta* **1754,** 253–262.

16. Mody, N., Leitch, J., Armstrong, C., Dixon, J. and Cohen, P. (2001) Effects of MAP kinase cascade inhibitors on the MKK5/ERK5 pathway. *FEBS Lett.* **502,** 21–24.

17. Davies, S. P., Reddy, H., Caivano, M., and Cohen, P. (2000) Specificity and mechanism of action of some commonly used protein kinase inhibitors. *Biochem. J.* **351,** 95–105.

18. Kuroki, M. and O'Flaherty, J. T. (1997) Differential effects of a mitogen-activated protein kinase kinase inhibitor on human neutrophil responses to chemotactic factors. *Biochem. Biophys. Res. Commun.* **232,** 474–477.

19. Mocsai, A., Banfi, B., Kapus, A., et al. (1997) Differential effects of tyrosine kinase inhibitors and an inhibitor of the mitogen-activated protein kinase cascade on degranulation and superoxide production of human neutrophil granulocytes. *Biochem. Pharmacol.* **54,** 781–789.

20. Tanoue, T. and Nishida, E. (2003) Molecular recognitions in the MAP kinase cascades. *Cell Signal.* **15,** 455–462.

21. Deshayes, S., Morris, M. C., Divita, G., and Heitz, F. (2005) Cell-penetrating peptides: tools for intracellular delivery of therapeutics. *Cell Mol. Life Sci.* **62,** 1839–1849.
22. Barr, R. K., Boehm, I., Attwood, P. V., Watt, P. M., and Bogoyevitch, M. A. (2004) The critical features and the mechanism of Inhibition of a kinase interaction motif-based peptide inhibitor of JNK. *J. Biol. Chem.* **279,** 36,327–36,336.
23. Bogoyevitch, M. A., Barr, R. K., and Ketterman, A. J. (2005) Peptide inhibitors of protein kinases-discovery, characterisation and use. *Biochim. Biophys. Acta* **1754,** 79–99.
24. Iles, K. E., Dickinson, D. A., Watanabe, N., Iwamoto, T., and Forman, H. J. (2002) AP-1 activation through endogenous H(2)O(2) generation by alveolar macrophages. *Free Radic. Biol. Med.* **32,** 1304–1313.
25. Martyn, K. D., Kim, M. J., Quinn, M. T., Dinauer, M. C., and Knaus, U. G. (2005) p21-activated kinase (Pak) regulates NADPH oxidase activation in human neutrophils. *Blood* **106,** 3962–3969.
26. Choi, M., Rolle, S., Wellner, M., et al. (2003) Inhibition of NF-κB by a TAT-NEMO-binding domain peptide accelerates constitutive apoptosis and abrogates LPS-delayed neutrophil apoptosis. *Blood* **102,** 2259–2267.
27. Gordon, J. (1991) Use of vanadate as protein-phosphotyrosine phosphatase inhibitor *Methods Enzymol.* **201,** 477–482.
28. Hancock, C. N., Macias, A., Lee, E. K., Yu, S. Y., Mackerell, A. D. Jr., and Shapiro, P. (2005) Identification of novel extracellular signal-regulated kinase docking domain inhibitors. *J. Med. Chem.* **48,** 4586–4595.

9

Transduction of Proteins into Intact Neutrophils

Tieming Zhao and Gary M. Bokoch

Summary

Neutrophils and related phagocytic leukocytes are notoriously difficult to transfect, making the introduction of proteins into these cells for biological studies problematic. We describe here two methods that have been successfully used to introduce proteins into intact primary human neutrophils while maintaining normal functional responses. The first utilizes a lipid-based reagent that transports proteins into intact neutrophils. This method is quick, easy, and is capable of transducing greater than 90% of the neutrophils in the population being studied. The second method involves the addition of a sequence derived from the HIV TAT protein to the protein to be introduced into the neutrophil. This requires both molecular biology to generate the initial construct as well as special procedures for protein isolation and renaturation. However, it also results in highly effective functional protein delivery into human neutrophils.

Key Words: Neutrophils; TAT proteins; GTPases; transduction; cell signaling; protein expression.

1. Introduction

To participate effectively in host defense, neutrophils must respond to chemotactic factors (*N*-formylated bacterial peptides, C5a, leukotriene B4, interleukin 8, etc.) and cytokines (tumor necrosis factor [TNF]-α, etc), leave the blood stream, and migrate across an endothelial cell layer into the tissues to contain and eliminate the infectious agents. Neutrophils represent the prototypic chemotactically sensitive cell, responding to chemotactic signals initiated through the binding of bacterial peptides and other chemokines to G protein-coupled receptors with speeds of up to 30 µm/min. Once at inflammatory sites, these phagocytic cells engulf the invading microorganisms and kill them through the generation of reactive oxygen species and the release of granule contents into the phagosome.

From: *Methods in Molecular Biology, vol. 412: Neutrophil Methods and Protocols*
Edited by: M. T. Quinn, F. R. DeLeo, and G. M. Bokoch © Humana Press Inc., Totowa, NJ

These intricate and highly coordinated events involve many signal transduction pathways and the action of small GTPases of the Rho family *(1)*.

Unfortunately, it has been difficult to study chemotactic signaling mechanisms in primary neutrophils because these differentiated cells are technically difficult to manipulate and transfect. Differentiatable myeloid cell lines used in some studies often do not respond as effectively as primary neutrophils, and may utilize different signaling mechanisms to regulate chemotactic responses. We describe here two methods that we have used to successfully introduce mutant proteins into primary human neutrophils for the purpose of studying signal transduction initiated by neutrophil chemotactic and phagocytic receptors.

The first method makes use of the commercial BioPORTER® protein delivery reagent, a unique lipid formulation that allows direct translocation of proteins into living cells (*see* www.genlantis.com). The BioPORTER reagent lipid captures proteins and transports them inside the neutrophils. We have found this method to be quick, easy, and effective, transducing the protein of interest into more than 90% of the neutrophils present. Most of our use of this method has involved transduction of Rho GTPases.

The second method makes use of the addition to the protein of interest of a portion of the HIV-TAT protein to transduce the protein to the cell interior. Various systems for generation of TAT constructs have been described *(2)*, and we have made use of the system popularized by Dowdy and colleagues *(3–5)*. Although cationic TAT peptides were originally speculated to cross plasma membranes directly, recent work has shown that this is a misconception. Uptake of TAT peptides appears to largely be due to endocytosis, with subsequent escape of a portion of the peptide to the cytoplasm *(6)*. However, we have found this method to work well for introducing certain proteins into human neutrophils.

2. Materials

2.1. Introduction of Proteins
into Human Neutrophils Using BioPORTER Reagent

1. Bioporter reagent (BP502401), available from Gene Therapy Systems, San Diego, CA.
2. Krebs-Ringer HEPES buffer: 118 mM NaCl, 4.8 mM KCl, 25 mM HEPES, 1.2 mM KH$_2$PO$_4$, 1.2 mM MgSO$_4$, 5.5 mM glucose, and 1 mM CaCl$_2$, adjust pH to 7.4.
3. Phosphate-buffered saline (PBS) buffer: 137 mM NaCl, 2.7 mM KCl, 10 mM Na$_2$HPO$_4$, 1.8 mM KH$_2$PO$_4$ (adjust to pH 7.4 as necessary).
4. Bath sonicator FS6 (Fisher Scientific).
5. Labquake shaker (Labindustries).

2.2 Delivery of TAT Fusion Proteins into Human Neutrophils

1. Buffer Z: 8 M urea, 100 mM NaCl, 20 mM HEPES, pH 8.0.

2. Imidazole/Buffer Z: dissolve to 20 m*M*, 50 m*M*, 100 m*M*, 250 m*M*, 500 m*M*, and 1 *M* of imidazole in Buffer Z, respectively.
3. 60% Isopropanol/20 m*M* HEPES
4. Dialysis tubing: SnakeSkin Pleated dialysis tubing with molecular weight cutoff of 7000 (Pierce).
5. 5-inch polystryrene chromatography columns (Evergreen Scientific).
6. Ni-NTA Agarose beads (Qiagen).
7. Centricon concentrators (Amicon).
8. Ultrasonic processor Sonicator (Heat Systems).

3. Methods

3.1. Introduction of Proteins into Human Neutrophils Using BioPORTER Reagent

In the BioPORTER method, three critical parameters are the concentrations of BioPORTER, of the proteins to be introduced, and the neutrophils themselves. These parameters are described in the method detailed here, which has been used successfully to introduce Rac2 GTPase and various mutations into human neutrophils *(7,8)*. However, for other proteins, these conditions and concentrations should be optimized accordingly to achieve effective transduction into cells.

1. Commercially available BioPORTER reagent is in the form of dry film. Add 250 μL of chloroform into each tube containing the dry film and vortex for 20 s. Transfer 1–10 μL solution to the bottom of one 1.5-mL microfuge tube (*see* **Note 1**) and let the chloroform evaporate completely for at least 4 h, preferably overnight in the tissue culture hood without UV light. Can be stored at −20°C.
2. To prepare the BioPORTER/protein complexes, dilute purified target protein in PBS buffer to the concentration of 100 μg/mL (*see* **Note 2**). Add 100 μL of the protein solution into the microfuge tube containing the dried BioPORTER reagent, and sonicate gently for 30 s in a bath sonicator.
3. Wash neutrophils isolated from human blood with Krebs-Ringer HEPES buffer and resuspend cells in the same buffer at 3×10^6 cells/900 μL (*see* **Note 3**).
4. Pipet 900 μL of cells into the microfuge tube of BioPORTER/protein complexes. This brings the final concentration of BioPORTER to 1–10 μL/mL (*see* **Note 4**) and final protein concentration to 10 μg/mL (*see* **Note 5**). Place the tube on a Labquake shaker at eight cycles per minute at room temperature for 2 h (*see* **Note 6**).
5. Centrifuge cells for 5 min at 300*g*, then aspirate off supernatant. Repeat twice more to remove external reagent/protein. Resuspend cells in the desired buffer and cell density for subsequent experiments.
6. To determine the efficiency of protein delivery, subject lysate of neutrophils with or without BioPORTER treatment to Western blot analysis of the introduced protein. The amount of cell lysate used for Western blot varies for different proteins (*see* **Note 7**).

7. Alternative methods of evaluation can also be used, e.g., immunostaining (*see also* **Subheading 3.2., step 22**).

3.2. Delivery of TAT-Fusion Proteins into Human Neutrophils

In the TAT fusion protein method, it is important to obtain purified TAT fusion protein and keep it active. Much of the TAT fusion protein produced will be insoluble and stored in inclusion bodies. This acts to protect the fusion protein and leads to high protein yields.

1. Clone the cDNA coding for the target protein into the pTAT-HA bacterial expression vector (*see* **Note 8**) *(4)*.
2. Transform the plasmid into an *Escherichia coli* BL21 strain. Spread *E. coli* on an Luria-Bertani (LB) plate and incubate at 37°C overnight (*see* **Note 9**).
3. Prepare 1 L of LB medium in a 2-L flask. Remove 40 mL to a 150-mL flask. Autoclave both flasks and add ampicillin to the final concentration of 100 µg/mL when flasks are cool.
4. Remove 2 mL of LB medium and inoculate with a colony from the LB plate. Incubate the tube in an incubator shaker at 250 cycles per minute at 37°C for the rest of the day.
5. Inoculate 40 mL medium in the 150-mL flask with the 2 mL *E. coli*. Culture on an incubator shaker at 250 cycles per minute at 37°C, overnight.
6. The next morning, add ampicillin to 960 mL LB medium in the 2-L flask to the final concentration of 100 µg/mL.
7. Inoculate the 960 mL LB medium with 40 mL of overnight culture and grow at 37°C until OD600 reaches between 0.6–0.8 (about 3 h).
8. Add isopropyl-β-D-thio-galactopyranoside (IPTG) to the final concentration of 1 mM (or as optimized) to induce protein expression. Grow at 37°C for another 5 h.
9. Centrifuge at 6000g for 15 min at 4°C. Remove supernatant and resuspend the pellet in 10 mLof 20 mM Imidazole/Buffer Z (*see* **Note 10**).
10. Sonicate four times at 30 s pulse (strength 4, 75% power) or until turbid. Centrifuge at 20,000g for 15 min at 4°C.
11. Equilibrate 2 mL of Ni-NTA agarose beads in a 5-inch chromatography column with 20 mL of 20 mM Imidazole/Buffer Z at room temperature. Load *E. coli* cell lysate on the column at room temperature (*see* **Note 11**).
12. Wash the column with 100 mL of 20 mM Imidazole/Buffer Z to remove contaminating protein, followed by 100 mL of 60% isopropanol/20 mM HEPES and then 100 mL of 20 mM HEPES to remove endotoxin.
13. Elute TAT fusion protein from the column with 4 mL of 50 mM Imidazole/Buffer Z, followed by 4 mL of 100 mM Imidazole/Buffer Z, 4 mL of 250 mM Imidazole/Buffer Z, 4 mL of 500 mM Imidazole/Buffer Z, and 4 mL of 1 M Imidazole/Buffer Z. Collect each elution fraction separately.
14. Apply 5 µL of each fraction to sodium dodecyl sulfate (SDS)-polyacrylamide gel electrophoresis (PAGE), and stain with Coomassie Blue to check for the TAT pro-

tein. Immunoblotting is also recommended. Pool the fractions that contain pure TAT fusion proteins.

15. To remove urea, add TAT protein to dialysis tubing and dialyse in 1 L of 20 m*M* HEPES buffer at 4°C four times for 1 h, 2 h, 3 h, and overnight, respectively. Change to fresh HEPES buffer after each dialysis (*see* **Note 12**).

16. To concentrate TAT fusion protein, prepare a Centricon concentrator from Amicon with the correct molecular weight cutoff for your protein of interest. Centrifuge the TAT fusion protein dialysate at 100,000*g* at 4°C for 15 min to remove protein aggregates. Apply 2 mL of supernatant to the Centricon concentrator, centrifuge for 30 min at 1000*g* or 5000*g*, depending on the type of concentrator used, and remove the flow-through. Repeat this several times until the desired volume is reached (*see* **Note 13**). Invert concentrator and centrifuge at 1000*g* for 2 min to transfer concentrated TAT fusion protein into a retention vial.

17. Do a standard BCA (or other) assay to determine protein concentration and either use fresh or store at −80°C.

18. Remove protein aggregates formed during storage by centrifugation at 100,000*g* at 4°C for 15 min immediately before incubation with neutrophils. Measure the protein concentration using a standard BCA (or other) protein assay.

19. Wash neutrophils isolated from human blood with Krebs-Ringer HEPES buffer. Resuspend cells in the same buffer at a concentration of 2×10^7 cells/mL (*see* **Note 14**). Add 1 mL of cells to a 1.5-mL microfuge tube.

20. Add TAT fusion protein to the final concentration of 500 n*M* (*see* **Note 15**) and place the tube on a Labquake shaker at eight cycles per minute. Incubate at 37°C for 30 min.

21. Centrifuge cells for 5 min at 300*g*, then aspirate off the supernatant. Resuspend cells in desired buffer and at desired concentration for subsequent experiments (*see* **Note 16**).

22. To determine the efficiency of protein delivery, subject neutrophils with or without TAT fusion protein treatment to extensive washes, and analyze cells by immunoblotting (*see* **Note 17**).

23. As desired, label your TAT fusion protein with 5–25 µg fluorescein isothiocyanate (FITC) (Molecular Probes) in 300 µL for 2 h in the dark. Then load onto a gel fitration column (S-6, S-12, S-200) in phosphate-buffered saline (PBS) and collect appropriate fractions. Add the purified TAT-FITC fusion protein to approx 1×10^6 cells in media, and check for uptake by microscopy or flow cytometry.

4. Notes

1. The amount of BioPORTER reagent should be optimized depending on the plate size where the subsequent experiments take place. According to the manufacture, the starting volume for experiments subsequently carried out in 1 well in 96-well plates is 1 µL, 24-well plate 2.5 µL, 12-well plate 5 µL, 6-well plate 10 µL, 60-mm dish 20 µL, 100-mm dish 35 µL. We have successfully performed experiments with 1–10 µL of BioPORTER in 96-, 24-, 12-, and 6-well plates, respectively.

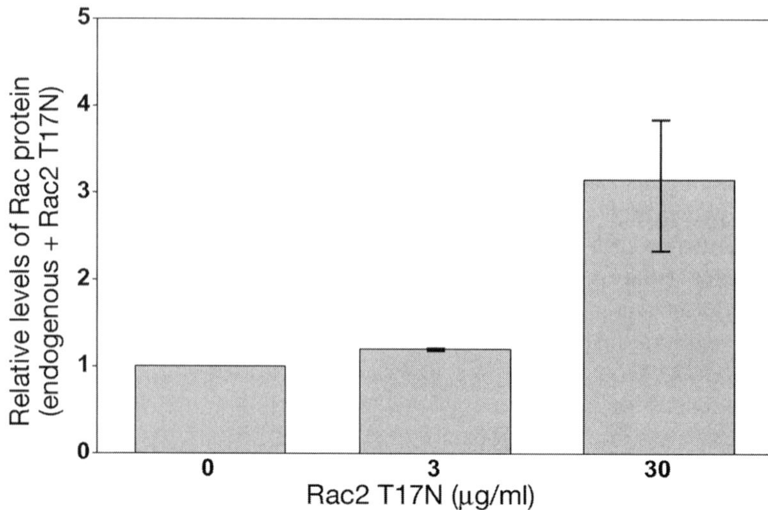

Fig. 1. Expression of exogenous Rac2 in human neutrophils with BioPORTER re-agent. The graph shows the level of expression of Rac2T17N relative to endogenous Rac2 in human neutrophils after incubation of cells with various concentrations of GTPase protein using BioPORTER reagent (*see* **ref. 7**) (Mean ± standard error of $n = 3$).

2. This concentration (100 µg/mL) is optimized for wild-type Rac2 and Rac2 mutants *(7,8)*. However, we have successfully performed experiments with concentrations from 30–300 µg/mL (3–30 µg/mL in the incubation with neutrophils; **step 4**). A concentration of 100 µg/mL can be used as a starting concentration for optimization of other proteins.

3. We used concentrations as high as 2×10^7 cells/mL successfully. However, though not very often, cells may aggregate at that concentration. This concentration is for human neutrophils: the concentration for murine neutrophils has not been determined.

4. Neutrophil viability is not affected when the BioPORTER concentration is lower than 3 µL/mL. However, in some experiments visible aggregation of neutrophils was observed when BioPORTER concentration was higher than 3 µL/mL.

5. For Rac2 concentrations lower than 10 µg/mL in the incubation with neutrophils, 1 µL/mL of BioPORTER is usually sufficient. For Rac2 concentrations of 10–30 µg/mL, 3 µL/mL of Bioporter is needed (*see* **Fig. 1**).

6. Keeping the incubation at room temperature instead of 37°C effectively reduced neutrophil aggregation. Incubation time generally varies from 1 to 4 h. Although effect of constitutively active Rac2 is apparent after 1 h incubation, more protein is delivered into the cells with longer incubation time. As a result of the short half life of isolated neutrophils, the incubation time will vary according to the length of subsequent experiments.

7. For example, 10 µg is needed for Rac2.

8. The pTAT-HA vector *(4)* contains a 6-histidine tag followed by the 11 amino acid TAT protein transduction domain flanked by glycine residues ensuring free bond

rotation, a hemaglutinin (HA) tag and a multicloning site. The cDNA of the target protein should be cloned into the multicloning site.

9. General safety precautions should be taken when producing, storing, and handling TAT fusion proteins. We recommend following NIH Biosafety Level 2 (BSL-2) guidelines for the safe handling of recombinant proteins.

10. *E. coli* can be saved at −80°C and resuspended until ready to purify.

11. Save some of the flow through for SDS-PAGE analysis. Always check your flow-through by Coomassie Blue stain or immunoblot. If abundant TAT fusion protein is found in the flow-through, decrease the concentration of imidazole in loading buffer Z. In contrast, if a high background of contaminating bacterial proteins is observed, increase the concentration of imidazole in loading buffer Z. The higher the imidazole in the load, the cleaner the peak fractions will be.

12. If significant protein precipitation is observed, add 10% glycerol to the dialysis buffer to help stabilize the protein. Alternatively, ion-exchange chromatography can be used.

13. We have stored TAT fusion proteins at 1 mg/mL at 4°C for several weeks, and at −80°C for indefinite periods. TAT fusion protein aggregates during storage, and this tends to happen more at higher protein concentrations.

14. Many experiments fail because of neutrophil aggregation, probably resulting from priming of neutrophils during the isolation process or from high cell concentration. Although it is not possible to completely prevent aggregation, it is less likely to occur at this cell concentration. We also noticed that keeping cells in larger volumes and/or leaving calcium out of the buffer (until necessary) helps prevent cell aggregation.

15. TAT fusion proteins were observed entering neutrophils at as low as 50 n*M* concentration.

16. Transduction of TAT fusion protein is concentration-dependent. Therefore, in the experiments that require long incubation times, not washing or only briefly washing the cells maintains chemical equilibrium and prevents TAT protein from slowly transducing back out of the cells. An example of the use of TAT fusion proteins to modulate human neutrophil NADPH oxidase activity *(9)* is shown in **Fig. 2**.

17. It has been reported that uptake of TAT protein involves endocytosis, and that the majority of TAT proteins that enter cells were degraded, with only traces reaching the cytosol *(6)*. This was sufficient to elicit biological effects, however *(5,6,9)*.

Acknowledgments

The authors thank Dr. Steven F. Dowdy, University of California at San Diego, for permission to describe here the TAT transduction methods he has developed. We thank Dr. Carl Nathan, Cornell University Medical School for the use of his TAT fusion proteins for the experiment described in **Fig. 2**. We acknowledge advice from Dr. Kersi Pestonjamasp on the use of BioPORTER in human neutrophils. This work was supported by National Institutes of Health grants GM39434 and GM44428 (to G. M. B). T. Z. is supported by a fellowship from the Arthritis Foundation, San Diego chapter.

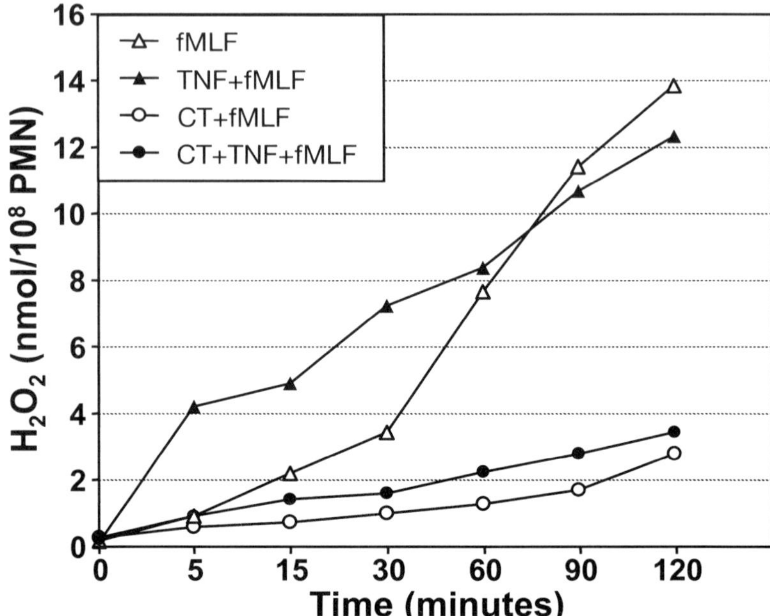

Fig. 2. Introduction of TAT-CT, a dominant negative Pyk2 construct, into human neutrophils. Human neutrophils were transduced with 500 n*M* TAT-CT or mock treated. The cells were then assayed for H_2O_2 formation with time under adherent conditions ± pretreatment with 25 ng/mL tumor necrosis factor (TNF)-α and stimulation with 0.1 µ*M* fMLP. The results shown demonstrate that the dominant negative TAT-CT enters the cells and inhibits the action of TNF_γ and fMLP (*see* **ref.** *8*).

References

1. Bokoch, G. M. (2005) Regulation of innate immunity by Rho GTPases. *Trends Cell Biol.* **15,** 163–171.
2. Joloit, A. and Prochiantz, A. (2004) Transduction peptides: from technology to physiology. *Nat. Cell Biol.* **6,** 189–196.
3. Schwarze, S. R., Hruska, K. A., and Dowdy, S. F. (2000) Protein transduction: unrestricted delivery into all cells? *Trends Cell Biol.* **10,** 290–295.
4. Becker-Hapak, M., McAllister, S. S., and Dowdy, S. F. (2001) TAT-mediated protein transduction into mammalian cells. *Methods* **24,** 247–256.
5. Han, H., Fuortes, M., and Nathan, C. (2003) Critical role of the carboxyl terminus of proline-rich tyrosine kinase (Pyk2) in the activation of human neutrophils by tumor necrosis factor: separation of signals for the respiratory burst and degranulation. *J. Exp. Med.* **197,** 63–75.
6. Richard, J. P., Melikov, K., Vives, E., et al. (2003) Cell-penetrating peptides-A reevaluation of the mechanism of cellular uptake. *J. Biol. Chem.* **278,** 585–590.

7. Gardiner, E. M., Pestonjamasp, K. N., Bohl, B. P., Chamberlin, C., Hahn, K. M., and Bokoch, G. M. (2002) Spatial and temporal analysis of Rac activation during live neutrophil chemotaxis. *Current Biol.* **12,** 2029–2034.
8. Zhao, T. and Bokoch, G. M. (2005) Critical role of proline-rich tyrosine kinase 2 in reversion of the adhesion-mediated suppression of reactive oxygen species generation by human neutrophils. *J. Immunol.* **174,** 8049–8055.
9. Zhao, T., Benard, V., Bohl, B. P., and Bokoch, G. M. (2003) The molecular basis for adhesion-mediated suppression of reactive oxygen species generation by human neutrophils. *J. Clin. Invest.* **112,** 1732–1740.

Optical Methods for the Measurement and Manipulation of Cytosolic Free Calcium in Neutrophils

Esther J. Hillson, Sharon Dewitt, and Maurice B. Hallett

Summary

The measurement and manipulation of cytosolic free Ca^{2+} permits the investigation of the mechanisms of generation of the Ca^{2+} signal and cellular responses to these Ca^{2+} signals within living neutrophils. The optical methods most applicable to neutrophils, which will be discussed here, are (1) the use of fluorescent indicators of Ca^{2+} and (2) photoactivation of reagents involved in Ca^{2+} signaling. Both of these synthetic agents can be loaded into neutrophils as lipid-soluble esters or can be microinjected into the cell. In this chapter, we will outline some of the techniques that have been used to monitor, visualize, and manipulate Ca^{2+} in neutrophils.

Key Words: Calcium signaling; cytosolic calcium flux; fluorescent calcium indicator dye; photoactivation; calcium imaging.

1. Introduction

The fluorescent Ca^{2+} chelator probes developed by Tsein and colleagues *(1,2)* have opened up the study of cytosolic free Ca^{2+} in small cells, including neutrophils. The ester method of loading the probe into the cytosol of neutrophils works successfully. These indicators can be synthesized or purchased with the ester groups, which both "mask" the Ca^{2+} binding part of the molecule and makes them lipid-soluble. Thus, the ester derivative readily crosses the plasma membrane and enters the neutrophil cytosol. In the cytosol, esterases cleave the ester bond to generate the acid form of the probe, which becomes entrapped within the cell, as it is hydrophilic and thus unable to easily cross the plasma (or other) membrane. It is the acid form of the probe that is also the Ca^{2+}-sensing form. There are two potential problems with this approach: (1) the accumulation of indicator in organelles, and (2) partial (rather than full) hydrolysis of the probe, which generates products with increased hydrophilicity, but that are insensitive

From: *Methods in Molecular Biology, vol. 412: Neutrophil Methods and Protocols*
Edited by: M. T. Quinn, F. R. DeLeo, and G. M. Bokoch © Humana Press Inc., Totowa, NJ

to Ca^{2+} *(3)*. However, these problems are easily avoided in the protocols given and the measurement methodologies would detect such problems. It is, however, important to limit the amount of probe loaded in this way because (1) as the indicator is a Ca^{2+} chelator, it will buffer cytosolic free Ca^{2+} changes and "blunt" the very response that is being measured and (2) there is a toxicity associated with the products of de-esterification of the probe (namely formaldehyde and H^+ ions), which can cause a reduction in ATP levels in neutrophils *(4)* and may have other toxic effects on the cell, including stimulation of cell aggregation. Some of these problems can be overcome by microinjection of nonesterified forms of the probes or high-molecular-weight forms that do not accumulate in organelles *(5,6)*, thereby circumventing the problems of location and toxicity associated with ester loading. Simple-lipid-assisted micro (SLAM)-injection works well with neutrophils and is the method of choice *(7)*.

In this chapter, we outline some of the techniques that have been used to monitor, visualize, and manipulate Ca^{2+} in neutrophils. Further details of some of these methods are given in **ref. 5**. A discussion of other methods not discussed here, such as the expression of luminescent and fluorescent indicators of Ca^{2+} are given in **ref. 8**.

2. Materials

1. Fluorescent Ca^{2+} probe: the properties of fluorescent Ca^{2+} probes suitable for use in neutrophils are presented in **Table 1**. These reagents and those for manipulation of cytosolic free Ca^{2+} are available from Invitrogen/Molecular Probes, Sigma, Calbiochem, Teflabs, or their agents (*see* **Notes 1** and **2**).
2. SLAM micropipets (Warner Instruments).

3. Methods

3.1. "Ester Loading" Procedure for Neutrophils

It is crucial that the loading medium contain Ca^{2+} (e.g., 1–2 mM) because as the Ca^{2+} chelator is generated within the cytosol, it will bind Ca^{2+} (the extent depending on its k_d) (*see* **Note 3**). Without additional extracellular Ca^{2+} to replace this, Ca^{2+} will be displaced from Ca^{2+} stores within the cell (or worse, not replaced at all). Another important point in the method is the 1000-fold dilution of the ester stock solution. This gives an acceptably low concentration the solvent dimethylsulfoxide (DMSO). DMSO can have toxic effect on neutrophils; therefore, it is important that it is kept to a minimum. At 0.1% (i.e., 1/1000), the effects of DMSO are minimal.

3.1.1. Dye Loading

1. Dissolve ester in dry DMSO (*see* **Note 1**) to give a stock (1–5 mg/mL). Store at –20°C.

Table 1
Fluorescent Ca^{2+} Probes Useful in Neutrophils

	kd (Ca^{2+} nM)	Excitation (nm)	Emission (nm)	Mode
Calcium Crimson	205	588	611	S
Calcium Green-1	189	506	534	S
Calcium Green-2	574	506	531	S
Calcium Green-5N	3300	506	531	S
Calcium Orange	328	554	575	S
FFP-18	400	340 & 380	505	R
Fluo3	864	506	526	S
Fluo3FF	62,000	506	526	S
Fluo4	345	494	516	S
Fura Red	133	420 & 480	640	R
Fura2	224	340 & 380	505	R
FuraFF	38,000	340 & 380	504	R
Indo-1	250	340	410 & 485	R
Mag-Fura-2	25,000	340 & 380	504	R
Mag-Fura-5	28,000	340 & 380	504	R
MOMO	nd	~495	~515	S
Quin2	114	340	490	S
Rhod-2	1000	553	576	S

All of these probes are available as cell-permeant esters. The mode of use is given as single wavelength (S) or the ratio of two wavelengths (R). nd, not determined.

2. Optional: mix 2.5 µL pluronic (25% w/v in DMSO) with 5 µL of ester stock. This step may assist transfer into the neutrophils, pluronic being a "dispersing agent" that assists in keeping the ester in solution.
3. Add 1 µL of ester to 1 mL of neutrophil suspension of $1–50 \times 10^6$ cell/mL in Ca^{2+}-containing medium (*see* **Note 4**).
4. Incubate neutrophils for 20–60 min at room temperature or 37°C.
5. Resuspend neutrophils in fresh medium. Loaded cells can be stored on ice to reduce leakage of Ca^{2+} probe.

3.1.2. Verification of Dye Loading

1. Check that fluorescence at 360 nm excitation (505 nm emission) is significantly higher than in nonloaded cells.
2. Record excitation or emission spectrum (quartz cuvet or optics) of loaded neutrophils to ensure that conversion of ester to its acid is complete (e.g., Fura2 ester and the acid have clearly different spectra; *see Molecular Probes Handbook*).
3. Treat neutrophils with either digitonin (150 mM) or Ca^{2+} ionophore (nonfluorescent ionomycin or Br A23187) and check that the spectral change is consistent with Ca^{2+} saturation for the probe, e.g., for fura2, the excitation spectrum peaks at 340 nm (*see* **Note 5**).

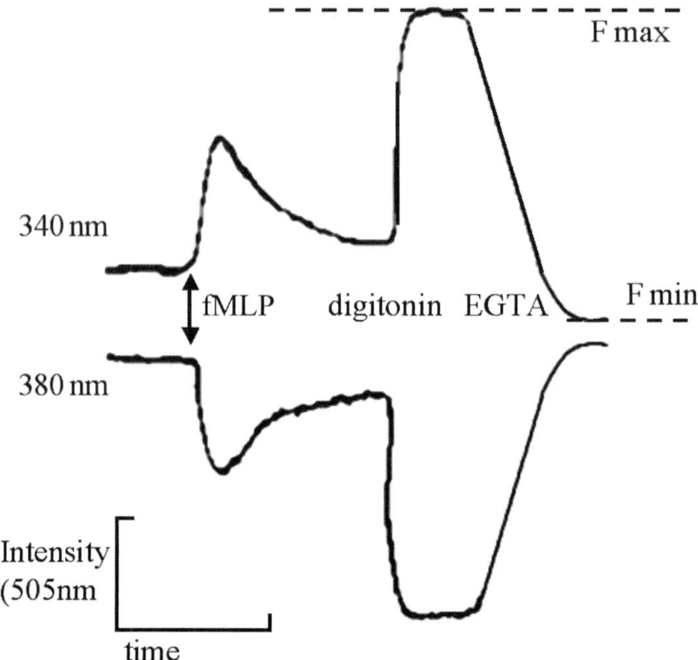

Fig. 1. A typical experiment in which cytosolic free Ca^{2+} concentration is monitored by Fura2. The emission at 505 nm is recorded for excitation at 340 nm and 380 nm as indicated. A stimulus is added that produces the "Ca^{2+} signature." The addition of digitonin and then EGTA is for calibration. For a single-wavelength probe, the values of Fmax and Fmin are used. For a dual-wavelength probes (as here), Rmax and Rmin are calculated from the two signals.

4. Add EGTA (20 mM) and check that the spectral change is consistent with zero saturation for the probe, e.g., for fura2, excitation peak is shifted toward 380 nm with the two spectra crossing at approx 360 nm.
5. Observe the fluorescence microscopically (e.g., Zeiss filters LP420/G365 or similar) to check loading has no obvious nonuniformity (i.e., in granules or phagosomes).

3.2. Measurement Protocols

3.2.1. Fluorometry

1. Add cell suspension to quartz or ultraviolet (UV) transmissible cuvet. Stir continuously to ensure that cells remain in the illumination beam. Set desired temperature (e.g., 37°C).
2. Set fluorometer to illuminate and record emission at the appropriate wavelengths. Record single- or dual-wavelength data as appropriate (*see* **Table 1**).
3. Once baseline data are constant (i.e., after warm-up time), add stimulus while recording (*see* **Fig. 1**).

4. Add ionomycin (1–4 μM) or digitonin (150 mM) at the end of the experiment to saturate the probe (*see* **Note 5**).
5. Once saturation has occurred, add EGTA (20 mM) (*see* **Fig. 1**).
6. Using the Kd parameters listed in Table 1, calculate the cytosolic free Ca^{2+} concentration throughout the time course using **eq. 1** or **2**, as appropriate (*see* **Fig. 1**).

3.2.2. Microfluorometery

By optically coupling a wavelength changer to the input port and a photomultiplier tube to the output port of a fluorescent microscope, the procedure given above can be used with individual cells. In order to reduce background signal from areas of the field not occupied by the cells (or by other cells), a pinhole in the focal plane can be useful. This could be either fixed, so that the cell is moved to the pinhole by the microscope stage, or moveable. However, as neutrophils, by their nature, tend to move as part of their response, it is preferable to replace the photomultiplier tube with an intensified charged coupled device (CCD) camera.

1. Set up a "virtual mask" by binning the data from the region of the image that includes the cell of interest (*see* **Note 6**). This approach has several advantages: (1) the cell is visualized and so that artifacts caused by cell movement are immediately apparent and (2) as the masks are electronic, several "masks" can be defined that read out the ratio values of more than one individual cell in the field.
2. Use ionomycin to provide the maximum and minimum signals when performing microfluorometry (*see* **Note 7**).

3.2.3. Flow Cytometry

Measurement of cytosolic free Ca^{2+} concentration within individual cells as a cell population is possible using flow cytometry. Time course of Ca^{2+} changes can be achieved by addition of stimulus to the population as it passes through the machine. Caution must be exercised in the interpretation of these data as, although individual cells are being interrogated, no single cell is followed through the time course and the same changes are merely population average changes. These often do not reflect changes at the single-cell level. For example, unless stimulus-induced Ca^{2+} spikes are synchronized within the population, these will not be observed by flow cytometry. With a UV laser cytometer, Indo-1, which can be used ratriometrically (*see* **Table 1**), is the probe of choice. However, single-wavelength indicators such as fluo3 and calcium green have been successfully used with conventional 488-nm lasers.

3.3. Calculation of Cytosolic Free Ca^{2+}

Single-wavelength dyes may be used for confocal microscopy or flow cytometry, but are not advised for conventional fluorometry or imaging. For these latter procedures, ratio dyes are preferable. The ratio of the two signals will be independent of the concentration (or amount) of probe (*see* **Notes 1** and **2**).

1. Obtain F_{max} and F_{min} at the end of the experiment by permeabilizing the cell (with ionophores) in the presence of Ca^{2+} and then chelating the Ca^{2+} with EGTA.
2. For a single wavelength indicator, calculate the cytosolic free Ca^{2+} concentration using **eq. 1**:

$$Ca^{2+} = K_d \cdot (F - F_{min}) / (F_{max} - F) \tag{1}$$

   where F is the fluorescent signal through the experiment, and F_{min} and F_{max} are the minimum and maximum obtainable signals from the indicator in the presence and absence of saturating amounts of Ca^{2+}, respectively.
3. For ratio dyes, the free Ca^{2+} required to give this signal ratio can be calculated using **eq. 2**:

$$Ca^{2+} = K_d \cdot \beta \cdot (R - R_{min}) / (R_{max} - R) \tag{2}$$

   where $\beta = Sf2/Sb2$, where Sxy is the emission signal at wavelength y ($y = 1$ for 340 nm and 2 for 380 nm) from the fully Ca^{2+}-saturated indicator ($x = b$), totally Ca^{2+}-free indicator ($x = f$), or variable Ca^{2+} saturation in the cell during the experiment ($x = v$), and $R = Sv1/Sv2$, $R_{max} = Sb1/Sb2$, and $R_{min} = Sf1/Sf2$.
4. Obtain R_{max} and R_{min} values at the end of the experiment by the addition of ionomycin or digitonin (to allow Ca^{2+} saturation), followed by EGTA (to remove Ca^{2+} from the probe) (*see* **Note 4**). R_{max} and R_{min} values are essential for the calculation.
5. The characteristic "Ca^{2+} signature" can be confirmed to originate solely from a change in Ca^{2+} by noting that the sum, $ASv1 + Sv2$ (where $A = (Sf2 - Sb2)/(Sb1 - Sf1)$), will be constant at all Ca^{2+} concentrations (**6**).

3.4. Imaging Ca²⁺ in Individual Neutrophils

3.4.1. Ratiometric Imaging

1. The fluorescent intensity at any point within the two-dimensional microscopic image of the cell will be proportional to the amount of probe in that "line of sight." Thus, it will be brighter at the center of the cell, where it is thicker, than at the edge, where the cell may become progressively thinner. Thus, it is helpful to use a ratiometric probe and take a ratio of two images to provide a "Ca^{2+} map" of the cytosolic free Ca^{2+} (*see* **Note 8**).
2. To take the ratio of two images, the changing excitation wavelength must be synchronized with the acquisition of images, often by a spinning filter wheel, optical chopper, or rapid-changing monochromator. These are commercially available from many sources (e.g., PTI, Cairn Instruments, etc.).
3. Imaging is best achieved by coupling the wavelength changer to the "fluorescent input" of the microscope and either a digital camera or a video camera to the "output."
4. Record a digitized signal to provide an array of pixels, each with a value that corresponds to the intensity of the fluorescent image in that region of the field being recorded.

Table 2
Imaging Speeds for Leica Resonant Scanning Confocal Microscope

X dimension (pixels)	Y dimension (pixels)	Scanning Mode	Time (ms/frame)
512	512	Uni	282
		Bi	141
512	256	Uni	215
		Bi	107
512	128	Uni	107
		Bi	53
512	64	Uni	62
		Bi	31
512	32	Uni	34
		Bi	**17**

5. After background subtraction, calculate the ratio, and use a look-up table (LUT) to provide color on the image corresponding to the cytosolic free Ca^{2+} concentration (Ca^{2+} map).
6. With an intensified video camera, the speed of acquisition is probably 25–30 frames/ s, so that one ratio image would take about 80 ms (*see* **Note 9**).

3.4.2. Confocal Imaging

The single-wavelength probes (e.g., fluo3 and calcium green) can only confidently be used with confocal imaging, because unlike conventional microscopy, confocal microscopy images only an optical section through the cell of a defined thickness. Therefore, the problems associated with cell thickness artifacts are eliminated. This optical sectioning enables Ca^{2+} to be monitored within the nucleus, through the cell perpendicular (xz-plane) to the normal viewing plane, or at any predefined locus *(9)*.

3.5. Rapid Ca²⁺ Imaging in Neutrophils

Many of the Ca^{2+} signals in neutrophils have time scales that can be measured over 10–100 s. However, it is becoming more evident that these Ca^{2+} events may be composed of faster events occurring on the millisecond time scale *(10)*. Repeated confocal laser scanning of a single line (*xt* scanning) through individual fluo3-loaded neutrophils can be used to accumulate data at a rate of at least 80 lines/s, giving a time resolution of greater than 12.5 ms with events in the cell distinguishable to about 0.1–0.2 µm lateral resolution *(11)*. Conventional confocal laser scanning in both x and y directions is necessarily slower, but resonant scanning in the x direction can generate useful images at 17.5 ms/frame (**Table 2**) and a rotating Nipkow disk (a series of pinpoints in the rotating disk) scans multiple laser beams across the field at up to 360 frames/s (3 ms/frame).

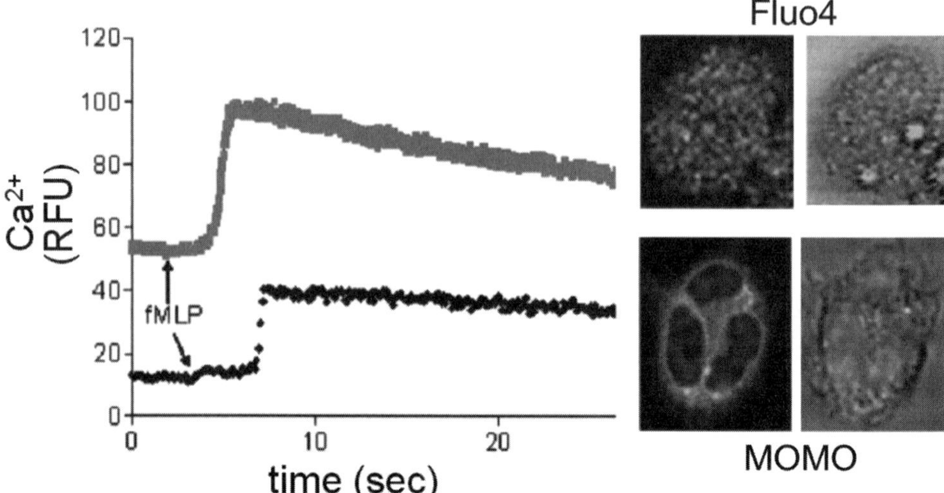

Fig. 2. Ca^{2+} trace generated from Fluo4- and MOMO-loaded neutrophils. The upper trace shows a typical signal from a fluo4-loaded neutrophil, with the distribution of fluo4 shown in the upper image pair and the lower trace shows a typical signal from a MOMO-loaded neutrophil, with the distribution of MOMO shown in the lower image pair. Note that MOMO locates to both the cytosol and the nuclear boundary but not within the nuclear lobes. The traces show "raw" fluorescence changes (relative fluorescence units) in response to fMLP (1 μM).

3.6. Near Membrane Ca^{2+} in Neutrophils

An analog of fura2, FFP-18, with a long hydrophobic tail accumulates in the membranes rather than the cytosol, and has been used to monitor near plasma membrane Ca^{2+} in neutrophils *(12,13)*. However, it is excited in the UV and does not readily permit confocal imaging. More recently, an equivalent "near membrane" probe excitable by visible light, MOMO (**Table 1**), has become available. This probe is more hydrophilic and partitions into the nuclear membrane of neutrophils (**Fig. 2**) and is excluded from the within the nuclear lobes. It is useful for locating Ca^{2+} signals, which originate near the nuclear boundary.

3.7. Manipulating Cytosolic Ca^{2+} in Neutrophils by Photolysis

1. Cytosolic free Ca^{2+} can be manipulated within neutrophils on demand by photo-release of "caged Ca^{2+}" (e.g., nitr-5), "caged Ca^{2+} chelator" (e.g., diazo-2), or caged IP_3 at defined times. These three "caged" compounds can be loaded into cells from their available acetoxymethyl-esters.
2. Other (nonesterified) caged compounds can also be introduced into cells by SLAM-injection *(7)*. The "caged" compound is inert until photolysis. In the case of "caged

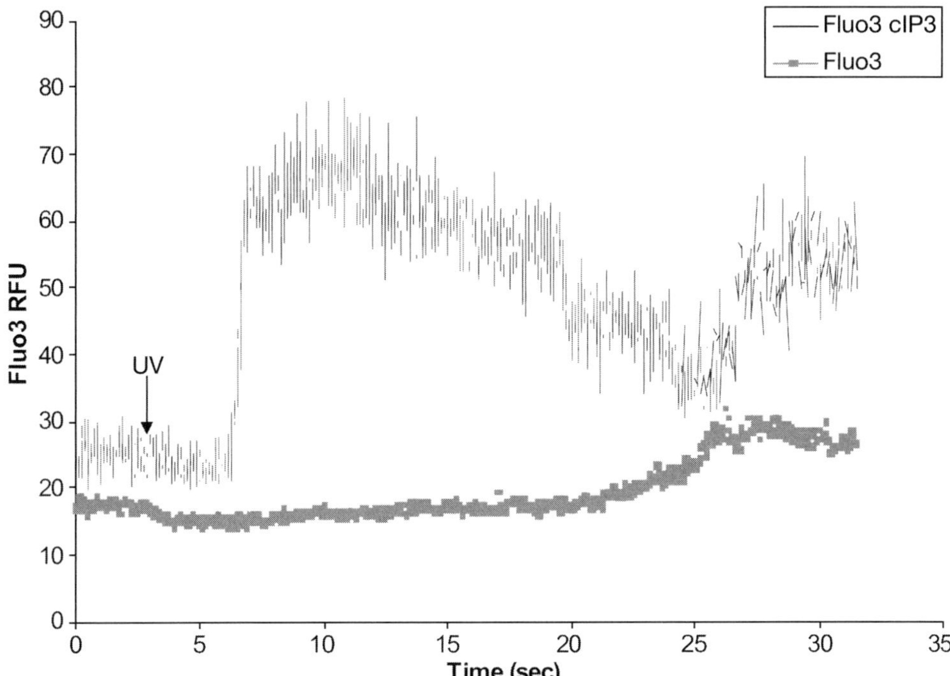

Fig. 3. Effect of Uncaging IP$_3$ on Neutrophil Cytosolic Free Ca^{2+} and the "Ultraviolet Exposure Artefact". The traces show the fluorescent signals from two fluo3-loaded neutrophils, the upper (black) trace from a cell also loaded with caged IP$_3$, and the lower trace from a cell only loaded with fluo3. The immediate and transient fluo3 signal observed on ultraviolet exposure was observed only in the cell with caged IP$_3$, whereas both cells have "late" rise in fluorescent signal at about 25 s. Although it is clear the late fluorescent signal cannot be due to uncaging of IP$_3$ and is hence artifactual, the initial signal is consistent with an IP$_3$ uncaging. This experiment illustrates the need to perform careful controls when using photolysis of caged compounds in neutrophils.

Ca^{2+}," the affinity of the chelator for Ca^{2+} changes dramatically on photolysis at 360 nm. In this way, the cytosolic free Ca^{2+} concentration can be elevated on photolysis *(14)*.

3. Load caged-IP$_3$ AM into neutrophils in the same way as described for fluorescent probes (**Subheading 3.1.**). The IP$_3$ concentration within the cytosol is then stepped up on photolysis (**Fig. 3**).
4. The efficiency of the uncaging procedure can be monitored using caged fluorescein as an easily quantifiable output.
5. Care must be taken to perform appropriate controls in neutrophils, as we have found that after UV exposure, a fluo3 or fluo4 signal often occurs, which is unrelated to the presence of caged IP$_3$ (**Fig. 3**).

6. Activation should be carefully controlled for by performing sham photolysis (i.e., no caged compound) or photolysis of presumed biologically inert compounds (e.g., fluorescein).

4. Notes

1. There are now a myriad of commercially available Ca^{2+} probes, many of which are available as ester derivatives suitable for cytosolic loading into neutrophils (*see* **Table 1**). When selecting which probe to use, it is important to consider its Ca^{2+} affinity. As in other cell types, neutrophils have resting cytosolic free Ca^{2+} concentrations of near 100 nM. On stimulation, this rises transiently to near 1 μM. As the fluorescent signal depends on the binding of Ca^{2+} to the probe, the Ca^{2+} dissociation constant, *kd*, will define the range over which the probe can be usefully employed. A probe with a kd of 300 nM will be only approx 25% saturated in the resting cell, and thus, 75% of its dynamic range will be available to monitor a rise in cytosolic free Ca^{2+}. It will, however, be difficult to measure cytosolic free Ca^{2+} concentrations above 1–3 μM, as the probe will be more than 90% saturated with Ca^{2+}. In contrast, a probe with a k_d of 1 μM would be more useful for higher Ca^{2+} changes. The k_d for Ca^{2+} will also determine its Ca^{2+} buffering effect within the cytosol. Fortunately, neutrophils have high endogenous Ca^{2+} buffering, with estimates ranging from 1:1000–1:3000 *(6,15)*. An intracellular fura2 concentration of 25–50 μM is thus estimated to increase the Ca^{2+} buffering by only about 10%. With probes of higher k_d such as fluo3 or fluo4, the buffering effect would be even less.

2. There are two types of fluorescent Ca^{2+} probes: (1) those that change their signal at two wavelengths, either on excitation or emission (ratiometric dyes) and (2) those that change their signal at only one wavelength (*see* **Table 1**). The single-wavelength, nonratiometric indicators, such as fluo3 or fluo4, can also be excited at visible wavelengths (such as 488 nm, a wavelength produced by laser light). Techniques involving confocal microscopy are powerful for locating the Ca^{2+} change within neutrophils and also can give information on fast time scales (1–10 ms). However, as the indicators produce only a single intensity change on binding Ca^{2+}, caution must be exercised in interpreting an increase in intensity as being solely due to an increase in cytosolic free Ca^{2+} concentration. For example, if a brightly loaded organelle moves into the confocal imaging plane, this would give an increased fluorescent signal unrelated to cytosolic free Ca^{2+} concentration. For this reason, it is recommended that Ca^{2+} measurements be initially performed using ratiometric indicators where there is certainty that the ratio change would be caused by a Ca^{2+} signal. The excitation spectrum of perhaps the most widely used indicator, fura2, shifts on binding Ca^{2+}, so that there is significant fluorescence from both the Ca^{2+}-bound and the Ca^{2+}-free forms of the probe. This permits monitoring of both the Ca^{2+}-free and the Ca^{2+}-bound forms of the indicator. By monitoring two wavelengths, one on either side of the isoemissive point (usually 340 nm and 380 nm), a re-assuring fluorescent "Ca^{2+} signature" is observed, with the signal at one wavelength rising and the signal at the other wavelength declining.

3. De-esterification that is catalyzed enzymatically in the cell will also occur spontaneously (at a slow rate). If this occurs before presentation to the cells, the probe either will not be able to enter the cells or enter the cells in the partially de-esterified, but in the non-Ca^{2+}-sensitive, form. Hydrolysis in the stored stock solution is slowed by reduced temperature and the exclusion of water. As DMSO is hydroscopic and will readily absorb water from the air, standard laboratory-grade DMSO is not recommended. Dry DMSO can be purchased in sealed containers in smaller volumes (Sigma). Another precaution that can reduce the water content of the solution is to be sure that when taking the container from the freezer (at $-20°C$) to let it warm up to room temperature before opening it. If this is not done, water from the air may condense on the inner surface of the container, contaminate the stock solution, and increase ester hydrolysis.

4. Extracellular Ca^{2+} must be present during the time that the Ca^{2+} chelating probe is loaded into the cells, so that it can replace Ca^{2+}, which will be bound to the probe. Otherwise, Ca^{2+} will be removed from intracellular sites. However, provided these precautions are taken, Ca^{2+} chelating probes have little obvious (adverse) effect on Ca^{2+} signaling. However, omission of extracellular Ca^{2+} during loading has been used as a deliberate strategy for "depleting cell Ca^{2+}" in order to establish an intracellular role for this ion in a particular cell activity.

5. Digitonin is often preferred when measuring cytosolic free Ca^{2+} concentration in a cell population in suspension, as it ensures that all fura2 gains access to high Ca^{2+} concentration in the extracellular medium. If ionomycin is used (e.g., during Ca^{2+} imaging), it is important that sufficient ionomycin is added to produce a truly maximal fluorescence signal, as it is possible to elevate cytosolic free Ca^{2+} with ionophores to concentrations that are less than micromolar. Under these latter conditions, the ionomycin signal will not correspond with R_{max}, and the cytosolic free Ca^{2+} concentration cannot be correctly calculated. The problem can be reduced by also increasing the extracellular Ca^{2+} concentration during the addition of ionomycin.

6. The information in a single pixel is limited by the noise associated with the detection system of the imager. A technique called binning is often used to collect the data from a pixel as well as its neighbors, and to treat the combined signal as though arising from a larger single pixel. Binning of information on neighboring pixels will increase the signal but decrease the spatial resolution. A statistical approach can be used to determine whether small areas within the cell truly have raised Ca^{2+}. Areas of interest in the image (with n pixels) can be chosen for comparison of their cytosolic free Ca^{2+} concentration. The areas will have statistically significant cytosolic free Ca^{2+} concentrations when n $>$ $2[\sigma(Z_{2\alpha}+Z_{2\beta})/\delta]^2$ or $\delta >$ $[\sqrt{(2/n)}][\sigma(Z_{2\alpha}+Z_{2\beta})]$, where n is the number of pixels in each of the two areas of the image, σ is the standard deviation of the distribution of Ca^{2+} values in the pixel arrays, Zx is the standard normal deviate exceeded with probability x, β is the significance level for the test, $(1 - \beta)$ is the power of the test, and δ is the difference in cytosolic free Ca^{2+} concentrations *(16)*. It can be seen that the ability to detect small and localized changes in Ca^{2+} depends on both the magnitude of the Ca^{2+} change and the area it occupies. Detection of both very small and very localized

Ca^{2+} changes is thus difficult and ultimately limited by the image noise, i.e., the variance (standard deviation) of individual pixel values.

7. Do not use digitonin to determine F_{max} in microfluorometry, as the Ca^{2+} indicator will be lost from the neutrophils (*see* **Note 3**). When this happens, the indicator will no longer be within the defined area and the information gathered will not be meaningful. Do not use digitonin to determine F_{max} in microfluorometry, as the Ca^{2+} indicator will be lost from the neutrophils (*see* **Note 3**). When this happens, the indicator will no longer be within the defined area and the information gathered will not be meaningful.

8. The major problems associated with fluorescent imaging are photobleaching and image noise. Photobleaching arises where excessive excitation results in the destruction of the fluorescent molecules and hence a reduction in the emission intensity (bleaching). Each fluorescent molecule emits about 10^4–10^5 photons/molecule before photolysis. With ratiometric methods, photobleaching is less of a problem, as the ratio will remain constant during the bleaching provided that the pairs of images are taken close together in time (when no significant bleaching has occurred). With nonratiometric confocal Ca^{2+} imaging, bleaching during the time of the experiment can be avoided by attenuating the laser light and increasing the detector (photomultiplier) sensitivity so that the minimum usable emission intensity is employed.

9. The time taken to change wavelengths can be minimized by using fast filter wheels, optical choppers, or rapid changing monochromators. However, the image quality is often poor at high speed and it is usually necessary to average a number of frames, thus increasing the signal but decreasing the time resolution, or to bin pixels to increase the signal (however, with decreased spatial resolution; *see* **Note 5**). The need to do either is reduced by using image intensification, but in practice, useful ratio images can rarely be acquired faster than about 1 s^{-1}. The speed barrier is overcome by rapid confocal imaging (*see* **Subheading 3.6.**). Other problems associated with Ca^{2+} imaging include photobleaching of the probe (*see* **Note 8**) and light-induced activation of the neutrophil under view (*see* **Subheading 3.7.** and **Note 10**). More details of Ca^{2+} imaging artifacts and their solutions may be found in **ref. 2**.

10. Ratiometric dyes require excitation near the UV region, which stimulates fluorescence from endogenous molecules such as NADPH within neutrophils, and consequently the signal:noise ratio of the Ca^{2+} probe is reduced.

References

1. Pozzan, T., Lew, D. P., Wollheim, C. B., Tsien, R. Y., and Rink, T. J. (1983) Is cytosolic ionized calcium regulating neutrophil activation? *Science* **221**, 1413–1415.
2. Hallett, M. B., Davies, E. V., and Pettit, E. J. (1996) Fluorescent methods for measuring and imaging the cytosolic free Ca^{2+} in neutrophils. *Methods: A Companion to Methods Enzymol.* **9**, 591–606.
3. Scanlon, M., Williams, D. A., and Fay, F. S. (1987) A Ca^{2+}-insensitive form of fura2 associated with polymorphoinuclear leukocytes-assessment and accurate Ca^{2+} measurement. *J. Biol. Chem.* **262**, 6308–6312.

 4. Al-Mohanna, F. A. and Hallett, M. B. (1988) The use of fura 2 to determine the relationship between intracellular free Ca^{2+} and oxidase activation in rat neutrophils. *Cell Calcium* **8,** 17–26.
 5. Hallett, M. B., Hodges, R., Cadman, M., et al. (1999) Techniques for measuring and manipulating free Ca^{2+} in the cytosol and organelles of neutrophils. *J. Immunol. Meth.* **232,** 77–88.
 6. Laffafian, I. and Hallett, M. B. (1998) Lipid-assisted microinjection: introducing material into the cytosol and membranes of small cells. *Biophys. J.* **75,** 2558–2563.
 7. Laffafian, I. and Hallett, M. B. (2000) Gentle micro-injection for myeloid cells using SLAM. *Blood* **95,** 3270–3271.
 8. Tepikin, A. V. (2000) *Calcium Signalling: A Practical Approach, 2nd ed.* Oxford Univ. Press, Oxford, UK, p. 230.
 9. Pettit, E. J. and Hallett, M. B. (1996) Localised and global cytosolic Ca^{2+} changes in neutrophils during engagement of CD11b/CD18 integrin visualised using confocal laser scanning reconstruction. *J. Cell Sci.* **109,** 1689–1694.
10. Kindzelskii, A. L. and Petty, H. R. (2003) Intracellular calcium waves accompany neutrophil polarization, formylmethionylleucylphenylalanine stimulation, and phagocytosis: a high speed microscopy study *J. Immunol.* **170,** 64–72.
11. Pettit, E. J. and Hallett, M. B. (1995) Early Ca^{2+} signalling events in neutrophils detected by rapid confocal laser scanning. *Biochem. J.* **310,** 445–448.
12. Davies, E. V. and Hallett, M. B. (1996) Near membrane Ca^{2+} changes resulting from store release in neutrophils: detection by FFP-18. *Cell Calcium* **19,** 355–362.
13. Davies, E. V. and Hallett, M. B. (1998) High micromolar Ca^{2+} beneath the plasma membrane in stimulated neutrophils. *Biochem. Biophys. Res. Commun.* **248,** 679–683.
14. Pettit, E. J. and Hallett, M. B. (1998) Release of "caged" cytosolic Ca^{2+} triggers rapid spreading of human neutrophils adherent via integrin engagement. *J. Cell Sci.* **111,** 2209–2215.
15. von Tscharner, V., Deranleau, D. A., and Baggiolini, M. (1986) Calcium fluxes and calcium buffering in human neutrophils. *J. Biol. Chem.* **261,** 10,163–10,168.
16. Armitage, P. and Berry, G. (1987) *Statistical Methods in Medical Research, 2nd ed.* Blackwell Scientific, Boston, MA, pp. 181–182.

11

Analysis of Electrophysiological Properties and Responses of Neutrophils

Deri Morgan and Thomas E. DeCoursey

Summary

The past decade has seen increasing use of the patch clamp technique on neutrophils and eosinophils. The main goal of these electrophysiological studies has been to elucidate the mechanisms underlying the phagocyte respiratory burst. NADPH oxidase activity, which defines the respiratory burst in granulocytes, is electrogenic because electrons from NADPH are transported across the cell membrane, where they reduce oxygen to form superoxide anion (O_2^-). This passage of electrons comprises an electrical current that would rapidly depolarize the membrane if the charge movement were not balanced by proton efflux. The patch clamp technique enables simultaneous recording of NADPH oxidase-generated electron current and H^+ flux through the closely related H^+ channel. Increasing evidence suggests that other ion channels may play crucial roles in degranulation, phagocytosis, and chemotaxis, highlighting the importance of electrophysiological studies to advance knowledge of granulocyte function. Several configurations of the patch clamp technique exist. Each has advantages and limitations that are discussed here. Meaningful measurements of ion channels cannot be achieved without an understanding of their fundamental properties. We describe the types of measurements that are necessary to characterize a particular ion channel.

Key Words: Proton current; ion channels; pH; zinc; phagocyte; respiratory burst; NADPH oxidase; electrophysiology; patch clamp.

1. Introduction

Ion channels are a diverse group of proteins, representatives of which are found in all plasma membranes. Channels allow the regulated passage of ions across membranes, which otherwise have extremely low permeability to ions. Ion channels perform a wide variety of functions in a large number of cellular processes. The development of the voltage-clamp method revolutionized the study of ion channels in excitable cells such as neurons and muscle *(1)*. The

From: *Methods in Molecular Biology, vol. 412: Neutrophil Methods and Protocols*
Edited by: M. T. Quinn, F. R. DeLeo, and G. M. Bokoch © Humana Press Inc., Totowa, NJ

patch clamp technique *(2)* dramatically extended the power of the voltage-clamp approach by making it possible to record current through individual ion channels and to faithfully record currents in the membranes of even the smallest cells. Knowledge of the presence and functions of ion channels in nonexcitable cells, such as epithelial cells and immune cells, has mushroomed from virtually nothing a quarter of a century ago to thousands of papers. Researchers now have heated arguments about the functions performed in these cells by ion channels that were not even known to exist a few decades ago. The first patch clamp study on human granulocytes described nonselective cation channels activated by increased intracellular free calcium *(3)*. Subsequently, the patch clamp technique has been used to explore proton, chloride, and potassium currents in neutrophils and eosinophils *(4–8)*, as well as in many other nonexcitable cells.

The phagocyte respiratory burst results in the production of large amounts of O_2^- via NADPH oxidase activation *(9)* (*see* **Note 1**). The patch clamp technique has proven to be a powerful way to examine the phagocyte NADPH oxidase, and various configurations of this technique have been used to examine the respiratory burst measured as electron current in real time *(7,10,11)*. These techniques have been used to examine the dependence of NADPH oxidase activity on temperature, pH_o and pH_i, NADPH concentration, and membrane voltage *(12–15)*. The patch clamp approach also enables recording current through voltage-gated proton channels (H^+ channels) during oxidase activation (*see* **Note 2**). Signaling pathways that regulate and coordinate the activity of NADPH oxidase and proton channels are an area of active investigation. A variety of evidence also implicates Cl^- channels in a host of neutrophil functions such as volume regulation, phagocytosis, chemotaxis, and host defense *(16,17)*. Thus, studies of granulocyte ion channels will undoubtedly continue to reveal the mechanisms of important physiological responses of these cells.

Although conceptually simple, the patch clamp technique is a complicated experimental method. Collecting meaningful data requires appropriate experimental design that is based on a genuine understanding of the properties of ion channels. We hope to introduce the reader to the patch clamp technique and provide an overview of the apparatus required, the general methods involved, and the application of this method to specific electrogenic transporters known to be present in human granulocytes. **Figure 1** shows a schematic diagram of the patch clamp setup and circuit. The essence of the patch clamp technique is the formation of an electrically-tight seal between the tip of a pipet and the plasma membrane of a cell (*see* **Note 3**). The glass pipet is filled with a solution that may mimic a physiological solution and makes contact with the electrode, shown as a wire. Electrical currents that pass through the membrane are converted into voltage, which is amplified and sent to the data acquisition software.

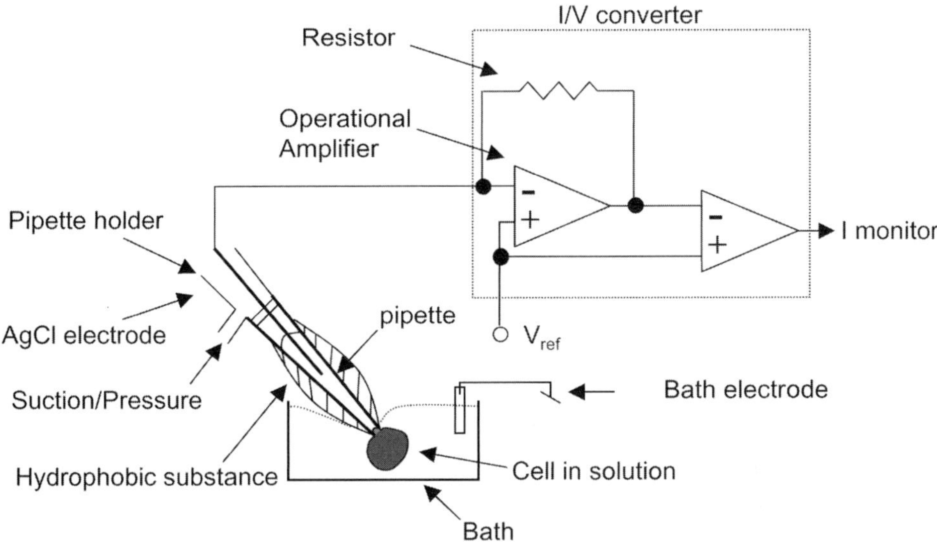

Fig 1. A simplified diagram of a cell being studied by a patch clamp is shown *(60, 63)*. A cell in a physiological bath solution is sealed to a glass pipet that is filled with a solution that makes contact with the electrode (usually a chlorided silver wire). The small currents due to ion movements across the membrane are measured as a voltage drop across a resistor. The glass pipet is coated with a hydrophobic substance to reduce the pipet/bath solution capacitance.

The patch clamp technique exists in several configurations that are distinguished by the size and orientation of the membrane through which current flow is measured. Whole-cell or perforated-patch configurations allow recording from the entire plasma membrane; in cell-attached or excised patch configurations, only the channels in a small patch of membrane are studied. In the perforated-patch configuration, the cytosol is intact; in whole-cell configuration the cytoplasm is replaced by the pipet solution. The choice of experimental solutions depends on the experimental goals (*see* **Note 4**). We give examples of solutions that have been used to study granulocytes, but more importantly, we discuss strategies of solution design.

2. Materials

2.1. Patch Clamp Equipment

The equipment comprising a patch clamp setup is listed below. Nearly all of the apparatus can be purchased commercially. It is beyond the scope of this chapter to discuss the relative merits of products of different manufacturers.

1. Inverted microscope.

  2. Perfusion bath and tubing.
  3. Pipet holder and headstage.
  4. Micromanipulators.
  5. Bath temperature control.
  6. Pressure/vacuum unit.
  7. Vibration isolation table.
  8. Faraday cage.
  9. Patch clamp amplifier.
 10. Computer with relevant data acquisition software.
 11. Interface between computer and patch clamp amplifier.
 12. Bath electrode.
 13. Patch pipet.

2.2. Bath (Reference) Electrode

  1. Silver wire.
  2. Epoxy resin.
  3. Agar powder.
  4. Glass or plastic tubing (3 mm bore).
  5. Soldering iron and solder.
  6. Bleach or NaOCl.
  7. Hot plate.
  8. Syringe with wide-bore needle.

2.3. Patch Pipet

  1. Glass capillary tubes.
  2. Pipet puller.
  3. Sylgard 184® (Dow Corning Corp.).
  4. Microscope.
  5. Heating filament attached to a power supply.
  6. Heat gun.
  7. Pipet solution (*see* **Subheading 2.4.**).

2.4. Solutions

Patch clamp experiments involve exchanging solutions to vary the environment of the cell or membrane patch. The composition of the solutions is predicated on the specific channel of interest and the type of information that is to be extracted from the experiment. For example, the most fundamental measurement of a channel suspected of being K^+ selective would be to measure V_{rev} in solutions with quite different K^+ concentrations (*see* **Note 5**). Pipet solutions that are intended for whole-cell measurements are often made to approximate cytosolic constituents, with high $[K^+]$ and very low $[Ca^{2+}]$.

2.4.1. Proton Channel Solutions

1. To isolate proton currents from other conductances that might be present in the cell membrane, common cations and anions such as Na^+, K^+, and Cl^- may be replaced with larger, less permeant ions such as tetramethylammonium (TMA^+), tetraethylammonium (TEA^+), or *N*-methyl-D-glucamine$^+$ and methanesulfonate ($MeSO_3^-$), aspartate$^-$, glutamate$^-$, isethionate ($HOCH_2CH_2SO_3^-$), cyanate (CN^-), or sulfate (SO_4^{2-}) *(5,18)*. Occasionally Cs^+ is used as an impermeant cation *(19)*, because Cs^+ blocks many K^+ channels.

2. Large H^+ currents tend to deplete intracellular protonated buffer, increasing pH_i. Because proton channels do not inactivate, any decay or "droop" of proton current during maintained depolarization indicates that depletion is occurring and that pH_i is increasing significantly *(20)*. To prevent or minimize such pH_i changes, a large concentration of buffer is required in the pipet solution *(21)*. Many studies have been done with approx 100 m*M* buffer, and concentrations as high as 200 m*M* buffer with no other cations have been used *(14)* (*see* **Note 6**). **Table 1** shows examples of solutions that have been used to study H^+ channels *(18,22,23)*.

3. Buffers should be chosen that have a pK_a near the desired pH of the solution. One consideration is that many buffers bind divalent cations, which creates a problem because inhibition by Zn^{2+} or Cd^{2+} is a frequently studied characteristic of proton currents. Because tricine binds Zn^{2+} with high affinity *(24)*, it removes Zn^{2+} from solution, so a different buffer would be preferred. Phosphate also binds zinc avidly *(25)*. The metal binding constants for several buffers were measured by Cherny and DeCoursey *(24)*.

4. Examples of buffers that have been used in solutions spanning a wide range of pH are as follows (numbers in parentheses are pK_a at 20°C): Homopiperazine-*N,N'*-bis-(2-ethanesulfonic acid) (Homopipes) (4.61); 2-(*N*-morpholino) ethanesulfonic acid (MES) (6.15); Bis-2-(hydroxyethyl)imino-tris(hydroxymethyl)methane (Bis Tris) (6.5); Piperazine-*N-N*-bis(2-ethanesulfonic acid) (PIPES) (6.7); *N-N*-bis (Hydroxyethyl)-2-aminosulfonic acid (BES) (7.1); *N*-(2-hydroxyethyl)-piperazine-*N*-(2-ethanesulfonic acid) (HEPES) (7.55); Tris-(hydroxymethyl) amino methane-hydrochloride (Tris) (8.3); 2-(*N*-cyclohexamino)-ethane sulfonic acid (CHES) (9.3); and 3-(cyclohexamino)-1-propanesulfonic acid (CAPS) (10.37).

2.4.2. Potassium Channel Solutions

Solutions for studying K^+ channels usually include K^+. $[K^+]$ can be varied, while maintaining constant ionic strength and osmolality, by replacing K^+ with Na^+. **Table 2** shows solutions that were used to study K^+ channels in granulocytes *(8)*.

2.4.3 Chloride Channel Solutions

It is possible to replace Cl^- with larger, less permeant ions. **Table 3** shows examples of bath and pipet solutions that have been used to study Cl^- channels *(5,18)*.

Table 1

Examples of Solutions Used for Proton Channel Characterization

pH	MeSO$_3$	CsCl	HCl	Base	EGTA	CaCl$_2$	MgCl$_2$	Buffer
Pipet Solutions								
5.5	96			80 TMA	1		2	100 MES
5.5	70		10	TEA		2	1	100 MES
6	80			130 TMA	1		2	100 MES
6	70		10	TEA		2	1	100 MES
6.5	128			75 TMA	1		2	100 BIS-TRIS
6.5	70		10	TEA		2	1	100 PIPES
7	128			115 TMA	1		2	100 HEPES
7	70		10	TEA		2	1	100 HEPES
7.5		75		50 CsOH				100 HEPES
8		75		44 CsOH				100 TRIS

pH	MeSO$_3$	CsCl	HCl	Base	BAPTA	EGTA	CaCl$_2$	MgCl$_2$	Buffer
Bath Solutions									
5.5	96			80 TMA		1		2	100 MES
6	70		10	TEA			1		100 MES
6.5	70		10	TEA	5		2		100 PIPES
6.5		35		132 CsOH	10				100 PIPES
7	70		10	TEA	5		2		100 HEPES
7.2		75		30 CsOH	10		11.4		100 HEPES

The table lists examples of solutions used for pipet and bath solutions when recording proton currents in whole cell configuration. Numbers indicate concentration in mM. Where no numbers are given the base was used to titrate the solution to the correct pH. TMA$^+$, tetramethylammonium; TEA$^+$, tetraethylammonium; MES, 2-(N-morpholino) ethanesulfonic acid; Bis Tris, Bis-2-(hydroxyethyl)imino-tris(hydroxymethyl) methane; PIPES, Piperazine-N-N-bis(2-ethanesulfonic acid); BES, N-N-bis(Hydroxyethyl)-2-aminosulfonic acid; HEPES, N-(2-hydroxyethyl)-piperazine-N-(2-ethanesulfonic acid); Tris, Tris-(hydroxymethyl) amino methanehydrochloride.

Table 2
Examples of Solutions Used for K⁺ Channel Characterization

pH	NaCl	KCl	CaCl$_2$	MgCl$_2$	HEPES	Glucose	NaOH
Pipet Solutions							
7	137.6	5	2	1	10	6	2.4
7	126.6	16	2	1	10	6	2.4
7	117.6	25	2	1	10	6	2.4
7	42.6	100	2	1	10	6	2.4
7	2.6	140	2	1	10	6	2.4

pH	NaCl	KCl	CaCl$_2$	MgCl$_2$	Buffer	BAPTA	EDTA	MgATP	Na$_2$ATP	HCl	KOH
Bath Solutions											
7	5	97.6	0.7		10 HEPES	10		1			42.6
7	3	97.6	1.6		10 HEPES		10		1	2	41.5
5.5		102	0.06	0.69	10 MES	10			2.5	3.6	38
8	4	97.2	0.83	0.76	10 Taps	10			0.5		42.6

The table lists examples of solutions used for pipet and bath solutions when recording potassium currents in whole cell configuration. Numbers indicate concentration in m*M*. *See* **Table 1** for definitions.

Table 3
Examples of Solutions Used for Cl⁻ Channel Characterization

pH	NaCl	Na_2SO_4	Na-Isethionate	TMA Cl	KCl	$CaCl_2$	$MgCl_2$	Mannitol	Base	Buffer
Pipet solutions										
7				145		2				20 BES
7.2	140				5	2	2			10 HEPES
7.2			140			2	2			10 HEPES
7.4	150				5	2	1		NaOH	10 HEPES
7.4	150				5	2	1	75	NaOH	10 HEPES

pH		$TMA\ MeSO_3$	NaCl	K-Asp	KCl	EGTA	$MgCl_2$	Buffer
Bath solutions								
5.5		96				1	2	100 MES
7.2			5		140	0.2	2	10 HEPES
7.2			5	110	30	0.2	2	10 HEPES

The tables list examples of pipet and bath solutions used for recording chloride currents in whole cell configuration. Numbers indicate concentration in m*M*. Where no numbers are given, the base was used to titrate the solution to the correct pH. *See* **Table 1** for definitions.

Table 4
Examples of Solutions Used for Perforated Patch Recordings

Solution	TMA MESO$_3$	(NH4)$_2$SO$_4$	MgCl$_2$	CaCl$_2$	EGTA	Buffer	pH	pH$_i$
Pipet solution	100	25	2	1.5	1	BES 10	7	N/A
Bath 1	100	25	2	1.5	1	MES 10	5.5	5.5
Bath 2	100	25	2	1.5	1	MES 10	6.5	6.5
Bath 3	100	25	2	1.5	1	BES 10	7.5	7.5
Bath 4	100	25	2	1.5	1	Tricine 10	8.5	8.5
Bath 5	100	2.5	2	1.5	1	BES 10	7	6
Bath 6	100	2.5	2	1.5	1	Tricine 10	8.5	7.5
Bath 7	100	2.5	2	1.5	1	HEPES	7	6

The table lists examples of solutions used for pipet and bath solutions when recording proton currents in perforated patch configuration. Numbers indicate concentration in mM. The pipet solution also contains 1 mg/mL amphotericin, and is the same for all bath solutions. pH$_i$ is calculated from **eq. 7** (*see* **Note 7**). *See* **Table 1** for definitions.

2.4.4. Perforated Patch Solutions

Perforated patch recordings are achieved when a pore-forming antibiotic, such as amphotericin or nystatin, is included in the pipet solution. The antibiotic forms pores that allow only small monovalent ions (e.g., Na$^+$, K$^+$, Cl$^-$, or NH$_4^+$) to carry current between the pipet and the cell. In order to control pH$_i$, a proton donor/acceptor system such as an ammonium or methylamine gradient can be used *(26)* (*see* **Note 7**). **Table 4** shows solutions that were used to change the pH symmetrically and asymmetrically *(15)*. The pH gradient can also be increased in increments by lowering [NH$_4^+$]$_o$ from 50 mM to 15, 9, 3, and 1 mM *(26)*, although lower [NH$_4^+$]$_o$ controls pH$_i$ less effectively *(10,26)*.

3. Methods

3.1. Patch Clamp Setup

In a patch clamp experiment, the pipet is positioned against the cell membrane, and suction is applied to the inside of the pipet until a seal of a very high resistance forms between the membrane and the glass. Hamill et al. *(2)* called the seal a "gigaohm seal" or "giga seal," because the resistance between the cell membrane and pipet is preferably in the gigaohm range. The pipet usually contains a physiological salt solution to mimic the microenvironment of the cell membrane, with a wire electrode immersed in the pipet solution.

The typical patch clamp setup consists of an inverted microscope with a small bath placed on the stage. The inverted microscope allows room for the bath, pipet electrode, electrode manipulators, perfusion apparatus, and temperature controller/detector. The bath has a discontinuous solution inlet and outlet to enable exchange of bath solutions. It is impossible to perform normal patch-clamp measurements without exchanging bath solutions. A continuous perfusion system forms a conduction pathway that can act as an antenna that introduces a great deal of noise. A discontinuous perfusion system has a junction to avoid this noise source.

Patch-clamp measurements are often performed at "room temperature," which may vary by a few degrees or more dramatically, depending on the building. Both proton and electron currents have unusually strong temperature dependence *(13,27)*; consequently, small temperature changes produce dramatic effects. Some designs for temperature control involve placing the chamber on a U-shaped piece of thermally conductive metal like copper, with Peltier devices sandwiched between the copper and a metal heat sink *(28)*. When this arrangement is used, care must be taken when decreasing the temperature, because contraction of the U-shaped metal lifts the chamber relative to the pipet, smashing it. Before lowering the temperature, it is advisable to lift the pipet and cell well above the chamber.

A reference electrode is placed in the bath and the pipet is attached to micromanipulators that enable positioning the pipet in the bath. Fine movements are required to target the pipet accurately. Both hydraulic (manually controlled) and motorized micromanipulators are used. The main requirement besides fine control is absence of drift. Cells are small and drift of cellular dimensions can abruptly end an experiment by pulling the pipet away from the cell or by smashing the pipet against the chamber.

The microscope, electrodes, and manipulators are placed on a vibration isolation table to prevent electrode movement. Vibrations from the floor can cause movement artifacts that can break the pipet tip. A commonly used vibration isolation table is an air table in which the legs are supported on pressurized air cylinders. Other tables are made of large, heavy slabs of marble or granite and are placed in room corners to minimize floor vibration. Locating the table in a cellar can also reduce movement artifacts.

The signal from the pipet electrode is amplified by the head-stage amplifier, lowpass filtered and sent to an analog to digital (A/D) converter, and finally to the computer where it is recognized by data acquisition software. Extraneous electrical noise from power cords to the microscope light, headstage, electrical thermometer, and micromanipulator motors can be shielded by surrounding them with braided wire. A Faraday cage (a wire screen cage) surrounding the entire microscope or just the microscope stage may be necessary to suppress electrical noise from overhead lights and other environmental electrical sources.

Because of the sensitivity of the headstage amplifier, it is important that all metal near the electrode be connected to ground to prevent the detection of line frequencies. It is important to avoid ground loops.

3.2. Electrode Preparation

The electrodes require manual preparation. The bath (reference) electrode can be used for long periods of time, but the patch pipet should be made immediately before use.

3.2.1. Bath (Reference Electrode)

The bath electrode provides the reference point from which voltage is measured. It is important that at each point in the recording apparatus where dissimilar materials meet, at least one current carrier can move freely in each direction. If this is not the case, electrode polarization may occur, in which large sustained currents decay because an opposing electromotive force develops at the offending interface. The most commonly used electrode in patch clamping is the silver-silver chloride electrode, made of a silver wire coated with AgCl. Current flowing into the solution or agar bridge is carried by Cl^-. A change in the bath chloride concentration will result in a voltage offset as a result of a liquid junction potential. Using an agar bridge between the AgCl and the bath solution keeps the concentration of chloride constant near the electrode and minimizes diffusion of the solution in the bridge into the chamber. Because patch-clamp chambers often have small volume (*e.g.*, <1 mL), the use of saturated KCl in the bridge (typically used with conventional microelectrodes) is avoided to prevent diffusion of these ions into the bath, which would change their concentrations. Bath electrodes are available for purchase but are often constructed on-site.

1. Cut the silver wire to a length of about 1.5 inches and solder it to the end of the wire intended to connect the bath to the setup.
2. Immerse the silver wire in bleach for several hr or overnight (*see* **Note 8**).
3. Cut a piece of glass or plastic tubing approx 1–2 inches in length, depending on its arrangement in the chamber, and epoxy the tube over the chlorided silver wire. Ensure that the epoxy completely encases the solder and any exposed, unchlorided silver wire. The tube should extend past the wire by 0.5 inch.
4. Heat 4% agar/Ringer's solution (w/v) in a beaker until the solution begins to boil, then remove from heat.
5. While the Ringer/agar solution is still hot, take up the solution with the syringe and inject the agar solution into the tube containing the silver wire.
6. Allow the electrode to cool and the agar to set.
7. When the electrode is not in use, it should be immersed in the solution used in the bridge and connected to ground.

3.2.2. Patch Pipet

The glass patch pipet contains an electrode, commonly a chlorided silver wire that contacts the pipet solution. The pipet is manufactured by a pipet puller that produces a tip opening a few microns or less in diameter. Commercially available pipet-pulling instruments heat a filament that surrounds the pipet while pulling on either end. The filament may be alternately heated and cooled by an adjustable computer program until the capillary draws out and breaks at the center. The pipet is heat polished to optimize its shape and resistance *(29)*. Lower tip resistance increases the fidelity of recording from whole cells (*see* **Note 9**). For small cells with currents of a few hundred picoamperes, a pipet resistance of 2–10 MΩ is typical and reasonable. Cells with larger currents require lower resistance pipets. Conversely, for recording tiny currents or a small number of channels in excised patches, the tip resistance is not critical. Pipets are best made immediately before use and should not be kept overnight.

1. Place the glass capillary tube (*see* **Note 10**) in the micropipet puller, and make two pipets.
2. Coat the pipet tip is coated with Sylgard, approaching very near the tip (*see* **Note 11**). Cure the Sylgard by heating with a heat gun.
3. Place the pipet on a microscope stage near a heating filament (*see* **Note 12**), and pass current through the filament to melt the pipet tip. The current can be adjusted empirically with a variable transformer. This process is observed through the microscope.
4. Switch off the current through the filament when the tip is rounded and the bore has narrowed to the desired diameter. If the microscope optics permits seeing the tip opening directly, direct observation is the easiest way to judge the progress of heat polishing (*see* **Note 13**).
5. Back-fill the pipet with a pipet solution, taking care to ensure that no air bubbles are present in the tip of the pipet. Tiny bubbles near the tip can be seen with a dissecting microscope. Bubbles can be dislodged by flicking the pipet shank with a fingertip, or by rubbing the serrated surface of a forceps against the shank of the pipet.
6. Mount the pipet in the pipet holder, place over the chlorided silver wire electrode, and tighten. When placed in the bath solution, the pipet resistance can be measured.

3.3. Seal Formation

The essence of the patch clamp technique is formation of a high-resistance seal (usually 10–100 GΩ or more) between the cell membrane and the pipet tip. This high resistance seal allows recording with very little background noise and small leak currents (i.e., current that flows from the pipet into the bath through the seal, rather than through the membrane). The process of positioning the pipet against the cell with the micromanipulators is similar for all adherent cells but differs when the cells are nonadherent. Filtering the pipet and bath solutions (e.g., at 0.2 µm) may improve seal formation.

3.3.1. Adherent Cells

1. Mount the pipet in the electrode holder, taking care not to scratch the chlorided surface of the wire and tighten the pipet in place.
2. Apply positive pressure to the pipet interior and lower the pipet tip into the bath solution.
3. After the pipet tip enters the bath solution, adjust the offset current to zero.
4. Position the pipet directly above the cell and lower the pipet toward the cell.
5. When the pipet is near the cell, lower the pipet carefully until the cell is slightly deformed by the pressure of the pipet, and then immediately apply gentle negative pressure (~2–5 inches of water).
6. Monitor the access resistance by observing the current produced by a small voltage pulse.

3.3.2. Nonadherent Cells

1. Follow **Subheading 3.3.1., steps 1–4**.
2. When the pipet tip is near the cell, reduce the positive pressure to avoid pushing (or "blowing") the cell away and move the pipet tip near the cell.
3. Apply negative pressure and suck the cell onto the pipet.
4. It may not be possible to avoid pushing the cell away by reducing the positive pressure. If this is the case, the pipet can be aligned slightly to the side of the cell. Immediately after moving the pipet in line with the cell, apply negative pressure to pull the cell onto the pipet tip.

3.4. Patch-Clamp Configurations

There are several patch clamp configurations: (1) cell attached patch, (2) inside-out patch, (3) whole cell, (4) outside-out patch, and (5) perforated patch (**Fig. 2**). Each configuration has advantages and disadvantages that should be considered when deciding which is appropriate for the desired experiment.

3.4.1. Cell-Attached Patch

1. After the pipet has contacted the cell and a GΩ seal has been achieved, the configuration is cell-attached patch.
2. The currents recorded will reflect only the ion channels in the patch of membrane that spans the tip of the pipet. Single channel events may be detected, because the patch of membrane is very small.
3. Changing the bath solution will not directly affect channels in the patch, because the cell is still intact. Also because the cell is intact, there is very limited control over the conditions in the cytosol; thus, the pH, ionic composition, and membrane potential will be uncontrolled and unknown.
4. If the cell has a high membrane resistance, current flowing through a single open channel in the patch may change the membrane potential of the cell substantially, producing a decaying unitary current *(30)*.

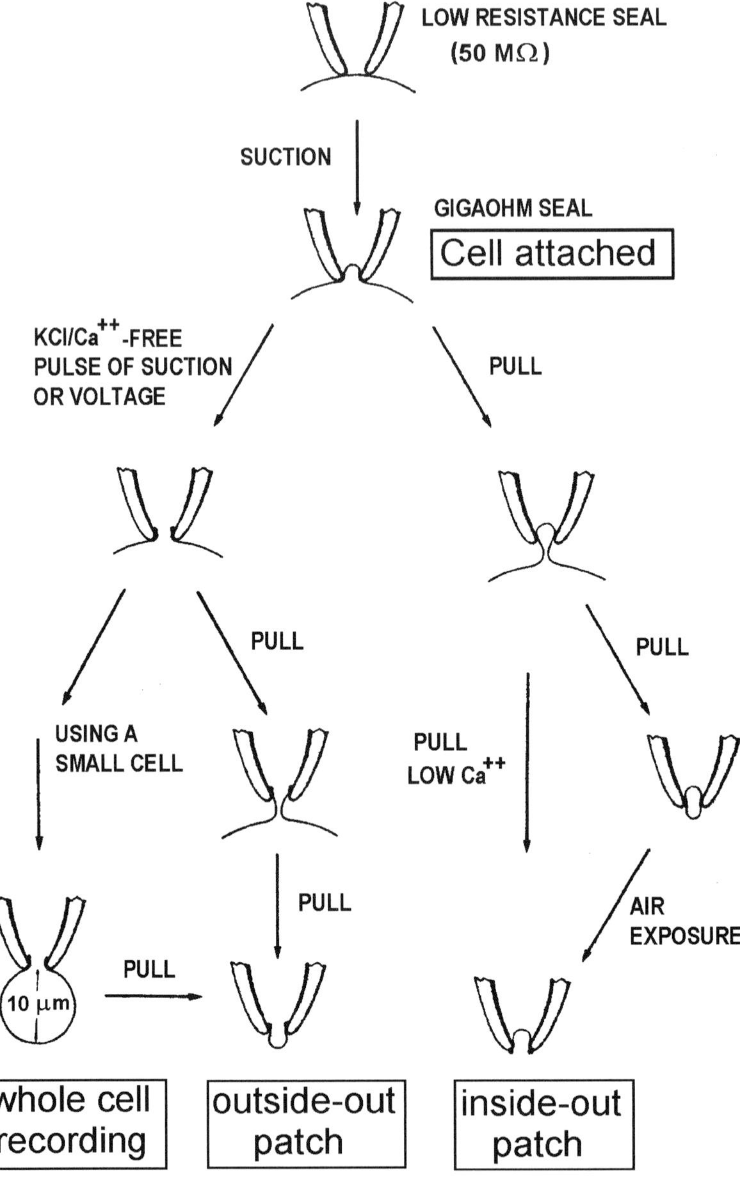

Fig. 2. The formation of the four main patch-clamp configurations is illustrated. *See* text for details. (From **ref. 2**.)

5. Because the pipet is extracellular, applied voltages (and recorded currents) need to be inverted to follow the convention of defining voltage as that inside the cell relative that outside. The actual membrane potential will be the sum of the command potential and the membrane potential of the cell, which is not known.

6. Sometimes cells are bathed in an isotonic K$^+$ solution with the intent of clamping the membrane potential near 0 mV, on the assumption that most cells have K$^+$ channels that will perform this function. The patch potential is then simply the inverse of the voltage applied to the pipet.

3.4.2. Inside-Out Patch

1. This configuration is achieved by pulling the pipet away from the cell after forming a cell-attached patch in the case of adherent cells, or by lifting the pipet briefly out of the bath solution for nonadherent cells.
2. It is best to excise the patch into solutions low in Ca^{2+}, because Ca^{2+} tends to "repair" membranes, producing a sealed vesicle on the end of the pipet, which is useless for recording.
3. As with the cell-attached patch, an inside-out patch permits recording single channels. The configuration allows complete control over the patch potential and the internal (bath) and external (pipet) solutions.
4. The bath solution (corresponding with the intracellular solution) can be exchanged easily during experiments.

3.4.3. Whole Cell

1. To form whole cell configuration from the cell-attached patch configuration, the membrane across the tip of the pipet is ruptured by suction or a brief pulse of high voltage, allowing access to the entire membrane of the cell.
2. Sometimes it is difficult to rupture the patch. Attaching a large (e.g., 50 mL) syringe to the tubing connected to the pipet interior and rapidly withdrawing the plunger is often effective. After membrane rupture, the pipet solution diffuses rapidly into the cell.
3. Whole-cell configuration enables recording macroscopic whole-cell currents that reflect all of the channels in the cell, with a defined internal solution. However, in this configuration the cytosol is removed, which may interrupt second messenger pathways that involve diffusible or unknown elements.

3.4.4. Outside-Out Patch

1. Outside-out patch configuration occurs when the pipet is pulled away from the cell after forming the whole-cell configuration.
2. Nonadherent cells tend to follow the pipet, but this may be remedied by pulling the cell away using a second pipet. The membrane ideally reforms across the tip of the pipet with the extracellular surface facing the bath solution.
3. This configuration is analogous to inside-out configuration, but with the opposite orientation, and has similar benefits and limitations, except that the external solution is readily varied.

3.4.5. Perforated Patch

1. This configuration is topologically like the cell-attached patch, except that the pipet solution contains a pore-forming molecule, such as ATP *(31)*, or an antibiotic like

amphotericin (~1 mg/mL) *(32)* or nystatin *(33)* that perforates the membrane and allows small ions to permeate. Larger pores that allow entry of larger molecules into the cell are formed by β-escin *(34)*.

2. Usually, the tip of the pipet is dipped into solution lacking the pore-forming molecule to fill the tip before back-filling the pipet. This enables the seal to be formed before the membrane becomes perforated.

3. The perforated-patch configuration gives electrical access to the cell, but prevents the loss of proteins and large molecules from the cytosol. One can record from all channels in the cell membrane, with the advantage that many signaling pathways remain intact *(31,33)*.

4. The perforated-patch configuration permits activation of NADPH oxidase in phagocytes *(10)*. In many other cells, it prevents the run-down of ionic currents that occurs in conventional whole-cell configuration *(35)*.

5. One problem with the perforated-patch configuration is that the patch may rupture spontaneously, producing whole-cell configuration *(13,36,37)*.

3.5. Electron Current Recordings

A recent application of the patch clamp technique is to record plasma membrane electron current (I_e) produced by the NADPH oxidase *(7)*. The enzyme is electrogenic and passes currents large enough to be recorded in human phagocytes (2–3 pA in neutrophils, 6–15 pA in eosinophils). Electron current can be measured in the perforated-patch configuration *(10)* or in excised, inside-out patches *(11)*.

3.5.1. Recording I_e in Whole Cells

Schrenzel et al. *(7)* first reported I_e in phagocytes studied in the whole cell configuration. Eosinophils were patched with a pipet solution consisting of 76 mM CsCl, 50 mM CsOH, 50 mM HEPES, pH 7.6, 10 mM TEACl, 1 mM MgCl$_2$ and 8 mM NADPH. The bath solution mimicked the pipet solution but lacked ATP or NADPH and was at pH 7.1. Under these conditions, the authors reported spontaneous appearance of I_e upon forming whole-cell configuration. The appearance of I_e without agonists suggests that these cells were activated spontaneously, perhaps by adhesion. In later experiments, I_e was observed with 25 μM GTPγs or 5 μM free Ca^{2+} in the pipet solution *(38)*.

3.5.2. I_e in Perforated Patch

The perforated patch configuration is highly suited to recording electron current. It is possible to record from the same cell before and after stimulation so that each cell is its own control. It appears to be beneficial to have glucose (1 mg/mL) in the bath solution to provide an energy source for the cells. Electron current can be recorded at −40 or −60 mV, voltages sufficiently negative to prevent activation of proton channels (at symmetrical pH) that would obscure the I_e.

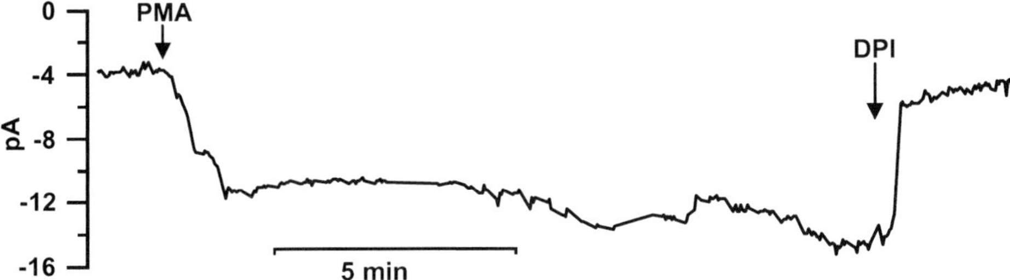

Fig. 3. A record from a human eosinophil in the perforated-patch configuration. The holding current is –60 mV. Pipet solution was 100 mM KMeSO$_3$, 1 mM MgCl$_2$, 1 mM EGTA, 25 mM (NH$_4$)$_2$SO$_4$, 10 BES, and 1 mg/mL amphotericin at pH 7.0. The bath solution was 100 mM TMAMeSO$_3$, 1 mM EGTA, 2 mM MgCl$_2$, 1.5 CaCl$_2$, 25 mM (NH$_4$)$_2$SO$_4$, and 10 mM BES, pH 7.0. At the times indicated by the arrows, the cell was stimulated with 60 nM phorbol myristate acetate and I$_e$ was inhibited by 12 µM diphenylene iodonium.

1. After forming a giga-seal, access to the cell increases over time as the amphotericin forms pores in the patch membrane. Depolarizing test pulses that activate H$^+$ current can be used to monitor the access resistance (*see* **Note 14**).
2. Begin recording when the H$^+$ current is stable.
3. The most reliable activator of NADPH oxidase is phorbol myristate acetate (PMA), which elicits I_e in almost all cells (*see* **Note 15**).
4. PMA produces a downward deflection of the holding current (I_{hold}) that usually stabilizes after a few minutes and persists for tens of minutes (**Fig. 3**).
5. If the increase in I_{hold} is due to NADPH oxidase activity, it should be sensitive to diphenylene iodonium (DPI), a well known inhibitor of the NADPH oxidase *(39)*. If a high concentration of DPI inhibits the presumed I_e only partially, then the DPI-insensitive current is likely due to appearance of a leak conductance.
6. During perforated patch recording, the patch sometimes ruptures spontaneously, producing whole cell configuration (*see* **Note 16**). When this occurs, I_e vanishes rapidly *(13)*, perhaps as a result of the diffusion of essential components from the cell into the pipet. I_e decreases several times faster when the patch is ruptured (τ = 5.6 s) than when I_e is inhibited by DPI (τ = 25.5 s) *(13)*.

3.5.3. I_e in Inside-Out Patch

Electron current can be recorded in the inside-out patch configuration *(11)*. A patch of membrane is excised from a stimulated granulocyte in which active oxidase complexes are already assembled. Exposing the internal surface of the patch membrane to NADPH enables electron flow across the membrane through the pre-activated NADPH oxidase complexes (*see* **Fig. 4** for a representative experiment) (*see* **Note 17**).

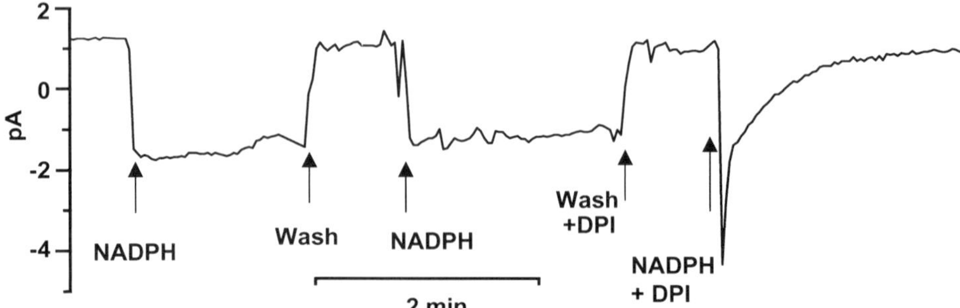

Fig. 4. I_e in an inside-out patch of membrane from a human eosinophil at −60 mV. The cell was stimulated with 60 n*M* PMA in cell-attached patch configuration 5 min before the patch was excised. Pipet solution was 200 m*M* HEPES, 2 m*M* MgCl, 2 m*M* EGTA, titrated to pH 7.5 with TMAMeSO₃. The bath solution was the same as the pipet solution with 50 µ*M* GTPγS and 3 m*M* MgATP added. The arrows indicate the addition of 2.5 m*M* NADPH or 12 µ*M* diphenylene iodonium.

1. Patch a granulocyte in the cell-attached patch configuration in Ringer solution.
2. Stimulate the cell by adding 60 n*M* PMA to the bath solution.
3. Exchange the bath to an intracellular solution (with low free Ca^{2+}).
4. Lift the pipet briefly out of the bath solution to produce an inside-out patch.
5. Add NADPH to the bath to elicit I_e.

3.6. Ion Channel Recording

3.6.1. Optimizing Conditions to Record Specific Types of Currents

1. If more than one type of ion channel or electrogenic transporter is present in a cell, as is almost always the case, then the currents that are recorded will be an additive mixture of the various component currents.
2. To focus on one particular type of channel, the recording conditions may be adjusted to maximize the conductance of interest and to eliminate extraneous currents. To record ionic currents, the permeant ion must be present in appropriate concentrations. To record K^+ currents, K^+ must be present, although most K^+ channels are also permeable to Rb^+, NH_4^+, and usually to a much smaller extent, other cations. To study currents carried by ions that are normally present at small concentrations, such as Ca^{2+} or H^+, it may be helpful to increase their concentration.
3. Eliminating extraneous currents can be achieved by several approaches:
   a. Eliminate all ions permeant through a channel. This may be difficult for channels that are not very selective. For example, many anion channels are notoriously promiscuous, allowing large anions, such as aspartate⁻ or methanesulfonate⁻, to permeate. Usually, replacing small physiological ions with larger ions of the same sign is sufficient to reduce the current (*see* **Note 18**).
   b. If using impermeant ions is not possible, inhibitors can be used. These should be as specific as possible, so that they do not affect the currents of interest (*see*

Note 19). Many channels have potent and specific inhibitors, often derived from venoms or toxins.

c. In some cases, the activation of a conductance can be prevented. To prevent activation of Ca^{2+}-activated channels, intracellular free Ca^{2+}, $[Ca^{2+}]_i$ can be buffered to low levels. Voltage-gated channels can be prevented from opening simply by avoiding the voltage range that activates the channels. For example, electron current can be recorded without interference from proton channels at -60 mV, because at symmetrical pH, proton channels do not open at this voltage. Most cells, including neutrophils *(5)*, have volume-regulated anion channels that are activated when osmotic imbalance occurs. Typically, hypotonic solutions cause cell swelling, which activates stretch- or swelling-activated channels. When one first ruptures the membrane patch to begin whole-cell recording, there is often a transient osmotic imbalance thought to be caused by the higher mobility of small anions in the pipet solution than that of the large anionic proteins in the cytoplasm. This diffusional imbalance produces both a Donnan potential *(30)* and transient activation of stretch-activated anion currents in neutrophils *(4,5)*. These stretch-gated currents usually go away after a few minutes *(18)*.

3.6.2. Requirements for Complete Characterization of an Ion Channel

3.6.2.1. SELECTIVITY: THE REVERSAL POTENTIAL MUST BE MEASURED

The selectivity of an ion channel is so important that channels are named for the ions they conduct. Potassium channels are selectively permeable to K^+, proton channels are selective for H^+, and so forth.

1. Selectivity can be determined only by studying the channel in various ionic conditions to determine which ions permeate. Permeability can be evaluated in two ways.
   a. Create conditions in which there is only a single ionic species (of one sign) on one side of the membrane, and determine whether that ion carries current in the correct direction. Although this approach seems straightforward, a prerequisite is a reliable method to determine whether the current is permeating the channel of interest. If the channel is voltage- and time-dependent, then the kinetics of the currents can be presumed to define the conductance, otherwise using blockers may be necessary. Identification of a channel by pharmacology alone is unacceptable for several reasons. First, truly specific inhibitors are rare. Second, pharmacology can be messy. Some drugs have multiple effects, such as changing the pH or altering metabolism. If the inhibitors are irreversible, or if they act slowly, then positive identification becomes tricky. For example, one published study incorrectly identified volume-regulated anion currents as Na^+ currents. The putative Na^+ currents consistently ran down after addition of an inhibitor, and the run-down was mistaken for slow and irreversible block.
   b. Selectivity is best determined by measuring the reversal potential, V_{rev}, and varying the species or concentration of ions in the bath. This approach can give precise information about relative permeability of a channel when many ions are measurably permeant. Addition or removal of impermeant ions should not

change V_{rev}, but changes in the concentration of the permeant ion should. The V_{rev} of voltage-gated proton currents does not change when the bath is changed among Cs^+, K^+, TMA^+, etc., although changes of a few millivolts are measured in some solutions which disappear when one corrects for liquid junction potentials. If one encounters a channel that indiscriminately conducts all cations or all anions (but does discriminate between ions of opposite charge), then the charge of the permeant ion can be identified by decreasing the ionic strength, replacing ions with uncharged molecules like glucose to maintain the osmolarity. Sometimes ions that are not permeant indirectly produce changes in V_{rev} that could be mistaken as an indication of permeability. For example, when the predominant cation in the bath solution is Na^+ or Li^+, compared with other cations, the measured V_{rev} of H^+ currents shifts in a positive direction *(40)*. Although such a shift would be expected if Na^+ were permeant, it is prevented by inhibiting Na^+/H^+-antiport and therefore is due to increased pH_i resulting from exchange of extracellular Na^+ for intracellular H^+ *(40)*.

2. How does one measure V_{rev}? The answer depends on the situation. One principle is that the idea is to determine the reversal potential of a conductance that is due presumably to a population of identical channels. However, all voltage-clamp data includes extraneous conductances that are observed as "leak" currents. If the conductance of interest is, for example, a non-voltage-gated (i.e., the conductance is "on" at all voltages) Ca^{2+}-activated K^+ conductance, then one might simply measure the voltage at which the current is zero to estimate V_{rev}. However, the measured V_{rev} will be incorrect unless the leak is corrected. Because leak conductances usually reverse near 0 mV, their effect is to decrease the absolute value of the measured V_{rev}. The larger the leak, the larger the error will be. If a specific inhibitor of the conductance of interest exists, one can subtract the current measured in the presence of blocker from that in its absence. The blocker-subtracted currents should reflect the properties of the conductance of interest. Two dangers exist. First, some drugs block channels in a voltage-dependent manner. In this case, the subtraction will give erroneous results unless block is essentially complete at all voltages. Second, the subtraction procedure introduces errors, because both the "leak" and conductance of interest may change with time, compromising the subtraction. One must recognize these limitations and accept that the results may have some error. One must be especially careful if two large numbers are subtracted to give a small result. If such a subtraction procedure results in V_{rev} values that follow the Nernst prediction within 0.1 mV, then one must be extremely suspicious that wish-fulfillment has entered into the data analysis process.

3. It is relatively easy to measure V_{rev} for channels that are voltage- and time-dependent. The two situations that may be encountered are illustrated in **Fig. 5**, when V_{rev} is outside (**Fig. 5A**) or within (**Fig. 5B**) the voltage range at which the conductance is activated. The classical method of estimating V_{rev} of voltage-gated channels, introduced by Hodgkin and Huxley *(1)*, is the "tail current" method (**Fig. 5A**). The membrane is held at a voltage at which the conductance is off, the "holding" potential, V_{hold}. A prepulse is applied to activate the conductance. Then the voltage

is stepped to various voltages at which most channels close, in this example, −20 through +10 mV in 10-mV increments. The object is to determine the voltage at which the resulting "tail currents" vanish; this is V_{rev}. Often tail currents decay exponentially, and if the initial amplitude is difficult to resolve because of capacity current interference, the current can be fitted by an exponential decay function, which can be extrapolated to the time at the start of the pulse. One could then plot the "instantaneous" current against voltage and interpolate to determine where the currents cross the voltage axis. This would give the wrong answer, +8 mV in our example, because one must always consider the leak current. The conductance of interest in **Fig. 5A** is the voltage-gated proton conductance. At V_{hold} there is 2 pA of inward current that is unrelated to proton current and therefore considered "leak." We are not interested in the properties of leak current, except in so far as we want to eliminate their effects. The leak current will be different at each test voltage where tail currents were recorded. The simplest way to eliminate the leak is to look not at the absolute current value at the start of the pulse, but at the amplitude of time-dependent tail current. Generic leak currents are usually time-independent, in contrast with proton currents. When we plot and interpolate the amplitudes of the tail current decay, we find that V_{rev} is actually +6 mV. The error due to leak was small here because the leak was small and reversed near V_{rev} for the proton conductance. Much larger errors have been published in reputable journals, including V_{rev} values that were described as being positive to voltages at which clearly outward tails currents decayed, because leak was not corrected. One presumption in the tail-current-decay-amplitude method is that the same number of channels is open at the end of each prepulse. In the example in **Fig. 5A**, one prepulse activated a smaller fraction of the total conductance, because the prepulse current was smaller. One can correct for this error retroactively by scaling the tail current according to the proportional decrease in prepulse current (in this example, the correction is negligible). Another presumption is that the tail current decays completely. In **Fig. 5A**, the tail current at +10 mV decayed only partially, resulting in maintained outward current. This can be confirmed by examining the secondary tail current when the voltage was finally returned to V_{hold}. There is a clear inward tail current after the pulse to +10 mV. One can determine whether complete decay can be expected at a given voltage by simply applying a pulse directly to that voltage to see if current is activated. If V_{rev} occurs at a voltage at which the conductance of interest is already activated, then a different approach must be used (discussed later).

4. Besides leak conductance, what other errors may occur? One systematic error arises from the necessity to activate the conductance in order to generate tail currents. The prepulse that activates the conductance also removes a large number of permeant ions from the cell. The result for proton currents is that the prepulse increases pH_i so that the measured V_{rev} tends to be more positive than the true value. This error can be minimized by using high buffer concentrations in the pipet solution and by keeping the prepulse small and brief. However, if the prepulse is too small, the tail currents will be small and difficult to interpret. Although the error

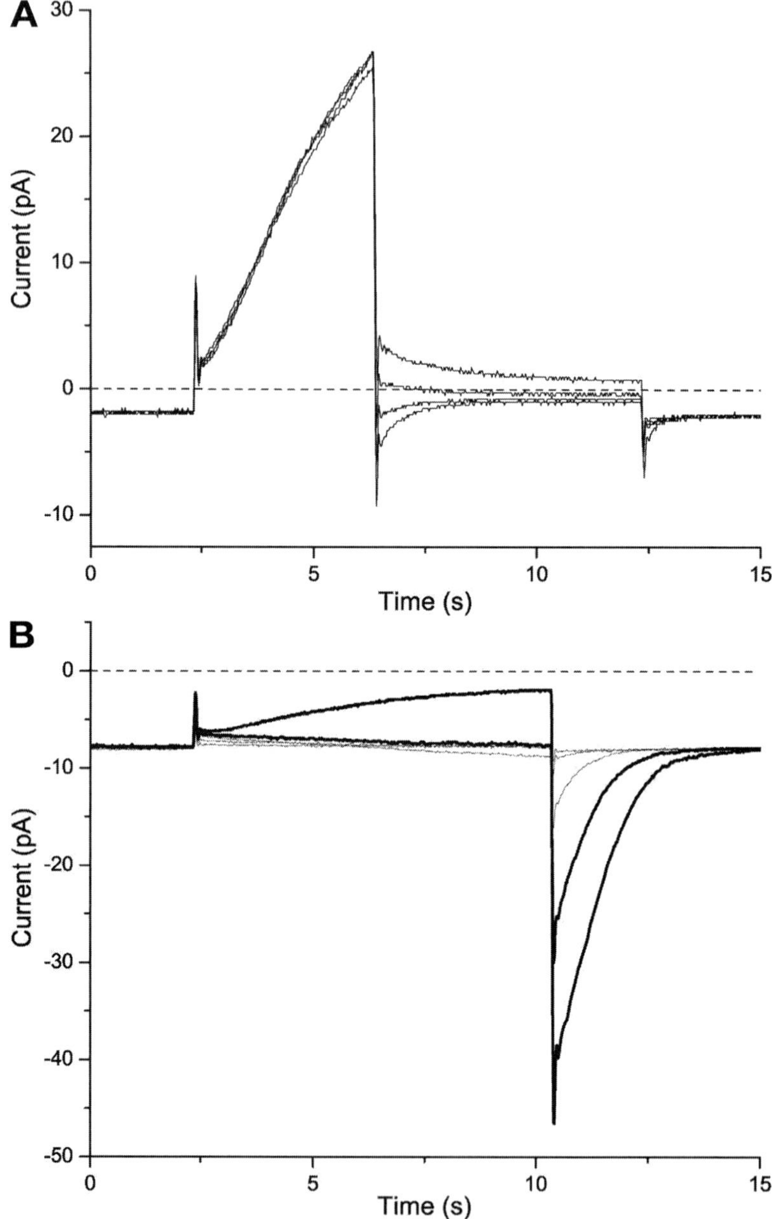

Fig. 5. Determining the reversal potential, V_{rev}, of a voltage-gated conductance, when V_{rev} is outside (**A**) or within (**B**) the voltage range at which the conductance is activated. (**A**) Tail current method illustrated by voltage-gated proton currents in a human eosinophil at pH_o 7.0 and pH_i 7.0. From a holding potential of −60 mV, where H^+ channels are closed, a prepulse to +80 mV activates the conductance. The membrane is then

cannot be eliminated, it can be recognized. Because of this error, it is more reliable to consider the change in V_{rev} measured when pH_o is varied, rather than the absolute value of V_{rev}, which will also have errors due to liquid junction potentials. If similar depletion of protonated buffer occurs during each V_{rev} determination, then the shift in V_{rev} will be the best indicator of selectivity.

5. Sometimes, the conductance of interest is activated at V_{rev} and therefore decaying tail currents do not occur. One solution is to cave in to error and simply measure the absolute current at various voltages, but this makes leak correction difficult. A simple and preferable alternative is to apply a family of depolarizing pulses, and observe the time dependent turn-on of current at various voltages, as illustrated in **Fig. 5B**. This human eosinophil was activated by PMA in perforated-patch configuration, and there is -8 pA of electron current at V_{hold}. That this large inward current is not leak can be deduced by noting the small spacing between the currents at the start of the four illustrated test pulses, before much proton current has turned on. During the pulse to -40 mV only leak current and I_e are evident, but the g_H is activated detectably at -30 mV, because a tail current appears upon repolarization. During the pulses to -20 and -10 mV, inward H^+ current turns on slowly during the pulses, and large inward tail currents are seen on repolarization. Finally, at 0 mV a large outward current turns on during the pulse. Despite the fact that this is clearly an outward current, its absolute value at the end of the pulse is still negative, because it is superimposed on the larger inward electron current. Only a computer program lacking the most rudimentary insight would declare that V_{rev} must be positive to 0 mV, simply because the total current at the end of the pulse is inward. It is obvious that V_{rev} for the voltage- and time-dependent proton conductance that we are interested in, falls between -10 mV and 0 mV (the two darker records), because the time-dependent current is inward and outward, respectively, in the two records. To determine V_{rev} we can first measure the amplitude of inward current at -10 mV (-1.2 pA) and the outward current at 0 mV ($+4.3$ pA) that turned on during the pulses. However, simply interpolating between these values on a current-voltage graph, for example, would give the incorrect V_{rev} value of -8 mV, because the g_H activated at these two voltages was a very different fraction of the total g_H. We can determine the relative g_H that was activated by each pulse from the amplitude of the tail current upon repolarization to -60 mV. We now scale the time-dependent current amplitudes according to the fraction of the g_H activated by the prepulse to arrive at the correct answer, -7 mV.

Fig. 5. (*Continued*) stepped to various voltages at which most channels close, -10 to $+20$ mV in 10-mV increments in this example. Bath solution TMAMeSO$_3$, pipet solution KMeSO$_3$, both with 50 mM NH$_4^+$ to control pH_i. (**B**) Determining V_{rev} from a family of depolarizing pulses. From $V_{hold} = -60$ mV, pulses were applied to -40 mV through 0 mV in 10-mV increments, with currents at -10 mV and 0 mV in bold. This human eosinophil was previously activated with PMA, and there is nearly -8 pA of electron current at V_{hold}. *See* text for details.

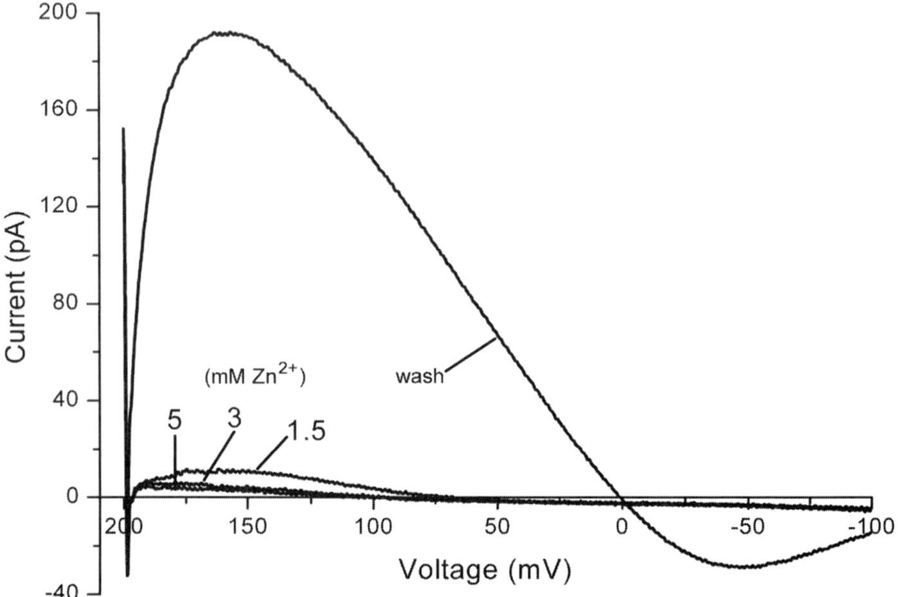

Fig. 6. Evaluation of V_{rev} using voltage ramps. Currents during voltage ramps (from +200 mV to −100 mV in 800 ms) in a human eosinophil studied in perforated-patch configuration with $KMeSO_4$ in the pipet solution and $TMAMeSO_4$ in the bath solution, both at pH 7.0 and both with 50 mM NH_4^+ to clamp pH_i near pH_o. The predominant conductance is due to voltage-gated proton channels, which are inhibited by Zn^{2+}. The sequence was 1.5 mM $ZnCl_2$, then 5 mM, then 3 mM, then washout with EGTA-containing bath solution. This cell was stimulated with phorbol myristate acetate and then diphenylene iodonium was added to inhibit NADPH oxidase.

6. Under some conditions, one may use a "quick-and-dirty" method to determine V_{rev}. One such method is to use voltage ramps (**Fig. 6**). Here the voltage in a human eosinophil was ramped rapidly from +200 mV to −100mV. Because H^+ current activation is faster at more positive voltages, nearly 200 pA of proton current was activated soon after the start of the ramp. The ramp was fast enough that the conductance was still active at 0 mV, then in the negative voltage range H^+ channels closed progressively. When the same ramp was applied after addition of high concentrations of Zn^{2+}, most, but not all of the proton current was abolished. Because a major effect of divalent cations is to shift the voltage dependence of H^+ channel activation to more positive voltages (*20,24,41,42*), the currents turned off at much more positive voltages. In the presence of Zn^{2+}, the current negative to +50 mV is almost entirely leak current. Estimating V_{rev} from voltage ramps suffers from at least three potential errors.

    a. A capacitive artifact produces an offset of the current recorded during a voltage ramp, typically of a few picoamperes, because the capacity current is proportional to dV/dt, which is constant during a voltage ramp. Even if one uses analog capacity correction, the correction is rarely perfect, and can change over time. This offset may be negligibly small, but can be problematic if the currents of interest are comparably small, as may occur in excised patches. The offset increases with the speed of the ramp. A quick way to test whether the offset is a problem is to apply the same ramp but in the opposite direction. If the currents do not superimpose, there is an offset problem. In the ramps in **Fig. 6**, the leak current intersects the X-axis near +70 mV. This may be the result of a small offset, because one expects leak currents to reverse near 0 mV, although this question was not studied systematically. However, the best estimate of V_{rev} in this experiment is not the point at which the control currents pass through the voltage axis (the zero-current potential), but rather the voltage at which the currents in the absence and presence of Zn^{2+} intersect, just negative to 0 mV.

   b. Significant numbers of the channel of interest must be open when the voltage passes V_{rev} to estimate V_{rev}. For example, voltage-gated proton channels generally open only positive to V_{rev} and thus V_{rev} can be estimated only by ramping from positive to negative voltages at a rate slow enough to allow H^+ channels time to open, but fast enough so that they do not all close by the time the voltage passes E_H. Although it seems obvious that one can measure a meaningful V_{rev} only if the channels of interest are open at that voltage, papers have been published in highly reputable journals in which this condition was not met. Naïve authors sometimes simply plot the total current and declare that V_{rev} is the voltage at which this current crosses the voltage axis, ignoring the fact that they have simply determined V_{rev} of the leak current.

   c. It is more difficult to separate the conductance of interest from leak or other conductances during ramps than during voltage pulses (discussed previously).

7. A second "quick-and-dirty" way to estimate V_{rev} of a voltage-gated channel is to apply a single voltage pulse that activates the conductance, and then repolarize to a voltage at which the channel closes. One can interpolate between the current at the end of the pulse and that at the beginning of the tail current to estimate V_{rev} *(43)*. This method assumes that the instantaneous current-voltage relationship is linear, i.e., that the open channel conductance is ohmic. **Figure 7** illustrates that the instantaneous current-voltage relationship for voltage-gated proton channels is not linear at symmetrical pH, but the rectification is weak enough that the error is not large if the two test voltages are close together. Leak must also be corrected. In some situations, the complete tail current method may be too time-consuming, and a rough estimate is better than none at all. An advantage of this method is that the two measurements are made at essentially the same time. Thus, although depletion would affect the result, the measurement is an accurate reflection of V_{rev} at that instant. Using the example in **Fig. 5B**, from the outward current at 0 mV, +4.3 pA, and the inward tail current at −60 mV, −34 pA, we interpolate to arrive at −7 mV as an estimate of V_{rev} which is identical with the result calculated by the more laborious formal methods described above.

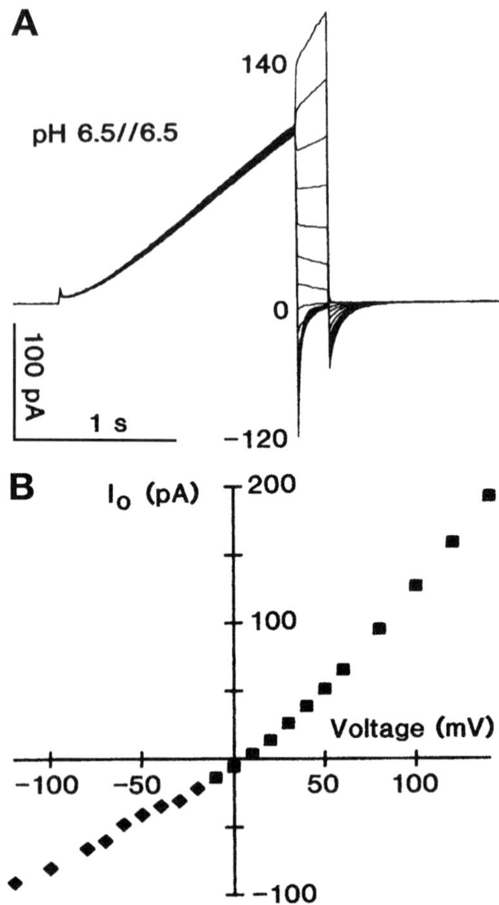

Fig. 7. Measurement of the instantaneous current-voltage relationship of proton channels. In **A**, currents during repeated voltage pulses are superimposed. The first step, the "prepulse," is to a voltage at which the g_H is activated, and then the membrane is stepped to a range of voltages. Ideally, the same number of channels is open at the end of each prepulse, and therefore the current at the start of the test pulse **(B)** reflects any rectification of the open channel current. If one measured the current-voltage relationship of a single open channel, the result should have the same shape as that plotted in **B**. If any gating that occurs during the test pulse, typically channels closing at negative voltages, is so rapid that the initial current cannot be resolved, one can fit the decaying current, usually to an exponential decay function, and extrapolate back to the start of the pulse. (From **ref. *73*.**)

3.6.2.2. Rectification

Rectification means that the current flowing through a single open channel is not a linear function of the applied driving force (the applied voltage relative to V_{rev}).

1. Rectification can be determined from instantaneous current-voltage curves, like that in **Fig. 7**. Errors can arise if the current at the end of the prepulse is not the same each time, which this can be corrected by scaling, as already described. If the tail currents decay rapidly, their amplitude can be difficult to determine.

2. Another way to detect rectification is to measure single-channel currents directly at different voltages. The preferred method is to use repeated voltage ramps, and then average segments of ramps containing either one or no open channels, and subtract one from the other. Distinct advantages of this method include: (a) it is efficient, (b) by superimposing records, one can immediately see whether there has been a change in the leak current, (c) one can resolve tiny currents near V_{rev} more reliably than during sustained voltages pulses, and (d) it is much easier to confirm which channel is which when there are multiple types of channels in the patch.

3.6.2.3. ACTIVATION BY LIGANDS OR VOLTAGE

1. A crucial property of ion channels is the factor that causes them to open.

2. Some channels are ligand-gated (i.e., they open when a particular substance binds). For example, endplate channels at the neuromuscular junction open when acetylcholine binds. Ca^{2+}-activated channels open when $[Ca^{2+}]_i$ increases above resting levels.

3. Many channels are voltage-gated. The voltage dependence of a channel can be described by applying a family of voltage pulses that span a voltage range in which the channel progresses from being closed to having a high probability of being open (P_{open}). Ideally, the current is measured after it reaches a steady-state level. In practice, this ideal is often not achieved, especially for proton currents that activate slowly and may decay during prolonged large pulses due to H^+ current-induced pH_i increases. One can "correct" the current measured before steady-state by fitting the current with a rising exponential, $I(t) = I_{max}(1 - e^{-t/\tau})$, and extrapolating to infinite time. This obviously assumes that an exponential describes the ideal activation behavior, which may not be the case, but may give a reasonable estimate. In this process, one also obtains the time constant (τ or τ_{act}) of activation of the current.

4. In most cells, proton currents activate with a sigmoidal time course, and thus a simple exponential does not accurately describe the kinetics. However, attempts to discover a simple model that fits the currents better have not been productive. Our philosophy is that if the main part of the current can be reasonably well described by a single exponential, then this provides a τ_{act} value that gives a rough indication of the rate at which most of the channels open, and it is a parameter that can be easily understood and communicated.

5. The resulting current-voltage relationship can be converted to a conductance-voltage relationship by dividing the current at each voltage (V) by the driving force ($V - V_{rev}$). The conductance ideally saturates at large voltages at which P_{open} is maximal. The single-channel correlate is the P_{open}-voltage relationship. Both of these relationships tend to be sigmoidal and are often fitted with a Boltzmann function (using proton channels as an example):

$$\frac{g_H}{g_{H,\text{max}}} = \frac{1}{1 + e^{-(V - V_{1/2})/k}} \tag{1}$$

or

$$\frac{P_{\text{open}}}{P_{\text{max}}} = \frac{1}{1 + e^{-(V - V_{1/2})/k}} \tag{2}$$

The midpoint of the curve, at which the conductance is half maximal, is $V_{1/2}$ (in mV) and the slope factor k (in mV) indicates the steepness of the voltage dependence. A smaller value of k means steeper voltage dependence.

3.6.2.4. SINGLE CHANNEL CONDUCTANCE

Single channel conductance (γ) is important because it determines how much current flows through the open channel. The total current across a cell membrane is described by:

$$I_{\text{total}} = (V - V_{\text{rev}})\, \gamma\, N\, P_{\text{open}} \tag{3}$$

where $(V - V_{\text{rev}})$ is the driving force, N is the number of channels in the membrane, and P_{open} is the probability of any channel being open.

1. When ion channels open and close, the current increases and decreases in steps of identical amplitude, which sum together when many channels open to form a macroscopic current in which the unitary events cannot be clearly detected.
2. For many ion channels, γ is large enough that single channel currents can be measured directly.
3. Detecting single channel currents is usually done using excised patches of membrane, because it is desirable that either one channel or no more than a few channels are present.
4. There is additional noise when a channel is open, and the total noise increases with each open channel so that it is difficult to resolve unitary events when more than a few channels are open at one time. If multiple types of channels are present, identification becomes very difficult.
5. Single channel currents can be recorded at constant voltage, during pulses, or during voltage ramps. If there is only one channel in a patch, one can hold the membrane at each voltage for long times and determine mean open and closed times as well as P_{open} for that voltage. P_{open} can be efficiently determined by converting the data to an amplitude histogram and comparing the integrals when one or no channel is open.
6. Giving pulses is useful for voltage-gated channels to derive a sense of whether the channel behaves as expected from macroscopic currents.
7. Voltage ramps are an excellent way to determine the reversal potential and the voltage dependence of the conductance, i.e., its rectification. Under most circumstances, the open channel current is not ohmic, rather the conductance changes with voltage or rectifies. A voltage ramp is applied repeatedly, and the currents are ana-

lyzed afterwards. The average of current segments with no open channels can be subtracted from the average of segments with exactly one channel open to produce the open-channel current-voltage relationship.

8. If the unitary conductance is too small to measure directly, it can be deduced from current fluctuations, or noise. Under stationary conditions (without time-dependence of I_{total}, P_{open} or driving force, etc.), the single channel current (i) can be calculated from the variance, σ^2 according to (**44**):

$$i = \sigma^2 /[I_{total} (1 - P_{open})] \tag{4}$$

The value used for σ^2 must have background noise subtracted, which can be determined from identical measurements in the presence of complete pharmacological blockade. If this is not feasible, one can determine background noise at subthreshold voltages, in the case of voltage-gated channels. The only parameter that is difficult to determine is P_{open}. The macroscopic conductance of a depolarization-activated channel may saturate at large positive voltages, but P_{open} at saturation may be much less than 1.0. However, by collecting data at different voltages, it is possible to deduce P_{open} (Fig. 12 of **ref. 45**). One can also collect data near $V_{threshold}$ where P_{open} is close to 0, and the error in $(1 - P_{open})$ is negligible; this approach, however, produces estimates only within a narrow voltage range. For noise analysis, it is important that the data are collected using an appropriate sampling rate and appropriate filtering. These issues are beyond the scope of this chapter and have been discussed elsewhere at length (**45,46**).

9. The transporters involved in the phagocyte respiratory burst have small conductance or transport rates. Proton channels have a conductance of approx 15 fS at physiological pH (**45**), which corresponds with currents of a few femtoamperes, on the order of approx 10,000 H^+/s, depending on the driving voltage. Store-operated Ca^{2+} (CRAC) channels may be present in neutrophils (**47**) and have similarly small conductance (**48**). In comparison, the NADPH oxidase transports only approx 300 e^-/s (**49**). It seems teleologically reasonable that in small cells, and especially in the much smaller phagosome, low transport rates would make the transporters more amenable to regulation and fine-tuning of their activity. A single large-conductance channel would produce a drastic and abrupt change in potential and ion concentrations, and thus would be unwieldy (**50**).

4. Notes

1. The NADPH oxidase complex is assembled from two membrane-bound subunits (gp91*phox* and p22*phox*) and several cytosolic subunits (p67*phox*, p47*phox*, Rac, and p40*phox*) (**9**). The NADPH oxidase works by transporting electrons from cytoplasmic NADPH across the membrane, where they reduce O_2 to $O_2^{\cdot-}$. This electron movement can be recorded as an electrical current (**7,10**). Because the electron transport is not directly coupled to a compensatory charge movement, it produces a separation of charge that leads to sustained depolarization of the membrane (**51**) beyond 0 mV in intact granulocytes (**52–56**). If this charge movement is not compensated, extreme depolarization results (**50,56**) that by itself directly prevents

electron flux across the membrane *(12)*. Almost all of this charge compensation is mediated by H^+ channels *(4,9,25,50,51,55–58)*.

2. H^+ channels are membrane proteins that are gated by both voltage and pH. They open with depolarization or an outward pH gradient (defined as $\Delta pH = pH_o - pH_i$). Increased pH_o and/or decreased pH_i cause the channels to open at lower voltages. Proton channels have two distinct gating modes, "resting" and "activated," that are observed before and after stimulation. Resting H^+ channels open almost exclusively positive to the Nernst potential for protons (E_H) and therefore extrude protons from cells. The activated mode occurs concurrently with NADPH oxidase activation, and is manifested as an increased proton conductance, faster channel opening on depolarization, slower channel closing, and a negative shift in the threshold voltage at which the channels open.

3. The patch clamp allows the recording of small currents and the detection of single channel currents in the cell membrane. Recording small currents with the patch clamp technique requires a large signal-to-noise ratio. Noise in the patch clamp set-up arises from the electronic circuitry, the pipet and holder assembly, the seal and cell membrane, and extraneous mechanical or electrical sources *(59,60)*. Noise interferes with the recordings and should be minimized.

4. A patch-clamp experiment typically consists of a number of voltage pulses applied to a cell or membrane patch. Changes in pH, ion concentration, temperature, osmolality, and drug concentration can be exploited to reveal the properties of the channels present. Channels are characterized by the stimulus that causes them to open, their ion permeability, opening and closing kinetics, conductance, and pharmacological sensitivity.

5. The current that passes through a channel is described by Ohm's law:

$$V = IR \tag{5}$$

where V is the potential difference in volts (V), I is current in amperes (A) and R is resistance in Ohms (Ω). The reciprocal of resistance is conductance (G, measured in Siemens, S) and indicates the facility of current flow.

The selectivity of an ion channel is determined by measuring the reversal potential (V_{rev}) and comparing it with the equilibrium potential or Nernst potential (E_X) of the ion (X) of interest. The Nernst equation shows that E_X depends on the ion concentrations inside and outside the cell *(44)*:

$$E_X = (RT/zF) \ln([X]_i / [X]_o) \tag{6}$$

where R is the gas constant, T is the absolute temperature, z is the charge of ion X, F is Faraday's constant, and $[X]_i$ and $[X]_o$ are the concentrations (brackets indicate concentration) of the ion inside and outside the cell, respectively. The term RT/zF is 25.26 mV for a monovalent cation at 20°C. For a channel that is perfectly selective for a particular ion, the reversal potential of the solution is identical to the Nernst potential.

6. For proton current recording, it is advisable to use a high intracellular buffer concentration to maximize control over pH. Reducing extracellular buffer from 100

mM to 1 mM had very little effect on voltage-gated proton currents *(21)*, because the bath solution represents an effectively infinite sink for protons. Intracellular buffer concentration was more critical, with distinct limitation of H$^+$ current when buffer was reduced from 100 to 10 mM even in excised, inside-out patches *(21)*. Several whole-cell patch-clamp studies in which pH$_i$ was determined have revealed that including 5–10 mM buffer in the pipet solution does not adequately control pH$_i$, compared with higher buffer concentrations, e.g., 100–120 mM *(19,41,61,62)*.

7. The high membrane permeability of the uncharged form of these molecules and the impermeability of charged form provides a virtually unlimited reservoir of intracellular proton donor/acceptor. At equilibrium, the concentration of the uncharged form will be nearly equal on both sides of the membrane due to its freely diffusible nature. By changing pH$_o$ and the concentration of the protonated donor, pH$_i$ can be established and maintained. In the case of ammonium, the relationship is give by *(26)*:

$$pH_i = pH_o - \log \left([NH_4^+]_i / [NH_4^+]_o \right) \tag{7}$$

The internal NH$_4^+$ concentration is fixed by the pipet solution and pH$_i$ can be varied by changing the NH$_4^+$ concentration in the bath. It is possible to lower pH$_i$ with respect to pH$_o$ by decreasing the external NH$_4^+$ concentration. Decreasing [NH$_4^+$]$_o$ to one-tenth [NH$_4^+$]$_i$ will ideally lower pH$_i$ by one unit. In practice, the control of pH$_i$ appears to be better at higher pH$_o$, perhaps because the neutral NH$_3$ is more abundant (the pK$_a$ of ammonium is 9.4).

8. A more traditional method is to pass current alternately in both directions through a silver wire that is immersed in a Cl$^-$-containing solution (e.g., 0.1 M HCl), to deposit a fine coat of AgCl. Sintered Ag/AgCl electrodes can also be purchased.

9. The pipet resistance (R_{pip}) is in series with the membrane resistance of the cell (R_m), and the applied voltage ($V_{command}$) will be divided according to these relative resistances. The result is that the actual voltage across the cell membrane (V_{true}) is less than the command voltage applied: $V_{true} = V_{command} [R_m/(R_m + R_{pip})]$. It should be noted that the resistance of the cell membrane decreases radically when a pulse that opens ion channels is applied, and therefore the series resistance error increases drastically when a large conductance is activated. Most patch-clamp amplifiers have a "series resistance compensation" circuit that can be adjusted to compensate some fraction of the series resistance. One manifestation of uncompensated series resistance is non-exponential decay of tail currents that normally decay exponentially.

10. Many types of glass have been evaluated for use as pipets *(29)*. Glass with a lower softening temperature is easier to fire polish after pulling and forms blunt pipets that have low access resistance. Harder glass with a higher softening temperature tends to form long, gradually tapering pipets with high resistance; hence are a poor choice for whole-cell recording. Hard glass may have other desirable properties, such as lower noise or better sealing proclivity *(63)*. Quartz pipets have been used to minimize noise during single channel recording *(64)*. Some types of glass, especially soft glasses used for whole-cell recording, contain heavy metals that can leach into the solution, affecting the recorded currents *(65,66)*. Such effects

can be controlled by including a divalent metal chelator such as EGTA in the pipet solution.

11. A pipet filled with solution and placed into a bath solution acts as a capacitor. During a step change in voltage there is a brief transient of current due to charging the pipet (fast) and cell membrane (slow) capacity. The pipet capacitance can be reduced by coating the part of the pipet that will be immersed with a hydrophobic substance, such as Sylgard® (Dow Corning Corp.), thus minimizing contact of the bath solution and the pipet and increasing the distance between the two conductors (the fluid in the bath and pipet) to reduce the capacitance *(29)*. When recording currents, it is necessary to null capacity transients using the analog capacity compensation controls of the patch clamp amplifier.

12. The wire filament is often coated with glass to prevent vaporization and deposition of metal on the pipet tip. Positioning a stream of air blowing toward the filament creates a steep temperature gradient that enables better control over the fire-polishing process.

13. If the tip opening cannot be seen, the tip resistance can be estimated by a "bubble test" in which a 10-mL syringe is attached to the base of the pipet via a tube. With the tip immersed in methanol and the syringe filled with 10 mL of air, apply pressure until bubbles first start to appear, and then note the amount of air left in the syringe. In one series of measurements, the resistance of the pipet filled with an isotonic KF solution and placed in a Ringer-containing bath was 3–5 MΩ for a bubble test number of 3, 5–8 MΩ for a bubble test number of 2, and 8–17 MΩ for a bubble test number of 1. Actual results may differ, depending on tip geometry and other factors. If the tip is too large, further polishing can be performed.

14. A more traditional way to monitor the access resistance is to apply a subthreshold test pulse and determine the time constant of decay of the capacity transient (τ_c). The decay becomes faster as the access resistance (R_a) decreases, because $\tau_c = R_a C_m$, where C_m is the capacity of the cell *(32)*.

15. Unlike arachidonic acid (AA), PMA does not perturb the cell membrane. Unlike f-Met-Leu-Phe, PMA needs no priming.

16. The intactness of the patch membrane can be assessed by including a fluorescent dye (e.g., Lucifer yellow) in the pipet solution *(13)*. The cell will fluoresce if the patch has ruptured. The intactness of the patch can also be assessed by using a dye visible by light microscopy (e.g., eosin) in the pipet solution *(37)*.

17. Petheö et al. *(11)* reported that the I_e in this method is susceptible to run-down over time, which was reduced by addition of GTPγS and ATP to the bath (internal) solution. In our hands, I_e often persists in inside-out patches with minimal run-down, without addition of anything to the bath besides saline solution. A speculative explanation for this difference is that our patches might include more cytoskeleton or other cytoplasmic structures that conceivably could influence oxidase stability.

18. Large, impermeant ions (methanesulfonate⁻, aspartate⁻, tetramethylammonium⁺, *N*-methyl-D-glucamine⁺) diffuse more slowly than the smaller physiological ions they replace, and thus significant liquid junction potentials may be present at the

tip of the pipet *(67,68)*. Contrary to popular opinion, liquid junction potentials are not corrected by nulling the current at the start of the experiment. If one wants to determine the absolute potential accurately, it is necessary to correct for liquid junction potentials. In many situations, such accuracy is not required, but if one wants to quantitate relative selectivity, for example, accuracy is desirable.

19. The apparent specificity of inhibitors often decreases with the passage of time. For example, charybdotoxin *(69)* and iberiotoxin *(70)* were originally thought to be specific for BK channels, but both were later found to also block other Ca^{2+}-activated K^+ channels *(71)* and charybdotoxin also blocks the purely voltage-gated Kv1.3 channel *(72)*.

Acknowledgments

This work was supported in part by the Heart, Lung and Blood Institute of the National Institutes of Health (research grants HL52671 and HL61437).

References

1. Hodgkin, A. L. and Huxley, A. F. (1952) Currents carried by sodium and potassium ions through the membrane of the giant axon of *Loligo. J. Physiol.* **116,** 449–472.

2. Hamill, O. P., Marty, A., Neher, E., Sakmann, B., and Sigworth, F. J. (1981) Improved patch clamp technique for high-resolution current recording from cells and cell-free membrane patches. *Pflügers Arch.* **391,** 85–100.

3. von Tscharner, V., Prod'hom, B., Baggiolini, M., and Reuter, H. (1986) Ion channels in human neutrophils activated by a rise in free cytosolic calcium concentration. *Nature* **324,** 369–372.

4. DeCoursey, T. E. and Cherny, V. V. (1993) Potential, pH, and arachidonate gate hydrogen ion currents in human neutrophils. *Biophys. J.* **65,** 1590–1598.

5. Stoddard, J. S., Steinbach, J. H., and Simchowitz, L. (1993) Whole cell Cl⁻ currents in human neutrophils induced by cell swelling. *Am. J. Physiol.* **265,** C156–C165.

6. Gordienko, D. V., Tare, M., Parveen, S., Fenech, C. J., Robinson, C., and Bolton, T. B. (1996) Voltage-activated proton current in eosinophils from human blood. *J. Physiol.* **496,** 299–316.

7. Schrenzel, J., Serrander, L., Bánfi, B., et al. (1998) Electron currents generated by the human phagocyte NADPH oxidase. *Nature* **392,** 734–737.

8. Tare, M., Prestwich, S. A., Gordienko, S., et al. (1998) Inwardly rectifying whole cell potassium current in human blood eosinophils. *J. Physiol.* **506,** 303–318.

9. Babior, B. M. (1999) NADPH oxidase: an update. *Blood* **93,** 1464–1476.

10. DeCoursey, T. E., Cherny, V. V., Zhou, W., and Thomas, L. L. (2000) Simultaneous activation of NADPH oxidase-related proton and electron currents in human neutrophils. *Proc. Natl. Acad. Sci. USA* **97,** 6885–6889.

11. Petheö, G. L., Maturana, A., Spat, A., and Demaurex, N. (2003) Interactions between electron and proton currents in excised patches from human eosinophils. *J. Gen. Physiol.* **122,** 713–726.

12. DeCoursey, T. E., Morgan, D., and Cherny, V. V. (2003) The voltage dependence of NADPH oxidase reveals why phagocytes need proton channels. *Nature* **422,** 531–534.

13. Morgan, D., Cherny, V. V., Murphy, R., Xu, W., Thomas, L. L., and DeCoursey, T. E. (2003) Temperature dependence of NADPH oxidase in human eosinophils. *J. Physiol.* **550,** 447–458.

14. Petheö, G. L. and Demaurex, N. (2005) Voltage- and NADPH-dependence of electron currents generated by the phagocytic NADPH oxidase. *Biochem. J.* **388,** 485–491.

15. Morgan, D., Cherny, V. V., Murphy, R., Katz, B., and DeCoursey, T. E. (2005) pH dependence of PMA activated electron current in human neutrophils. *J. Physiol.* **569,** 419–431.

16. Menegazzi, R., Busetto, S., Dri, P., Cramer, R., and Patriarca, P. (1996) Chloride ion efflux regulates adherence, spreading and respiratory burst of neutrophils stimulated by tumor necrosis factor-α (TNF) on biological surfaces. *J. Cell Biol.* **135,** 511–522.

17. Moreland, J. G., Davis, A. P. Bailey, G., Nauseef, W. M., and Lamb, F. S. (2006) Anion channels, including ClC-3, are required for normal neutrophil oxidative function, phagocytosis, and transendothelial migration. *J. Biol. Chem.* **281,** 12,277–12,288.

18. Cherny, V. V., Henderson, L. M., and DeCoursey, T. E. (1997) Proton and chloride currents in Chinese hamster ovary cells. *Membr. Cell Biol.* **11,** 337–340.

19. Demaurex, N., Grinstein, S., Jaconi, M., Schlegel, W., Lew, D. P., and Krause, K.-H. (1993) Proton currents in human granulocytes: regulation by membrane potential and intracellular pH. *J. Physiol.* **466,** 329–344.

20. DeCoursey, T. E. and Cherny V. V. (1994) Voltage-activated hydrogen ion currents. *J. Membr. Biol.* **141,** 203–223.

21. DeCoursey, T. E. and Cherny, V. V. (1996) Effects of buffer concentration on voltage-gated H^+ currents: does diffusion limit the conductance? *Biophys. J.* **71,** 182–193.

22. Gordienko, D. V., Tare, M., Parveen, S., Fenech, C. J., Robinson, C., and Bolton, T. B. (1996) Voltage-activated proton current in eosinophils from human blood. *J. Physiol.* **496,** 299–316.

23. Schrenzel, J., Lew, D. P., and Krause, K.-H. (1996) Proton currents in human eosinophils. *Am. J. Physiol.* **271,** C1861–C1871.

24. Cherny, V. V. and DeCoursey, T. E. (1999) pH-dependent inhibition of voltage-gated H^+ currents in rat alveolar epithelial cells by Zn^{2+} and other divalent cations. *J. Gen. Physiol.* **114,** 819–838.

25. Femling, J. K., Cherny, V. V., Morgan, D., et al. (2006) The antibacterial activity of human neutrophils and eosinophils requires proton channels but not BK channels. *J. Gen. Physiol.* **127,** 659–672.

26. Grinstein, S., Romanek, R., and Rotstein, O. D. (1994) Method for manipulation of cytosolic pH in cells clamped in the whole cell or perforated patch configurations. *Am. J. Physiol.* **267,** C1152–C1159.

27. DeCoursey, T. E. and Cherny, V. V. (1998) Temperature dependence of voltage-gated H^+ currents in human neutrophils, rat alveolar epithelial cells, and mammalian phagocytes. *J. Gen. Physiol.* **112,** 503–522.

28. Chabala, L. D., Sheridan, R. E., Hodge, D. C., Power, J. N., and Walsh, M. P. (1985) A microscope stage temperature controller for the study of whole-cell or single-channel currents. *Pflügers Arch.* **404,** 374–377.

29. Rae, J. L. and Levis, R. A. (1984) Patch voltage clamp of lens epithelial cells: theory and practice. *Mol. Physiol.* **6,** 115–162.

30. Fenwick, E. M., Marty, A., and Neher, E. (1982) A patch clamp study of bovine chromaffin cells and of their sensitivity to acetylcholine. *J. Physiol.* **331,** 577–597.

31. Lindau, M. and Fernandez, J. M. (1986) IgE-mediated degranulation of mast cells does not require opening of ion channels. *Nature* **319,** 150–153.

32. Rae, J., Cooper, K., Gates, P., and Watsky, M. (1991) Low access resistance perforated patch recordings using amphotericin B. *J. Neurosci. Methods* **37,** 15–26.

33. Horn, R. and Marty, A. (1988) Muscarinic activation of ionic currents measured by a new whole cell recording method. *J. Gen. Physiol.* **92,** 145–159.

34. Fan, J. S. and Palade, P. (1998) Perforated patch recording with β-escin. *Pflügers Arch.* **436,** 1021–1023.

35. Falke, L. C., Gillis, K. D., Pressel, D. M., and Misler, S. (1989) 'Perforated patch recording' allows long-term monitoring of metabolite-induced electrical activity and voltage-dependent Ca^{2+} currents in pancreatic islet B cells. *FEBS Lett.* **251,** 167–172.

36. Chung, I. and Schlichter, L. C. (1993) Criteria for perforated-patch recordings: ion currents versus dye permeation in human T lymphocytes. *Pflügers Arch.* **424,** 511–515.

37. Strauss, U., Herbrik, M., Mix, E., Schubert, R., and Rolfs, A. (2001) Whole-cell patch-clamp: true perforated or spontaneous conventional recordings? *Pflügers Arch.* **442,** 634–638.

38. Bánfi, B., Schrenzel, J., Nüsse, O., et al. (1999) A novel H^+ conductance in eosinophils: unique characteristics and absence in chronic granulomatous disease. *J. Exp. Med.* **19,** 183–194.

39. Robertson, A. K., Cross, A. R., Jones, O. T. G., and Andrew, P. W. (1990) The use of diphenylene iodonium, an inhibitor of NADPH oxidase, to investigate the antimicrobial action of human monocyte derived macrophages. *J. Immunol. Methods* **133,** 175–179.

40. DeCoursey, T. E. and Cherny, V. V. (1994) Na^+-H^+ antiport detected through hydrogen ion currents in rat alveolar epithelial cells and human neutrophils. *J. Gen. Physiol.* **103,** 755–785.

41. DeCoursey, T. E. (1991) Hydrogen ion currents in rat alveolar epithelial cells. *Biophys. J.* **60,** 1243–1253.

42. Byerly, L., Meech, R., and Moody, W. Jr. (1984) Rapidly activating hydrogen ion currents in perfused neurones of the snail, *Lymnaea stagnalis. J. Physiol.* **351,** 199–216.

43. Humez, S., Fournier, F., and Guilbault, P. (1995) A voltage-dependent and pH-sensitive proton current in *Rana esculenta* oocytes. *J. Membr. Biol.* **147,** 207–215.

44. Hille, B. (2001) *Ion Channels of Excitable Membranes, 3rd edition*. Sinauer, MA.

45. Cherny, V. V., Murphy, R., Sokolov, V., Levis, R. A., and DeCoursey, T. E. (2003) Properties of single voltage-gated proton channels in human eosinophils estimated by noise analysis and direct measurement. *J. Gen. Physiol.* **121,** 615–628.

46. Neher, E. and Stevens, C. F. (1977) Conductance fluctuations and ionic pores in membranes. *Annu. Rev. Biophys. Bioeng.* **6,** 345–381.

47. Demaurex, N., Monod, A., Lew, D. P., and Krause, K.-H. (1994) Characterization of receptor-mediated and store-regulated Ca^{2+} influx in human neutrophils. *Biochem. J.* **297,** 595–601.

48. Zweifach, A. and Lewis, R. S. (1993) Mitogen-regulated Ca^{2+} current of T lymphocytes is activated by depletion of intracellular Ca^{2+} stores. *Proc. Natl. Acad. Sci. USA* **90,** 6295–6299.

49. Cross, A. R., Higson, F. K., Jones, O. T. G., Harper, A. M., and Segal, A. W. (1982) The enzymic reduction and kinetics of oxidation of cytochrome b_{-245} of neutrophils. *Biochem. J.* **204,** 479–485.

50. Murphy, R. and DeCoursey, T. E. (2006) Charge compensation during the phagocyte respiratory burst. *Biochim. Biophys. Acta* **1757,** 996–1011.

51. Henderson, L. M., Chappell, J. B., and Jones, O. T. G. (1987) The superoxide-generating NADPH oxidase of human neutrophils is electrogenic and associated with an H^+ channel. *Biochem. J.* **246,** 325–329.

52. Geiszt, M., Kapus, A., Nemet, K., Farkas, L., and Ligeti, E. (1997) Regulation of capacitative Ca^{2+} influx in human neutrophil granulocytes. Alterations in chronic granulomatous disease. *J. Biol. Chem.* **272,** 26,471–26,478.

53. Jankowski, A. and Grinstein, S. (1999) A noninvasive fluorometric procedure for measurement of membrane potential. *J. Biol. Chem.* **274,** 26,098–26,104.

54. Bankers-Fulbright, J. L., Gleich, G. J., Kephart, G. M., Kita, H., and O'Grady, S. M. (2003) Regulation of eosinophil membrane depolarization during NADPH oxidase activation. *J. Cell Sci.* **116,** 3221–3226.

55. Rada, B. K., Geiszt, M., Káldi, K., Tímár, C., and Ligeti, E. (2004) Dual role of phagocytic NADPH oxidase in bacterial killing. *Blood* **104,** 2947–2953.

56. Demaurex, N. and Petheõ, G. L. (2005) Electron and proton transport by NADPH oxidases. *Philos. Trans. R. Soc. Lond. B. Biol. Sci.* **360,** 2315–2325.

57. Henderson, L. M., Chappell, J. B., and Jones, O. T. G. (1987) The superoxide-generating NADPH oxidase of human neutrophils is electrogenic and associated with an H^+ channel. *Biochem. J.* **886,** 425–433.

58. DeCoursey, T. E. (2004) During the respiratory burst, do phagocytes need proton channels or potassium channels or both? *Science's STKE* **2004,** pe21(http://stke.sciencemag.org/cgi/content/full/sigtrans;2004/233/pe21?ijkey=lPoLtn7p09qzw&keytype=ref&siteid=sigtrans).

59. Rae, J. L. and Levis, R. A. (1984) Patch voltage clamp of lens epithelial cells: theory and practice. *Mol. Physiol.* **6,** 115–162.

60. Sigworth, F. J. (1995) Electronic design of the patch clamp, in *Single-Channel Recording, 2nd edition* (Sakmann, B., and Neher, E. eds.), Plenum, NY, pp. 95–127.
61. Byerly, L. and Moody, W. J. (1986) Membrane currents of internally perfused neurons of the snail, *Lymnaea stagnalis*, at low intracellular pH. *J. Physiol.* **376,** 477–491.
62. Kapus, A., Romanek, R., Qu, A. Y., Rotstein, O. D., and Grinstein, S. (1993) A pH-sensitive and voltage-dependent proton conductance in the plasma membrane of macrophages. *J. Gen. Physiol.* **102,** 729–760.
63. Levis, R. and Rae, J. L. (1992) Constructing a patch clamp setup. *Meth. Enzymol.* **207,** 14–66.
64. Levis, R. A. and Rae, J. L. (1993) The use of quartz patch pipettes for low noise single channel recording. *Biophys. J.* **65,** 1666–1677.
65. Cota, G. and Armstrong, C. M. (1988) Potassium channel "inactivation" induced by soft-glass pipettes. *Biophys. J.* **55,** 107–109.
66. Rojas, L. and Zuazaga, C. (1988) Influence of the patch pipette glass on single acetylcholine channels recorded from *Xenopus* myocytes. *Neurosci. Lett.* **88,** 39–44.
67. Neher, E. (1992) Correction for liquid junction potentials in patch clamp experiments. *Methods Enzymol.* **207,** 123–131.
68. Ng, B. and Barry, P. H. (1995) The measurement of ionic conductivities and mobilities of certain less common organic ions needed for junction potential corrections in electrophysiology. *J. Neurosci. Methods* **56,** 37–41.
69. Miller, C., Moczydlowski, E., Latorre, R., and Phillips, M. (1985) Charybdotoxin, a protein inhibitor of single Ca^{2+}-activated K^+ channels from mammalian skeletal muscle. *Nature* **313,** 316–318.
70. Galvez, A., Gimenez-Gallego, G., Reuben, J. P., et al. (1990) Purification and characterization of a unique, potent, peptidyl probe for the high conductance calcium-activated potassium channel from venom of the scorpion *Buthus tamulus*. *J. Biol. Chem.* **265,** 11,083–11,090.
71. Hermann, A. and Erxleben, C. (1987) Charybdotoxin selectively blocks small Ca-activated K channels in *Aplysia* neurons. *J. Gen. Physiol.* **90,** 27–47.
72. Sands, S. B., Lewis, R. S., and Cahalan, M. D. (1989) Charybdotoxin blocks voltage-gated K^+ channels in human and murine T lymphocytes. *J. Gen. Physiol.* **93,** 1061–1074.
73. Cherny, V. V., Markin, V. S., and DeCoursey, T. E. (1995) The voltage-activated hydrogen ion conductance in rat alveolar epithelial cells is determined by the pH gradient. *J. Gen. Physiol.* **105,** 861–896.

12

Analysis of Neutrophil Apoptosis

Emma L. Taylor, Adriano G. Rossi, Ian Dransfield, and Simon P. Hart

Summary

Neutrophil-derived granule enzymes, oxidants, and mediators have been implicated in the pathogenesis of a variety of inflammatory diseases. Neutrophil apoptosis is associated with the loss of expression of adhesion molecules and greatly reduced responsiveness to external stimuli, so that these cells become functionally isolated from their environment. In contrast with necrosis, apoptosis is associated with preservation of plasma membrane integrity, so that release of harmful neutrophil contents is limited, and the inert neutrophils are phagocytosed by local macrophages. Furthermore, phagocytosis of apoptotic neutrophils by human macrophages in vitro suppresses release of macrophage-derived pro-inflammatory mediators. In this way, by downregulating neutrophil functions and triggering "silent" clearance by phagocytes, apoptosis provides a mechanism for the safe disposal of potentially destructive inflammatory cells. Many of the molecular events involved in the apoptosis pathway have been identified and several complementary methods may be employed to identify and quantitate neutrophil apoptosis. This chapter will discuss analysis of neutrophil morphology, DNA fragmentation, membrane changes, mitochondrial alterations, caspase activation, and phagocytosis of apoptotic neutrophils by macrophages.

Key Words: Apoptosis; caspase; necrosis; DNA fragmentation; phagocytosis.

1. Introduction

Once recruited in large numbers to a site of tissue damage, neutrophils die locally by undergoing apoptosis, which is the process of programmed cell death in vertebrates. The removal of apoptotic neutrophils is thought to be an important step in preventing the release of granule contents and chemotactic factors into the extracellular fluid, thereby halting further injury during resolution of inflammation. The characteristic morphological features of apoptosis in human neutrophils were first reported by Savill et al. in 1989 *(1)*. Until that time, it

From: *Methods in Molecular Biology, vol. 412: Neutrophil Methods and Protocols*
Edited by: M. T. Quinn, F. R. DeLeo, and G. M. Bokoch © Humana Press Inc., Totowa, NJ

was widely believed that neutrophils died by undergoing necrosis, the classical mode of cell death that follows irreversible cell injury *(2)*, but which is associated with disruption of the plasma membrane and release of intracellular contents, which in the case of the neutrophil may promote further tissue injury. Savill et al. demonstrated that when human neutrophils were aged in culture for 10 h at 37°C a proportion of cells displayed densely condensed nuclei and vacuolated cytoplasm. Electron microscopy of the same cell population revealed chromatin condensation, nucleolar prominence, vacuolated cytoplasm, and intact cytoplasmic granules. In contrast to apoptosis in other cell lineages, substantial cell shrinkage, membrane blebbing, and formation of apoptotic bodies are not prominent features of apoptotic neutrophils when observed by light microscopy. Neutrophils are terminally differentiated cells and undergo constitutive apoptosis, so that after culture in vitro for 20 h, typically >50% of the cells will have undergone apoptosis. In vitro aging of neutrophils has been successfully used to demonstrate the pro-apoptotic or anti-apoptotic effects of numerous mediators, bacterial products, and pharmaceutical agents *(3–5)*. Potential apoptosis-regulating agents are incubated with a suspension of freshly isolated human neutrophils and apoptosis is assessed over time. Early time points (e.g., 4 h) are used to look for induction of apoptosis, and later time points (e.g., >20 h) for assessing inhibition of apoptosis. It should be appreciated that the methods of neutrophil isolation and culture can profoundly influence rates of constitutive apoptosis and the response to external apoptosis-modulating reagents. Human neutrophil isolation may involve multiple steps including leukocyte sedimentation (using Percoll, Hypaque-Ficoll, gelatine, etc.), many centrifugation steps, temperature changes, and red cell lysis. Each step has the potential to influence neutrophil responsiveness and apoptosis. Furthermore, contamination of neutrophil preparations, even with low numbers of monocytes or eosinophils, can influence rates of neutrophil apoptosis. In order to overcome this variable, some researchers have recommended neutrophil isolation protocols based on antibody-dependent depletion of contaminating cells *(6)*, which in turn requires careful quality-control steps. Other variables that must be taken into account when investigating human neutrophil apoptosis include the pH of media, glucose concentration, oxygen tension, and cell density during culture *(7,8)*.

Substantial research activity has identified many of the molecular events involved in the apoptosis pathway, which in turn have led to a variety of different techniques that may be used to identify and quantify apoptotic neutrophils in the laboratory. The inevitable consequence of initiation of the apoptosis pathway is disassembly of the intracellular machinery which is mediated principally by effector caspases, a family of mammalian cysteine proteases which cleave specific aspartate-containing sites in a variety of target proteins. Caspases exist constitutively in the cell as zymogens, and upon an initiating pro-apoptotic stimulus they are sequentially activated in a tightly regulated, irreversible cascade. Caspases have been classified into initiators (caspases 2, 8, 9, 10, 12) and effectors (caspases 3, 6, 7, 14) *(9)*. Initiation of the biochemical pathways of apoptotic cell death can broadly be defined as occurring via either intrinsic or extrinsic mechanisms, although it is now recognized that a certain amount of cross-talk occurs between the two. The intrinsic pathway is recruited in response to cell damage or stress, and involves changes at the level of the mitochondria including processing and activation of pro-apoptotic Bcl-2 family member proteins (e.g., Bax), release of cytochrome c from mitochondria, formation of the apoptosome (composed of cytochrome c, APAF-1, and procaspase 9), and activation of caspase 9. The extrinsic pathway of apoptosis results from engagement of cell surface death receptors (e.g., Fas [CD95], tumor necrosis factor [TNF] receptor) leading to formation of death-inducing signaling complex (DISC), and involves cleavage and activation of procaspase 8 as the initiator caspase. Both mechanisms result in downstream processing and activation of caspase 3, which is the predominant effector caspase responsible for the cleavage of cell components *(10)*. Normal cell membranes exhibit marked phospholipid asymmetry, with phosphatidylcholine and sphingomyelin predominantly on the external layer and most of the phosphatidylethanolamine and phosphatidylserine on the inner layer. Apoptosis is associated with loss of normal asymmetrical distribution of aminophospholipids, and in particular phosphatidylserine becomes exposed on the exterior of the cell, where it may mediate binding of opsonins and permit recognition of apoptotic cells by phagocytes *(11)*. Activation of an endonuclease during apoptosis leads to internucleosomal digestion of DNA into monomers or oligomers or nucleosomal DNA of 200 base pairs and multiples thereof, and a regular "ladder-type" pattern of oligonucleosomal DNA fragmentation on agarose gel electrophoresis is commonly considered to define apoptosis. Apoptotic neutrophils are recognized and ingested by macrophages *(11)*, and it has been shown that macrophage phagocytic capacity for apoptotic neutrophils may be influenced by cytokines *(12)*, prostaglandins *(13)*, glucocorticoid hormones *(14)*, and interaction of surface adhesion molecules with neighboring cells and extracellular matrix components *(15)*. Furthermore, apoptotic cells may become opsonised, whereby naturally occurring soluble opsonins bind to the cell surface and initiate phagocytosis *(16)*.

Neutrophil apoptosis may be assessed in the laboratory by analysis of cell morphology by microscopy, assessing plasma membrane alterations, measuring loss of mitochondrial membrane potential, looking for evidence of caspase activation, and gel electrophoresis of genomic DNA. Methods for quantifying macrophage phagocytosis of apoptotic neutrophils are also described in this chapter.

2. Materials

2.1. Cytocentrifuge Preparations for Light Microscopy

1. Cytocentrifuge chambers, filter cards, glass slides, and coverslips.
2. Methanol, Diff-Quik™ stains (Baxter Healthcare), and distrene-80, plasticizer, xylene.
3. Iscove's modified Dulbecco's medium (IMDM) (Invitrogen).

2.2. Plasma Membrane Alterations

1. 96-well U-bottom flexible polyvinyl chloride plate (Becton Dickinson Labware).
2. Annexin binding buffer: 140 mmol NaCl, pH 7.4, 20 mM HEPES, 4 mM CaCl$_2$.
3. Stock solution of 1 mg/mL propidium iodide in phosphate-buffered saline (PBS) (Sigma-Aldrich) (*see* **Note 1**).
4. Cytometry buffer: cation-free PBS containing 0.1% bovine serum albumin. Store at 4°C.
5. For CD16 staining: 3G8 (anti-FcγRIIIB) F(ab')$_2$ fragments (Ancell, Bayport, MN); fluorescein isothiocyanate (FITC)-conjugated F(ab')$_2$ goat anti-mouse immunoglobulin (Ig) (Dako); nonbinding murine IgG$_1$ control (Dako).

2.3. Mitochondrial Permeability

Mit-E-Ψ mitochondria permeability detection kit (Biomol International): contains lyophilized Mit-E-ψ reagent, to be reconstituted in 125 μL of dimethylsulfoxide (DMSO) per vial then stored at −20°C (avoid repeated freeze–thawing), and 10X assay buffer, to be diluted before use. Once diluted, assay buffer can be stored at 4°C for up to 7 d.

2.4. Western Blotting for Caspases

1. Tris-buffered saline (TBS): dissolve 8 g of NaCl, 0.2 g of KCl, and 3 g Trizma base in 800 mL of distilled water. Adjust the pH to 7.4 with HCl and add distilled water to 1 L. Autoclave aliquots and store at room temperature.
2. Protease inhibitor buffer (*see* **Note 2**): to 0.3 mL of TBS add 0.5 mL of protease inhibitor cocktail for use with mammalian cell and tissue extracts (1:50 dilution in TBS) (Sigma-Aldrich), supplemented with 4-(2-aminoethyl)-benzenesulfonylfluoride (AEBSF; 20 μL of 400 mM stock in H$_2$O), aprotinin (20 μL of 0.15 μM stock in H$_2$O), leupeptin (20 μL of 20 mM stock in H$_2$O), pepstatin A (40 μL of 0.75 μM stock in methanol), sodium orthovanadate (20 μL of 1 M stock, pH 10, boiled), benzamidine (20 μL of 0.5 M stock), levamisole (20 μL of 2 M stock) and β-glycerophosphate (60 μL of 3.33 M stock in H$_2$O) (all from Sigma-Aldrich).
3. 10% NP-40 in TBS.
4. Sample Buffer (for 10 mL of 4X Sample Buffer): 5.5 mL of 50% glycerol, 2 mL of 20% sodium dodecyl sulfate (SDS), 2.5 mL of 0.5 M Tris-HCl (pH 6.8), 20 μL of 1% (w/v in ethanol) bromophenol blue. For reducing buffer, add 100 μL of β-mercaptoethanol in a fume hood.

5. 12% acrylamide separating gel (*see* **Note 3**): 2.5 mL of 1.5 *M* Tris-HCl (pH 8.8), 4 mL of 30% acrylamide/bis-acrylamide solution (37.5:1 ratio), 3.3 mL of dH$_2$O, 100 µL of 10% (w/v) SDS. Add 30 µL of *N,N,N',N'*-tetramethylethylenediamine (TEMED) and 60 µL of 10% (w/v) ammonium persulphate to initiate polymerization.
6. dH$_2$O-saturated butanol: mix approx 75 mL of butan-2-ol with approx 25 mL of distilled water and shake vigorously. Allow to separate, and store at room temperature. Use the top layer.
7. 4.5% acrylamide stacking gel: 2.5 mL of 0.5 *M* Tris-HCl (pH 6.8), 1.5 mL of 30% acrylamide/bis-acrylamide solution (37.5:1 ratio), 5.8 mL of dH$_2$O, 100 µL of 10% (w/v) SDS. Add 30 µL of TEMED and 60 µL of 10% (w/v) ammonium persulphate to initiate polymerization.
8. Running buffer: 25 m*M* Tris, 192 m*M* glycine pH 8.3, 0.1% SDS.
9. Transfer buffer: 25 m*M* Tris, 192 m*M* glycine pH 8.3, 20% methanol.
10. Hybond-P polyvinylidene difluoride (PVDF) membrane (Amersham Biosciences).
11. Benchmark™ prestained molecular weight standards (Invitrogen).
12. Blocking buffer: TBS containing 0.1% Tween 20 (polysorbat 20)® and 5% Marvel dried milk powder.
13. Primary antibody: anti-caspase-3 (active) polyclonal rabbit antibody (Biovision Research Products).
14. Secondary antibody: horseradish peroxidase (HRP)-conjugated polyclonal goat anti-rabbit Ig (Dako).
15. Enhanced chemiluminescence (ECL) reagents (Amersham Biosciences).

2.5. Fluorimetric Caspase Kit

Homogeneous Caspases Assay Kit (Roche Diagnostics) (*see* **Note 4**): contains 1X incubation buffer, stock caspase substrate solution (500 µ*M* DEVD-R110 in DMSO), positive control (lysate from apoptotic camptothecin-treated U937 cells), and R110 standard for calibration curve construction (1 m*M* in DMSO).

2.6. Caspase Profiling Plate

1. ApoAlert™ Caspase Profiling Plate (Clontech): contains 96-well microplate with immobilized substrates for caspase 2 (VDVAD-AMC), caspase 3 (DEVD-AMC), caspase 8 (IETD-AMC), and caspase 9 (LEHD-AMC) in 24 wells each, lysis Buffer, 2X Reaction Buffer, 100X dithiothreitol (DTT) solution, and inhibitors of caspases 2, 3, 8, and 9.
2. 10% NP-40 in TBS.

2.7. Gel Electrophoresis for DNA Fragmentation

1. Wizard® genomic DNA purification kit (Promega).
2. 0.5X TBE running buffer: Make a 5X stock (pH 8.3) with 54 g Trizma base, 27.5 g boric acid, and 20 mL 0.5 *M* ethylenediamine tetraacetic acid (EDTA) (pH 8.0) in 1 L of dH$_2$O.

3. SeaKem LE agarose (Cambrex) for DNA electrophoresis (*see* **Note 5**).
4. 10 mg/mL ethidium bromide solution (Sigma-Aldrich).
5. 6X blue/orange loading dye (Promega).

2.8. Propidium Iodide Intercalation into DNA

1. 96-well flat-bottom flexible polyvinyl chloride plate (Becton Dickinson Labware).
2. 70% ethanol in distilled H_2O.
3. 0.5 mg/mL solution of RNase A (Sigma-Aldrich).
4. Solution of 0.1 mg/mL propidium iodide in PBS (Sigma-Aldrich).

2.9. Terminal Deoxynucleotidyltransferase-Mediated dUTP Nick End Labeling Staining

1. 96-well flat-bottom flexible polyvinyl chloride plate (Becton Dickinson Labware).
2. *In Situ* Cell Death Detection Kit, Fluorescein (Roche Diagnostics): contains 10X enzyme solution (terminal deoxynucleotidyltransferase [TdT]) in storage buffer and 1X labeled nucleotide mixture in reaction buffer. This protocol also requires PBS (wash buffer), 3% H_2O_2 in methanol (blocking solution), 4% paraformaldehyde in PBS at pH 7.4 (fixation buffer; freshly prepared), and 0.1% Triton X-100 in 0.1% sodium citrate (permeabilization buffer; freshly prepared).

2.10. Plate-Based Phagocytosis Assay

1. 2.5% glutaraldehyde (Sigma-Aldrich) in PBS.
2. 0.1 mg/mL dimethoxybenzidine (o-diansidine) (Sigma-Aldrich) in PBS, made up fresh from 1 mg/mL frozen stock.
3. 30% hydrogen peroxide solution (Sigma-Aldrich).
4. Costar 48-well TC-treated microplates (Fisher Scientific).

2.11. Flow Cytometry-Based Phagocytosis Assay

1. 5-chloromethylfluorescein diacetate (CMFDA; CellTracker™ Green; Molecular Probes).
2. 0.25% trypsin/1 m*M* EDTA solution (Invitrogen).

3. Methods

3.1. Analysis of Neutrophil Morphology by Light Microscopy

1. Suspend freshly-isolated peripheral blood neutrophils (at least 97% purity) at 4×10^6 cells/mL in IMDM supplemented with 10% autologous serum (*see* **Note 6**).
2. Add 135 µL of neutrophil suspension to wells of a 96-well, flat-bottom flexible plate. To each well, add 15 µL of apoptosis-modifying agents (10X concentration) or buffer control, cover with a lid, and incubate at 37°C in a 5% CO_2 incubator for the desired length of time.
3. Load a cytospin chamber with 100 µL of aged neutrophil suspension.
4. Cytocentrifuge at 300 rpm (30*g*) for 3 min.
5. Air-dry for 5 min.

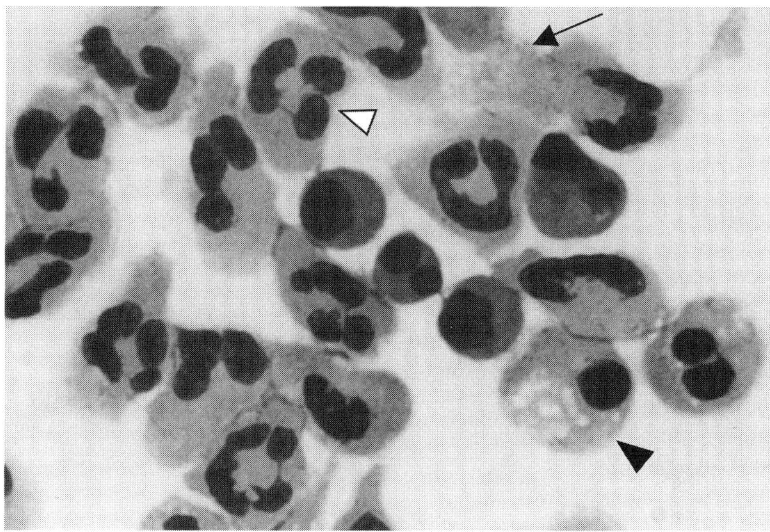

Fig. 1. A cytocentrifuge preparation of human neutrophils aged for 22 h in vitro. Nonapoptotic neutrophils exhibit characteristic multilobed nuclei and an open pattern of chromatin (open arrowhead). Apoptotic neutrophils show condensation of nuclear material into one to three dark-staining nuclear bodies with prominent nucleoli, and often demonstrate cytoplasmic vacuolation (closed arrowhead). Late apoptotic neutrophils appear as "ghosts" with little or no nuclear material (closed arrow).

6. Fix in methanol for 1 min. Drain.
7. Stain in Diff Quik™ solution 1 or equivalent acid dye for 1 min. Drain.
8. Stain in Diff Quik solution 2 or equivalent basic dye for 1 min. Drain and rinse with distilled water.
9. Allow to dry, mount with a drop of DPX, and apply a coverslip. View using a light microscope with a ×40 or ×100 objective and count >200 cells per slide. Apoptotic neutrophils can be identified by their characteristic condensation of nuclear material, prominent nucleoli, and frequently vacuolation of the cytoplasm (**Fig. 1**) (*see* **Note 7**).

3.2. Annexin V Staining

Apoptosis is associated with the loss of phospholipid asymmetry and the exposure of phosphatidylserine on the outer surface of the plasma membrane. Annexin V is a protein that binds specifically to phophatidylserine in the presence of Ca^{2+}, and its fluorescent derivatives can be used to identify apoptotic cells using a simple and rapid flow cytometric method.

1. Suspend freshly isolated peripheral blood neutrophils (at least 97% purity) at 4×10^6 cells/mL in IMDM supplemented with 10% autologous serum (*see* **Note 6**).

2. Add 135 μL of neutrophil suspension to wells of a 96-well, flat-bottom flexible plate.
3. Add 15 μL of apoptosis-modifying agents (10X concentration) or buffer control to each well.
4. Cover plate, and incubate at 37°C in a 5% CO_2 incubator for the desired length of time.
5. Resuspend cells by gentle pipetting and transfer 100 μL of the cell suspension to a 96-well, U-bottom flexible plate (*see* **Note 8**) and centrifuge at 200*g* for 2 min at room temperature. Discard the supernatants.
6. Vortex the plate for 5 s to disrupt the cell pellet.
7. Resuspend the cells in 50 μL of recombinant human Annexin V-FITC (Caltag) freshly diluted 1:50 in annexin binding buffer (*see* **Note 9**).
8. Incubate for 5 min at room temperature. No wash is required.
9. Co-label membrane permeable necrotic or late apoptotic cells by staining with the membrane impermeant nuclear dye propidium iodide (PI; 5 μg/mL) 30 s prior to running the sample.
10. Analyze by flow cytometry using FL-1/FL-2 channel analysis following appropriate compensation (*see* **Note 10**).
11. Live cells are annexin V-negative and PI-negative; apoptotic cells are annexin V-positive and PI-negative; necrotic cells are both annexin V-positive and PI-positive (**Fig. 2**).

3.3. Loss of CD16 Expression

Neutrophil apoptosis is associated with a marked downregulation of FcγRIII (CD16) *(21)*. A simple flow cytometric method using an anti-CD16 monoclonal antibody will reliably discriminate apoptotic from non-apoptotic neutrophils in a mixed cell population.

1. Suspend freshly isolated peripheral blood neutrophils (at least 97% purity) at 4×10^6 cells/mL in IMDM supplemented with 10% autologous serum (*see* **Note 6**).
2. Add 90 μL of neutrophil suspension to wells of a 96-well, flat-bottom flexible plate.
3. Add10 μL of apoptosis-modifying agents (10X concentration) or buffer control to each well.
4. Cover plate, and incubate at 37°C in a 5% CO_2 incubator for the desired length of time.
5. Resuspend cells by gentle pipetting and transfer 100 μL of the cell suspension to a 96-well U-bottom flexible plate (*see* **Note 8**) and centrifuge at 200*g* for 2 min at 4°C. Discard the supernatants.
6. Vortex the plate for 5 s to disrupt the cell pellet.
7. Incubate neutrophils with 20 μg/mL monoclonal murine IgG_1 anti-CD16 (clone 3G8) or nonbinding control IgG_1 in 50 μL cytometry buffer on ice for 30 min.
8. Add 75 μL of cytometry buffer. Centrifuge at 200g for 2 min at 4°C. Discard the supernatants and vortex the plate for 5 s to disrupt the cell pellet.

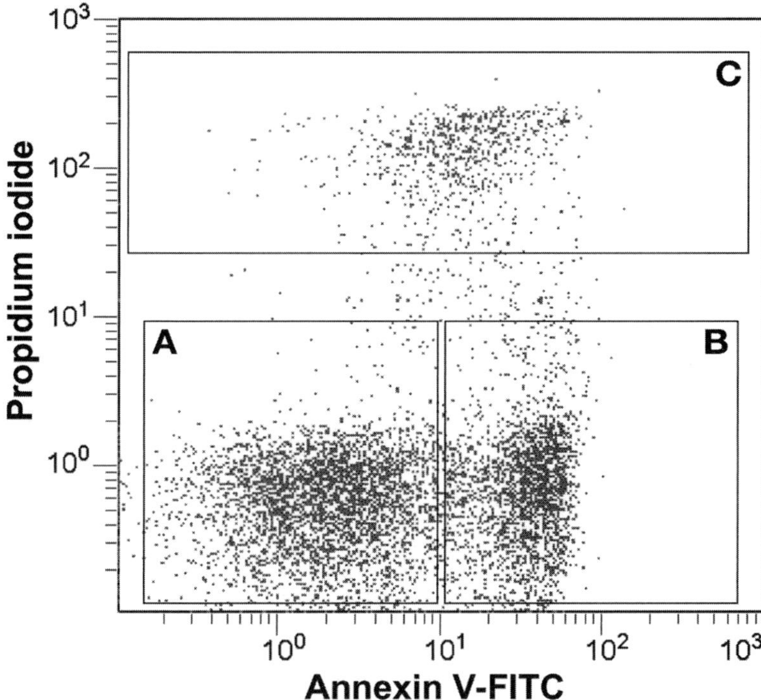

Fig. 2. Annexin V staining. An example two-colour flow cytometry dot plot of aged human neutrophils stained with annexin V-fluorescein isothiocyanate (FL-1, *x*-axis) and propidium iodide (FL-2, y-axis) is shown. The regions indicated represent **(A)** nonapoptotic neutrophils, **(B)** apoptotic neutrophils, and **(C)** late apoptotic or necrotic neutrophils.

9. Add 125 µL of cytometry buffer. Centrifuge at 200*g* for 2 min at 4°C. Discard the supernatants and vortex the plate for 5 s to disrupt the cell pellet.
10. Add 50 µL of 1:40 F(ab')$_2$ goat anti-mouse Ig-FITC in cytometry buffer on ice for 30 min.
11. Repeat **steps 8** and **9**. Resuspend the cells in 50 µL of cytometry buffer.
12. Analyse cells using a flow cytometer. Apoptotic neutrophils have low FL-1 channel fluorescence; non-apoptotic neutrophils have high FL-1 fluorescence (*see* **Note 11**).

3.4. Measurement of Mitochondrial Membrane Potential Using JC-1.

Upon activation of apoptosis-promoting proteins such as Bid, Bax, and Bak, the potential across the mitochondrial membrane ($\Delta\Psi_M$) is lost and pores are formed which allow the passage of proteins from the mitochondria into the cytosol, leading to activation of caspase 9 and subsequently caspase 3 *(22)*. Changes in neutrophil mitochondrial membrane potential may be measured using JC-1 (5,5',6,6'-tetrachloro-1,1',3,3'-tetra-ethyl-benzimidazocarbo-cyaniniodide; Molecular Probes), a cationic dye which exhibits membrane potential-dependent accumulation in mitochondria indicated by a fluorescence emission shift from green (525 nm) to red (590 nm). In viable cells with intact mitochondrial membrane potential, JC-1 is able to enter mitchondria and polymerise, where it fluoresces in the red (FL-2) channel. However, when $\Delta\Psi_M$ is lost during apoptosis, JC-1 is unable to penetrate the mitochondrial membrane and it remains in monomeric form in the cytosol and fluoresces in the green (FL-1) channel. Apoptosis-associated mitochondrial depolarisation is indicated by a decrease in the red/green fluorescence intensity ratio *(23)* (*see* **Note 12**). Although not described here, the change in fluorescence of JC-1 associated with apoptosis can also be quantified flow cytometrically.

1. This protocol assumes the use of a Mit-E-Ψ mitochondria permeability detection kit (Biomol International UK, Exeter, UK)
2. Suspend freshly isolated peripheral blood neutrophils (at least 97% purity) at 2×10^6 cells/mL in IMDM supplemented with 10% autologous serum (*see* **Note 6**).
3. Dispense 1 mL of neutrophil suspension (2×10^6 cells) into 2-mL round-bottomed polypropylene tubes and incubate ± apoptosis-modifying agents at 37°C for the desired length of time (*see* **Note 13**).
4. Centrifuge at 220g for 5 min and discard the supernatants.
5. Gently resuspend each cell pellet in 1 mL of freshly prepared Mit-E-Ψ (JC-1) solution and incubate at 37°C for 10 min.
6. Add 2 mL of assay buffer added to each tube and centrifuge at 220g for 5 min. Discard the supernatants.
7. Resuspend the cell pellets in 1 mL of assay buffer and centrifuge at 220g for 5 min. Discard the supernatants.
8. Resuspend the cell pellets in 1 mL of assay buffer.
9. Transfer duplicate aliquots of 100 µL from each sample to a black 96-well plate.
10. Measure fluorescence using a fluorescence microtiter plate reader (excitation of 485 nm; emission of 590 nm for red fluorescence and 530 nm for green fluorescence).
11. Calculate the ratio of red:green fluorescence.

3.4. Caspase Activation

Caspase 3 is the archetypal effector caspase which has received the most attention in studies of apoptosis. Caspase 3 is synthesized as an inactive zymogen (32 kDa) that is cleaved in cells undergoing apoptosis into large (17 kDa) and small (12 kDa) subunits that associate to form the active enzyme. Western blotting can be used to observe the increase in active caspase or disappearance of inactive procaspase on treatment with an apoptotic stimulus. For studying caspase 3, several antibodies are appropriate for Western blot analysis; a rabbit polyclonal antibody that detects only the large (p17) cleaved fragment of active

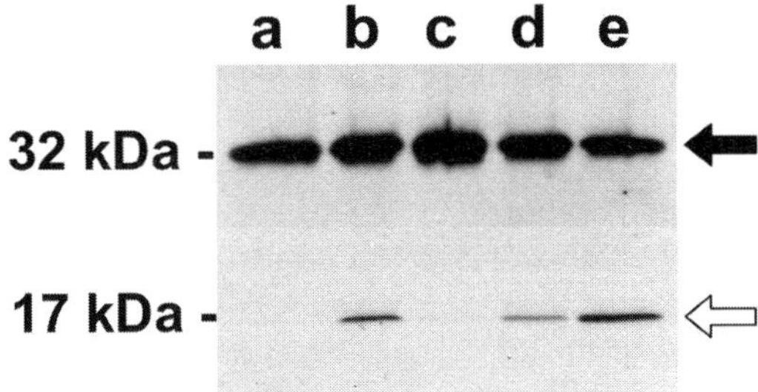

Fig. 3. Western blotting for caspase 3. Human neutrophils were incubated in Iscove's modified Dulbecco's medium + 10% autologous serum at 37°C under the following conditions: (**A**) medium only (2 h); (**B**) 10 ng/mL tumor necrosis factor (TNF)-γ (2 h); (**C**) 0.1 μg/mL gliotoxin (2 h); (**D**) 0.1 μg/mL gliotoxin + 10 ng/mL TNF-γ (2 h); (**E**) medium only (20 h). Cells were lysed and subjected to sodium dodecyl sulfate-polyacrylamide gel electrophoresis and Western blotting and probed for whole caspase 3 (closed arrow) and cleaved active caspase 3 (open arrow).

caspase 3, a mouse monoclonal antibody that detects only the procaspase (p32) form, and a mouse monoclonal antibody and a rabbit polyclonal antibody to detect both the p32 zymogen and the p17 active subunit (Biovision Research Products). An example of Western blotting for active caspase 3 is shown in **Fig. 3**.

3.4.1. Western Blotting for Caspases

1. These instructions assume the use of a minigel apparatus such as the Atto Corporation Dual Mini Slab Chamber (GRI Ltd, Braintree, UK) or the mini-PROTEAN electrophoresis system (BioRad Laboratories).
2. Suspend freshly-isolated peripheral blood neutrophils (at least 97% purity) at 5×10^6 cells/mL in IMDM supplemented with 10% autologous serum (*see* **Note 6**).
3. Dispense 1 mL of neutrophil suspension (5×10^6 cells) into 2-mL round-bottomed polypropylene tubes and incubate ± apoptosis-modifying agents at 37°C for the desired length of time (*see* **Note 13**).
4. Prepare and pour a 12% acrylamide gel. Overlay with approx 0.5 mL of dH_2O-saturated butanol whilst polymerization occurs (allow 30 min). Rinse off butanol with dH_2O and pour a 4.5% acrylamide stacking gel on top and insert comb. Allow the stacking gel to polymerise, then assemble the electrophoresis apparatus and remove the comb.
5. Centrifuge the neutrophil suspensions at 220*g* for 5 min at 4°C. Discard the supernatants.
6. Resuspend the cell pellets in protease inhibitor buffer (90 μL per 5×10^6 cells) and incubate on ice for 10 min (*see* **Note 2**).
7. Add 10% NP-40 in TBS (10 μL per 5×10^6 cells), immediately vortex for 5 s, and incubate for a further 10 min on ice.

8. Centrifuge at 20,000*g* for 20 min at 4°C, and retain the detergent-soluble supernatant containing the cytosolic and membrane fractions.
9. Transfer the supernatants to fresh tubes on ice and add 33 µL (per 100 µL) 4X sample buffer. Heat at 95°C for 5 min. Samples may be stored at −20°C prior to SDS-polyacrylamide gel electrophoresis (PAGE).
10. Load samples (20 µL per lane for a minigel) onto a 12% polyacrylamide gel alongside molecular weight standards and run at 80 V for approx 15 min to allow samples to migrate through the stacking gel. Increase the voltage to 160 V through the separating gel until the dye front reaches the bottom of the gel.
11. Transfer proteins from the gel to the PVDF membrane at 70 V for 1 h at room temperature.
12. Wash the membrane in TBS/0.1% Tween 20 for 5 min.
13. Block the membrane in 10 mL of 5% Marvel dried milk powder in TBS/0.1% Tween 20 at room temperature on a rocking platform for 30 min.
14. Wash the membrane in TBS/0.1% Tween 20 for 5 min.
15. Incubate with the primary antibody (anti-caspase-3 [active] polyclonal rabbit antibody) at 1 µg/mL in TBS/0.1% Tween 20 containing 5% Marvel dried milk powder. Incubate in a covered hydrophobic tray (just large enough to accommodate the blot) for 2 h at room temperature on a rocking platform. Use sufficient volume to just cover the blot.
16. Wash membrane three times in TBS/0.1% Tween 20 each for 5 min at room temperature on a rocking platform.
17. Incubate with the secondary antibody (HRP-conjugated polyclonal goat anti-rabbit Ig) diluted 1:2500 in TBS/0.1% Tween 20 containing 5% Marvel dried milk powder. Incubate in a covered hydrophobic tray (just large enough to accommodate the blot) for 2 h at room temperature on a rocking platform. Use sufficient volume to just cover the blot.
18. Wash membrane three times in TBS/0.1% Tween 20 each for 5 min at room temperature on a rocking platform.
19. Develop using ECL according to the manufacturer's instructions.

3.4.2. Fluorometric Homogeneous Caspase Assay

Apoptosis usually occurs via caspase-dependent processes, and an increase in caspase activity gives a general indication of the occurrence of apoptosis. Commercial kits are available in which total caspase activity can be measured (homogeneous caspase assay). However, these kits do not dissect out precisely which caspases are active. They utilize cleavage of a fluorescence-conjugated, modified nonspecific caspase substrate (e.g., VAD-fmk, DEVD) to generate a fluorescent product (e.g., FITC, rhodamine 110), and fluorescence correlates with the extent of total caspase activity.

1. These instructions assume the use of a Homogeneous Caspases Assay Kit (Roche Diagnostics).

2. Suspend freshly isolated peripheral blood neutrophils (at least 97% purity) at 10^6 cells/mL in IMDM supplemented with 10% autologous serum (*see* **Note 6**).
3. Plate out neutrophils (10^5 per well; *see* **Note 14**) and incubate at 37°C ± apoptosis-modifying agents for the required time in a black 96-well microplate (total volume 100 µL).
4. Dilute stock caspase substrate 1:10 in incubation buffer. Add 100 µL of freshly prepared caspase substrate to each well, plus duplicate wells containing medium alone (negative control) and duplicate wells containing positive control lysate.
5. Cover the plate and incubate at 37°C for at least 1 h.
6. Measure fluorescence with a plate reader (excitation, 470–500 nm; emission, 500–560 nm).

3.4.3. Caspase Profiling Assay

Fluorometric assays for specific caspases work in a similar way to the homogeneous caspase assays, but exploit a certain degree of specificity between the substrates of individual caspases or groups of caspases. Fluorescence-conjugated substrates specific for certain caspases (e.g., 2, 3, 8, and 9) are immobilized in a 96-well plate. When cell lysates are added to the wells and incubated with the substrates, the amount of fluorescence generated correlates with the activation of that particular caspase. They can therefore be used to study particular pathways of apoptosis by looking for activity of a caspase protease that is specific for a particular pathway or cell type, e.g., caspase 8 in death receptor-mediated death.

1. These instructions assume the use of the ApoAlert™ Caspase Profiling Plate (*see* **Notes 15** and **16**).
2. Suspend freshly-isolated peripheral blood neutrophils (at least 97% purity) at 2×10^6 cells/mL in IMDM supplemented with 10% autologous serum (*see* **Note 6**).
3. Dispense 1 mL of neutrophil suspension (2×10^6 cells) into 2-mL round-bottomed polypropylene tubes and incubate ± apoptosis-modifying agents at 37°C for the desired length of time (*see* **Note 13**).
4. Centrifuge at 220g for 5 min at 4°C. Discard the supernatants.
5. Resuspend the cells in 400 µL ice-cold 1X lysis buffer and incubate for 10 min on ice.
6. Meanwhile, add 10 µL of DTT per 1 mL of 2X reaction buffer, then add 50 µL to each well of the 96-well caspase profiling plate.
7. Cover the plate with film and incubated for 5 min at 37°C.
8. Vortex the neutrophil lysates, then add 50 µL from each lysate to duplicate wells of each caspase substrate.
9. Cover the plate with film and incubate for 2 h at 37°C.
10. Measure fluorescence using a plate reader (excitation, 380 nm; emission, 460 nm).

3.5. Gel Electrophoresis of DNA

A characteristic event of apoptosis is the endonuclease-mediated cleavage of DNA at regular intervals along its length, thus generating single-nucleosome

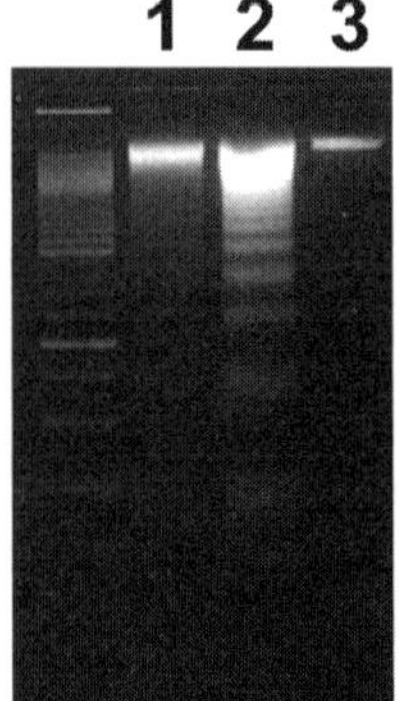

Fig. 4. Internucleosomal DNA fragmentation in neutrophils. Neutrophils were incubated for 8 h in the presence of phosphate-buffered saline (control, lane 1), 0.1 µg/m gliotoxin (inducer of neutrophil apoptosis, lane 2), or 0.2 m*M* dibutyryl-cAMP (inhibitor of neutrophil apoptosis, lane 3). DNA was extracted and run on a 2% agarose gel containing ethidium bromide and the characteristic laddering pattern of apoptosis visualised under ultraviolet light.

fragments of around 180 bp, or oligonucleosomal fragments at multiples thereof, following earlier large-scale (50–200 kbp) degradation *(26)*. Such ordered fragmentation produces discretely sized lengths of DNA, which produce a distinct "laddering" pattern on electrophoresis, in contrast to necrotic cell death in which DNA is cleaved randomly, thus producing a smear on a DNA gel.

1. Suspend freshly-isolated peripheral blood neutrophils (at least 97% purity) at 5×10^6 cells/mL in IMDM supplemented with 10% autologous serum (*see* **Note 6**).
2. Dispense 1 mL of neutrophil suspension (5×10^6 cells) into 2-mL round-bottomed polypropylene tubes and incubate ± apoptosis-modifying agents at 37°C for the desired length of time (*see* **Note 13**).
3. Extract the genomic DNA using a Wizard Genomic DNA Purification Kit.
4. Run the DNA (23 µL of DNA plus 7 µL of loading dye) on a 2% agarose gel containing ethidium bromide (2.5 µg/mL) in TBE buffer at 10 V/cm.
5. Visualize gel under ultraviolet illumination. DNA from apoptotic cells exhibits a characteristic ladder pattern (**Fig. 4**).

3.6. Hypodiploid DNA Content

DNA fragmentation also leads to an apparent reduction in nuclear DNA content of ethanol-permeabilised cells, so that staining with a DNA-intercalating dye such as propidium iodide allows detection of a "hypodiploid" cell population. This technique works particularly well with neutrophils, because they are terminally differentiated cells which do not undergo proliferation, and consequently generate only two peaks when DNA content is measured: diploid (viable) cells and hypodiploid (apoptotic) cells.

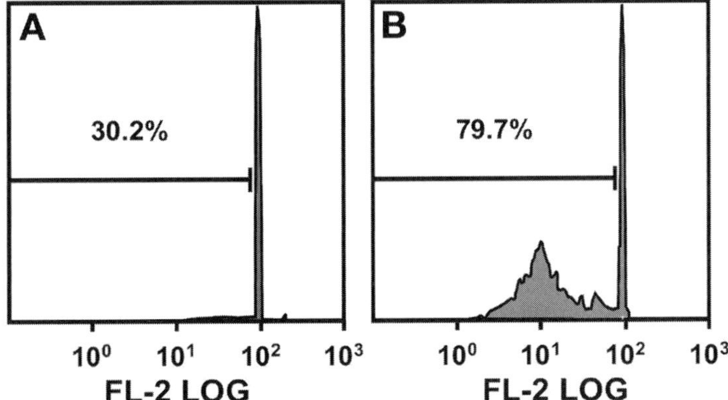

Fig. 5. Propidium iodide intercalation into DNA of neutrophils. Neutrophils were incubated at 37°C for (**A**) 8 h or (**B**) 20 h, then assessed by flow cytometry to determine the percentage of permeabilised cells with hypodiploid DNA content. After 8 h in culture 30.2% of the cells had undergone apoptosis associated with internucleosomal DNA fragmentation. After 20 h in culture, 79.7% of the cells had undergone apoptosis.

1. Suspend freshly isolated peripheral blood neutrophils (at least 97% purity) at 5×10^6 cells/mL in IMDM supplemented with 10% autologous serum (*see* **Note 6**).
2. Add 90 μL of neutrophil suspension to wells of a 96-well, flat-bottom flexible plate. To each well add 10 μL of apoptosis-modifying agents (10X concentration) or buffer control, cover with a lid, and incubate at 37°C in a 5% CO_2 incubator for the desired length of time.
3. Harvest cells by pipetting and transfer to 96-well, U-bottomed flexiwell plate (*see* **Note 8**).
4. Centrifuge the plate at 200*g* for 2 min at 4°C and discard the supernatants. Vortex the plate for 5 s to disrupt the cell pellets.
5. Resuspend the cells in 70% ethanol and incubate on ice for 10 min to permeabilize them.
6. Centrifuge the plate at 200*g* for 3 min at 4°C and discard the supernatants.
7. Wash the cells three times by adding 100 μL of PBS per well, centrifuging the plate at 200*g* for 3 min at 4°C, and discarding the supernatants.
8. Resuspend the cells in 60 μL of RNase A (0.5 mg/mL) followed by 60 μL of propidium iodide (0.1 mg/mL)
9. Incubate at room temperature for 25 min in the dark.
10. Transfer the cells to flow cytometry tubes, add 100 μL of PBS, and analyze by flow cytometry (FL2 channel) to determine the percentage of cells with hypodiploid DNA content (**Fig. 5**).

3.7. TdT-Mediated dUTP Nick End Labeling Staining for DNA Breaks

The presence of DNA strand breaks can be assessed by enzymatic methods, because DNA breaks create acceptor sites for enzymes such as TdT. Addition

of TdT together with fluorescein-12-2'deoxyuridine-5'-triphosphate is used
to reveal DNA fragmentation in the TdT-mediated dUTP nick end labeling
(TUNEL) technique.

1. These instructions assume the use of the *In Situ* Cell Death Detection Kit.
2. Prepare freshly-isolated peripheral blood neutrophils (at least 97% purity) resuspended at 2×10^7 cells/mL in IMDM supplemented with 10% autologous serum
   (*see* **Note 6**).
3. Add 90 µL of neutrophil suspension to wells of a 96-well, flat-bottom flexible plate.
   To each well, add 10 µL of apoptosis-modifying agents (10X concentration) or
   buffer control.
4. Cover plate, and incubate at 37°C in a 5% CO_2 incubator for the desired length of
   time.
5. Transfer 100 µL of neutrophil suspension to a 96-well, U-bottom flexible plate (*see*
   Note 8) and centrifuge at 200*g* for 2 min at 4°C. Discard the supernatants.
6. Wash the cells three times by adding 100 µL of PBS per well, centrifuging the plate
   at 200*g* for 3 min at 4°C, discarding the supernatants, and vortexing the plate for 5 s.
7. Add 100 µL of fixation solution to each well.
8. Incubate on a shaker for 60 min at room temperature.
9. Add 200 µL of PBS to each well then centrifuge the plate at 200*g* for 10 min at 4°C
   and discard the supernatants.
10. Resuspend the cells in permeabilization solution for 2 min on ice.
11. Add 50 µL of nucleotide mixture to the two negative control wells. The TUNEL
    reaction mixture is then made up by mixing the enzyme solution (50 µL) with the
    remaining 450 µL of nucleotide mixture.
12. The two positive control wells are treated for 10 min at room temperature with
    DNase I to introduce DNA strand breaks.
13. Wash twice in PBS (200 µL per well) then resuspend in TUNEL reaction mixture
    (50 µL per well).
14. Cover the plate and incubate at 37°C for 60 min in the dark.
15. Wash twice in PBS (200 µL per well) then transfer to flow cytometry tubes for
    analysis of fluorescence levels (FL-1).

3.8. Plate-Based Assay for Phagocytosis of Apoptotic Neutrophils

Macrophage phagocytosis of apoptotic neutrophils may be assessed using
minor modifications of a serum-free phagocytosis assay first described by
Newman et al. in 1982 *(1,27,28)*. This method uses adherent human monocyte-
derived macrophages, which are most efficient at ingesting apoptotic cells, but
it has also been used successfully with murine peritoneal and bone marrow-
derived monocytes and human apoptotic neutrophils.

1. This method assumes the use of adherent macrophages in Costar 48-well TC-treated
   microplates.
2. Suspend 10^8 freshly-isolated peripheral blood neutrophils (at least 97% purity) in
   20 mL of IMDM supplemented with 10% autologous serum (*see* **Note 6**).

3. Dispense the neutrophil suspension into a 75-cm^2 cell culture flask and stand the flask on its end in an incubator for 20 h at 37°C in a 5% CO_2 atmosphere.

4. Harvest the neutrophil suspension into a 50-mL conical polypropylene tube and wash twice in warm IMDM (50 mL volume per wash) by centrifuging at 220g for 5 min and discarding the supernatant. After each wash, gently resuspend the neutrophil pellet in 1 mL of warm IMDM using a plastic pipet to avoid clumping of the cells.

5. Count the cells using a haemocytometer and finally resuspend the aged neutrophils at 4×10^6 cells/mL in warm (37°C) IMDM (no serum).

6. Rinse the macrophage monolayer with warm (37°C) IMDM to remove nonadherent cells.

7. Overlay the macrophage monolayer with 0.5 mL (2×10^6 cells) of the suspension of aged neutrophils in IMDM (serum-free), and incubate for 60 min at 37°C in a 5% CO_2 atmosphere.

8. Wash each well with ice-cold PBS (without cations). Use an inverted microscope to check that noningested neutrophils have been largely removed.

9. Repeat the wash as necessary, continually checking by microscopy to ensure that the macrophage monolayer is not disrupted.

10. Fix in 2.5% glutaraldehyde for 30 min and rinse off with PBS.

11. Stain for myeloperoxidase (MPO) with 0.1 mg/mL dimethoxybenzidine and 0.03% (v/v) hydrogen peroxide in PBS for about 60 min at room temperature.

12. Count the percentage of macrophages (MPO-negative) that have phagocytosed one or more apoptotic neutrophils (MPO-positive) by examination with an inverted microscope of at least five fields (minimum 400 cells), and record as the mean percent phagocytosis of duplicate or triplicate wells (**Fig. 6**).

3.9. Flow Cytometry-Based Phagocytosis Assay

This modification of the plate-based phagocytosis assay utilizes a fluorescent chloromethyl dye that diffuses freely through plasma membranes to label the cytoplasm of live neutrophils. This method avoids the need for vigorous washing of the macrophage monolayer following interaction with aged neutrophils and eliminates potential observer bias when counting macrophages. It is less laborious than the plate-based counting method and so is particularly suitable for investigating a large number of conditions. However, we recommend that new investigators validate this method against the "gold standard" plate-based counting assay in their own laboratories (*29*).

1. This method assumes the use of adherent macrophages in 48-well, TC-treated microplates.

2. Suspend 10^8 freshly isolated peripheral blood neutrophils (at least 97% purity) in 5 mL of warm IMDM in a 50-mL Falcon® conical polypropylene tube.

3. Add 10 µL of 10 mM CMFDA (20 µM final concentration), mix gently, and incubate at 37°C for 15 min.

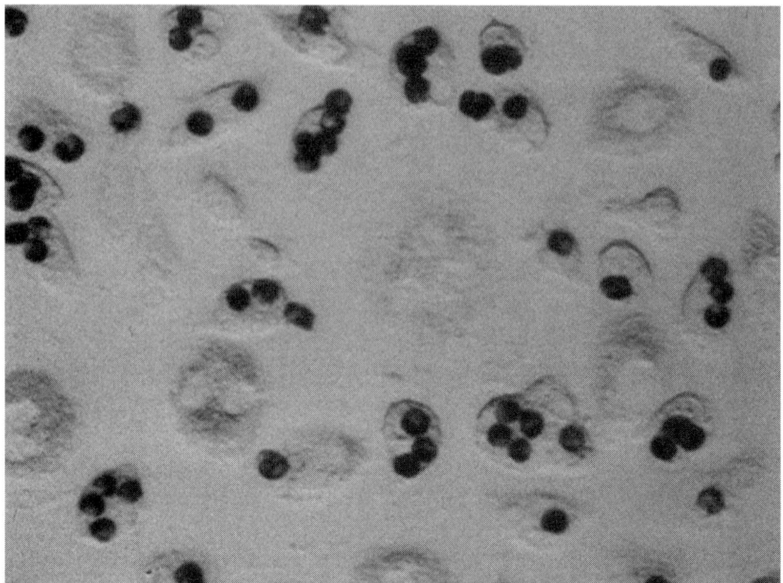

Fig. 6. Phagocytosis assay with human monocyte-derived macrophages and aged neutrophils. Following 60 min incubation at 37°C, the macrophage monolayer was washed with cold phosphate-buffered saline, fixed with glutaraldehyde, and stained for myeloperoxidase (MPO). Ingested neutrophils are MPO-positive (dark reaction product), whereas macrophages are MPO-negative.

4. Dilute the labeled neutrophils by adding 15 mL of warm IMDM supplemented with 10% autologous serum (see **Note 6**).
5. Dispense the neutrophil suspension into a 75-cm^2 cell culture flask and stand the flask on its end in an incubator for 20 h at 37°C in a 5% CO_2 atmosphere.
6. Harvest the neutrophil suspension into a 50-mL Falcon conical polypropylene tube and wash twice in warm IMDM (50 mL per wash) by centrifuging at 220g for 5 min and discarding the supernatant. After each wash, gently resuspend the neutrophil pellet in 1 mL of warm IMDM using a plastic pipet to avoid clumping of the cells.
7. Count the cells using a hemocytometer and finally resuspend the aged neutrophils at 4×10^6 cells/mL in warm (37°C) IMDM (no serum).
8. Rinse the macrophage monolayer with warm (37°C) IMDM to remove nonadherent cells.
9. Overlay the macrophage monolayer with 0.5 mL (2×10^6 cells) of the suspension of labelled aged neutrophils in IMDM (serum-free), and incubate for 60 min at 37°C in a 5% CO_2 atmosphere.
10. After a 60 min incubation at 37°C, aspirate the neutrophil suspension from the wells.
11. Detach the macrophages by incubation with 300 μL of 0.25% trypsin/1 mM EDTA solution (Invitrogen) for 20 min at 37°C followed by 20 min at 4°C (*see* **Note 17**).

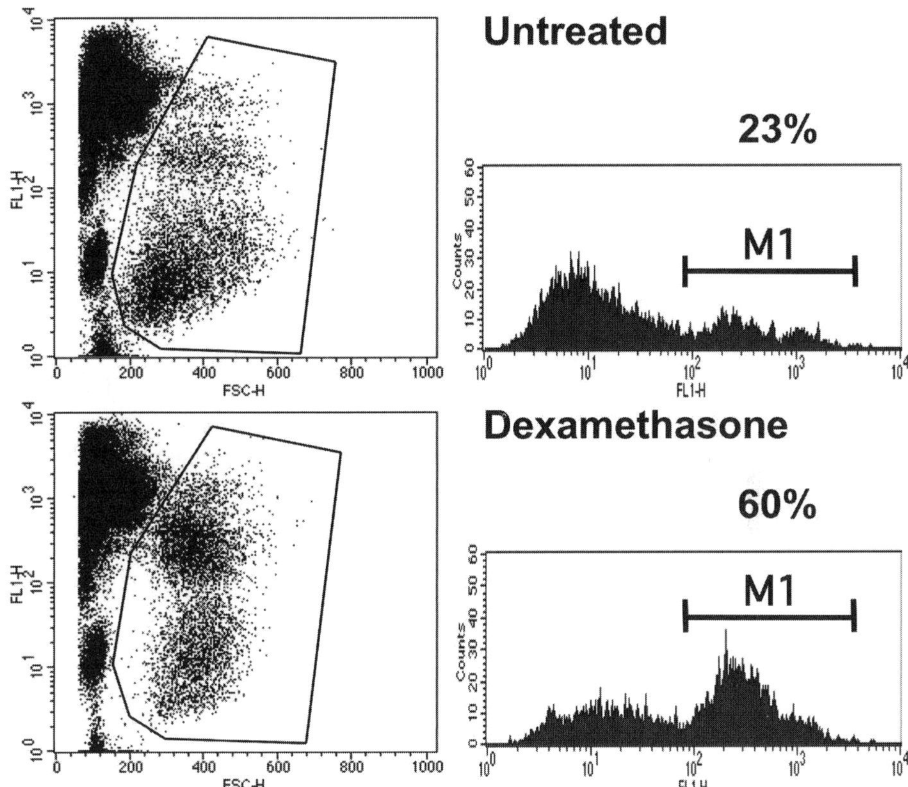

Fig. 7. Flow cytometry-based phagocytosis assay. Aged human neutrophils were labeled with 5-chloromethylfluorescein diacetate (CMFDA) and incubated with monocyte-derived macrophages cultured in medium alone (Untreated) or in the presence of 1 μ*M* dexamethasone as indicated. Macrophages were detached and analysed by flow cytometry. Macrophages were distinguished by their typical high forward-scatter (FSC) characteristics, and macrophages which had ingested CMFDA-labeled apoptotic neutrophils were FL1-positive. Dexamethasone substantially increased the percentage of macrophages that phagocytosed labeled apoptotic neutrophils.

12. Harvest the detached macrophages by forceful pipetting. Analyze the entire 300 μL of sample (unfixed) immediately using a FACSCalibur flow cytometer (Becton Dickinson), aiming to collect at least 5000 events in the macrophage gate.
13. Apoptotic cells and macrophage populations are identified by their distinct laser-scatter properties. The number of the FL-1-positive events in the macrophage gate is divided by the total number of macrophages to obtain the proportion of macrophages that have internalised apoptotic cells. (**Fig. 7**).
14. Treatment of monocytes for 5–7 days with 1 μ*M* dexamethasone can be used to increase the percentage of macrophages that phagocytose apoptotic neutrophils (**Fig. 7**).

4. Notes

1. 1 µ*M* TO-PRO3 (Molecular Probes) may also be used to identify necrotic cells, and
   has the advantage that fluorescence is measured in the FL-4 channel, which requires
   minimal compensation when used alongside FL-1.

2. Neutrophil granules contain high concentrations of proteases. Extreme care must
   be taken to keep all samples on ice during the preparation of neutrophil lysates in
   order to prevent the protein of interest being degraded before blotting. Higher
   protease inhibitor concentrations are required for neutrophils than for other cell
   types. The protocol describes high protease inhibitor concentrations that we have
   used successfully in our laboratory; however, some experimentation with regard
   to manipulation of concentrations may be required. Recent evidence demonstrates
   that pre-incubation of neutrophils with 1 m*M* phenylmethylsulfonylfluoride (PMSF;
   Sigma-Aldrich) for up to 1 h prior to lysis may be required to prevent neutrophil
   protease-mediated cleavage of intracellular proteins following addition of lysis
   buffer *(17)*.

3. The 12% gel recommended in this protocol is based on blotting for cleaved caspase
   3 (14–21 kDa). However, gel percentages may be altered according to the size of
   the protein of interest. In general, low-percentage gels are used to blot for high-
   molecular-weight proteins, and high-percentage gels for low-molecular-weight
   proteins. Therefore, the composition of the gel may have to be altered depending
   on the size of the protein of interest. Some proteins that are important in apoptosis
   are of similarly low size to cleaved caspase 3, e.g., Bax (21 kDa) and truncated
   Bid (15 kDa), and therefore a similar percentage gel will probably be appropriate,
   whereas other proteins such as procaspase 8 (55 kDa) and procaspase 3 (32 kDa)
   are larger and may require a lower percentage gel for optimal resolution.

4. The protocol described is based on the homogeneous caspase kit available from
   Roche Diagnostics, and represents the simplest method for fluorimetric detection
   of total caspase activity. However, other commercial kits are available, which may
   be based on either the same or slightly different protocol.

5. We have found that SeaKem LE agarose is effective for separation of DNA fragments. It forms strong gels with high electrophoretic mobility, and shows very low
   background levels when stained with ethidium bromide.

6. We culture neutrophils in IMDM containing 10% serum. We routinely use autologous platelet-rich plasma-derived serum, prepared from platelet-rich plasma (harvested after centrifugation of citrate-anticoagulated blood) by recalcification with
   20 m*M* CaCl$_2$ for 1 h at 37°C in glass tubes. However, bovine serum may also be
   used. Alternatively, neutrophils may be cultured in the absence of serum with a
   small amount of added protein (e.g., 0.5% [w/v] serum albumin), but apoptosis will
   proceed more rapidly. Cell-saver pipet tips should be used when pipetting neutrophils to minimize cell damage.

7. Some investigators add extra serum to the cells in the cytospin chamber to avoid
   artifacts caused by cell breakage during centrifugation. We have observed that adding serum reduces the effectiveness of deposition of neutrophils that have progressed

to a late stage of apoptosis (late apoptotic or postapoptotic cells), so potentially underestimating the true rate of neutrophil apoptosis. Late apoptotic neutrophils appear as cell ghosts with little or no evidence of nuclear staining, having undergone "nuclear evanescence" *(18–20)*. Furthermore, late apoptotic cell ghosts in a cell population may be overlooked because the lack of nuclear material means that they stain very faintly with Romanowsky stains such as Diff Quik, and in the past late apoptotic cells may have been gated out as "debris" when analyzed by flow cytometry.

8. The flexible plate is best trimmed and mounted in a standard 96-well tissue culture plate for support. Alternatively, for small numbers of samples, 2-mL round-bottomed polypropylene vials may be used.

9. Dilutions prior to flow cytometry should be performed using Annexin binding buffer, because in the absence of Ca^{2+} Annexin V will rapidly dissociate from PS on the apoptotic cell surface.

10. Flow cytometers interpret the laser forward-scatter (FS) signal in different ways. For example, the Coulter Epics XL measures integral (area) FS signals, whereas the Becton-Dickinson FACSCalibur records height signals from the FS pulse. This difference significantly affects how late apoptotic neutrophils appear in a standard FS vs side-scatter dot plot. On a Coulter Epics XL, late apoptotic neutrophils appear as a distinct low-FS population that can be gated separately from the population of nonapoptotic/early apoptotic neutrophils. This has the advantage that a nuclear stain such as propidium iodide is not routinely required to identify the membrane permeable late apoptotic cells. On a FACSCalibur cytometer, the late apoptotic neutrophils appear within a single population along with the nonapoptotic and early apoptotic cells. Digital cytometers such as the FACSVantage analyze the area of the FS pulse similarly to the Coulter machine.

11. Eosinophils present in the granulocyte population also have low levels of CD16 expression, so eosinophil counts (by analysis of cell morphology in cytocentrifuge preparations or CD16 expression) should be performed at baseline.

12. Another way of analyzing mitochondrial changes associated with apoptosis is Western blotting of the Bcl-2 family proteins that are important in the control of the intrinsic (mitochondrial) apoptotic pathway. The expression profile of Bcl-2 family proteins in peripheral blood neutrophils has been analyzed and shows expression of Bak, Bad, Bcl-w, and Bfl-1 in these cells, but relatively little Bcl-2, Bcl-xL, Bik, and Bax *(24)*. Therefore, the more abundant proteins should be considered first for analysis for roles in neutrophil apoptosis, although some proteins, such as Bax, may be transcriptionally upregulated in response to certain apoptotic stimuli under the control of p53, and may therefore still be important in neutrophil apoptosis. The Western blotting protocol is identical to that described for caspases.

13. For short-term culture, tubes may be incubated in a shaking heated block set to 37°C. For longer incubation periods, a hole is made in the top of each polypropylene round-bottomed tube, and the tubes are placed in an incubator in a 5% CO_2 atmosphere at 37°C.

14. The protocol for this kit recommends a cell number of 4×10^4 cells per well in a volume of 100 µL. However, in assays involving neutrophils cell numbers may be increased in order to amplify the signals obtained from the assay particularly as neutrophils are so abundant. On the other hand, cell density may affect the rate of spontaneous apoptosis of these cells during culture, with a high cell density promoting survival, especially at densities of 8×10^6 cells/mL and above *(25)*. Therefore, some experimentation may be required to manipulate the cell density, volume, and time of culture in order to gain the best results with this assay.

15. Colorimetric assays for single-caspase activity are also available. These are similar to the fluorometric assays, and follow a similar protocol, but instead utilize cleavage of a chromophore (e.g., p-nitroanalide) from caspase substrates as a measure of caspase activation. Development of colour can be monitored using a spectrophotometer or microplate reader (405 nm wavelength light), and activity quantified by comparison with a calibration curve constructed using known standards.

16. This protocol applies to any cell type, as caspase activation is a general event of apoptosis and is not unique to neutrophils. However, differences may exist between the expression profiles of the various caspases in different cell types, and this must be borne in mind when selecting an assay to use in neutrophils. Caspases 1, 3, 4, and 7–10 are expressed in neutrophils *(24)*. In contrast, it has been reported that caspase 2 is absent from peripheral blood neutrophils, although it is expressed in HL-60 cells.

17. Treatment with trypsin-EDTA may lead to clumping of cells leading to blockage of the flow cytometer's sample intake nozzle. Clumping may be minimised by adding 50 µL of bovine serum to each well following incubation with trypsin-EDTA.

Acknowledgments

This work was funded by a Wellcome Trust Research Fellowship (E. T.) and a Medical Research Council Clinician Scientist Fellowship (S. P. H.). The Western blot demonstrating activation of caspase 3 (**Fig. 3**) was kindly provided by Tara Sheldrake. We are grateful to Shonna MacCall for her helpful comments.

References

1. Savill, J. S., Wyllie, A. H., Henson, J. E., Walport, M. J., Henson, P. M., and Haslett, C. (1989) Macrophage phagocytosis of aging neutrophils in inflammation: programmed cell death in the neutrophil leads to its recognition by macrophages. *J. Clin. Invest.* **83,** 865–875.

2. Hurley, J. V. (1983) Terminations of acute inflammation I. Resolution, in *Acute Inflammation* (Hurley, J. V., ed.), Churchill-Livingstone, London, pp. 109–117.

3. Ward, C., Chilvers, E. R., Lawson, M. F., et al. (1999) NF-kappaB activation is a critical regulator of human granulocyte apoptosis in vitro. *J. Biol. Chem.* **274,** 4309–4318.

4. Taylor, E. L., Megson, I. L., Haslett, C., and Rossi, A. G. (2001) Dissociation of DNA fragmentation from other hallmarks of apoptosis in nitric oxide-treated neutrophils: differences between individual nitric oxide donor drugs. *Biochem. Biophys. Res. Commun.* **289,** 1229–1236.

5. Walker, A., Ward, C., Taylor, E. L., et al. (2005) Regulation of neutrophil apoptosis and removal of apoptotic cells. *Curr. Drug Targets Inflamm. Allergy* **4,** 447–454.

6. Sabroe, I., Jones, E. C., Usher, L. R., Whyte, M. K. B., and Dower, S. K. (2002) Toll-like receptor (TLR)2 and TLR4 in human peripheral blood granulocytes: a critical role for monocytes in leukocyte lipopolysaccharide responses. *J. Immunol.* **168,** 4701–4710.

7. Dransfield, I. and Rossi, A. G. (2004) Granulocyte apoptosis: who would work with a 'real' inflammatory cell? *Biochem. Soc. Trans.* **32,** 447–451.

8. Rossi, A. G., Ward, C., and Dransfield, I. (2004) Getting to grips with the granulocyte: manipulation of granulocyte behaviour and apoptosis by protein transduction methods. *Biochem. Soc. Trans.* **32,** 452–455.

9. Wyllie, A. H. (1997) Apoptosis: an overview. *Br. Med. Bull.* **53,** 451–465.

10. Zimmermann, K. C., Bonzon, C., and Green, D. R. (2001) The machinery of programmed cell death. *Pharmacol. Ther.* **92,** 57–70.

11. Savill, J., Dransfield, I., Gregory, C., and Haslett, C. (2002) A blast from the past: clearance of apoptotic cells regulates immune responses. *Nat. Rev. Immunol.* **2,** 965–975.

12. Ren, Y. and Savill, J. S. (1995) Proinflammatory cytokines potentiate thrombospondin-mediated phagocytosis of neutrophils undergoing apoptosis. *J. Immunol.* **154,** 2366–2374.

13. Rossi, A. G., McCutcheon, J. C., Roy, N., Chilvers, E. R., Haslett, C., and Dransfield, I. (1998) Regulation of macrophage phagocytosis of apoptotic cells by cAMP. *J. Immunol.* **160,** 3562–3568.

14. Liu, Y., Cousin, J. M., Hughes, J., et al. (1999) Glucocorticoids promote nonphlogistic phagocytosis of apoptotic leukocytes. *J. Immunol.* **162,** 3639–3646.

15. McCutcheon, J. C., Hart, S. P., Canning, M., Ross, K., Humphries, M. J., and Dransfield, I. (1998) Regulation of macrophage phagocytosis of apoptotic cells by adhesion to fibronectin. *J. Leukoc. Biol.* **64,** 1–8.

16. Hart, S. P., Smith, J. R., and Dransfield, I. (2004) Phagocytosis of opsonized apoptotic cells: roles for 'old-fashioned' receptors for antibody and complement. *Clin. Exp. Immunol.* **135,** 181–185.

17. Kato, T., Sakamoto, E., Kutsuna, H., Kimura-Eto, A., Hato, F., and Kitagawa, S. (2004) Proteolytic conversion of STAT3alpha to STAT3gamma in human neutrophils: role of granule-derived serine proteases. *J. Biol. Chem.* **279,** 31,076–31,080.

18. Hebert, M. J., Takano, T., Holthofer, H., and Brady, H. R. (1996) Sequential morphologic events during apoptosis of human neutrophils. Modulation by lipoxygenase-derived eicosanoids. *J. Immunol.* **157,** 3105–3115.

19. Ren, Y., Stuart, L., Lindberg, F. P., et al. (2001) Nonphlogistic clearance of late apoptotic neutrophils by macrophages: efficient phagocytosis independent of beta 2 integrins. *J. Immunol.* **166,** 4743–4750.

20. Hart, S. P., Alexander, K. M., MacCall, S. M., and Dransfield, I. (2005) C-reactive protein does not opsonize early apoptotic human neutrophils, but binds only membrane-permeable late apoptotic cells and has no effect on their phagocytosis by macrophages. *J. Inflamm. (Lond)* **2,** 5.

21. Dransfield, I., Buckle, A.-M., Savill, J. S., McDowall, A., Haslett, C., and Hogg, N. (1994) Neutrophil apoptosis is associated with a reduction in CD16 (FcgRIII) expression. *J. Immunol.* **153,** 1254–1263.

22. Harada, H. and Grant, S. (2003) Apoptosis regulators. *Rev. Clin. Exp. Hematol.* **7,** 117–138.

23. Martin, M. C., Dransfield, I., Haslett, C., and Rossi, A. G. (2001) Cyclic AMP regulation of neutrophil apoptosis occurs via a novel protein kinase A-independent signaling pathway. *J. Biol. Chem.* **276,** 45041–45050.

24. Santos-Beneit, A. M. and Mollinedo, F. (2000) Expression of genes involved in initiation, regulation, and execution of apoptosis in human neutrophils and during neutrophil differentiation of HL-60 cells. *J. Leukoc. Biol.* **67,** 712–724.

25. Hannah, S., Nadra, I., Dransfield, I., Pryde, J. G., Rossi, A. G., and Haslett, C. (1998) Constitutive neutrophil apoptosis in culture is modulated by cell density independently of beta2 integrin-mediated adhesion. *FEBS Lett.* **421,** 141–146.

26. Nagata, S. (2005) DNA degradation in development and programmed cell death. *Annu. Rev. Immunol.* **23,** 853–875.

27. Newman, S. L., Henson, J. E., and Henson, P. M. (1982) Phagocytosis of senescent neutrophils by human monocyte-derived macrophages and rabbit inflammatory macrophages. *J. Exp. Med.* **156,** 430–442.

28. Hart, S. P., Dougherty, G. J., Haslett, C., and Dransfield, I. (1997) CD44 regulates phagocytosis of apoptotic neutrophil granulocytes, but not apoptotic lymphocytes, by human macrophages. *J. Immunol.* **159,** 919–925.

29. Jersmann, H. P., Ross, K. A., Vivers, S., Brown, S. B., Haslett, C., and Dransfield, I. (2003) Phagocytosis of apoptotic cells by human macrophages: Analysis by multiparameter flow cytometry. *Cytometry* **51A,** 7–15.

IV

Neutrophil Adhesion and Chemotaxis

13

Optical and Fluorescence Detection of Neutrophil Integrin Activation

Ulrich Y. Schaff, Melissa R. Sarantos,
Harold Ting, and Scott I. Simon

Summary

Neutrophils are among the first cells to respond to acute inflammation through a multi-step process initiated by selectin mediated rolling, which transitions to an integrin/inter-cellular adhesion molecule-dependent arrest and transmigration across endothelium. A conformational shift in the CD11/CD18 adhesion receptor on neutrophils is a critical determinant of the efficiency of recruitment on inflamed endothelium. For instance, β_2-integrin expression level is upregulated up to 10-fold by fusion of cytoplasmic granule pools of CD11b/CD18 (Mac-1). Furthermore, a rapid increase in affinity and membrane clustering of CD11a/CD18 (LFA-1) is necessary for efficient deceleration and arrest in shear flow. We present methods here to quantify the changes in receptor expression and affinity that support neutrophil adhesive phenotypes. Techniques involving real-time fluorescence flow cytometry and parallel plate rheometry coupled with light microscopy are presented.

Key Words: Selectins, integrins; ICAM-1; adhesion molecules; neutrophil; shear stress.

1. Introduction

Neutrophil rolling and transition to arrest on inflamed endothelium are dynamically regulated by the affinity of the β_2-integrin for binding intercellular adhesion molecule (ICAM)-1. Conformational shifts are thought to regulate molecular affinity and adhesion stability. Also critical to adhesion efficiency is membrane redistribution of active LFA-1 into dense submicron clusters where multimeric interactions strengthen adhesion *(1)*. How conformational shifts in LFA-1 effect changes in affinity that influence the nature of leukocyte rolling and adhesion remain critical questions in inflammation biology.

From: *Methods in Molecular Biology, vol. 412: Neutrophil Methods and Protocols*
Edited by: M. T. Quinn, F. R. DeLeo, and G. M. Bokoch © Humana Press Inc., Totowa, NJ

                                                                 Schaff et al.

Strong evidence links conformation of the ligand binding region (the I domain) of LFA-1 to specific affinity in binding ICAM-1 *(2,3)*. Site-directed amino acid point mutations in the I domain have revealed distinct structural conformations that correlate to a range of LFA-1 affinity states. ICAM-1 equilibrium binding constants can be increased four orders magnitude through these mutations, ranging between low (i.e., 1600 μM), intermediate (i.e., 9 μM), and high-affinity (i.e., 0.15 μM) *(3)*. Neutrophil adhesion receptors are able to rapidly regulate their affinity for ligand and concentrate at precise membrane regions within the area of contact with inflamed endothelium.

Ligand binding and receptor clustering at these sites can in turn trigger intracellular signaling that promote stable adhesion and neutrophil migration. We describe methods for quantifying changes in integrin affinity, imaging of activation epitopes on integrins, and real-time fluorescence imaging of neutrophils in shear flow. These methods allow a direct correlation of receptor affinity and neutrophil recruitment.

2. Materials

2.1. Detection of Changes in β_2-integrin Affinity

1. Amino microsphere latex beads (D = 6 μm, 2.7% solids-latex) (Biosciences, Piscataway, NJ).
2. Phosphate-buffered saline (PBS) without Ca^{2+} or Mg^{2+} (Gibco/Invitrogen).
3. Sulfo-SMCC maleimide-NHS-ester crosslinker (Pierce), sulfhydryl linker SATA (Pierce, Rockford, IL).
4. Dimethylsulfoxide (DMSO) (Sigma-Aldrich), ethylenediamine tetraacetic acid (EDTA) (Sigma-Aldrich, St. Louis, MO).
5. 3500 MWCO dialysis membrane (Pierce Rockford, IL).
6. Alexafluor488 (Molecular Probes).
7. Graphpad Prism version 4.0 for Windows (Graphpad Software Inc., San Diego, CA).
8. Recombinant human LFA-1 with an inserted leucine zipper, anti-leucine zipper monoclonal antibody (324C), and dimeric ICAM-1/Ig (ICOS Corporation, Bothell, WA).

2.2. Immunofluorescence Imaging of Neutrophil Surface Epitopes

1. Primary antibody conjugated directly to fluorophore in absence of sodium azide.
2. 35 mm diameter glass coverslips coated with an endothelial monolayer.
3. Poly(dimethyl siloxane) (PDMS) microfluidic flow chambers and reservoirs.
4. Fluorescent microscope with excitation and emission filter wheels, a high-sensitivity camera, and excitation/brightfield shutters.
5. HEPES-buffered saline: 10 mM KCl, 110 mM NaCl, 10 mM glucose, 1 mM MgCl$_2$, and 30 mM HEPES, pH 7.4, 290 mOsm.
6. 25% Human serum albumin (HSA).
7. L-Ascorbic acid (Sigma).

8. Hanks balanced salt solution (HBSS; Gibco).
9. Syringe pump with 1-mL syringe.
10. 20 cm of PE 20 tubing.

3. Methods

3.1. Detection of Changes in β_2-integrin Affinity

We describe here methodology for measuring changes in affinity state of LFA-1 by measuring rates of binding and dissociation of ICAM-1 in a cell-free system on beads and on neutrophils *(4)*.

1. Pipet 200 μL of amino microsphere latex beads into a 1.5-mL Eppendorf tube. Add 1.0 mL PBS without Ca^{2+} or Mg^{2+} and centrifuge the tube at 1400 rpm ($160g$) for 5 min or until a pellet is visible.
2. Remove supernatant and repeat procedure 3.1 two more times. After the last wash, resuspend beads in 200 μL PBS.
3. Mix beads using an Eppendorf Thermomixer R (Brinkmann Instruments) with 125 μ*M* sulfo-SMCC maleimide-NHS-ester crosslinker (Pierce) for 2 h at 4°C.
4. Add a sulfhydryl group to anti-leucine zipper antibody 324C by mixing 10 μL of SATA solution (13 mg/mL in DMSO) to 100 μL of antibody (at 3.95 mg/mL) at 450 rpm for 30 min at room temperature.
5. Dialyze the solution (3500 molecular weight cutoff membrane) in 1.5 L PBS containing 1 m*M* EDTA.
6. After 6 h of dialysis, remove the buffer, add fresh buffer, and allow dialysis to proceed overnight.
7. Centrifuge the crosslinker-modified beads and remove the 200 μL of PBS.
8. Remove the 324C-SH from the dialysis membrane and use this volume to resuspend the amino-crosslinker beads. Mix the beads and 324C-SH (450 rpm) for 2 h at 4°C. Centrifuge the beads and resuspend in 200 μL PBS.
9. Measure the concentration of the beads on a Coulter Multisizer II. The final concentration should be approx 10^7 beads/mL. Store the beads at 4°C. They are stable for approx 3 mo.
10. Directly prior to experimentation, mix recombinant LFA-1 heterodimer with an inserted leucine zipper between the α and β subunits (20 μg/mL) with 10^5 beads/ 100 μL PBS at 37°C, 450 rpm for 30 min. The bivalent binding structure of the antibody will allow for clustering of the LFA-1 into a dimer-like configuration presented on the bead surface. The cell-free system offers the opportunity to prescribe LFA-1 distribution in a manner that shifts the binding spectrum towards bivalent driven adhesion of ICAM-1 (**Fig. 1**).
11. Centrifuge the beads with antibody captured LFA-1 and resuspend in 100 μL PBS and incubate (37°C, 450 rpm for 30 min) with activator $MgCl_2$ (3 m*M*).
12. Collect data from the beads with activated LFA-1 on a FACScan flow cytometer where fluorescently labeled (Alexafluor488) ICAM-1 (20 μg/mL) is added to the beads directly prior to reading and bead fluorescence is detected in real time.

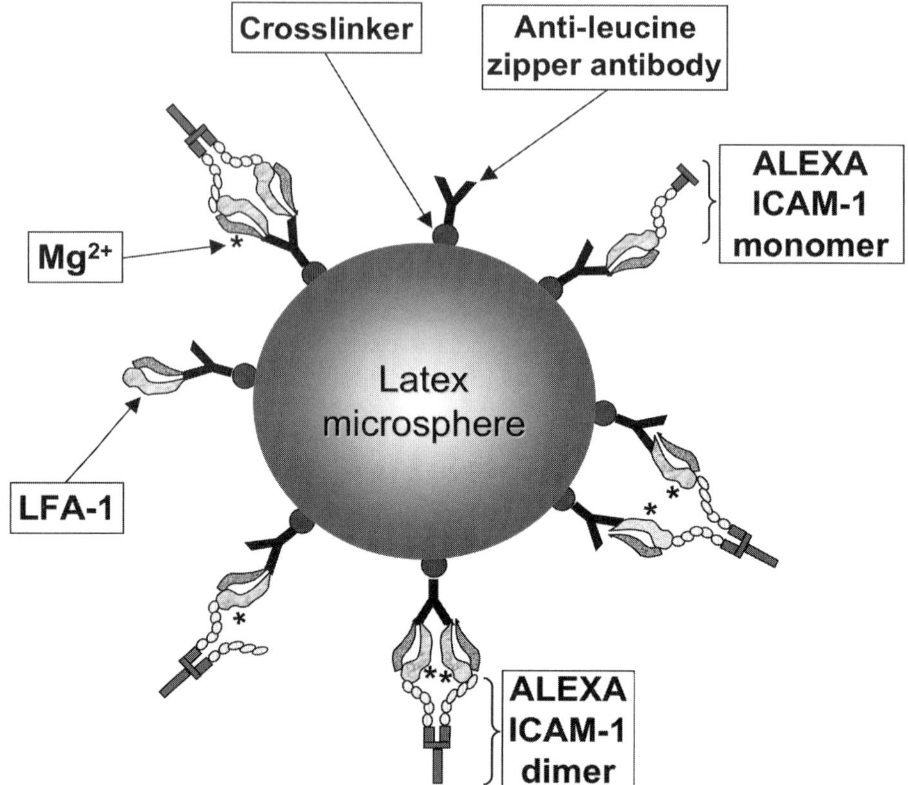

Fig. 1. LFA-1 bead assembly. Bead with antibody bound LFA-1 binding monomeric and dimeric intercellular adhesion molecule-1.

> After 7 min, add inhibitor to the sample without stopping the data collection of the cytometer. Let data collection proceed for an additional 15 min or until fluorescence has reached background level.

13. To measure kinetic rates for ICAM-1 binding to freshly isolated neutrophils, repeat **step 8**, replacing beads with neutrophils and running the experiment in HEPES-buffered saline solution (+0.1% HSA and 1.5 m*M* Ca^{2+}) instead of PBS.

14. Average the fluorescence over 10-s intervals and plot vs time (**Fig. 2**).

15. Analyze the curves using Graphpad Prism version 4.0 for Windows. Obtain constants k_{off} and k_{ob} (observed rate constant) by performing one phase exponential decay ($Y = specific\ binding * e^{-kT} + nonspecific\ binding$) and one phase exponential association [$Y = Y_{max}(1 - e^{-kT})$] curve fits of the real time data, respectively. Calculate the k_{on} as: $k_{on} = (k_{ob} - k_{off})\ /\ [ICAM\text{-}1\text{-}Alexa]$. Calculate data points by taking the average fluorescence over 5–25 s depending on the rate of change. Calculate K_D, the equilibrium dissociation constant, as $k_{off}\ /\ k_{on}$.

16. Determine statistical significance ($p \leq 0.05$) by a one-way analysis of variance (ANOVA) with a Newman-Keuls Multiple Comparison Post Test.

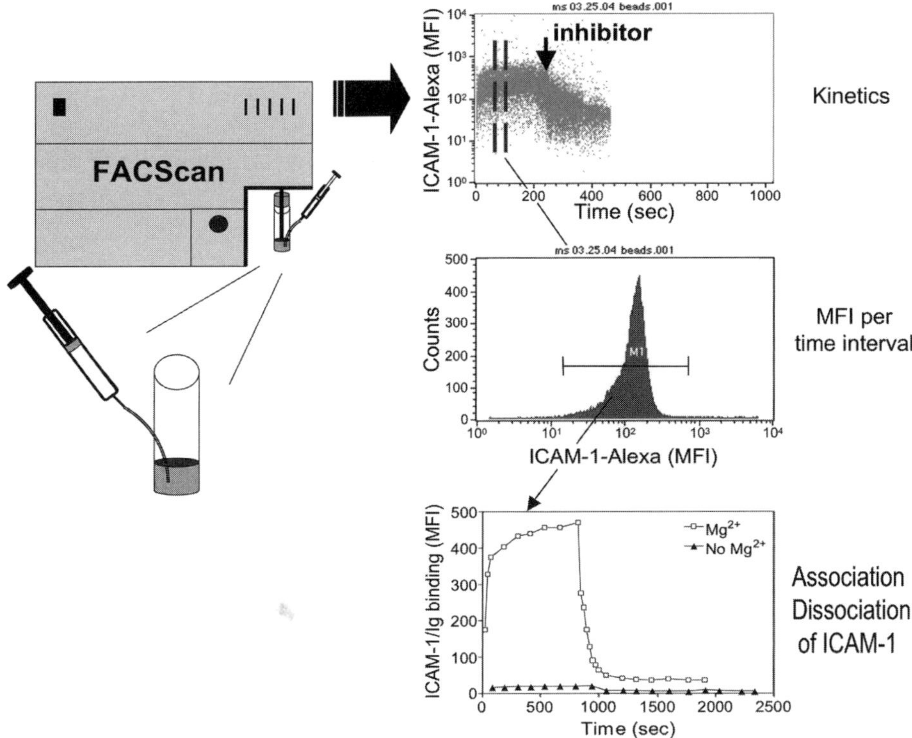

Fig. 2. Experimental set-up and time course analysis of fluorescent intercellular adhesion molecule-1 binding to LFA-1 on beads.

3.2. Immunofluorescence Imaging
of Surface Epitopes on Neutrophils in Shear Flow

Immunofluorescent labeling of fixed cells has long been utilized to visualize the subcellular location of specific proteins. However, real-time alterations in the distribution of adhesion molecules and formation of high density clusters is a critical stage in the transition from leukocyte rolling to arrest on inflamed endothelium *(5,6)*. Therefore, we developed the following protocol for labeling and measuring the distribution of a given leukocyte surface protein in an in vitro model of recruitment to endothelium (*see* **Note 1**).

Although a standard parallel plate flow chamber may be substituted, our protocol takes advantage of the intrinsic low flow rate of a PDMS microfluidic device designed to mimic the geometry of a small blood vessel *(7)*. Taking these considerations into account, we have developed the following protocol to quantify both the formation of high-density adhesion molecule clusters on a neutrophil surface and the concurrent adhesion and arrest of these neutrophils on an inflamed endothelial surface.

3.2.1. Preparation of Monolayers

1. Grow substrate cell monolayers on glass coverslips to permit the use of high numerical aperture oil objectives. Growing endothelial monolayers on a glass coverslip requires a bioadhesive coating such as crosslinked gelatin (*see* **Note 2**).
2. Wash monolayers extensively with HBSS to remove serum before assembling any type of flow chamber on the coverslip.

3.2.2. Preparation of Flow Channel

1. For a removable vacuum-sealed channel, soak all PDMS devices overnight in deionized water at 4°C, dissolving all small bubbles.
2. Before interfacing the PDMS with a cultured cell monolayer, briefly dip the PDMS in a beaker containing approx 100 mL of HBSS to avoid osmotic stress.
3. Connect the PDMS device to its vacuum leads, and invert it such that drops of HBSS remain on the channel engraved surface.
4. Use needle nosed tweezers to place a monolayer coated coverslip on top of the wetted PDMS and initialize the vacuum using a stopcock.
5. Use a stage heater to maintain experimental conditions at 37°C. A heated and humidified enclosure provides optimal experimental conditions. For oil objectives, begin heating at least 1 h in advance of the experiment in order to allow optics to equilibrate at the experimental temperature.
6. To prevent desiccation of channels, place drops of HBSS on top of all access holes. Use blunt tweezers to hold down the PDMS and press fit a fluid-filled reservoir and pump fluidic tubes into prepunched holes in the PDMS device.
7. Draw fluid through the microchannels into the pump tubing using the syringe pump. Set the volumetric flow rate of the syringe pump based on the desired fluid shear stress and **eq. 1**.

$$\tau = \frac{6\mu Q}{Wh^2} \tag{1}$$

Where τ is shear stress, μ is fluid viscosity, Q is fluid flow rate, w is channel width, and h is channel height.

3.2.3. Real-Time Fluorescence Imaging

1. Dilute neutrophils to a concentration of 1×10^6 cells/mL in HEPES buffered saline solution containing 10 mg/mL HSA and 20 µg/mL of the appropriate labeling and control antibody. Incubate for 5 min at room temperature.
2. Centrifuge the labeled neutrophils, and resuspend the pellet to a concentration of 1×10^6 cells/mL in HEPES-buffered saline containing 0.5 mM ascorbic acid, 1.5 mM calcium, and 1 mg/mL HSA.
3. Use these washed, labeled neutrophils as the sample for a microfluidic flow chamber by adding them to the inlet reservoir. Draw the sample into the flow chamber at the desired shear rate.
4. Optimize camera settings for the labeling conditions by taking single images of the first neutrophils that adhere to the substrate. When imaging moving cells, it is important to use the minimum exposure time necessary to obtain a clear image.

5. Once a number of neutrophils adhere to the underlying substrate, take a sequence of alternating fluorescent and brightfield images using excitation/emission filter wheels (*see* **Note 3**).
6. The distribution of the labeled neutrophil surface epitopes may be processed and analyzed using Image Pro (Mediacybernetics), MetaMorph (Molecular Devices), or National Institutes of Health (NIH) image analysis software. Defining clusters of epitopes as regions with signal intensity three standard deviations above the background intensity, one may quantify the number, size, and location of high-density adhesion molecuses (i.e., β_2 integrins) on the neutrophil surface.

3.2.4. Quantifying Density of Cell Adhesion

It may be useful to correlate topographic redistribution of adhesion receptors with the kinetics of adhesion and efficiency of neutrophil recruitment to the endothelium under a given experimental condition. The surface density of rolling, arrested, and transmigrating neutrophils on the substrate may be used as an index of the efficiency of recruitment to the endothelium (*see* **Notes 4–6**).

1. Identify neutrophils by their phase-bright appearance in the same focal plane as the endothelial monolayer.
2. Adhesive surface densities are quantified by counting the number of rolling, arrested, or transmigrating neutrophils in an average of multiple (>5) microscopic fields and normalizing this number to the surface area contained in each field.
3. Arrested neutrophils are defined as those neutrophils that move less than one cell diameter in 30 s.
4. Rolling neutrophils are defined as having moved more than one cell diameter along the direction of shear flow.
5. Transmigrating neutrophils are defined as neutrophils that become phase-dark and significantly shape-change (aspect ratio >1.5), moving more than one cell diameter independently of the direction of shear flow.

4. Notes

1. There are several biological and technical challenges to performing immunofluorescent measurements in real time on moving cells. In order to avoid induced clustering of surface epitopes, one must use primary rather than robust secondary antibody labeling. Long exposure to high-intensity light excitation can alter both leukocyte and endothelial cell function as a result of phototoxicity *(8)*.
2. The motion of rolling or migrating leukocytes severely constrains the practical exposure time for image capture. The combination of these factors requires that the fluorescent or confocal microscope be optimized for maximal efficiency of specific light collection, with collection rates typically faster than five frames per second. Because real-time immunofluorescence requires high concentrations of primary labeling agents, it is also important to minimize the rate of sample consumption and dead volume in the flow chamber used to simulate blood shear flow.
3. To minimize autofluorescence of metabolic products that may interfere with leukocyte imaging, grow monolayers in phenol-red free medium.

4. When a microscopic field becomes unsuitable for further imaging as a result of photobleaching, move the stage to another region of the flow channel.
5. These data may be acquired concurrently with fluorescence imaging, and may be optimized by switching to a lower magnification phase contrast objective after **Subheading 3.1., steps 1–9** (at a cost of lower fluorescence resolution).
6. To ensure that a recombinant adhesion molecule such as ICAM-1 or E-selectin can be imaged upon binding, it is useful to employ multimeric forms of these molecules. This will result in a 5- to 10-fold longer lifetime of binding. Several companies now sell IgG and IgM chimeras of cell adhesion molecules and these can be directly conjugated to a fluorophore of choice. A second option is conjugation of recombinant molecules to quantum dots of 5- to 25-nm diameters. These are very bright and stable particles with a variety of conjugation chemistries available to ligate recombinant molecules in monomeric and polymeric valence.

Acknowledgments

This work was supported by NIH grants AI47294. (S. I. S.), AI060555 (U.Y.S), and by an National Science Foundation Center for Biophotonics Science and Technology.

References

1. Lum, A. F., Green, C. E., Lee, G. R., Staunton, D. E., and Simon, S. I. (2002) Dynamic regulation of LFA-1 activation and neutrophil arrest on intercellular adhesion molecule 1 (ICAM-1) in shear flow. *J. Biol. Chem.* **277,** 20,660–20,670.
2. Lupher, M. L. Jr., Harris, E. A., Beals, C. R., Sui, L. M., Liddington, R. C., and Staunton, D. E. (2001) Cellular activation of leukocyte function-associated antigen-1 and its affinity are regulated at the I domain allosteric site. *J. Immunol.* **167,** 1431–1439.
3. Shimaoka, M., Xiao, T., Liu, J. H., et al. (2003) Structures of the αL I domain and its complex with ICAM-1 reveal a shape-shifting pathway for integrin regulation. *Cell* **112,** 99–111.
4. Sarantos, M. R., Raychaudhuri, S., Lum, A. F., Staunton, D. E., and Simon, S. I. (2005) Leukocyte function-associated antigen 1-mediated adhesion stability is dynamically regulated through affinity and valency during bond formation with intercellular adhesion molecule-1. *J. Biol. Chem.* **280,** 28,290–28,298.
5. Green, C. E., Pearson, D. N., Camphausen, R. T., Staunton, D. E., and Simon, S. I. (2004) Shear-dependent capping of L-selectin and P-selectin glycoprotein ligand 1 by E-selectin signals activation of high-avidity β_2-integrin on neutrophils. *J. Immunol.* **172,** 7780–7790.
6. Kim, M., Carman, C. V., Yang, W., Salas, A., and Springer, T. A. (2004) The primacy of affinity over clustering in regulation of adhesiveness of the integrin $\alpha_L\beta_2$. *J. Cell Biol.* **167,** 1241–1253.
7. Saetzler, R. K., Jallo, J., Lehr, H. A., et al. (1997) Intravital fluorescence microscopy: impact of light-induced phototoxicity on adhesion of fluorescently labeled leukocytes. *J. Histochem. Cytochem.* **45,** 505–513.

14

Analysis of Neutrophil Polarization and Chemotaxis

Mary A. Lokuta, Paul A. Nuzzi, and Anna Huttenlocher

Summary

Neutrophil polarization and directed migration (chemotaxis) are critical for the inflammatory response. Neutrophil chemotaxis is achieved by the sensing of narrow gradients of chemoattractant and the subsequent polarization and directed migration toward the chemotactic source. Despite recent progress, the signaling mechanisms that regulate neutrophil polarization during chemotaxis have not been clearly defined. Here, we describe methods to analyze neutrophil polarization and asymmetric redistribution of signaling components induced by chemoattractant using immunofluorescence. Further, methods are described to dissect the role of specific signaling pathways during chemotaxis by the use of murine neutrophils from transgenic mouse models. Finally, methods for time-lapse microscopy and transwell assay for the analysis of neutrophil chemotaxis will also be discussed.

Key Words: Neutrophil; HL-60; polarization; chemotaxis; immunofluorescence; time-lapse microscopy.

1. Introduction

Neutrophils are key mediators of the immune response and are often the first responders to inflammatory stimuli *(1,2)*. They migrate rapidly up very shallow gradients of inflammatory mediators by the spatially restricted recruitment and activation of signaling molecules. Chemotaxis is achieved by two distinct processes: the actin-independent sensing of chemoattractant gradients and the subsequent actin-mediated neutrophil polarization, with the generation of a leading-edge pseudopod toward the source of chemoattractant *(3–9)*. The resting neutrophil is maintained in a rounded, non-adherent state and, in response to either a gradient or uniform concentration of chemoattractant, adopts a polarized morphology *(10)*. A hallmark of the polarized morphology is the asymmetric recruitment of both membrane and signaling components *(11)*. However, the mechanism by which this is achieved and how the responses may differ to uniform or gradients of different chemoattractants remains poorly understood. In fact, studies suggest that neutrophils display a hierarchical response to external

From: *Methods in Molecular Biology, vol. 412: Neutrophil Methods and Protocols*
Edited by: M. T. Quinn, F. R. DeLeo, and G. M. Bokoch © Humana Press Inc., Totowa, NJ

cues and are able to prioritize external stimuli. For example, a recent study demonstrates that neutrophils display preferential responses to end-target chemoattractants (*N*-formylmethionyl-leucyl-phenylalanine [fMLP]) as compared to intermediary chemoattractants (interleukin [IL]-8) by the activation of specific signaling pathways *(12–14)*. A major constraint to progress in understanding neutrophil motility has been the difficulty purifying resting, primary neutrophils from humans and mouse, and the limited tools available for manipulation. A powerful tool to study specific pathways of neutrophil function both in vitro and in vivo is by the study of neutrophils isolated from transgenic mouse models. Furthermore, analysis of neutrophils from patients can also provide important information about neutrophil motility and chemotaxis. Defects in neutrophil polarization and directed migration have been observed in patients with immune deficiency disorders, including leukocyte adhesion deficiency (LAD) and lazy leukocyte syndrome *(15–17)*, and more recently in a patient with chronic inflammatory disease *(18)*. Further, methods to analyze human and mouse neutrophil polarization using immunofluorescent detection of the actin *(19)* and microtubule cytoskeleton *(20)* and lipid membrane domains *(21)* are also discussed. Finally, analysis of neutrophil random and chemotactic migration using transwell assay and time-lapse microscopy will be presented *(22)*.

2. Materials

2.1. Purification of Human and Mouse Primary Neutrophils

2.1.1. Common Supplies

1. 50 mL, 17×120 mm and 15 mL, 17×120 mm screw-cap conical tubes (Sarstedt).
2. 18 G $\times$ 1.5 inch, 21 G $\times$ 1.5 inch, 22 G $\times$ 1.5 inch, and 30 G $\times$ 0.5 inch needles (Becton Dickinson).
3. Hemocytometer.

2.1.2. Buffers and Solutions

1. Dulbecco's phosphate-buffered saline (DPBS)/human serum albumin (HSA)/Hep: DPBS without Ca^{2+} or Mg^{2+} (Mediatech; *see* **Note 1**) containing 0.15% HSA (American Red Cross NDC 52769-251-05) and 10 U/mL preservative-free heparin.
2. ACK buffer: 155 mM NH_4Cl, 10 mM $KHCO_3$, and 127 µM EDTA made in sterile irrigation H_2O (siH_2O; Baxter Healthcare Corporation).
3. 70% EtOH: 700 mL of 100% ethyl alcohol USP and 300 mL of distilled and deionized (dd)H_2O.
4. Histopaque®-1077, Histopaque-1083, and Histopaque-1119 (all from Sigma).
5. 3% sterile, aged thioglycollate solution: dissolve 30 g of thioglycollate medium in 1 L of cold, distilled H_2O, and heat to boiling in order to completely dissolve the powdered medium. Autoclave 50- to 100-mL aliquots, and store at room temperature in the dark for at least a month prior to use.

6. USP: preservative-free sodium heparin at 1,000 IU/mL (American Pharmaceutical Partners Inc. NDC 63323-276-02).
7. Polymorphprep™ (Nycomed, Oslo, Norway).
8. Phosphate-buffered saline (PBS; Sigma #P3813) made up to 5X in 200 mL siH₂O (*see* **Note 1**).
9. DPBS without Mg^{2+} and Ca^{2+} (DPBS; Mediatech).

2.1.3. Purification of Murine Neutrophils From Peripheral Blood

1. Mice of at least 6–8 wk of age.
2. Gauze pads.
3. Halothane (Halocarbon Laboratories). Halothane is a hepatotoxin and should only be used in a fume hood in conjunction with an institution approved animal use protocol.
4. Insulin syringe (Becton Dickinson).

2.1.4. Purification of Murine Neutrophils From Bone Marrow

1. Mice of at least 6–8 wk of age.
2. Institution approved apparatus for euthanasia.
3. Forceps, straight blade surgical scissors, and delicate surgical scissors (Fisher).
4. 10-mL luer-lock™ syringe (Terumo Medical Corporation).
5. 100 × 15 mm polystyrene Petri dish (Fisher).
6. 70-μm nylon cell strainer (BD Biosciences).

2.1.5. Purification of Intraperitoneally Induced Murine Neutrophils

1. Mice of at least 6–8 wk of age.
2. 3-mL luer-lock syringe (Becton Dickinson).
3. Institution approved apparatus for euthanasia.
4. Forceps.
5. Straight-blade surgical scissors (Fisher).

2.2. Analysis of Neutrophil Polarization

2.2.1. Neutrophil Treatment for Immunofluorescent Staining

1. Polystyrene 24-well plate (Falcon).
2. 12-mm round cover glass (Fisher).
3. Fibrinogen (Sigma).
4. DPBS.
5. EGM-2MV (Cambrex Biosciences).
6. Neutrophil activators: fMLP (Sigma #F3506), recombinant complement factor 5a (C5a; Sigma #C5788), leukotriene B_4 (LTB_4; Calbiochem #434625), and/or recombinant IL-8 (Sigma #I1645).
7. 1.5-mL microfuge tubes.
8. Dulbecco's phosphate buffered saline with Mg^{2+} and Ca^{2+} (DPBS+; Mediatech; *see* **Note 1**).
9. Water-jacketed tissue culture incubator set to 37°C and 10% CO_2.

2.2.2. Cell Fixation and Staining for Actin or G_M3

1. PBS made up to 1X in ddH_2O (Sigma; *see* **Note 1**).
2. Fixative: 6.6% paraformaldehyde (Electron Microscopy Sciences), 0.05% glutaraldehyde (Sigma), and 0.25 mg/mL saponin (Alfa Aesar) in PBS. This reagent should be prepared fresh daily.
3. Quenching solution: 0.15 M glycine in PBS. This reagent should be prepared fresh daily.
4. The only difference in fixation procedures for actin and G_M3 is the blocking buffer. Actin blocking buffer: 10% heat inactivated fetal bovine serum (FBS) and 0.25 mg/mL saponin in PBS for the visualization of actin. G_M3 blocking buffer: 10% heat-inactivated FBS in PBS.
5. For actin staining: anti-β-actin mouse monoclonal antibody AC15 (Sigma #A1978) and fluorescein isothiocyanate (FITC)-conjugated sheep anti-mouse immunoglobulin (Ig)G (ICN) or rhodamine-conjugated phalloidin (Molecular Probes).
6. For G_M3 staining: anti-G_M3 (Seikagaku #370695) and FITC-conjugated Fab fragment goat anti-mouse IgM (Jackson Immunoresearch).
7. 2-(4-Amidinophenyl)-6-indolecarbamidine dihydrochloride (DAPI; Sigma) at 10 mg/mL in ddH_2O.
8. Glass microscope slides (Fisherbrand).
9. Mounting media such as Vectashield® (Vector).

2.2.3. Cell Fixation and Staining for Tubulin

1. Fixative: 0.7% glutaraldehyde (Sigma) in PBS (*see* **Note 1**).
2. Permeabilization buffer 1: 0.5% Triton X-100 in PBS.
3. Permeabilization buffer 2: 0.5% sodium dodecyl sulfate in PBS.
4. Reducing agent: 1.5 mg/mL sodium borohydride ($NaBH_4$) in PBS. Sodium borohydride is a powerful reducing agent which releases hydrogen gas when added to aqueous solutions. Add the $NaBH_4$ to PBS in a fume hood. Keep the lid of the vessel only loosely tightened to allow the vessel to vent for at least 1 hr prior to use. This reagent should be prepared fresh daily.
5. Blocking solution: 10% heat-inactivated FBS in PBS.
6. FITC-conjugated monoclonal anti-α-tubulin clone DM1A (Sigma #F2168).
7. DAPI.
8. Glass microscope slides (Fisherbrand).
9. Mounting media.

2.3. Analysis of Neutrophil Chemotaxis

2.3.1. Transwell Assay

1. 3-μm-pore transwell filters (Costar).
2. Fibrinogen.
3. EGM-2MV (Cambrex Biosciences). HyQ® CCM1™ is an acceptable serum-free alternative for primary neutrophils (Hyclone).
4. Neutrophil activators, such as fMLP, C5a, LTB_4, and/or IL-8.

5. 0.5 *M* ethylenediamine tetraacetic acid (EDTA), pH 7.4.

2.3.2. Live Imaging of Chemotaxing Cells

1. For nonfluorescence time-lapse microscopy: 35 × 10 mm non-tissue-culture plastic Petri dish (Falcon).
2. For fluorescence time-lapse microscopy: 35 × 10 mm non-tissue-culture plastic Petri dish with an 18-mm-diameter hole cut in the center of the dish using a drill press.
3. For fluorescence time-lapse microscopy: 22-mm-diameter microscope cover glass (Fisherbrand).
4. For fluorescence time-lapse microscopy: Norland Optical Adhesive 68 (Norland Products).
5. For fluorescence time-lapse microscopy: ultraviolet light source, such as UVP ultraviolet (UV) Transilluminator and/or a tissue culture hood UV light source.
6. For fluorescence time-lapse microscopy: fibronectin, purified according to Ruoslahti *(23)*.
7. For both fluorescence and nonfluorescence time-lapse microscopy: fibrinogen.
8. Appropriate culture medium for the cells. We routinely use EGM-2MV for primary neutrophils and Gey's medium for HL-60 cells *(24)*: 138 mM NaCl, 6 mM KCl, 1 mM Na$_2$HPO$_4$, 5 mM glucose, 20 mM HEPES, 1 mM CaCl$_2$, 1 mM MgSO$_4$ and 0.5% HSA (American Red Cross). HyQ CCM1 is an acceptable serum-free alternative for primary neutrophils (Hyclone).
9. Chemotactic agents, such as 12.5 mM IL-8, 2.5 µg/mL C5a, 25 µg/mL LTB$_4$, or 10 µM fMLP.
10. Inverted fluorescent microscope with both 20X air and 60X oil differential interference contrast (DIC) objectives and a cooled charge coupled device (CCD) camera.
11. A system to maintain sample temperature at 37°C, such as The Box (Life Imaging Services).
12. Software capable of capturing time-lapse and/or dual wavelength time-lapse image series, such as the MetaMorph® Imaging System (Universal Imaging Corporation).
13. Microinjection system, such as Eppendorf Femtojet® microinjector.
14. Microloader pipet tips and Femtotips® sterile injection capillary (Eppendorf).

3. Methods

3.1. Purification of Primary Neutrophils (see also *Chapters 2 and 3*)

There are multiple sources of murine neutrophils which differ from one another primarily by maturity. Elicited peritoneal neutrophils are activated cells and most responsive to stimuli. Peripheral blood neutrophils are fully mature and capable of responding to appropriate activators. They also exhibit a very low basal activation state. Bone marrow-derived neutrophils are likely to be at various stages of neutrophil differentiation with some immature neutrophils in the preparation. However, an important consideration is the number of neutrophils obtained from each compartment. For example, we typically get 6–12 × 10^5 peripheral blood neutrophils/mouse. This number is quite small and limits the downstream applications. Therefore, neutrophils from the other compartments may be more appropriate depending on the experimental conditions. The cell yields can vary greatly depending on the strain, age, and size of mouse being studied; however, using 8- to 12-wk-old wild-type mice, we typically obtain between 8 and 13×10^6 neutrophils per mouse from the bone marrow and 1-5×10^7 intraperitoneally induced neutrophils per mouse (*see* **Note 2**).

3.1.1. Purification of Murine Neutrophils From Peripheral Blood

1. Place one to two pieces of gauze into the bottom of a 50-mL conical tube. In a fume hood, add 500 µL Halothane or other institution-approved anesthetic directly to the gauze and recap the tube until use.
2. Restrain a mouse by holding it securely with one hand and placing it on a surface to which it can grasp firmly. Gently pull on its tail to encourage the mouse to grasp the surface firmly. With the thumb and forefinger, grasp the mouse behind both ears at the scruff of the neck close to the base of the skull. Then tuck the tail under the pinkie finger.
3. While restraining the mouse, hold its head just inside the conical tube containing the anesthetic until it is unresponsive to a *firm* pinch of its hind footpad. Continue holding the mouse while it is being anesthetized so that it does not crawl into the tube and inhale too much anesthetic. This is important because the heart must continue beating in order to obtain as much of the blood as possible.
4. Perform a cardiac puncture by laying the animal on its back with the head furthest away. Rinse the chest area with 70% EtOH. Placing counter-tension on the abdomen and using an insulin syringe containing 100 µL of heparin, enter the chest just below the xiphoid process at approx 25–30° from the chest and angle toward the heart.
5. If no blood is visible when pulling on the plunger, pull the syringe out slightly without removing it from the skin and try approaching the heart from a different angle.
6. Once blood appears, still holding the syringe, slowly withdraw the plunger, continuously drawing blood.
7. If blood flow into the syringe stops, try to reposition the needle. One should obtain 0.8–1.2 mL of blood per mouse *(25)*.
8. Add at least an equal volume of DPBS/HSA/Hep to the blood, and overlay mixture onto a discontinuous gradient of (starting from the bottom) 6 mL of Histopaque-1119, 2 mL of Histopaque-1083, and 1.5 mL of Histopaque-1077 in a 15-mL conical tube. Up to 5 mL of diluted blood may be overlayed onto this gradient.
9. Centrifuge gradient for 30 min at 700*g* and room temperature with the brake off. After the spin, there will be two leukocyte bands.
10. Aspirate the uppermost layer, which lies between the Histopaque-1077 and the Histopaque-1083 and contains the monocytic cells.
11. Transfer the neutrophil layer, which lies between the Histopaque-1119 and the Histopaque-1083 layers, to a 50-mL conical tube.
12. Rinse these cells with DPBS/HSA/Hep, and centrifuge at 500*g* for 10 min. The brake may be applied once the centrifuge has slowed to below 200*g* (*see* **Note 3**).

13. Aspirate, rinse, and resuspend neutrophils in 25 mL of ACK buffer for 3 min in order to lyse the red blood cells.
14. Bring the volume to 50 mL with DPBS/HSA/Hep, and centrifuge at 500*g* for 10 min.
15. **Steps 13** and **14** may be repeated if too many red blood cells remain (*see* **Note 4**).
16. Resuspend the neutrophil pellet in 100–200 µL of DPBS per mouse.
17. Count cells by hemocytometer and visually assess their purity (*see* **Note 5**).

3.1.2. Purification of Murine Neutrophils From Bone Marrow

1. Euthanize the mouse, and quickly remove the long bones of the limbs.
2. Using the delicate scissors, quickly clean the tissue from the surface of the bones.
3. Cut the tips off both ends of the long bones. Flush the marrow compartment of the bones into an iced Petri dish by inserting a 30-G needle on a 10-mL syringe filled with ice-cold DPBS/HSA/Hep. Marrow from the same strain can be pooled into a single dish.
4. After harvesting marrow from all of the mice, gently disperse the marrow cells with an 18-G needle on a 10-mL syringe, and transfer into a 50-mL conical tube.
5. Bring volume to 50 mL with DPBS/HSA/Hep, and centrifuge at 500*g* for 5 min (*see* **Note 3**).
6. Resuspend cell pellet in 5 mL of DPBS/HSA/Hep, filter through a 70-µm cell sieve, and overlay onto a discontinuous gradient of (starting from the bottom) 1 mL of Histopaque-1119 and 5 mL of Histopaque-1083.
7. Centrifuge tubes (700*g*, 30 min) at room temperature with the brake off.
8. Remove all but the last 1.5–2 mL, and add 25 mL of ACK buffer directly to this bottom fraction. Transfer mixture to a 50-mL conical tube, and incubate at room temperature for 3 min.
9. Bring the volume to 50 mL with DPBS/HSA/Hep, and spin at 500*g* for 5 min to rinse.
10. Resuspend cell pellet in 0.5–1 mL of DPBS per mouse, and count using a hemocytometer (*see* **Note 5**).

3.1.3. Purification of Intraperitoneally Induced Murine Neutrophils

1. Inject 2 mL of aged thioglycollate into the peritoneal space with a 22-G needle on a 3-mL syringe. To do this, immobilize the mouse with one hand and, with the needle bevel facing upward, inject into the lateral abdominal fur crease at approximately the level of the knee *(26)*.
2. After 4 h, euthanize the mouse.
3. Lay the animal on its back with its head furthest away, and clean the abdominal area with 70% EtOH.
4. Taking care to maintain the integrity of the membrane, remove the skin to expose the peritoneal membrane. To remove the fur and skin, use a blunt pair of forceps and pull upward at centerline of the abdomen. Clip a small hole through the skin. Keeping the blunt side of the straight blade scissors down gently insert the tip of the scissors and use them to separate the skin from the peritoneal membrane. Reflect the skin back around the abdomen.
5. Inject 3 mL of ice cold DPBS/HSA/Hep into the peritoneal cavity with a 21-G needle on a 3-mL syringe.

6. While holding the upper body of the mouse stationary, grasp the mouse at the base of the tail and shake or massage the abdomen vigorously for one full minute.
7. Lay the mouse on its side and insert the needle with the bevel facing the back of the mouse.
8. Slowly withdraw the fluid from the peritoneal space, taking care not to suck tissues into the needle. To facilitate this, keep a small amount of tension on the membrane to form a pouch into which the lavage fluid can drain before being removed.
9. The fluid should be cloudy but relatively free of red blood cells. Approximately 2.5 mL of lavage fluid should be recovered.
10. Quickly transfer the lavage fluid to a 50-mL conical tube on ice.
11. Bring volume to 50 mL with cold DPBS/HSA/Hep, and centrifuge at $500g$ for 5 min at 4°C (*see* **Note 3**).
12. Resuspend cell pellet in 5–10 mL of ACK for 3 min as needed to lyse the red blood cells.
13. Bring to 50 mL with cold DPBS/HSA/Hep, and centrifuge at $500g$ for 5 min at 4°C.
14. Resuspend cell pellet in 1–2 mL of DPBS/HSA/Hep per mouse, count cells on the hemocytometer, and resuspend to $\leq 5 \times 10^6$ cells/mL (*see* **Note 5**).

3.1.4. Purification of Human Peripheral Blood Neutrophils

Neutrophils are easily activated, and special steps must be taken to prevent the cells from becoming activated during the purification (*see* **Notes 3–5**). It is important to ensure that the neutrophils are quiescent and have a rounded, non-polar morphology. We routinely visualize the cells after the purification process and will only proceed with the planned experiments if most of the neutrophils are in the rounded, quiescent morphology. When working with human neutrophils, a detailed Human Subjects Protocol must be approved prior to starting the experiment.

1. Draw human peripheral venous blood from a healthy donor into a syringe containing preservative-free heparin to a final concentration of 100 IU/mL.
2. Neutrophil purification should be performed within 2 h of the blood draw. The use of glass vacutainers and tubes should be avoided as the neutrophils will adhere to the glass surface.
3. Gently overlay 5–7.5 mL of blood onto an equal volume of Polymorphprep™ in a 15-mL conical tube.
4. Centrifuge gradients at $500g$ for 25 min and 18–22°C with the brake turned off. Prior to the completion of the spin increase to $700g$ for an additional 10 min with the brake off (*see* **Note 4**).
5. Upon completion of the spin, two leukocyte bands should be visible. The mononuclear band is the uppermost of the two and is directly below the plasma layer. The granulocyte band is between the clear and the pink media. The red cells usually pellet below the gradient.
6. Remove the plasma and top, mononuclear cell layer. The plasma may be retained if an autologous source of plasma is required for a downstream application.

7. Remove the neutrophil layer, and combine neutrophils from up to 6 of these gradient tubes into a 50-mL conical tube.
8. Bring the volume to 50 mL with 1X DPBS and centrifuge at 500g for 10 min to remove residual Polymorphprep media (*see* **Note 3**).
9. Remove and discard supernatant.
10. Resuspend the cell pellet in 8 mL of siH_2O for 30 s to lyse the red blood cells, and immediately add 2 mL of 5X PBS. Start timing the incubation as soon as the siH_2O is added to the pellet; however, it is a good idea to gently pipet up and down once to resuspend the cell pellet in the siH_2O. (*see* **Note 6**). An additional lysis and spin may also be added, however, keep in mind that this increases the likelihood that the neutrophils may become activated (*see* **Note 5**).
11. Centrifuge at 400g for 5 min at room temperature.
12. Remove supernatant, and resuspend neutrophils in 1X DPBS at a concentration $\leq 5.0 \times 10^6$. Store protected from light at 4°C to prevent the cells from clumping and activating (*see* **Note 5**).

3.2. Cell Treatment for Immunofluorescenct Staining

1. Place a coverslip into as many wells of a 24-well plate as needed. It is a good idea to run each sample in duplicate.
2. Coat the coverslips with 10 µg/mL of fibrinogen in DPBS for 1 h at 37°C.
3. Rinse coverslips twice with 1X DPBS, and keep them in DPBS at 37°C until use.
4. Dilute neutrophils to 0.5×10^6 cells/mL in DPBS and pretreat if desired. If the desired treatment of the cells requires the addition of a reagent which is not particularly soluble in aqueous solutions, dilute the cells to 1×10^6 cells/mL. Add an equal volume of this cell dilution to a neutrophil culture media such as EGM-2MV that contains the reagent and proceed with the pretreatment prior to plating the cells.
5. Add 250 µL of DPBS+ to each of the wells.
6. Add 250 µL of cells with an appropriate activator to each of the wells, and incubate at 37°C and 10% CO_2. We routinely use a final concentration of 100 nM fMLP, 1.25 nM IL-8, 1.25 ng/mL C5a, or 25 ng/mL LTB_4 for 10 min 37°C and 10% CO_2, depending on the experiment (*see* **Note 7**).

3.3. Cell Fixation Techniques

It is important to note that no single fixation technique works for every antibody, and modifications of fixation protocols are often required to optimize a given antigen for optimal antibody recognition. In this chapter, we provide details for the examination of polarized actin *(27)*, G_M3-containing lipid rafts *(28)*, and tubulin *(29)* in neutrophils.

3.3.1. Cell Fixation and Staining for Actin (see **Fig. 1**)

1. Gently aspirate treatment media, and carefully add 500 µL of fixative directly to the wells for 15 min at room temperature.
2. Rinse coverslips twice with PBS, and add 500 µL of 0.15 M glycine for 15 min at room temperature.

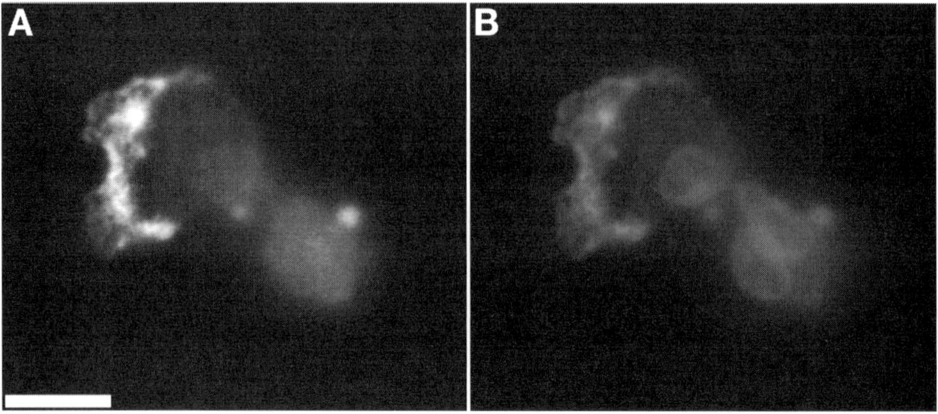

Fig. 1. Immunofluorescent staining for actin in a primary human neutrophil. Neutrophils were stimulated with 100 nM fMLP for 15 min at 37°C and 10% CO_2. Cells were fixed and stained with rhodamine-conjugated phalloidin and counterstained with DAPI, as described. Actin is shown in **A**, and the overlay is shown in **B**. Images were taken with a 100× oil-immersion objective. Bar, 10 µm. (color insert will appear after page 300)

3. Rinse coverslips twice with PBS, and add 500 µL of blocking buffer. It is best to block overnight at 4°C.
4. Transfer the coverslips to a piece of Parafilm placed in a humidified chamber. We include wetted Whattman paper under the Parafilm.
5. Add 100 µL of AC15 antibody diluted 1:1000 in blocking buffer per coverslip, and incubate for 30 min at room temperature.
6. Rinse the coverslips three times with PBS.
7. Add 100 µL of FITC-conjugated sheep-anti mouse IgG diluted 1:250 in blocking buffer, and incubate for 30 min at room temperature. The inclusion of DAPI at 1:10,000 with the secondary antibody is recommended for neutrophils. Rhodamine-conjugated phalloidin can also be used to stain the F-actin (*see* **Note 8**).
8. Rinse the coverslips three times with PBS. Leave the last wash on for 30 min.
9. Using 3–5 µL of mounting media on a glass slide, invert coverslip onto the mounting media, and allow it to dry at room temperature protected from light.
10. Seal the edges of the coverslips with nail polish, and allow them to dry thoroughly at room temperature protected from light before trying to visualize the samples.

3.3.2. Cell Fixation and Staining for G$_M$3 (see **Fig. 2**)

1. Follow **Subheading 3.3.1., steps 1–4**. Note that the blocking buffer for actin staining contains saponin whereas that for G$_M$3 does not.
2. Add 50 µL of anti-G$_M$3 antibody diluted 1:50 in blocking buffer to each coverslip, and incubate for 1 h at room temperature.
3. Rinse the coverslips three times with PBS.

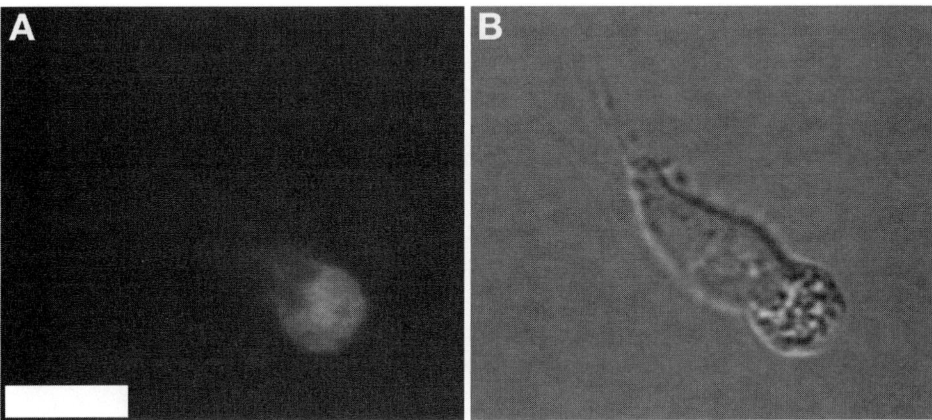

Fig. 2. Immunofluorescent staining for G_M3 in a primary human neutrophil. Neutrophils were stimulated with 100 n*M* fMLP for 15 min at 37°C and 10% CO_2. Cells were fixed and stained with anti-G_M3 antibody, as described. A differential interference contrast image of the cells was also captured to show cell shape and position of immunofluorescent staining. G_M3 is shown in **A**, and the overlay is shown in **B**. Images were taken with a 60× oil-immersion objective. Bar, 10 µm. (color insert will appear after page 300)

4. Add 100 µL of FITC-conjugated goat-anti mouse IgM diluted 1:200 in blocking buffer, and incubate for 30 min at room temperature. The inclusion of DAPI at 1:10,000 with the secondary antibody is recommended for neutrophils (*see* **Note 8**).
5. Rinse the coverslips three times with PBS. Leave the last wash on for 30 min.
6. Using 3–5 µL of mounting media on a glass slide, invert coverslip onto the mounting media and allow it to dry at room temperature protected from light.
7. Seal the edges of the coverslips with nail polish and allow them to dry thoroughly at room temperature protected from light before trying to visualize the samples.

3.3.3. Cell Fixation and Staining for Tubulin (see **Fig. 3**)

1. Gently aspirate treatment media, and carefully add 500 µL of 0.7% glutaraldehyde directly to the wells for 15 min at room temperature.
2. Rinse coverslips twice with PBS, add 500 µL of 0.5% Triton X-100, and incubate for 15 min at room temperature.
3. Rinse coverslips twice with PBS, add 500 µL of 0.5% SDS, and incubate for 15 min at room temperature.
4. Rinse coverslips twice with PBS, add 1 mL of 1.5 mg/mL $NaBH_4$, and incubate for 30 min at room temperature.
5. Rinse coverslips twice with PBS, and add 500 µL of blocking buffer. It is best to block overnight at 4°C.
6. Transfer the coverslips to a piece of Parafilm placed in a humidified chamber. We include wetted Whattman paper under the Parafilm.

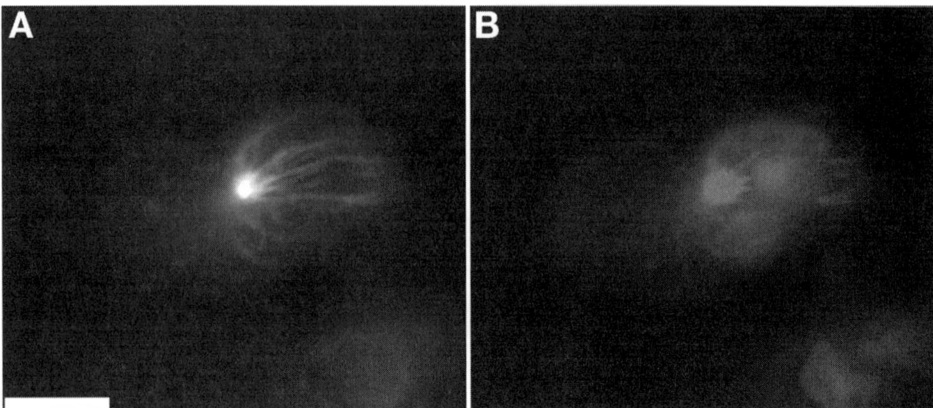

Fig. 3. Immunofluorescent staining for tubulin in a primary human neutrophil. Neutrophils were treated as described in **Subheading 3.2.** and stimulated with 100 n*M* fMLP for 15 min at 37°C 10% CO_2. Cells were then fixed and stained with fluorescein isothiocyanate-conjugated anti-tubulin antibody and counterstained with DAPI as described in **Subheading 3.3.3.** Actin is shown in **A** and the overlay is shown in **B**. Images were taken with a 100× oil-immersion objective. Bar, 10 µm. (color insert will appear after page 300)

7. Add 100 µL of FITC-conjugated tubulin diluted 1:200 in blocking buffer to each coverslip, and incubate for 30 min room temperature. The inclusion of DAPI at 1:10,000 with the tubulin antibody is recommended for primary neutrophils (*see* **Note 8**).
8. Using 3–5 µL of mounting media on a glass slide, invert coverslip onto the mounting media and allow it to dry at room temperature protected from light.
9. Seal the edges of the coverslips with nail polish and allow them to dry thoroughly at room temperature protected from light before visualizing the samples.

3.4. Analysis of Chemotaxing Cells

Numerous techniques have been used to examine chemotaxis including chemotaxis under agarose *(30,31)*, Dunn chemotactic chambers *(32)*, and microfluidic devices *(33,34)*. In this chapter we will describe modifications of a widely used Boyden Millipore filter or transwell assay. Further, live time-lapse microscopy of chemotaxing cells is also discussed.

3.4.1. Transwell Assay

The transwell assay consists of two chambers separated by a membrane of defined pore size. Neutrophils are placed in the top chamber and upon completion of the assay, the number of cells which migrate through the membrane is determined, indicating the amount of migration *(35)*. This can be used to examine cell migration under a variety of conditions. Random migration can be assayed by the addition of an inflammatory mediator in both the top and bottom chambers. Chemotactic migration can be assayed by placing the chemoattractant only in the bottom chamber. The transwell assay has the advantage of being able to simultaneously perform replicates and does not require expensive equipment to perform. The disadvantages include the difficulty in distinguishing between enhanced random cell motility and directed cell migration (i.e., chemokinesis vs. chemotaxis) and the relatively steep gradient that must be formed in order to see chemotaxis *(36,37)*.

1. Coat two 3-µm-pore, 12-mm-diameter transwells for each experimental condition with 2.5 µg/mL of fibrinogen. To coat, add 600 µL in the bottom well and 200 µL on top of the membrane, and incubate at 37°C for 1 h (*see* **Note 9**).
2. Wash transwells twice with DPBS. It is important to thoroughly remove all DPBS and have the membranes completely dry prior to the start of the assay. Otherwise, media will leak from one compartment to the next, and the gradient will not be properly formed. These can be dried in a laminar flow hood for at least 2–3 h and up to overnight.
3. For duplicate samples, bring 8×10^5 cells to 400 µL in appropriate medium for each experimental condition and pretreat if desired (*see* **Note 10**).
4. Add 600 µL of medium with or without chemoattractant to the bottom chamber and 200 µL of this cell preparation to the top of each of two filters. For chemoattractants we routinely use 1.25 n*M* IL-8, 1.25 ng/mL C5a, 10–100 n*M* fMLP, or 25 ng/mL LTB$_4$ (*see* **Note 7**). It is a good idea to include a media-alone control to ensure increases in migration are truly being assayed.
5. Incubate the samples for 1–3 h at 37°C and 10% CO_2.
6. Add 60 µL of 0.5 *M* EDTA to the bottom chamber, and place the plate at 4°C for 10–15 min. This step helps to dislodge the neutrophils from the bottom of the filter.
7. Remove the filters from the well, and count the numbers of cells which migrated into the bottom chambers. The results can be expressed by cell number per milliliter.

3.4.2. Live Nonfluorescent Imaging of Chemotaxing Cells (see *Fig. 4*)

This assay has the advantage of being able to visualize individual cells, track them, and determine their speed and directional persistence *(38)*. The primary disadvantage to this technique is the requirement for relatively expensive, specialized equipment.

1. Coat 35×10 mm non-tissue-culture plastic Petri dishes with 2 mL of 2.5 µg/mL fibrinogen in DPBS for 1 h at 37°C.
2. Rinse dishes twice with DPBS, and keep moist with DPBS at 37°C until use.
3. Pre-equilibrate culture medium, such as EGM-2MV, to 10% CO_2 and 37°C.
4. Plate 1×10^5 cells in 3 mL of culture media, and incubate at 10% CO_2 and 37°C for 30 min.
5. Place 5–10 µL of chemoattractant in a 0.5-mL microfuge tube, and spin at 14,000*g* for 5 min to clarify (*see* **Note 11**).

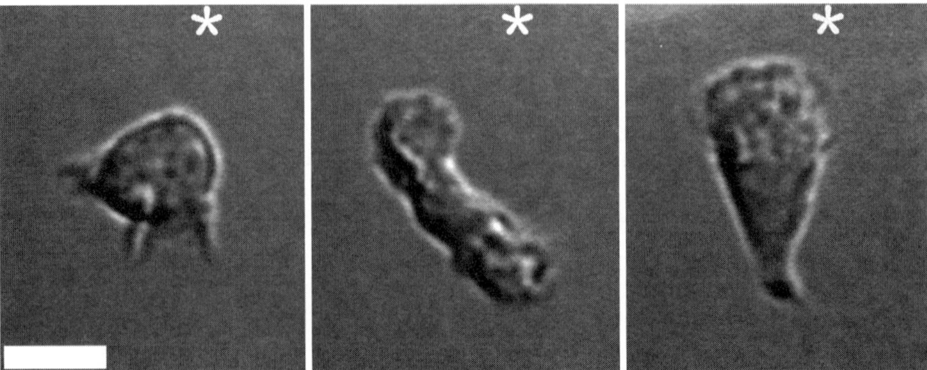

Fig. 4. Differential interference contrast images of a chemotaxing differentiated HL-60 cell. Differentiated HL-60 cells were plated onto glass bottom plates in Gey's medium, as described. The Femtotip® was loaded with 45 μM C5a and set with a back pressure of 12 hPa. Images shown were captured at 2-min intervals with a 60× oil-immersion objective. The asterisk indicates the direction of the opening of the Femtotip in each image. Bar, 10 µm.

6. Using the microloader pipet tips, load approx 2–3 µL chemoattractant into the tip of the Femtotip, taking care to ensure that there are no air pockets within the chemoattractant (*see* **Note 11**).
7. Install the Femtotip on the microinjector, and apply a back pressure of 10 hPa for neutrophils or 12 hPa for HL-60 cells.
8. Place the cell sample on the microscope stage with the 20× DIC objective in place, and drop the Femtotip until it just breaks the top surface of the medium. Then find the tip of the Femtotip through the eyepieces.
9. After locating the tip of the Femtotip, focus slightly below the plane of the tip, then drop the Femtotip just to the level of focus, but not beyond. Repeat this process, taking care to note when the cells come into view in the background. When the cells become visible on the dish surface, proceed cautiously so as not to touch the tip of the Femtotip to the surface of the dish.
10. Bring the cells into focus, being sure to adjust the Köhler illumination and optimize all the settings on the image acquisition software.
11. Bring the Femtotip down close to, but not touching the surface of the dish and move to a clean, nonpolarized field of cells. Begin capturing as quickly as possible, as the neutrophils will respond quite rapidly (*see* **Note 12**).
12. Images can be captured every 1–15 s, depending on what is being assayed.

3.4.2. Live Fluorescent Imaging of Chemotaxing Cells

This assay can be used to examine the subcellular localization of fluorescently tagged proteins of interest during chemotaxis. Details regarding the generation of stable HL-60 cell lines expressing fluorescently-tagged proteins are given elsewhere *(35)*.

1. Several days prior to the experiment, prepare glass bottom plates by applying four to five drops of optical adhesive to the underside of the Petri dishes just around the edge of the 18-mm hole (*see* **Note 13**).
2. The drops should each be approx 2–3 mm in diameter. Using a pipet tip, spread adhesive around the hole while trying to avoid the cut edge of the hole.
3. Apply the 22-mm cover glass to the adhesive on the underside of the Petri dish. Make sure that the cover glass makes firm contact with the adhesive and that no gaps exist, because media could leak through onto the microscope.
4. The adhesive must be activated by exposure to an ultraviolet light source with maximum absorption at 350–380 nm. The cure time is dependent on the wavelength and strength of the light source and the distance from it. Expose both sides to the light source making sure to remove the lid during the exposure (*see* **Note 14**).
5. On the day of the experiment, coat the cover glass with 100 µg/mL of fibronectin and 2.5 µg/mL of fibrinogen in 750 µL of DPBS for 1 h at 37°C.
6. Rinse twice with DPBS, and keep moist with DPBS at 37°C until use.
7. Resuspend 5×10^5 differentiated HL-60 cells in 3 mL of Gey's medium that has been pre-equilibrated to 37°C and 5% CO_2, and incubate for 10 min at 37°C and 5% CO_2 to plate cells.
8. Place 5–10 µL of chemoattractant in a 0.5-mL microfuge tube, and spin at 14,000*g* for 5 min to clarify (*see* **Note 11**).
9. Apply a very thin coating of immersion oil to the entire bottom surface of the cover glass. It important to ensure that this coating is thin to avoid contact with a 20× non-oil objective. Once this is applied, do not set the dish down on anything.
10. Place the dish on the stage, and find the cells with the 20× objective, taking care not to touch the oil on the bottom of the dish.
11. Using the microloader pipet tips, load approx 2–3 µL of chemoattractant into the tip of the Femtotip, taking care to ensure that there are no air pockets within chemoattractant (*see* **Note 11**).
12. Install the Femtotip on the microinjector, and apply a back pressure of 12 hPa.
13. Lower the Femtotip until it just breaks the top surface of the medium. Then find the tip of the Femtotip through the eyepieces by focusing well above the plane of the cells.
14. After locating the tip of the Femtotip, drop the Femtotip just to the level of focus but not beyond by focusing below the plane of the tip. Repeat this process, taking care to note when the cells come into view in the background. The tip must be relatively close to the dish and in the center of the field if it is to be found quickly with the higher objective.
15. Using the proper procedure for the microscope, switch to the 60× objective and bring the cells into focus. The Femtotip should be visible, although it will be out of focus. Adjust the Köhler illumination and optimize the settings on the image acquisition software to acquire both DIC and fluorescent images.

16. Bring the Femtotip down close to, but not touching, the surface of the cover glass and move to a clean, nonpolarized field of cells. Begin capturing as quickly as possible, as the HL-60 cells will respond quite rapidly (*see* **Note 12**).
17. Images can be captured every 1–15 sec, depending on what is being assayed.

4. Notes

1. There are several types of phosphate-buffered saline used in this chapter, and it is important to distinguish between them. DPBS without Ca^{2+} or Mg^{2+} is referred to as DPBS. DPBS with Ca^{2+} and Mg^{2+} is designated DPBS+. Powdered PBS made up to 1X or 5X is referred to as PBS. In general, DPBS or DPBS+ are used whenever the cells are responsive, because these buffers are cell-culture-tested. The powdered PBS is used primarily for rinses during the staining procedures.
2. It should be noted that appropriate institutional approval of an animal protocol should be obtained prior to performing any of the mouse procedures. For additional information on the anatomy of mice, refer to **ref. *39***.
3. Neutrophils are susceptible to activation. Even the vibration caused by applying the brake during centrifugation can cause neutrophil activation. However, we have found no adverse effects during washes when a low brake is applied after the speed falls below 200*g*.
4. A hemocytometer is more useful for counting the neutrophil preparations than a Coulter Counter because the purity of the preparation can readily be assessed. An additional benefit is the observation that live neutrophils will rapidly adhere to and spread on the glass surface. This observation allows confirmation that the neutrophils are responsive.
5. Neutrophils are easily activated during the purification process. To improve the likelihood that the neutrophils are as quiescent as possible, the purification procedures should be performed as quickly as possible. Do not allow the cells to remain in the Polymorphprep beyond the length of the required spin. Do not allow the cells to remain pelleted for any significant length of time. The centrifuge should be very carefully balanced during all spins, because vibrations can induce neutrophil activation. Do not keep the cells at concentrations greater than 5×10^6 cells/mL, as this will cause them to activate. Neutrophils will bind avidly to glass, so the use of glass vessels, vacutainers, etc. is not recommended.
6. Occasionally there will be a significant amount of red cell contamination if the red blood cells fail to pellet completely. In this case, the siH_2O and 5X PBS can be scaled appropriately based on a visual assessment of the degree of red cell contamination.
7. Murine neutrophils will not recognize human IL-8 but chemotax nicely to C5a and LTB_4 and fairly well to fMLP. HL-60 cells do not respond well to IL-8 but will chemotax nicely to C5a and fMLP. However, they do not polarize well to fMLP.
8. Reagents containing fluorophores should be protected from the light. We wrap our humidified chambers with aluminum foil to protect them from the ambient light during the staining protocol. Care should also be taken to protect the samples from light once they are mounted onto the slides. Storage at −20°C is recommended

to protect the fluorescence signal. DAPI co-staining of neutrophils allows confirmation based on the nuclear shape that the cell of interest is a neutrophil.

9. Other extracellular matrix proteins can also be used to coat the membrane. If examining haptotaxis toward an adhesive substrate, coat only the bottom chamber and leave the top chamber dry during this incubation.

10. For primary neutrophils, we routinely use EGM-2MV and incubate them for 30 min 37°C prior to the start of the assay. An alternative serum-free media is HyQ CCM1 from Hyclone. For assaying HL-60 cells, we use Gey's medium. It is important to note that if neutrophils in DPBS are added to the EGM-2MV for the top, the medium for the bottom must be similarly diluted.

11. Chemoattractants must be spun to clarify prior to loading the Femtotip. Care must then be taken to avoid the debris pellet when withdrawing the chemoattractant from the tube in order to fill the Femtotip. Additionally, chemoattractants should be resuspended in the least viscous or particulate solution possible. Otherwise, the Femtotip will clog and no gradient of chemoattractant will be formed. Additionally, this step should not be completed before the samples are ready because evaporation from the Femtotip tip will occur and may clog it.

12. The distance from the Femtotip to the dish surface is crucial for the proper generation of a gradient. If the tip is too far from the dish surface, the gradient will be irregularly formed and will not accurately represent the capacity for chemotaxis of the given sample. If the tip is too close, the cells will adhere to and clog the needle tip. Make a concerted effort to ensure that this distance is carefully controlled between samples. Likewise, the angle of the tip will alter the shape of the gradient and should be standardized to allow for sample comparison. We have empirically determined an angle that gives us a reproducible gradient generation and have fashioned a cardboard cutout for a guide to allow consistency between experiments.

13. Keep in mind that glass-bottomed dishes will cause primary neutrophil adhesion and loss of the rounded, quiescent morphology.

14. Ensure that the adhesive has cured to avoid leaking. For appropriate cure times, *see* http://www.norlandprod.com.

Acknowledgments

This work was supported by the American Heart Association O255769N, 0225401Z, and National Institutes of Health 1P01AI50500-01.

References

1. Nathan, C. (2006) Neutrophils and immunity: challenges and opportunities. *Nat. Rev. Immunol.* **6,** 173–182.

2. Mollinedo, F., Borregaard, N., and Boxer, L. A. (1999) Novel trends in neutrophil structure, function and development. *Immunol. Today* **20,** 535–537.

3. Van Haastert, P. J. and Devreotes, P. N. (2004) Chemotaxis: signalling the way forward. *Nat. Rev. Mol. Cell Biol.* **5,** 626–634.

4. Postma, M., Bosgraaf, L., Loovers, H. M., and Van Haastert, P. J. (2004) Chemotaxis: signalling modules join hands at front and tail. *EMBO Rep.* **5,** 35–40.

5. Iijima, M., Huang, Y. E., and Devreotes, P. (2002) Temporal and spatial regulation of chemotaxis. *Dev. Cell* **3,** 469–478.

6. Chung, C. Y., Funamoto, S., and Firtel, R. A. (2001) Signaling pathways controlling cell polarity and chemotaxis. *Trends Biochem. Sci.* **26,** 557–566.

7. Meili, R., Sasaki, A. T., and Firtel, R. A. (2005) Rho Rocks PTEN. *Nat. Cell Biol.* **7,** 334–335.

8. Meili, R. and Firtel, R. A. (2003) Two poles and a compass. *Cell* **114,** 153–156.

9. Weiner, O. D., Rentel, M. C., Ott, A. et al. (2006) Hem-1 complexes are essential for Rac activation, actin polymerization, and myosin regulation during neutrophil chemotaxis. *PLoS. Biol.* **4,** e38.

10. Dekker, L. V. and Segal, A. W. (2000) Perspectives: signal transduction. Signals to move cells. *Science* **287,** 982–3, 985.

11. Weiner, O. D., Servant, G., Welch, M. D., Mitchison, T. J., Sedat, J. W., and Bourne, H. R. (1999) Spatial control of actin polymerization during neutrophil chemotaxis. *Nat. Cell Biol.* **1,** 75–81.

12. Heit, B., Tavener, S., Raharjo, E., and Kubes, P. (2002) An intracellular signaling hierarchy determines direction of migration in opposing chemotactic gradients. *J. Cell Biol.* **159,** 91–102.

13. Foxman, E. F., Kunkel, E. J., and Butcher, E. C. (1999) Integrating conflicting chemotactic signals. The role of memory in leukocyte navigation. *J. Cell Biol.* **147,** 577–588.

14. Foxman, E. F., Campbell, J. J., and Butcher, E. C. (1997) Multistep navigation and the combinatorial control of leukocyte chemotaxis. *J. Cell Biol.* **139,** 1349–1360.

15. Hogg, N., Henderson, R., Leitinger, B., McDowall, A., Porter, J., and Stanley, P. (2002) Mechanisms contributing to the activity of integrins on leukocytes. *Immunol. Rev.* **186,** 164–171.

16. Pinkerton, P. H., Robinson, J. B., and Senn, J. S. (1978) Lazy leucocyte syndrome —disorder of the granulocyte membrane? *J. Clin. Pathol.* **31,** 300–308.

17. Miller, M. E., Oski, F. A., and Harris, M. B. (1971) Lazy-leucocyte syndrome. A new disorder of neutrophil function. *Lancet* **1,** 665–669.

18. Lokuta, M. A., Cooper, K. M., Aksentijevich, I., Kastner, D. L., and Huttenlocher, A. (2005) Neutrophil chemotaxis in a patient with neonatal-onset multisystem inflammatory disease and Muckle-Wells syndrome. *Ann. Allergy Asthma Immunol.* **95,** 394–399.

19. Eddy, R. J., Pierini, L. M., and Maxfield, F. R. (2002) Microtubule asymmetry during neutrophil polarization and migration. *Mol. Biol. Cell* **13,** 4470–4483.

20. Ding, M., Robinson, J. M., Behrens, B. C., and Vandre, D. D. (1995) The microtubule cytoskeleton in human phagocytic leukocytes is a highly dynamic structure. *Eur. J. Cell Biol.* **66,** 234–245.

21. Kindzelskii, A. L., Sitrin, R. G., and Petty, H. R. (2004) Cutting edge: optical microspectrophotometry supports the existence of gel phase lipid rafts at the lamellipodium of neutrophils: apparent role in calcium signaling. *J. Immunol.* **172,** 4681–4685.

22. Nuzzi, P. A., Lokuta, M. A., and Huttenlocher, A. (2006) Analysis of neutrophil chemotaxis. *Methods Mol. Biol.* **370,** 23–35.
23. Ruoslahti, E., Hayman, E. G., Pierschbacher, M., and Engvall, E. (1982) Fibronectin: purification, immunochemical properties, and biological activities. *Methods Enzymol.* **82 Pt A,** 803–831.
24. Hauert, A. B., Martinelli, S., Marone, C., and Niggli, V. (2002) Differentiated HL-60 cells are a valid model system for the analysis of human neutrophil migration and chemotaxis. *Int. J. Biochem. Cell Biol.* **34,** 838–854.
25. Hoff, J. (2000) Methods of blood collection in the mouse. *Lab Animal* **29,** 47–53.
26. Miner, N. A., Koehler, J., and Greenaway, L. (1969) Intraperitoneal injection of mice. *Appl. Microbiol.* **17,** 250–251.
27. Eddy, R. J., Pierini, L. M., and Maxfield, F. R. (2002) Microtubule asymmetry during neutrophil polarization and migration. *Mol. Biol. Cell* **13,** 4470–4483.
28. Kindzelskii, A. L., Sitrin, R. G., and Petty, H. R. (2004) Cutting edge: optical microspectrophotometry supports the existence of gel phase lipid rafts at the lamellipodium of neutrophils: apparent role in calcium signaling. *J. Immunol.* **172,** 4681–4685.
29. Ding, M., Robinson, J. M., Behrens, B. C., and Vandre, D. D. (1995) The microtubule cytoskeleton in human phagocytic leukocytes is a highly dynamic structure. *Eur. J. Cell Biol.* **66,** 234–245.
30. Heit, B. and Kubes, P. (2003) Measuring chemotaxis and chemokinesis: the under-agarose cell migration assay. *Sci. STKE.* **2003,** L5.
31. Foxman, E. F., Kunkel, E. J., and Butcher, E. C. (1999) Integrating conflicting chemotactic signals. The role of memory in leukocyte navigation. *J. Cell Biol.* **147,** 577–588.
32. Wells, C. M. and Ridley, A. J. (2005) Analysis of cell migration using the Dunn chemotaxis chamber and time-lapse microscopy. *Methods Mol. Biol.* **294,** 31–41.
33. Rhoads, D. S., Nadkarni, S. M., Song, L., Voeltz, C., Bodenschatz, E., and Guan, J. L. (2005) Using microfluidic channel networks to generate gradients for studying cell migration. *Methods Mol. Biol.* **294,** 347–357.
34. Abhyankar, V. V., Lokuta, M. A., Huttenlocher, A., and Beebe, D. J. (2006) Characterization of a membrane-based gradient generator for use in cell-signaling studies. *Lab. Chip.* **6,** 389–393.
35. Nuzzi, P. A., Lokuta, M. A., and Huttenlocher, A. (2006) Analysis of neutrophil chemotaxis. *Methods Mol. Biol.* **96,** In press.
36. Zigmond, S. H. (1977) Ability of polymorphonuclear leukocytes to orient in gradients of chemotactic factors. *J. Cell Biol.* **75,** 606–616.
37. Zigmond, S. H. and Hirsch, J. G. (1973) Leukocyte locomotion and chemotaxis. New methods for evaluation, and demonstration of a cell-derived chemotactic factor. *J. Exp. Med.* **137,** 387–410.
38. Lokuta, M. A., Nuzzi, P. A., and Huttenlocher, A. (2003) Calpain regulates neutrophil chemotaxis. *Proc. Natl. Acad. Sci. USA* **100,** 4006–4011.
39. Popesko, P., Rajtova, V., and Horak, J. (2003) *A Colour Atlas of Anatomy of Small Laboratory Animals.* Elsevier.

15

Quantifying and Localizing
Actin-Free Barbed Ends in Neutrophils

Michael Glogauer

Summary

We describe here a permeablization method that retains coupling between *N*-formyl-methionyl-leucyl-phenylalanine (fMLP) receptor stimulation and barbed-end actin nucleation in neutrophils. Using fluorescently-tagged actin monomers, we are able to quantify and localize actin-free barbed ends generated downstream of chemoattractant receptors. Partial permeabilization of the neutrophils with the mild detergent *n*-octyl-β-glucopyranoside maintains signaling from membrane receptor to the actin cytoskeleton while allowing for the introduction of inhibitors and activators of signal transduction pathways implicated in regulating actin cytoskeleton dynamics. This is a useful assay for studying signal transduction to the actin cytoskeleton in neutrophils.

Key Words: Neutrophils; actin nucleation; free barbed ends; actin cytoskeleton.

1. Introduction

Neutrophils are recruited to sites of infection, inflammation, or injury through the process of chemotaxis, which involves directed cell migration up the concentration gradient of extracellular chemoattractants that include bacterial products such as *N*-formylated peptides (e.g., *N*-formylmethionyl-leucyl-phenylalanine [fMLP]) *(1)*. The chemotactic signal transduction pathways responsible for orienting neutrophils towards the chemoattractant source regulate the cytoskeletal machinery that drives cell motility *(2)*. Cell polarization and migration up a chemoattractant concentration gradient depends on a series of actin dependent polymerization–depolymerization cycles that result in extension of the leading lamella *(3)*. Actin assembly dynamics in cells are mediated through regulating the availability of the high-affinity barbed end of the actin filament *(4)*. Although it is easier to study signal transduction to the actin cytoskeleton in transfectable stromal cells than in terminally differentiated neutrophils, it is

From: *Methods in Molecular Biology, vol. 412: Neutrophil Methods and Protocols*
Edited by: M. T. Quinn, F. R. DeLeo, and G. M. Bokoch © Humana Press Inc., Totowa, NJ

becoming more evident that many signal transduction pathways have critical differences that are cell-type specific *(5)*. Although fibroblasts and neutrophils both use the same actin machinery to drive motility, neutrophils are able to move 10 times faster than fibroblasts. In order to study the steps linking receptor perturbation to actin assembly directly in primary neutrophils, we describe here an adaptation of an approach previously taken with another hematopoietic cell, the blood platelet, to analyze the steps linking receptor perturbation to actin assembly *(6)*. The principal experimental technique involves controlled partial permeablization of the plasma membrane so as to preserve essential signaling intermediates required to sustain the pathways, but to permit introduction of activators and inhibitors of signal transduction cascades. This technique has yielded interesting information on the actin cytoskeleton regulatory elements downstream of the fMLP receptor in neutrophils *(7)*. We describe here the technique for use with neutrophils which allows for both the quantification of free barbed ends nucleation sites and the visual localization of these nucleation sites downstream of chemoattractant receptors.

2. Materials

2.1. Permeablization and Quantification of Actin-Free Barbed Ends

1. Buffer A: 5 mM Tris-base, 0.2 mM $CaCl_2$, 0.2 mM ATP, 0.5 mM dithiothreitol (DTT), pH 8.0. Store at −20°C.
2. Pyrene-labeled rabbit skeletal muscle actin dissolved in cold Buffer A (10 mg/mL). Snap-freeze in liquid nitrogen, and store in single-use aliquots (5 μL each) at −80°C.
3. PHEM Buffer: 60 mM PIPES, 25 mM HEPES, 10 mM EGTA, 2 mM $MgCl_2$, pH 6.0. Store at 4°C.
4. Buffer B: 10 mM Tris-base, 0.1 mM EGTA, 2 mM $MgCl_2$, 100 mM KCl, 0.5 mM DTT, pH 7.0. Store at −20°C. Add the following protease inhibitors to buffer B just before use to give the final indicated concentration: 10 μM phallacidin, 42 nM leupeptin, 10 mM benzamidine, and 0.123 mM aprotinin. Add ATP solution, as described below.
5. ATP solution: 100 mM ATP, pH 7.0. Store at −20°C. Add to buffer B for a final concentration of 0.5 mM ATP immediately before use.
6. Permeablization buffer: 4% (w/v) of *n*-octyl-β-glucopyranoside (OG; Sigma) in PHEM buffer. Store at −20°C.
7. Hanks balanced salt solution (HBSS) with Ca^{2+} and Mg^{2+}: 0.14 M NaCl, 5.4 mM KCl, 1 mM Tris-HCl, 1.1 mM $CaCl_2$, 0.4 mM $MgSO_4$, 1 mM HEPES, pH 7.2. Store at 4°C.
8. 0.5 M DTT. Store at −20°C.
9. fMLP: prepare 9.1 nM fMLP (Sigma) in HBSS, make fresh.
10. Spectorfluorometer with excitation filter of 386 nm and emission filter of 388 nm.
11. 96-well microtiter plate, black flat-bottom, or appropriate tubes depending on the spectrofluorometer used.

2.2. Rhodamine-Labeled Actin Nuclei Staining

1. Rhodamine-labeled rabbit skeletal muscle: dissolve in cold Buffer A to a final concentration of 30 μM. Incubate on ice for 2 h and centrifuge for 20 min at 60,000g and 4°C.
2. 5% (w/v) bovine serum albumin (BSA): dissolve 5 g of BSA (Sigma) in 100 mL of double-distilled sterile water.
3. Permeabilization solution: 4% (w/v) of n-octyl-β-glucopyranoside in PHEM buffer. Store at −20°C.
4. Triton Permeablization Buffer: 0.1 %(v/v) Triton X-100 in PBS. Store at 4°C.
5. 100 nM fMLP in HBSS made fresh before each experiment.
6. Formaldehyde: Prepare 3.7 % (w/v) solution in phosphate-buffered saline (PBS) fresh for each experiment.
7. 10X PBS (Gibco): prepare 1X PBS by dilution of one part of 10X PBS with nine parts water.
8. Microscope cover slips (24 × 50 × 0.10 mm).
9. Polystyrene tissue culture chamber glass slides (Lab-Tek).
10. Alexa 488 Phalloidin (Molecular Probes).
11. Immuno-Fluore Mounting Medium (ICN Pharmaceuticals).

3. Methods

Neutrophils are difficult cells to work with because they are terminally differentiated, protease-rich cells that are easily activated during isolation. As a result, permeablization protocols can unleash the proteases, which result in the breakdown of the signaling proteins, complicating experimental analysis. In order to overcome this problem, we use a protocol that results in partial plasma membrane disruption while maintaining receptor signaling into the cytoplasmic compartment. The partial access to the cytoplasm allows for the introduction of recombinant proteins, signaling intermediates, and fluorescent actin monomers, which enable us to study and determine the signaling intermediates involved in the regulation of the actin cytoskeleton downstream of specific membrane receptors.

3.1. Measuring the Increase in Free Barbed Ends/Actin Nuclei After Neutrophil Activation

1. To reconstitute pyrene-labeled actin, take an aliquot (5 μL/each) from the −80°C freezer, dilute with cold buffer A, and incubate mixture on ice for 1 h.
2. Centrifuge sample for 20 min at 60,000g and 4°C to remove any F-actin. Pyrene-labeled actin monomers are ready for use.
3. Isolate human or murine neutrophils from peripheral blood as described in Chapters 2 and 3, and resuspend neutrophils at 5 × 10^6 cells/mL in HBSS.
4. Aliquot 1 × 10^6 neutrophils into microfuge tubes for each sample and pellet (400g for 5 min). Samples should include resting and fMLP-activated neutrophils.

5. Resuspend cells in 90 µL of HBSS, and add 10 µL fMLP (100 n*M* final concentration).
6. Incubate samples for 30 s at 37°C, and immediately permeabilize.
7. For neutrophil permeabilization, add 10 µL of permeabilization buffer to resting or fMLP-treated samples, and incubate for 15 s (*see* **Notes 1** and **2**).
8. Rapidly stop the permeabilization process by adding 185 µL of Buffer B to dilute the OG.
9. Transfer samples to a 96-well microtiter plate or cuvet/test tube.
10. Add pyrene-labeled actin (15 µL) to each well (1 µ*M* final concentration).
11. Read the fluorescence intensity over 600 s with a spectrophotometer or fluorescence microplate reader with excitation set at 366 nm and emission at 386 nm.
12. Calculate the increase in free barbed ends generated by comparing the fluorescence intensity slopes for the stimulated and control/unstimulated cells (**Fig. 1**).

3.2 Rhodamine-Labeled Actin Nuclei Staining

1. Coat eight-chamber tissue culture glass slides with 5% BSA at room temperature for 1 h.
2. Rinse the slides three times with PBS.
3. Isolate human or murine neutrophils from peripheral blood as described in Chapters 2 and 3.
4. Resuspend neutrophils at 1×10^6 cells/mL in HBSS containing 1% (w/v) gelatin, and aliquot 100 µL of cells per slide well.
5. Incubated the slides for 10 min at 37°C to allow for attachment of the cells.
6. Remove buffer and replace with 90 µL of Buffer B.
7. Add 10 µL of permeabilization buffer, and incubate for 15 s to partially permeabilize the cells (*see* **Note 2**).
8. Rapidly stop the permeabilization process by adding 185 µL of Buffer B to dilute the OG.

Fig. 1. (*Opposite page*) Measuring changes in free-actin barbed ends levels after activation in neutrophils. Wild-type (WT) and Rac2 null neutrophils were permeabilized and the actin polymerization curve was measured in a fluorescence spectrophotometer (excitation, 368 nm; emission, 388 nm) following the addition of pyrene-labeled actin monomers. Left panel: raw data plot showing the increase in fluorescence resulting from pyrene monomers polymerizing from free barbed ends in the permeabilized neutrophils. Verification that polymerization is originating from free barbed ends is carried out by adding cytochalasin B, which blocks 95% of the polymerization curve (data not shown). The slope of the fluorescence-polymerization curve is proportional to the number of free barbed ends. Right panel: the percent increase in free barbed ends is calculated from a minimum of five plots per slopes for each treatment group. The ratio of the fMLP activated group slope divided by the control (untreated) group slope gives the free barbed end increase as a percent of the control group. We see that while there is a threefold increase in free barbed ends in wild type neutrophils, cells lacking the small GTPase Rac2, which is known to regulate actin assembly, show no significant increase in free barbed ends after fMLP stimulation.

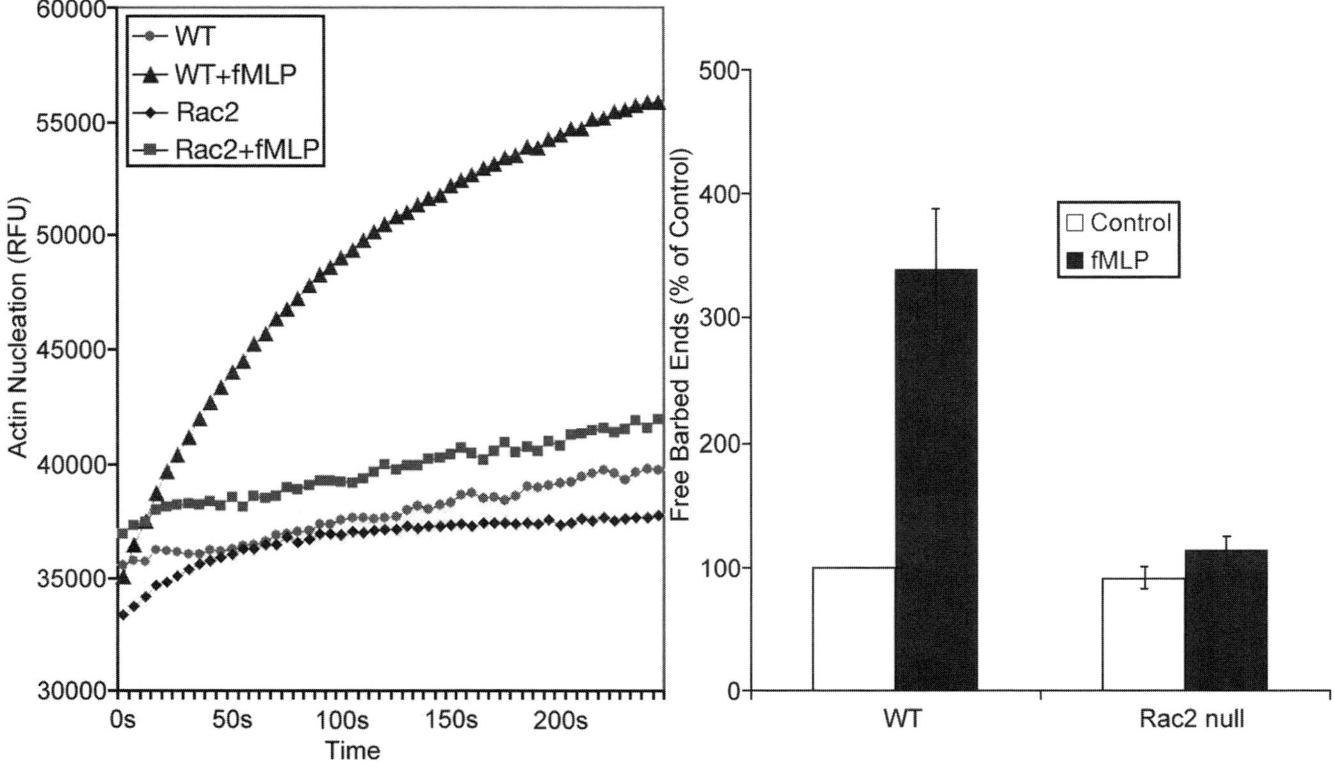

Actin Nucleation (RFU)
60000
55000
50000
45000
40000
35000
30000
WT
WT+fMLP
Rac2
Rac2+fMLP
0s
50s
100s
150s
200s
Time
Free Barbed Ends (% of Control)
500
400
300
200
100
0
Control
fMLP
WT
Rac2 null

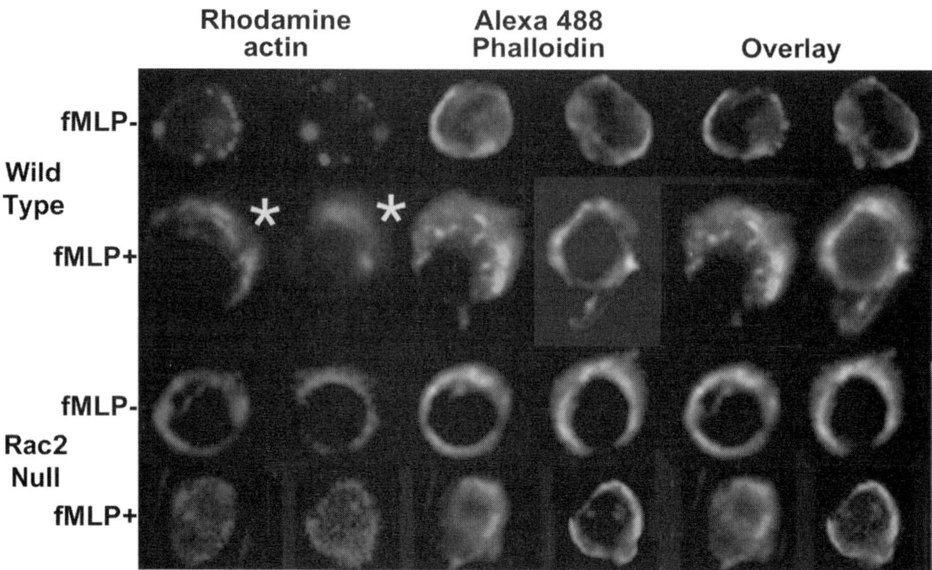

Fig. 2. Localizing free-actin barbed ends generated after fMLP stimulation in neutrophils. Wild-type and Rac2 null neutrophils were allowed to attach to bovine serum albumin-coated coverslips, and permeabilized as described above. Following the addition of fMLP, rhodamine-actin monomers were added to label the free barbed ends. After fixing the cells fluorescently tagged phalloidin, which binds to actin filaments was added. We note there is dramatic increase in free barbed ends at the leading edge (*) of the activated neutrophils. Also note that in the overlay (orange), the rhodamine tag is clearly at the advancing edge of the actin filaments (green). Neutrophils lacking the small GTPase Rac2, show no significant increase in free barbed ends or the development of an actin rich leading edge after fMLP stimulation. (color insert will appear after page 300)

9. Add 10 µL of 100 n*M* fMLP, and incubate for 3 min to stimulate the attached neutrophils. Control samples are treated the same with buffer instead of fMLP.
10. Add 10 µL of rhodamine-labeled actin in Buffer B (1 µ*M* final concentration) to each well, and incubate for 10 min at room temperature.
11. Rapidly wash the slides three times with PBS.
12. Add 3.7% formaldehyde, and incubate for 10 min at room temperature to fix the samples.
13. Wash the slides three times with PBS.
14. Add 100 µL of 0.1% Triton in PBS, and incubate for 4 min to completely permeablize the cells.
15. Wash the samples three times with PBS and then aspirated dry from one corner.
16. Add 100 µL of Alexa 488 Phalloidin diluted in PBS (1:40), and incubate for 20 min at room temperature in the dark.

17. Wash the samples five times for 2 min each with PBS and then aspirated dry from one corner.
18. Mount the samples by carefully removing the gasket (a razor blade may be required), and adding mounting medium and a coverslip. Use nail varnish at the edges of the coverslip to seal the sample.
19. Slides can be viewed immediately or stored in the dark at 4°C for up to 2 wk (*see* **Note 3**).
20. View the slides under phase contrast microscopy (to locate the cells and identify the focal plane) and under confocal microscopy. Excitation at 543 nm induces rhodamine fluorescence (red emission) for the new actin filaments, which originate from the free barbed ends following cell activation, whereas excitation at 488 nm induces Alexa 488 fluorescence (blue emission) to enable visualization of all actin filaments (those present before activation and after activation). Software can be used to overlay the fluorescence images. Examples of the signals for activated neutrophils are shown in **Fig. 2**.

4. Notes

1. Neutrophils should be used within 3 h of isolation. They should be kept on a gentle rotator at 4°C until they are used.
2. For OG permeabilization, pipet up and down thre times slowly to mix. This should take about 10–15 s. Add stop buffers immediately after mixing.
3. Air bubbles are undesirable in the mounting medium, and slow, careful application of the coverslip minimizes their appearance.

References

1. Niggli, V. (2003) Signaling to migration in neutrophils: importance of localized pathways. *Int. J. Biochem. Cell Biol.* **35,** 1619–1638.
2. Weiner, O. D., Servant, G., Welch, M. D., Mitchison, T. J., Sedat, J. W., and Bourne, H. R. (1999) Spatial control of actin polymerization during neutrophil chemotaxis. *Nat. Cell Biol.* **1,** 75–81.
3. Fenteany, G. and Glogauer, M. (2004) Cytoskeletal remodeling in leukocyte function. *Curr. Opin. Hematol.* **11,** 15–24.
4. Cicchetti, G., Allen, P. G., and Glogauer, M. (2002) Chemotactic signaling pathways in neutrophils: from receptor to actin assembly. *Crit. Rev. Oral Biol. Med.* **13,** 220–228.
5. Sun, C. X., Downey, G. P., Zhu, F., Koh, A. L., Thang, H., and Glogauer, M. (2004) Rac1 is the small GTPase responsible for regulating the neutrophil chemotaxis compass. *Blood* **104,** 3758–3765.
6. Hartwig, J. H., Bokoch, G. M., Carpenter, C. L., et al. (1995) Thrombin receptor ligation and activated Rac uncap actin filament barbed ends through phosphoinositide synthesis in permeabilized human platelets. *Cell* **82,** 643–653.
7. Glogauer, M., Hartwig, J., and Stossel, T. (2000) Two pathways through Cdc42 couple the N-formyl receptor to actin nucleation in permeabilized human neutrophils. *J. Cell Biol.* **150,** 785–796.

16

Measurement of Neutrophil Adhesion Under Conditions Mimicking Blood Flow

Mark A. Jutila, Bruce Walcheck, Robert Bargatze, and Aiyappa Palecanda

Summary

Neutrophil migration from blood into tissues is required for effective innate immune responses against infection. Adhesion of the neutrophil in blood to the vascular endothelium and eventual migration through the vessel wall and accumulation at the site of infection involves different classes of adhesion molecules. In vivo intravital microscopy studies show that different adhesion molecules mediate binding events under shear forces associated with blood flow vs binding events that take place under static conditions. To fully analyze the function of these adhesion molecules in vitro, assays must reflect the hemodynamic forces associated with blood flow. We outline two approaches used to study neutrophil adhesion under conditions that mimic blood flow.

Key Words: Adhesion; blood flow; shear; neutrophil; migration; inflammation; endothelial cell.

1. Introduction

Neutrophils are not typically found in healthy tissues, thus central to their role in countering infectious agents is their extravasation from the blood into sites of infection. Adhesion molecules expressed by neutrophils in the blood, the endothelial cell lining of vessels within the tissue, and other cells such as platelets, are required for effective extravasation *(1)*. In vivo measurements have shown that cells in the blood travel at surprisingly high velocities. Even in the reduced flow associated with the capillary and post-capillary beds, velocities of 500–5000 µm/s are common *(2)*. Years of work on defining the molecular basis for leukocyte/endothelial cell interactions demonstrate that specific classes of adhesion molecules preferentially mediate binding under shear versus static conditions *(1,3–6)*. For example, adhesive interactions between leukocytes

From: *Methods in Molecular Biology, vol. 412: Neutrophil Methods and Protocols*
Edited by: M. T. Quinn, F. R. DeLeo, and G. M. Bokoch © Humana Press Inc., Totowa, NJ

and endothelial cells mediated by the β_2 integrins (LFA-1, MAC-1) preferentially occur under low shear; whereas adhesion molecules of the selectin family (L-, E-, and P-selectin) effectively mediate adhesion under high shear conditions *(1,6)*. The adhesive events mediated by β_2 integrins and selectins are different, with selectin-mediated adhesion being defined as a tethering event, followed by rolling, and the β_2 integrin-mediated event being defined as "tight" adhesion.

Selectin- and integrin-mediated events occur sequentially and are the cornerstone of the multistep paradigm of leukocyte recruitment *(1)*, with the first step being shear dependent tethering, followed by rolling along the vessel wall. The second step involves an increase in adhesive activity of the leukocyte induced by activating events (adhesion molecule or chemotactic factor triggered). The third step is characterized by tight adhesion, followed by transendothelial migration. To accurately measure the function of adhesion molecules involved in neutrophil extravasation, assays must reflect the flow and corresponding shear forces associated with the vascular system.

Use of intravital microscopy is best to study physiologically relevant neutrophil adhesion, but these experiments require many animals, are costly, are not conducive to testing multiple adhesion molecules and/or agonists, and are technically challenging. As such, in vitro assays have been developed which reflect, in part, the dynamics of the vascular system and allow analysis of each step of the multistep model of leukocyte recruitment outlined previously. Two general approaches have been developed. In the first, endothelial cells, purified adhesion molecules or other adhesive substrates are immobilized on the internal surface of small glass capillary tubes, which are incorporated into a circulatory "loop" and flow established by use of a peristaltic pump (capillary tube flow assay). In the second approach, adhesive substrates are established on the internal surface of planar chambers, which are then connected to a syringe or peristaltic pump (parallel plate flow chamber). In both approaches, cells are passed over the adhesive substrate at a defined shear stress and use video microscopy and image analysis to analyze adhesive interactions. Flow assays are uniquely suited to investigate the binding kinetics of neutrophil adhesive events occurring very rapidly in a time scale shorter than that of most static adhesion assays. In addition, processes subsequent to the initial adhesion events can be studied, such as cell stabilization, spreading, transendothelial migration, and the distribution pattern of particular cell surface determinants on neutrophils and endothelial cells during their interactions.

In the sections that follow, we describe the capillary tube and parallel plate flow chamber assays. Examples of experiment design, setup, and data generation/presentation are presented. Detailed protocols are provided mainly for the capillary tube assay, which are focused on three analyses of a novel selectin inhibitor. For the parallel plate flow assay, more attention is given to experimental design, although procedures are provided for labeling neutrophils with 5(6)-carboxyfluorescein succinimidyl ester (CFSE) and using this as a means of analyzing these cells in whole blood.

2. Materials

2.1. Capillary Tube Flow Assay

1. Hank's balanced salt solution (HBSS) without $CaCl_2$ and $MgCl_2$ (Mediatech).
2. HBSS supplemented with 2 mM $CaCl_2$.
3. RPMI or Dulbecco's modified Eagle's medium (DMEM) medium (Gibco).
4. Fetal bovine serum (FBS), HEPES.
5. Mono-Poly Resolving Media: (M-PRM; MP Biomedicals) for human polymorpho-nuclear neutrophils (PMN) isolation.
6. 50-mL centrifuge tubes (FisherBiotech).
7. Sterile transfer pipets for layering of blood on M-PRM gradient (FisherBiotech).
8. 10-mL disposable syringe(s) for cell reservoir and infusion of reagents (Fisher Biotech).
9. Adhesive substrate, e.g., endothelial cells, transfected adherent cells, purified adhesion molecules, which is dictated by the specific experiment. For this chapter, use of recombinant human E-selectin-immunoglobulin (Ig)G chimera (R&D systems), human umbilical-cord endothelial cells (HUVECs; Cambrex Corp.), which are Factor VIII and low-density lipoprotein (LDL)-receptor positive (cultured in endothelial-cell growth media [EGM; Clonetics]) and Chinese hamster ovary (CHO) cells expressing recombinant E-selectin will be presented.
10. The human pro-monocytic cell line, U937 (American Type Culture Collection [ATCC]), has been used extensively to study human leukocytes. U937 cells express the E-, P- and L- selectin ligand, P-selectin glycoprotein ligand (PSGL), and are able to roll on selectins.
11. Recombinant human tumor necrosis factor (TNF)-α (R & D Systems).
12. Capillary tube flow system: a central feature of the capillary tube system is an artificial vessel that is created by growing endothelial cells, transfected cell lines expressing single or multiple adhesion molecules, or immobilizing purified adhesion molecules on the internal surface of small diameter glass capillary tubes. The "vessels" are integrated into a loop system in which fluid can be recirculated via a peristaltic pump. Cells are injected into the system, and their interaction with the endothelial cell monolayer is monitored by video-microscopy. Agents can be infused into the assay and their effect on leukocyte-endothelial cell interactions readily measured (**Fig. 1**).
13. Programmable peristaltic pump.
14. Tubing: silicone tubing (Masterflex, Cole Parmer #96400-14); tubing adapters: male (Cole Parmer #06359-25) and female (Cole Parmer, #06359-05); extension sets (Abbot Laboratories #4612-04); and three-way stopcock.
15. Capillary tubes: outer diameter (OD) = 0.0665 inch, inner diameter (ID) = 0.0555 inch, ordered as precut 2.5 cm in length (Drummond Scientific).

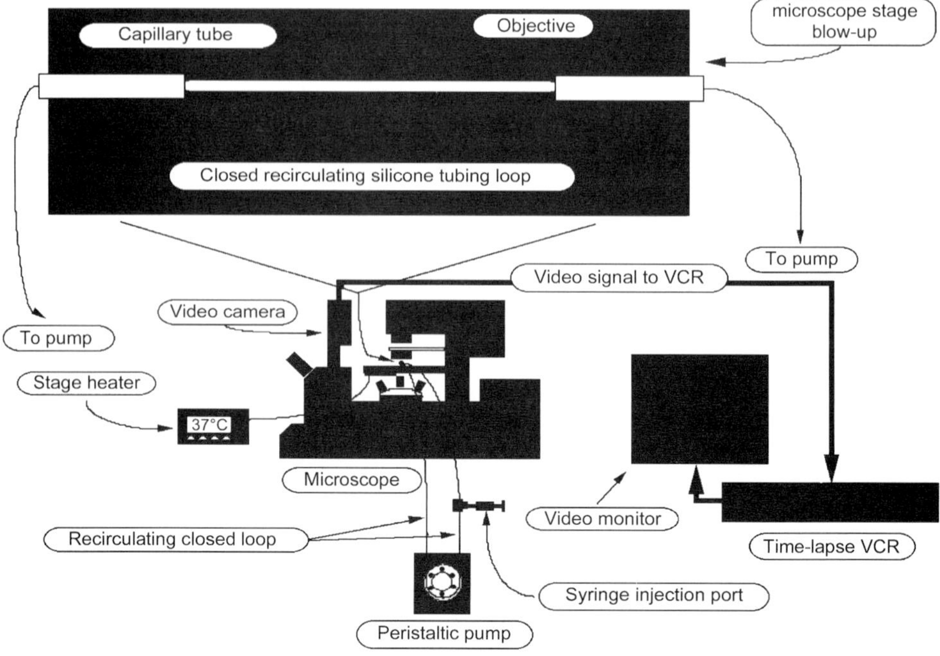

Fig. 1. Schematic of the in vitro capillary shear assay system.

16. Stop-cock valves or manifold to hold the disposable syringes (Cole Parmer).
17. EL-246 is a novel antibody that binds to a conserved epitope on both L- and E-selectin *(7)*.
18. Inverted microscope with video camera.
19. VHS deck or digital feed into computer.
20. Computer and image analysis software, e.g., National Institutes of Health (NIH) ImageJ (available on the Internet at http://rsb.info.nih.gov/nih-image).

2.2. Parallel Plate Flow Chamber Assay

1. 100-mL glass syringe for use on the syringe pump to draw neutrophils through a parallel plate flow chamber.
2. 10-mL disposable syringes for cell reservoir and infusion of reagents (FisherBiotech).
3. Three-way stopcock valves or manifold to mount the disposable syringes (Cole Parmer).
4. 35-mm Petri dishes (FisherBiotech). Tissue-culture-grade plastic tends to be stickier for neutrophils and can increase their firm attachment and spreading.
5. Substrate: endothelial cells (Clonetics), transfected adherent cells, purified adhesion molecules grown on or adsorbed to the 35-mm dish.
6. 2.5% bovine serum albumin (BSA) (FisherBiotech) in HBSS without $CaCl_2$ and $MgCl_2$ (Mediatech) to block the surface of the 35-mm dish.

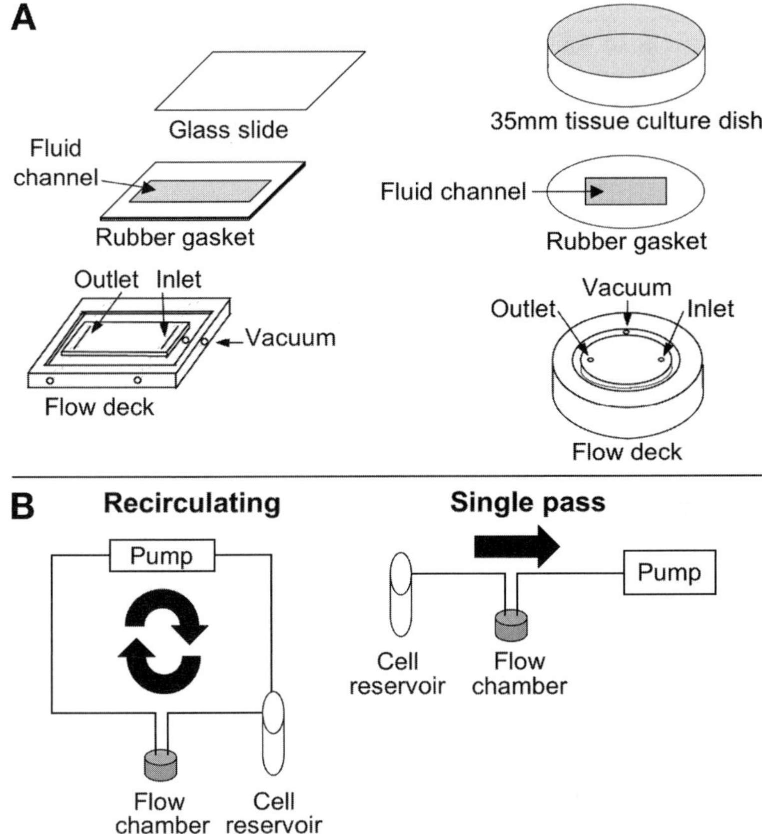

Fig. 2. Schematics of typical parallel plate flow chamber configurations **(A)** and the recirculating (left) and single pass (right) fluid flow paths **(B)**.

7. Ethylenediamine tetraacetic acid (EDTA) (FischerBiotech).
8. Dextran solution: 3% (w/v) dextran (molecular weight = 500,000) dissolved in HBSS without $CaCl_2$ and $MgCl_2$)
9. Ficoll-Hypaque (1.077 g/mL) (Mediatech).
10. Percoll solution: 30% (v/v) Percoll (Sigma) in HBSS without $CaCl_2$ and $MgCl_2$.
11. CFSE or 5(6)-carboxyfluorescein diacetate (CFDA) (Molecular Probes).
12. Parallel plate flow chamber: two commonly used parallel plate flow chamber designs are shown in **Fig. 2**. The basic setup of the parallel plate flow chamber consists of an acrylic flow deck (Glycotech) that is typically placed on a slide or in a 35-mm tissue culture dish. Flow chambers are transparent to allow visualization of cell adhesion events using real-time video microscopy.
13. Programmable syringe or peristaltic pump.
14. Inverted microscope with video camera.

15. VHS deck or digital feed into computer.
16. Specific inhibitors, which are dictated by the experiment (e.g., monoclonal antibodies, small molecule inhibitors, enzymes, etc.).
17. Computer and image analysis software, e.g., NIH ImageJ.

3. Methods

3.1. Capillary Tube Continuous Recirculating Shear Assay System

The capillary tube flow assay is performed in accord with protocols described in previous publications *(8–12)*. This simulated microenvironment is an excellent tool for evaluating the molecular, cellular, and vascular surface interactions associated with inflammatory and infectious diseases. The shear forces generated in the capillary tube are similar to the shear factors measured in blood vessels *(13)*. As a result of the extensive data set that can be collected, this experimental system and method allows for the discrimination of the diverse behaviors observed for individual leukocyte subsets under a variety of experimental conditions. Either leukocytes and/or endothelial cells can be activated in this system using a spectrum of cytokines or activating factors. Blocking of interactions with monoclonal antibodies, enzymatic treatment, or chemically defined compounds can be observed and accurately quantified. This assay is useful for testing the prophylactic (pre-adhesion treatment) and therapeutic (post-adhesion treatment) efficacy of potential new drug compounds that interfere with leukocyte trafficking. As examples of the utility of this system, we demonstrate the ability to test the efficacy of a unique selectin inhibitor on multiple ligand/receptor configurations (*see* **Note 1**).

3.1.1. Isolation of Human Neutrophils for Use in the Capillary Tube Shear Assay

1. Collect heparinized whole blood from healthy volunteers (*see* **Notes 2** and **3**).
2. Mix M-PRM well by inverting two to three times immediately before use, and layer 3 mL of M-RPM in sterile polypropylene tubes.
3. Layer 5 mL of whole blood on top of M-RPM medium using a sterile serological pipet.
4. Centrifuge at $300g$ for 35 min at 25°C without braking. Following centrifugation, four distinct layers should be seen. If there is not a distinct granulocyte layer (i.e., if there are still significant amounts of red blood cells [RBCs present in this layer], tubes can be centrifuged for an additional 5–15 min).
5. Collect the plasma layer using a sterile transfer pipet. Transfer plasma to a sterile 50-mL tube and store on ice.
6. Collect the mononuclear cell (lymphocyte and monocyte) layer using a sterile transfer pipet. Transfer mononuclear cells to a 50-mL tube and dilute to 45 mL with cold HBSS without $CaCl_2$ and $MgCl_2$.

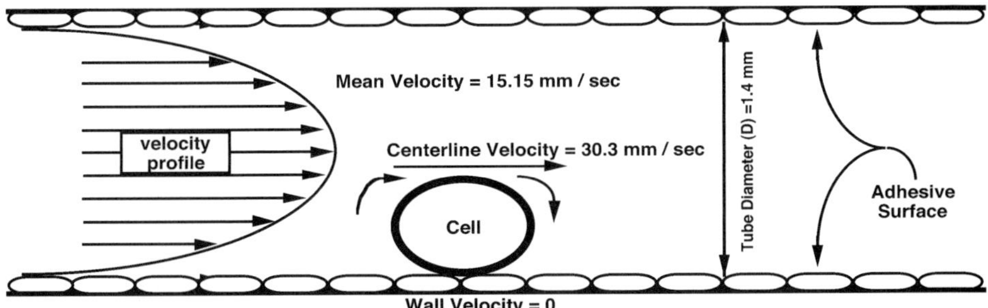

Fig. 3. Calculations for a typical in vitro capillary tube shear assay experiment. These assays provide detailed information on cell–cell or cell–substrate interactions. This information can include rolling speed, rolling behavior, number of cells binding per unit area vs time, and the measurement of the effects of signaling or adhesion modulators on these interactions.

7. Collect the neutrophil layer using a sterile transfer pipet. Transfer neutrophils to a 50-mL tube and dilute to 45 mL with cold HBSS (without $CaCl_2$ and $MgCl_2$).
8. Centrifuge the neutrophils at 250g for 10 min at room temperature.
9. Remove the supernatant and resuspend the pellet in 0.5 mL of HBSS (without $CaCl_2$ and $MgCl_2$). Mix gently, dilute to 3 mL with the same buffer, and place on ice.
10. Count cells and keep on ice until used in the assay.

3.1.2. The Capillary Tube-Based Shear Assay System

1. Attach the glass capillary tube containing an adhesive substrate (i.e., purified adhesion molecule, CHO cells expressing a recombinant adhesion molecule, or activated endothelial cells) to silicone tubing to form a closed loop in which media and cells are to be re-circulated (*see* **Note 4**).
2. Mount the tube on an inverted microscope (*see* **Fig. 1**).
3. Regulate flow using a variable speed peristaltic pump to simulate in vivo blood flow shear conditions (1–3 dynes/cm^2; **Fig. 3**).
4. Infuse leukocytes/cell lines into the system at 4×10^6 cells/mL in sterile HEPES-buffered (20 mM) DMEM or RPMI (pH 7.0), plus 2% FBS or human plasma.
5. Continuously monitor and videotape establishment of the rolling interaction for at least 8 min. During this time, control or experimental conditions are established and maintained.
6. At the 8-min time period, inject test or control compounds into the recirculating loop.
7. Record the leukocyte/endothelial cell interactions for an additional 8 min.

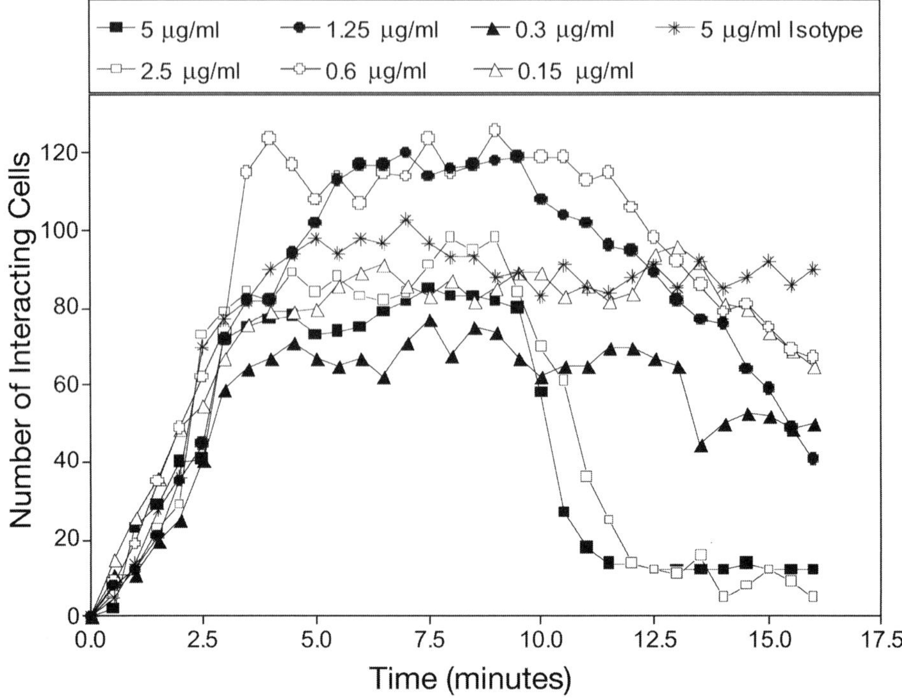

Fig. 4. E-selectin dependent rolling of U937 cells. Recombinant human E-selectin-IgG Fc chimera was immobilized on the internal surface of glass capillary tubes in the capillary tube assay. U937 cells were injected into the loop and cell interaction monitored at physiologic shear rates. Rolling interaction established for a period of 8 min. Different doses of an anti-E-/L-selectin monoclonal antibody EL246 was then injected into the loop and the U937-E-selectin interaction recorded for an additional 8 min. Each line represents analyses from separate tubes. Data are plotted as numbers of cells interacting at every 30-s interval in untreated and antibody-treated loops.

8. Determine the number of cells rolling on the substrate before and after the injection of adhesion modifiers by individual frame analysis of the video recording. In some experiments, control or test compounds are infused simultaneously with the interacting cells, and the interactions are monitored for a total of 10 min.
9. The data are generally presented as the number of rolling cells within the field of view vs time **(Fig. 4)**.

3.1.3. E-Selectin-Mediated Rolling of U937 Cells

1. For this assay, use recombinant E-selectin immunoglobulin (Ig)G chimera as a substrate to support rolling.
2. Rinse the assay capillary tubes three times with phosphate-buffered saline (PBS) on the day before the assay, and soak them in Petri dishes containing PBS until ready for use.

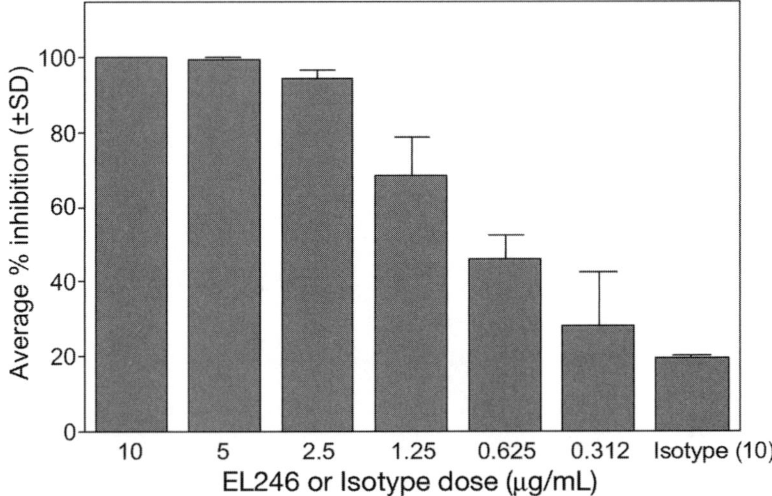

Fig. 5. EC50 of EL-246 to reverse/inhibit U937 rolling on recombinant E-selectin. U937 cell rolling on E-selectin chimera was established as described in **Fig. 3**. Monoclonal antibody EL-246 at different doses was injected into the capillary tube assay and the effect on the numbers of rolling cells recorded. The data are presented as percent inhibition {[100 − (average number of cells rolling between 6 and 8 min/average number rolling between 14 and 16 min) × 100]}. The data are averages of three independent experiments.

3. Remove PBS, and add 70–90 µL of goat anti-human IgG (10 µg/mL in PBS) to each tube.
4. Wrap the Petri dish containing the tubes in parafilm, and incubate overnight at 4°C.
5. On the day of the assay, remove the IgG solution from each tube, and rinse the tubes three times with PBS.
6. Add 70–90 µL of human E-selectin-Fc chimera to each tube (0.5 µg/mL in PBS) and incubate at room temperature for 1 h.
7. Incorporate the capillary tube into the shear system as described previously.
8. Establish flow to generate the physiological venular shear stress of 2 dynes/cm^2.
9. Inject U937 cells (4 × 10^6/loop) into the loop. U937 cells interact with the immobilized E-selectin and display a "rolling" behavior. The numbers of rolling cells increases from time of injection to approx 6 min (**Fig. 4**).
10. Inject EL246 or isotype control monoclonal antibody 8 min after cell infusion and monitor for an additional 8 min.
11. As shown in a representative experiment in **Fig. 4** and summarized in **Fig. 5**, EL246 completely reversed the rolling interaction at 5 µg/mL. The reversal was dose-dependent. EL-246 reversed the rolling of U937 on recombinant E-selectin with an IC$_{50}$ of approx 0.6 µg/mL (**Fig. 5**). The rolling interaction was unaffected by an isotype-matched irrelevant monoclonal antibody (**Fig. 5**).

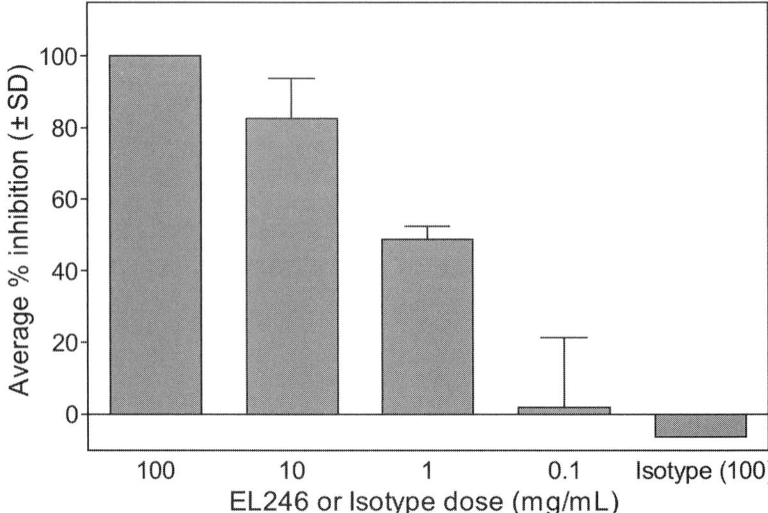

Fig. 6. IC50: EL246 inhibition of U937 cell binding to cell based E-selectin. Chinese hamster ovary cells transfected to express human E-selectin were grown on the internal surface of glass capillary tubes and the tubes incorporated into the capillary tube assay. U937 cells were pretreated with EL246 or isotype matched control monoclonal antibody prior to injection into the loop. The data are represented as average percent inhibition (± SD) from three independent experiments.

3.1.4. U937 Cell Rolling on CHO Cells Expressing Human E-Selectin

To more closely reflect the selectin-leukocyte interaction thought to occur in vivo, CHO cells transfected with human E-selectin cDNA can be used as the substrate in the capillary tube instead of purified adhesion molecules.

1. Grow CHO cells to confluency on the internal surface of the capillary tube, as determined by microscopic analysis. This is normally done by seeding the tubes the day before the assay and incubating them submerged in cRPMI (10% FBS, v/v) in Petri dishes at 37°C and 5–10% CO_2.
2. On the day of the assay, rinse the capillary tubes in assay medium (RPMI or DMEM, plus 2% BSA [w/v] or human plasma) and then incorporate into the shear system, as described previously.
3. Establish flow to generate the physiological venular shear stress of 2 dynes/cm².
4. Inject EL-246 or isotype control monoclonal antibody treated U937 cells (4×10^6/ loop) (*see* **Notes 5** and **6**).
5. Monitor for 8 min.
6. As shown a representative experiment in **Fig. 6**, monoclonal antibody EL-246 inhibited the binding of U937 cells to CHO-E-selectin transfectants. However, the IC_{50}

for inhibition was higher (>1 µg/mL) for inhibiting a cell-based adhesion event as opposed to reversal (**Fig. 6**) of binding to purified adhesion molecule (~0.6 µg/mL).

3.1.5. Human Neutrophil Rolling on Activated HUVECs

The critical adhesive interactions between leukocytes and endothelium involve both a selectin-mediated rolling followed by integrin-mediated sticking, both of which are prerequisite to transmigration *(1)*. The two in vitro shear assays described previously do not have the complete complement of adhesion molecules required for leukocyte–endothelial interactions that occur in vivo. Therefore, an assay can be set up wherein HUVECs are grown on the internal surface of glass capillary tubes to mimic a "blood vessel," and the endothelial cells are activated with TNF-α (1 ng/mL) for 4 h as a surrogate inflammatory stimulus. The 4-h activation induces the expression of E-selectin and the integrin ligands intercellular adhesion molecule (ICAM) and vascular cell adhesion molecule (VCAM)-1 on HUVECs.

1. Grow HUVECs on the internal surface of capillary tubes as described previously for CHO cells.
2. Observe the capillary tubes microscopically on the day of the assay to ensure that complete monolayers of endothelial cells are established.
3. Rinse the tubes in assay medium (RPMI or DMEM, plus 2% BSA [w/v] or human plasma), and incorporate them into the shear system, as described previously.
4. Establish flow to generate the physiological venular shear stress of 2 dynes/cm^2.
5. Inject EL-246 or isotype control monoclonal antibody treated PMN (4×10^6/loop) into the loop.
6. Monitor for 8 min. As described previously for the CHO transfectants, PMN roll briefly before sticking to the activated HUVECs.
7. As shown in a representative experiment, pretreatment with monoclonal antibody EL-246 inhibited the PMN binding to HUVECs in a dose-dependent manner (**Fig. 7**). An irrelevant isotype matched monoclonal antibody had no effect on PMN–HUVEC interaction.

3.2. Parallel Plate Flow Chamber Design

The initial design of the parallel plate flow chamber is based upon that described by Hochmuth and colleagues to study red blood cells *(14)*. The parallel plate flow chamber was used in early studies on neutrophils by Wilkinson et al. *(15)* and Forrestor et al. *(16)* to study their adhesive characteristics on absorbed plasma proteins. Lawrence et al. described one of the first parallel plate flow chamber assays to study neutrophil adhesion to endothelium *(17)*. Since these earlier studies, numerous researchers have utilized the parallel plate flow chamber and modified versions of it to examine the dynamics of neutrophil adhesion to various substrates, including endothelial cells, platelets, leukocytes, transfected cell lines, and purified molecules *(18–22)*.

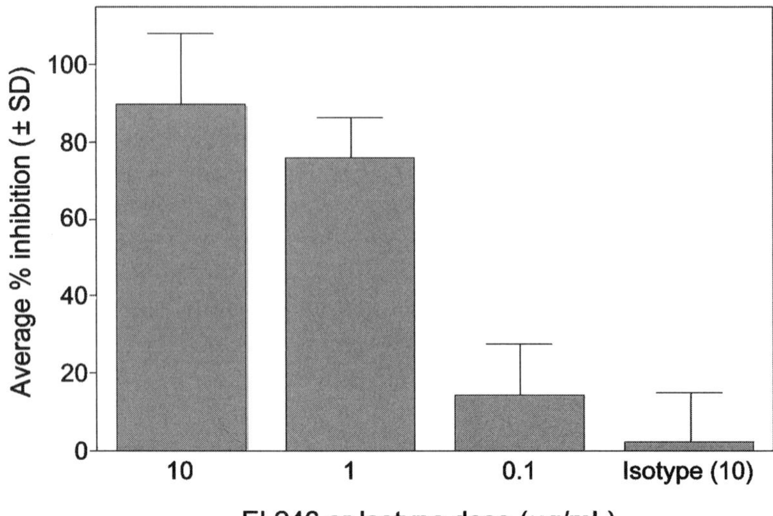

Fig. 7. IC50: EL-246 inhibition of polymorphonuclear neutrophils (PMN) binding to tumor necrosis factor (TNF)-α-activated human umbilical vein endothelial cells (HUVECs). HUVECs were grown on the internal surface of glass capillary tubes, activated with TNF-α (4 h prior to assay) and the tubes incorporated into the capillary tube assay. PMN were pretreated with EL246 or isotype control monoclonal antibody prior to injection into the loop. The data are represented as average percent inhibition (± SD) from three independent experiments.

3.2.1. Human Neutrophil Isolation

1. Collect heparin or sodium citrate anticoagulated venous blood from healthy donors (*see* **Notes 2** and **3**).
2. Add dextran solution to whole blood to achieve 1.5% dextran final concentration.
3. Allow the blood–dextran mixture to sediment for 45 min at room temperature.
4. Collect upper layer and layered onto Ficoll-Hypaque (1.077 g/mL).
5. Centrifuge at 300g.
6. Collect the neutrophil pellet and resuspend in HBSS without $CaCl_2$ and $MgCl_2$, containing 0.5 mM EDTA and 25 mM HEPES (*see* **Note 7**).
7. Resuspend neutrophils in HBSS containing 2 mM $CaCl_2$ and 25 mM HEPES immediately before use to permit selectin and integrin function.

3.2.2. Fluorochrome Labeling of Neutrophils

Flowing neutrophils, when mixed with other cells (e.g., RBCs) or depending on the substrate (e.g, attached neutrophils), can be difficult to visualize in order to examine their adhesive properties. Thus, we utilize cell dyes, such as CFSE, to label cells prior to their perfusion into the flow chamber *(23,24)*.

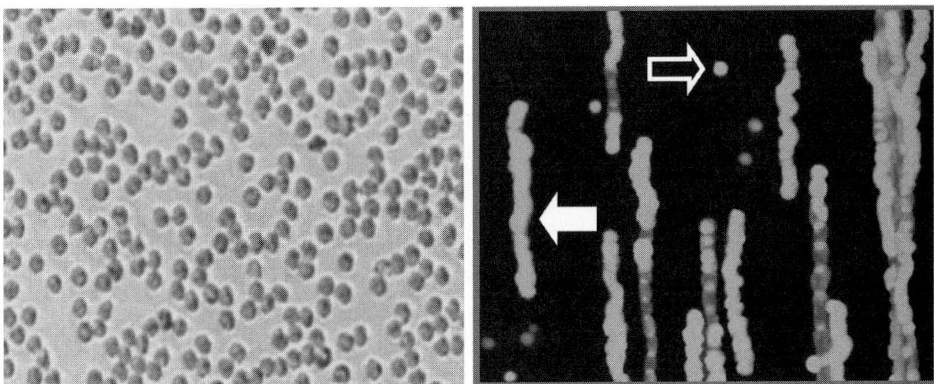

Fig. 8. Neutrophil rolling on a monolayer of neutrophils in the presence of red blood cells. Human neutrophils isolated from peripheral blood were allowed to form a monolayer on recombinant P-selectin-human IgG chimera (0.11 µg/mL) (R&D systems) substrate (left panel). Isolated neutrophils were labeled with CFSE, mixed with red blood cells at a physiological ratio, and perfused into the flow chamber (2 dynes/cm^2). The frames from a 15-s video segment (300× magnification) were superimposed to reveal tracks of labeled neutrophils rolling over the neutrophil monolayer (right panel, solid arrow). For comparison, labeled leukocytes that attached directly to the P-selectin substrate rolled a very short distance during the 15-s time window (right panel, open arrow). (Reproduced from **ref. 24**, with permission.)

1. Add 1 µL of CFSE (5 m*M* in dimethylsulfoxide [DMSO]) to 1 mL of neutrophils (4 × 10^6 cells), and incubate for 10 min at 37°C in the dark with occasional agitation.
2. Wash labeled neutrophils twice at room temperature with HBSS without CaCl$_2$ and MgCl$_2$.
3. Resuspend labeled neutrophils in HBSS containing 2 m*M* CaCl$_2$ and 25 m*M* HEPES immediately before use.
4. Visualize labeled neutrophil staining and morphology using fluorescent real-time microscopy (**Fig. 8**) (*see* **Note 8**).

3.2.3. Parallel Plate Flow Chamber Setup and Experimental Design

Two general types of experiments often performed using parallel plate flow chambers include detachment (shear resistance assay) and accumulation assays.

1. Prepare the substrate of interest on the surface of the slide or dish, as described previously for the capillary tube flow assay.
2. Place a silicon rubber gasket between the flow deck and slide/dish to form a fluid channel over the substrate (*see* **Fig. 2**).
3. Place the assembled flow chamber on the microscope stage and hold it together via vacuum pressure or carefully placed stage clips.
4. Connect the silicone tubing inlet and outlet ports to a cell reservoir and a pump (syringe or peristlatic), respectively (**Fig. 2B**) (*see* **Note 9**).

5. Establish flow to generate the physiological venular shear stress of 2 dynes/cm^2 (*see* **Note 10**).
6. For detachment assays, introduce neutrophils (5×10^6) first into the flow chamber, and allow them to settle onto the substrate of interest under static conditions *(23,25–27)*.
7. Directly count the number of input cells before and then after a particular magnitude and duration of flow.
8. Present data as number of leukocytes bound per unit area of visualized substrate (*see* **Note 11**).
9. For accumulation assays, establish flow to generate the physiological venular shear stress of 2 dynes/cm^2 (*see* **Note 10**).
10. Introduce neutrophils (5×10^6) into the continuous flow to create contact time constraints on cell adhesion.
11. Determine the rate of accumulation as a function of area and time to create an adhesion parameter of cells/area/time (*see* **Note 12**). Rolling velocities of tethered neutrophils can be determined as well (*see* **Note 13**).

4. Notes

1. Each of the approaches described here is expensive to setup, requiring video microscopy and image analysis systems. However, once the hardware is in place, the costs associated with the experimental analyses are no greater than any other immune function assay. The capillary tube system is available on a contract service basis (ProteoFlow™, LigoCyte Pharmaceuticals) for investigators that wish to examine shear dependent adhesion events, but do not want to spend the up-front costs associated with establishing their own system.
2. In performing cell adhesion experiments under flow, the condition of the cell preparation is important. The function of adhesion molecules on neutrophils are easily altered by activation events induced by improper cell preparation, contaminants (e.g., endotoxin) in reagents, glassware, etc. and events taking place in the animal or human source of the cells. Two main problems occur if the cells are too activated: (1) loss of key surface adhesion molecules, such as L-selectin and (2) cellular aggregation. If the cells become partially activated cell aggregation can be minimized by using calcium/magnesium free HBSS and a low amount of EDTA. We have presented two different methods for preparing human neutrophils for use in rolling assays, which we have found to be effective.
3. In preparing neutrophils from whole blood, residual platelets can bind to the neutrophils and complicate or confound interpretation of their binding to a substrate as a result of the expression of various platelet adhesion molecules. Platelet contamination can be greatly minimized by layering isolated neutrophils on 30% Percoll in HBSS without $CaCl_2$ and $MgCl_2$ and centrifuging at 48g, followed by one low-speed wash (48g) in HBSS without $CaCl_2$ and $MgCl_2$.
4. The circulation loop allows multiple infusions, via an injection port, of various antibodies/test compounds during the continuous recirculation of leukocytes across the

interactive surface of the capillary tube. An inverted microscope-video capture system (**Fig. 1**) is used to survey the entire length of the tube and high-resolution phase-contrast recording of the interactive field is performed for subsequent analysis.

5. Previous studies had shown that the interaction of U937 cells with E-selectin expressed on CHO cells is different from that on purified recombinant E- selectin. Specifically, U937 cells roll and then most of them stick to the transfected CHO cells (data not shown). The sticking is likely due to cross-reactivity of human integrins on the U937 cells binding to counter-receptors on the CHO cells. Because anti-selectin antibodies will not reverse an integrin-mediated adhesion, the U937 cells must be pretreated with monoclonal antibody EL-246 prior to injection into the loop.

6. Occasionally, neutrophils or cell lines almost immediately stick upon encountering a specific adhesive substrate. This is presumably due to rapid (or pre-) activation of the β_2 integrins on the leukocyte, which bind ICAM-1 or similar integrin ligands on the adhesive substrate. Once such events take place, they prohibit analysis of any role of selectins. Thus, to examine selectin function in these settings, pretreatment experiments, as described, can be performed or inhibitors of the integrin interactions can be used, which will not disrupt any potential selectin-mediated event.

7. EDTA prevents neutrophil aggregation during their short-term storage while preparing for the shear flow assays.

8. New optical tools have enabled real-time and confocal imaging of neutrophil–endothelial cell interactions in parallel flow chambers. This has allowed for examination of intriguing aspects of the spatial and temporal distribution of adhesion molecules and other cell-surface determinants during the various steps of neutrophil transendothelial cell migration *(22)*.

9. The fluid flow path used with the flow chamber is either single pass, allowing for a continuous exchange of media during the assay, or recirculating (**Fig. 2B**). Stopcock valves (or a manifold) can be added to the tubing line to be used for injection ports to infuse reagents of interest.

10. Fluid flow, which is controlled by the pump, determines the forces acting on the neutrophils. Fluid flow is laminar and simulates nonturbulent blood flow in the microcirculation. The wall shear rate that characterizes the fluid flow is a gradient of flow velocities and is defined as a change in velocity (distance/time) per change in radial displacement (distance)—hence, units of wall shear rate are reciprocal seconds (s^{-1}). In parallel plate flow chambers, the gradient of flow velocities is parabolic for a Newtonian fluid (fluid that flows like water) in that fluid velocity is lower near the flow chamber walls and increases with distance from the walls (for example, *see* **Fig. 3**). Wall shear rate is calculated from the geometry of the rubber gasket fluid channel and volumetric flow rate. The thickness of the rubber gasket and width of the fluid channel can be varied to allow convenient flow rates to be used, which can conserve cells and reagents. Wall shear stress is a critical force modulating leukocyte/substrate interactions and is defined as a shear force acting per unit area (pressure). Units of wall shear stress are dynes/cm^2 (dyn/cm^2).

11. The viscosity of the assay fluid impacts the shear force on the cells (discussed previously). Assays can be done in simple buffers, such as HBSS, or even whole blood. However, the effective shear forces imparted on cells in these fluids will be different, although the flow rates will be the same.

12. The accumulation assay has been a convenient approach to study the binding kinetics of specific receptor/ligand pairs *(28,29)*, including binding affinity and the rate constants of bond formation and bond breakage. Findings by this approach, however, can be confounded by neutrophils tethering not only to the intended substrate but to attached neutrophils, platelets, leukocyte microparticles, etc., resulting in nonlinear cell accumulation.

13. Near the flow chamber wall, nonadhering neutrophils will roll end-over-end, which is due to the gradient of fluid streamlines. This rolling velocity by nonattached cells is referred to as the critical velocity. A platelet, for instance, will have a lower critical velocity than a leukocyte because it is much smaller and occupies slower moving fluid streamlines closer to the wall. If the rolling velocity of a neutrophil adjacent to the flow chamber wall is lower than its critical velocity, then it can be assumed that adhesion-mediated tethering/rolling is occurring.

Acknowledgments

Much of the work summarized in this chapter was supported by funds from the NIH and US Department of Agriculture awarded to both Drs. Jutila and Walcheck, and LigoCyte Pharmaceuticals. The authors also thank Jill Graff for constructive comments and critique of the chapter.

References

1. Butcher, E. C. (1991) Leukocyte-endothelial cell recognition: three (or more) steps to specificity and diversity. *Cell* **67,** 1033–1036.

2. Ley, K., Pries, A. R., and Gaehtgens, P. (1988) Preferential distribution of leukocytes in rat mesentery microvessel networks. *Pflugers Arch.* **412,** 93–100.

3. Ley, K., Gaehtgens, P., Fennie, C., Singer, M. S., Lasky, L. A., and Rosen, S. D. (1991) Lectin-like cell adhesion molecule 1 mediates leukocyte rolling in mesenteric venules in vivo. *Blood* **77,** 2553–2555.

4. Abbassi, O., Kishimoto, T. K., McIntire, L. V., Anderson, D. C., and Smith, C. W. (1993) E-selectin supports neutrophil rolling in vitro under conditions of flow. *J. Clin. Invest.* **92,** 2719–2730.

5. Von Andrian, U. H., Chambers, J. D., McEvoy, L. M., Bargatze, R. F., Arfors, K.-E., and Butcher, E. C. (1991) Two-step model of leukocyte-endothelial cell interaction in inflammation: distinct roles for LECAM-1 and the leukocyte β_2 integrins *in vivo*. *Proc. Natl. Acad. Sci. USA* **88,** 7538–7542.

6. Lawrence, M. B. and Springer, T. A. (1991) Leukocytes roll on a selectin at physiologic flow rates: distinction from and prerequisite for adhesion through integrins. *Cell* **65,** 859–873.

7. Jutila, M. A., Watts, G., Walcheck, B., and Kansas, G. S. (1992) Characterization of a functionally important and evolutionarily well-conserved epitope mapped to the short consensus repeats of E-selectin and L-selectin. *J. Exp. Med.* **175,** 1565–1573.
8. Berlin, C., Bargatze, R. F., Campbell, J. J., et al. (1995) α4 integrins mediate lymphocyte attachment and rolling under physiologic flow. *Cell* **80,** 413–422.
9. Berg, E. L., McEvoy, L. M., Berlin, C., Bargatze, R. F., and Butcher, E. C. (1993) L-selectin-mediated lymphocyte rolling on MAdCAM-1. *Nature* **366,** 695–698.
10. Bargatze, R. F., Kurk, S., Watts, G., Kishimoto, T. K., Speer, C. A., and Jutila, M. A. (1994) In vivo and in vitro functional examination of a conserved epitope of L- and E-selectin crucial for leukocyte-endothelial cell interactions. *J. Immunol.* **152,** 5814–5825.
11. Bargatze, R. F., Kurk, S., Butcher, E. C., and Jutila, M. A. (1994) Neutrophils roll on adherent neutrophils bound to cytokine-induced endothelial cells via L-selectin on the rolling cells. *J. Exp. Med.* **180,** 1785–1792.
12. Jutila, M. A., Bargatze, R. F., Kurk, S., et al. (1994) Cell surface P- and E-selectin support shear-dependent rolling of bovine gamma/delta T cells. *J. Immunol.* **153,** 3917–3928.
13. Perry, M. A. and Granger, D. N. (1991) Role of CD11/CD18 in shear rate-dependent leukocyte-endothelial cell interactions in cat mesenteric venules. *J. Clin. Invest.* **87,** 1798–1804.
14. Hochmuth, R. M., Mohandas, N., and Blackshear, P. L., Jr. (1973) Measurement of the elastic modulus for red cell membrane using a fluid mechanical technique. *Biophys. J.* **13,** 747–762.
15. Wilkinson, P. C., Lackie, J. M., Forrester, J. V., and Dunn, G. A. (1984) Chemokinetic accumulation of human neutrophils on immune complex-coated substrata: analysis at a boundary. *J. Cell Biol.* **99,** 1761–1768.
16. Forrester, J. V. and Lackie, J. M. (1984) Adhesion of neutrophil leucocytes under conditions of flow. *J. Cell Sci.* **70,** 93–110.
17. Lawrence, M. B., McIntire, L. V., and Eskin, S. G. (1987) Effect of flow on polymorphonuclear leukocyte/endothelial cell adhesion. *Blood* **70,** 1284–1290.
18. Usami, S., Chen, H. H., Zhao, Y., Chien, S., and Skalak, R. (1993) Design and construction of a linear shear stress flow chamber. *Ann. Biomed. Eng.* **21,** 77–83.
19. Jones, D. A., Smith, C. W., and McIntire, L. V. (1996) *Methods for In Vitro Analysis of Leukocyte Adhesion Under Flow Conditions.* Balckwell Sciences, Inc., Cambridge.
20. Brown, D. C. and Larson, R. S. (2001) Improvements to parallel plate flow chambers to reduce reagent and cellular requirements. *BMC Immunol.* **2,** 9.
21. Lawrence, M. B. (2001) *In Vitro Flow Models of Leukocyte Adhesion.* Oxford University Press, Inc., New York.
22. Simon, S. I. and Green, C .E. (2005) Molecular mechanics and dynamics of leukocyte recruitment during inflammation. *Annu. Rev. Biomed. Eng.* **7,** 151–185.
23. St. Hill, C. A., Bullard, K. M., and Walcheck, B. (2005) Expression of the high-affinity selectin glycan ligand C2-O-sLeX by colon carcinoma cells. *Cancer Lett.* **217,** 105–113.

24. St Hill, C. A., Alexander, S. R., and Walcheck, B. (2003) Indirect capture augments leukocyte accumulation on P-selectin in flowing whole blood. *J. Leukoc. Biol.* **73,** 464–471.
25. van Kooten, T. G., Schakenraad, J. M., van der Mei, H. C., and Busscher, H. J. (1992) Development and use of a parallel-plate flow chamber for studying cellular adhesion to solid surfaces. *J. Biomed. Mater. Res.* **26,** 725–738.
26. Lawrence, M. B., Kansas, G. S., Kunkel, E. J., and Ley, K. (1997) Threshold levels of fluid shear promote leukocyte adhesion through selectins (CD62L,P,E). *J. Cell Biol.* **136,** 717–727.
27. Mattila, P. E., Green, C. E., Schaff, U., Simon, S. I., and Walcheck, B. (2005) Cytoskeletal interactions regulate inducible L-selectin clustering. *Am. J. Physiol. Cell. Physiol.* **289,** C323–332.
28. Alon, R., Hammer, D. A., and Springer, T. A. (1995) Lifetime of the P-selectin-carbohydrate bond and its response to tensile force in hydrodynamic flow. *Nature* **374,** 539–542.
29. Puri, K. D., Finger, E. B., and Springer, T. A. (1997) The faster kinetics of L-selectin than of E-selectin and P-selectin rolling at comparable binding strength. *J. Immunol.* **158,** 405–413.

17

Model Systems to Investigate Neutrophil Adhesion and Chemotaxis

Nancy A. Louis, Eric Campbell, and Sean P. Colgan

Summary

Polymorphonuclear neutrophil (PMN) recruitment from the blood stream into surrounding tissues, followed by migration through the tissue with triggered release of oxidative enzymes or eventual clearance from the epithelial surface, involves a regulated series of events central to acute responses in host defense *(1)*. Accumulations of large numbers of neutrophils within mucosal tissues are pathognomonic features of both acute and chronic inflammatory conditions including sepsis *(2,3)* and inflammatory bowel disease *(4)*, but the precise signals governing neutrophil adhesion and transmigration remain to be fully characterized. Previous chapters examine methods employed for both neutrophil isolation and study of the mechanisms underlying regulation of PMN rolling behavior. Here, we describe in vitro experimental models for the examination of PMN adhesion to endothelial and epithelial monolayers as well as the characterization of signals influencing neutrophil migration, both along acellular matrices and across endothelial and epithelial monolayers, in the physiologically relevant directions. Studies employing these model systems allow further elucidation of the mechanisms governing PMN adhesion and transmigration.

Key Words: Neutrophil; PMN; endothelia; epithelia; T84; HMEC; chemoattractant; under agarose; adhesion; myeloperoxidase; transmigration; Boyden chamber.

1. Introduction

The vast majority of inflammatory cells are recruited to, as opposed to resident at, inflammatory lesions *(5)* and therefore, polymorphonuclear neutrophil (PMN) recruitment and activation is the critical first response in many inflammatory processes. In noninflammatory conditions, PMN are sequestered in an inactive state within the vasculature until an inflammatory stimulus activates them for recruitment from the vasculature. Such stimuli trigger molecular changes, including the release of soluble chemotactic mediators, such as tumor necrosis factor (TNF)-α *(6)*, interleukin (IL)-8 *(7)*, and lipopolysaccharide (LPS) *(8)*,

From: *Methods in Molecular Biology, vol. 412: Neutrophil Methods and Protocols*
Edited by: M. T. Quinn, F. R. DeLeo, and G. M. Bokoch © Humana Press Inc., Totowa, NJ

accompanied by changes in endothelial apical surface protein and endothelial-bound chemokine expression resulting in a selectin-mediated change in PMN rolling behavior *(9)*. The subsequent series of events stimulates integrin-mediated PMN binding to the endothelium *(10)*, followed by exit from the vasculature via transendothelial migration across endothelial tight junctions via interactions with members of the junctional adhesion molecule (JAM) family *(11)* (practical methods employed to study the signals governing these interactions are detailed in earlier chapters).

Following their exit from the vasculature into the surrounding tissues, further PMN–tissue interactions trigger either proinflammatory responses characterized by targeted but potentially self-destructive degranulation *(12)*, with release of proteases and reactive oxygen species *(13)* or clearance of the PMN from the tissue via apoptosis *(14)*. Alternatively, PMN may continue to migrate through the tissue and across the epithelium. The current paradigm for transepithelial migration (TEM) of PMN across epithelial monolayers envisions a process consisting of sequential molecularly defined events, including initial CD11b/18-mediated interactions of PMN with members of the JAM family of proteins *(15)*. Subsequent CD47-mediated orchestration of PMN movement through the paracellular space is now a well accepted principle, involving CD47 interactions with signal regulatory protein (SIRP)α *(16)*. PMN then reach the apical epithelial surface where they are either activated by lumenal antigens or cleared into the lumen by a CD55-dependent mechanism *(17)*.

Here, we discuss the design and implementation of models to study PMN migration both along acellular organic support matrices and across endothelial or epithelial cell monolayers. These in vitro models permit the examination of the influence of soluble mediators on the properties of PMN migration, as well as further examination of the results of modulation of endothelial or epithelial cell surface protein expression on cell–cell interactions governing both PMN adhesion and migration.

2. Materials

2.1. Maintenance of Epithelial and Endothelial Cell Cultures

1. T84 epithelial cell media: 1:1 ratio of Dulbecco's modified Eagle's medium (DMEM; Gibco/BRL) and Hams F112 Medium, supplemented with 10% fetal bovine serum (Invitrogen), 15 m*M* HEPES buffer, pH 7.5, 14 m*M* NaHCO$_3$, 40 mg/L penicillin, 8 mg/L ampicillin, and 9 mg/L streptomycin.
2. Human microvascular endothelial cell (HMEC) media: DMEM supplemented with 10% fetal bovine serum; 15 m*M* HEPES buffer, pH 7.5, 14 m*M* NaHCO$_3$, 40 mg/L penicillin, 8 mg/L ampicillin, 9 mg/L streptomycin, 10 ng/mL epidermal growth factor (Collaborative Biomedical Products), and 1 µg/mL hydrocortisone (Sigma).
3. 0.05% Trypsin-ethylenediamine tetraacetic acid (EDTA) solution (Gibco/BRL).

4. Membrane permeable supports in various formats and pore sizes (Corning-Costar).
5. Collagen, Type IV (Sigma) (*see* **Note 1**).
6. Voltage clamp (Iowa Dual Voltage Clamps, Bioengineering, University of Iowa) interfaced with an equilibrated pair of calomel electrodes and a pair of Ag-AgCl electrodes.
7. Evohm (World Precision Instruments Inc.).

2.2. Under-Agarose Migration Assay

1. 60-mm tissue culture plates (Corning-Costar).
2. Hank's balanced salt solution (HBSS+) supplemented with 10 mM HEPES (both from Sigma).
3. Modified HBSS (without Ca^{2+} or Mg^{2+}) (HBSS–) supplemented with 10 mM HEPES (both from Sigma).
4. Agarose and gelatin (Fisher Scientific).
5. Cameo quick stain II (Cambridge Diagnostic Products, Ft. Lauderdale, FL).
6. 5% buffered glutaraldehyde solution (Fisher Scientific).

2.3. Neutrophil Adhesion and Transmigration

1. Histopaque 1077 (Sigma).
2. Gelatin (Fisher Scientific).
3. Ammonium chloride lysis buffer: mix 8.29 g of NH4Cl, 1 g of NaCO$_3$, 0.038 g of EDTA (disodium salt) in 1 L of distilled, deionized H_2O.
4. Acid citrate dextrose: mix 13.7 g of citric acid, 25 g of sodium citrate, and 20 g of dextrose in 1 L of distilled, deionized H_2O.
5. Citrate buffer: titrate 1 M sodium citrate with 1 M citric acid (both from Sigma) to a final pH of 4.2.
6. 2',7'-bis-(2-carboxyethyl)-5-(and-6)-carboxyfluorescein (BCECF; Molecular Probes/ Invitrogen). For stock, dissolve in dimethyl sulfoxide to achieve a final concentration of 1 µg/µL.
7. 2.2'-azino-*bis*-(3-ethylbenzothiazoline-6-sulfonic acid) (ABTS) reagent: add 50 mg ABTS to 100 mL of 100 mM citrate buffer and a final concentration of 0.03% H_2O_2 (Sigma).

2.4. Chemoattractants

2.4.1. N-Formylmethionyl-Leucyl-Phenylalanine

1. Prepare a 10 mM stock of *N*-formylmethionyl-leucyl-phenylalanine (fMLP) (Sigma) in dimethylsulfoxide (DMSO). Store at −20°C.

2.4.2. Bacterial Supernatant

1. Prepare fresh for each experiment.
2. Grow a culture of *Streptococcus faecalis* overnight in tryptone broth at 37°C on a rotating platform.
3. Centrifuging the bacterial culture at 1000g for 15 min at 4°C.
4. Filter-sterilize through a 0.45-µm filter prior to use.

2.4.3. Zymosan-Activated Serum

1. Zymosan-activated serum (ZAS) contains complement-derived chemotactic fragment and is prepared fresh for each experiment.
2. Prepare serum from pooled samples of coagulated blood obtained by venipuncture, as described *(18)*.
3. Centrifuge clotted blood at 2500*g* for 20 min.
4. Filter-sterilize the serum supernatant through a 0.45-µm filter and store at −20°C prior to use.
5. Before each experiment, thaw 3 mL of serum.
6. Add 25 mg zymosan (Sigma), and incubate the mixture at 37°C for 60 min with frequent mixing.
7. Destroy residual complement by incubating the mixture at 56°C for 30 min.
8. Centrifuge at 3000*g* for 15 min.
9. Collect the supernatant and store at 4°C prior to use.

3. Methods

The influence of soluble mediators on the mechanics of PMN movement can be studied in a reductionist in vitro model by examining the influence of individual mediators on PMN migration in relation to an organic, acellular support system serving as an experimental extracellular matrix. Studies of the migration of PMN across an agarose–gelatin matrix have allowed the determination of the relative efficacy of different chemotactic mediators *(19)* as well as the examination of the physical properties of PMN migration through the extracellular matrix *(20)* (*see* **Note 2**).

The majority of the work examining the soluble signals and cell–cell interactions governing PMN adhesion to and migration across cellular monolayers has been accomplished using Boyden chamber systems. First described in 1962 *(21)*, this in vitro model employs a two-chamber system to examine the ability of isolated human PMN to move in response to a chemotactic gradient across a monolayer of endothelial or epithelial cells plated on a permeable plastic support with pores of a sufficient size to allow the PMN to pass through. Use of this model system has been instrumental in defining the influence of soluble mediators *(22)*, as well as the regulated expression of surface proteins *(23)* on PMN adhesion and transmigration. Additionally, interactions between the PMN and the apical epithelial surface can be examined specifically through determination of the number of PMN remaining loosely adherent to the apical epithelial surface upon completion of a transmigration experiment *(9,24)*.

3.1. Under Agarose Migration

3.1.1. PMN Isolation and Labeling

1. PMN are purified from anti-coagulated whole venous blood by density centrifugation as described in previous chapters.

2. Centrifuge whole blood at $400g$ for 20 min at 25°C with no brake.
3. Remove plasma and mononuclear cells by aspiration.
4. Sediment red blood cells with 2% gelatin.
5. Remove residual red blood cells by lysis in ice-cold NH_4Cl buffer.
6. Resuspend isolated PMN ice-cold $HBSS^-$ to a concentration of 1.2×10^8 PMN/mL.
7. Add 1.0 µL of BCECF-AM stock per mL of cells (final concentration of 1 µM) and incubate for 15 min at 37°C.
8. Centrifuge labeled PMN at 14,000g in a microfuge and wash once with $HBSS^-$, pelleted again by centrifugation at 14,000g and resuspend to a final concentration of 1.25×10^6 PMN/mL.

3.1.2. Preparation of Agarose-Gelatin Chemotaxis Plates

1. Prepare migration plates daily and use within 2 h of preparation.
2. Dissolve agarose in HBSS+ by incubation in a boiling water bath for about 10 min. At the same time, dissolve gelatin in HBSS+ at 56°C for 10 min.
3. Mix equal volumes of 1% gelatin and 2% agarose at 56°C.
4. Add 6 mL of the mixture to each 15 mm × 60 mm tissue culture dish that is needed.
5. Allow the plates to further solidify at 4°C for about 45 min.
6. Eight sets of three wells each are radially punched by placing each plate over a template and punching wells using a 3-mm-diameter stainless steel bore (**Fig. 1**).
7. Remove the agarose:gelatin plugs with a Pasteur pipet attached to a suction flask assembly. The distance between the rims of the wells is 3 mm (**Fig. 1**).

3.1.3. Migration Assay

1. Each migration plate consists of eight rows of three wells radiating out from the center of the plate. Each row has an inner, an outer and a middle well. Four rows designated as test contain PMN in the middle well and the appropriate chemoattractant in the adjacent well. The other four rows on the same plate are designated control and contain PMN and HBSS+ in the adjacent well (**Fig. 1**).
2. Pre-incubate the migration plates with 10 µL of chemoattractant added to the inner wells of the test rows for 45 min at 37°C in a moist chamber.
3. Immediately after pre-incubation, empty all wells on all plates and evacuate residual moisture using a Pasteur pipet attached to a suction flask assembly.
4. Add 10 µL of HBSS+ to all outer row wells of all eight well sets.
5. Add fresh chemoattractant to all inner wells of test rows, as described above, thus creating a steeper chemoattractant gradient.
6. Add HBSS+ to all inner even wells.
7. Add 10 µL of PMN suspension (1×10^8/mL) to all middle row wells as a final step.
8. Incubate the plates in a moist chamber at 37°C for 120 min.
9. Immediately after incubation, stop PMN function by the addition of 5 mL of 5% buffered glutaraldehyde to each plate.
10. Store the plates overnight at 4°C.
11. Rinse away excess fixative with H_2O and carefully lift the agarose out of the plates.
12. Allow the cell plaques to dry onto the plates before staining.

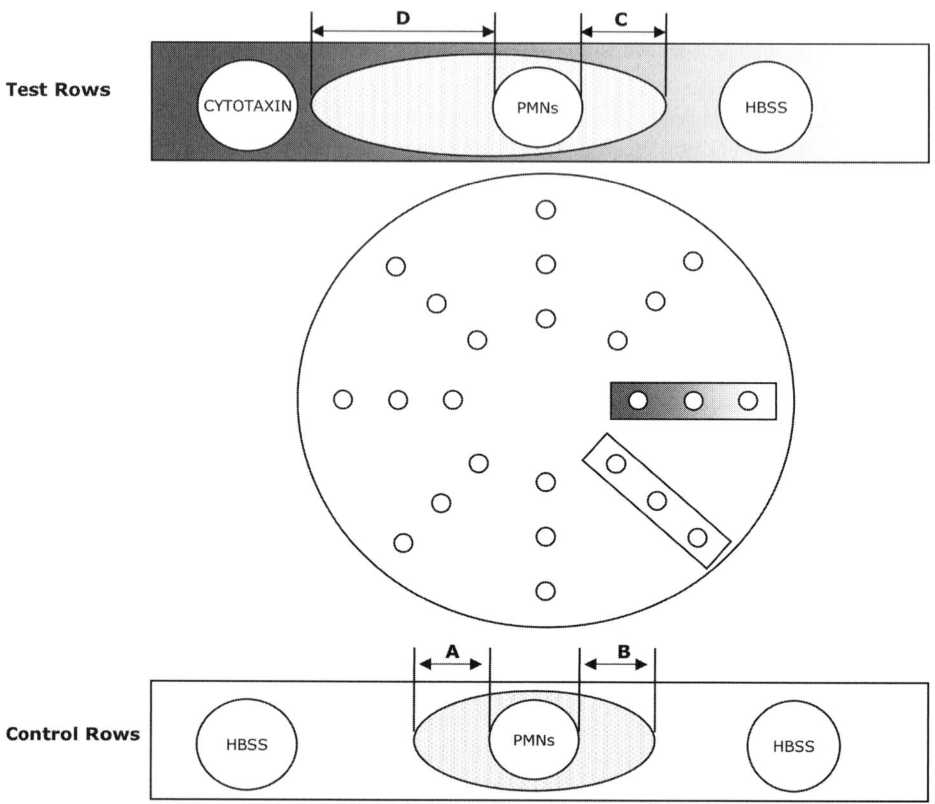

Fig. 1. Schematic overview of the organization of migration wells used in the under agarose assay. The plates contain 1% agarose and 0.5% gelatin; the wells are 3 mm in diameter. Each row (control and test) is alternated four times around the plate. Also indicated are the measurements taken from the plates. HBSS, Hank's balanced salt solution.

13. Stain the cells by adding 3 mL of Camco Quick Stain II for 5 min at room temperature with regular mixing.
14. Rinse the plates with deionized H_2O and air-dry.

3.1.4. Collection of Migration Data

1. PMN migration is determined by the leading front method *(25)*.
2. Measure the distance of migration using a micrometer scale placed in the ocular of a light microscope. Use the 10× objective.
3. Measurements are presented in migration units, with 1 mm equal to 0.96 migration units.
4. All plates are measured by the same person without prior knowledge of the experimental conditions (i.e., blindly).

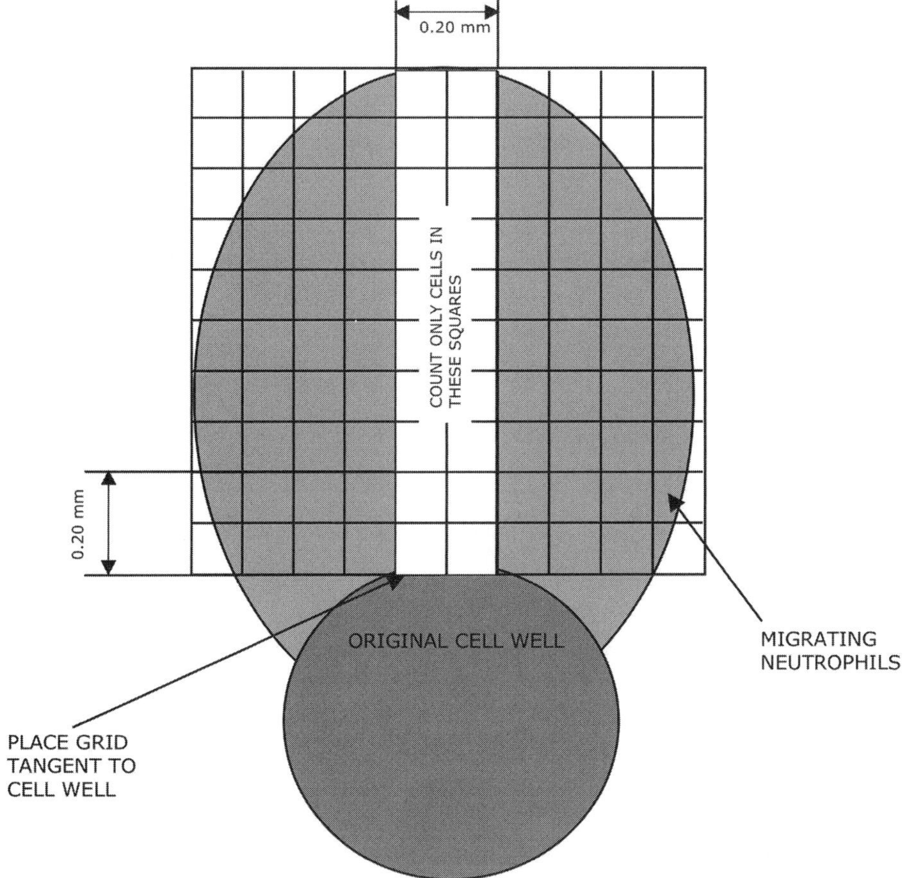

Fig. 2. Schematic representation of the cell count assay used to quantitate the number of leukocytes migrating into the migration pattern on under agarose plates.

5. The distances cells migrate toward the inner wells (A) and outer wells (B) containing buffer are represented as mean values of spontaneous migration (A+B/2). All values are obtained from the four rows of wells taken from each plate (**Fig. 1**).
6. The distance the leading front of cells moves toward the chemoattractant is referred to as directed cell migration (quantity D), while the distance the cells move away from the chemoattractant well is called random migration (quantity C).
7. These measurements can be used to determine chemotactic index (D/C), chemotactic differential (D − C), relative chemotactic differential [(D − C)/(A + B/2)], relative directed migration [D/(A + B/2)], and chemokinetic index [C/(A + B/2)].
8. Quantification of the number of migrated cells is obtained from the same wells as those used to measure the linear migration distances. Using a 10× objective and an ocular grid, count the number of PMN in a 0.20 mm^2 area (**Fig. 2**).

9. Only directed migration (D) pattern is quantified. For each well, evaluate the number of cells distributed in 28 different grid spaces.
10. Counts are not obtained for the grid space immediately adjacent to the cell well, as the cell density is far too great to attempt accurate quantitation.
11. In cases where a cell is positioned in two or more grids confluently, the cell is considered in the grid where the majority of the nucleus is present. Counts are done blindly by the same person.

3.2. PMN Adhesion to Endothelial Cell Monolayers

3.2.1. Plating of Epithelial or Endothelial Cells

1. Trypsinize HMECs using trypsin/EDTA solution (*see* **Note 3**).
2. Wash cells with phosphate-buffered saline and dilute in the appropriate medium to a final concentration of 1×10^5 cells/mL.
3. Plate HMECs in 24-well plates and grow to confluence (*see* **Notes 4–6**).

3.2.2. Adhesion to Endothelial Cell Monolayers

1. Activate HMECs for PMN adhesion by adding 1 µg/mL LPS (isolated from *Salmonella minnesota*) to the media and incubating for 24 h at 37°C.
2. Wash monolayers with HBSS+.
3. Cover cells with 400 µL of HBSS+ containing 200 n*M* fMLP (buffer at 37°C).
4. Add 400 µL of BCECF-labeled PMN to each well for a final concentration of 5×10^5 PMN/well.
5. Immediately following addition of PMN, centrifuge plates at 150*g* (no brake), for 4 min at room temperature to uniformly settle PMN onto the surface of the endothelium.
6. Collect a baseline reading of fluorescence (excitation 485 nm/emission 530 nm).
7. Incubate the plate for 5 min at 37°C.
8. Remove the supernatant by aspiration with a Pasteur pipet.
9. Add 1.0 mL of HBSS+ (buffer at 37°C) and read fluorescence again.
10. Incubate the plate at 37°C. At 5-min intervals, wash and read fluorescence as described in **steps 8** and **9** for a total of 4 readings.
11. Plot the data as the background-subtracted fluorescence vs the wash number. Increased PMN adherence is indicated by a slower decline in measured fluorescence over serial washes.
12. To determine the actual number of residually adherent PMN, add 1 mL of ice-cold HBSS+ and 50 µL of 10% Triton-X-100.
13. Incubate the plates for 20 min in a 4°C cold room on an orbital shaker set at 180 rpm.
14. The number of PMN in each well is determined by myeloperoxidase assay relative to standards generated from known concentrations of PMN from the same isolation (described under **Subheading 3.3.3.**).

3.3. PMN Transmigration Assay

In preparation for the study of physiological regulators of either transendothelial or transepithelial PMN migration, it is important to plate cells such that

PMN will migrate across the monolayers in the appropriate physiological direction. Thus, endothelial cells are plated within the upper well of the chamber such that PMN cross the endothelial monolayer from the apical to basolateral surface (**Fig. 3A**). Conversely, epithelial cells are plated and grown to confluence on the bottom surface of the upper chamber membrane such that PMN added to the upper chamber traverse the monolayer from the basolateral to apical direction (**Fig. 3B**).

3.3.1. Plating of Epithelial And Endothelial Cells

1. Trypsinize T84 or HMEC cells using trypsin/EDTA solution (*see* **Note 3**).
2. Wash cells with phosphate-buffered saline and dilute in the appropriate medium to a final concentration of 1×10^5 cells/mL.
3. Plate HMECs on the upper side (as *inserts,* **Fig. 3A**) and epithelial cells (e.g., T84 cells) on the lower side (as *inverts,* **Fig. 3B**) of polyethylene supports (*see* **Notes 4–6**).

3.3.2. Transmigration Assay

1. Prior to assays of transepithelial migration, wash confluent T84 cell monolayers with 200 µL warm HBSS+ in the apical (upper) chamber and 1 mL warm HBSS+ in the basolateral (lower) chamber to remove any secreted chemical mediators.
2. Because the initial transepithelial resistance (TER) has been observed to fluctuate after washing, monolayers should be allowed to recover at 37°C until the measured TER has equilibrated prior to beginning the transmigration experiment (*see* **Note 7**).
3. After equilibration of T84 monolayers, HBSS+ is removed from both upper and lower chambers and replaced with 150 µL of fresh prewarmed HBSS+ and 1.0 mL of 10^{-6} *M* fMLP in HBSS+, respectively (*see* **Note 8**).
4. Add the desired number of PMN to the upper chamber in a volume of 40 µL of HBSS–. For example, if applying 10^6 PMN, add 40 µL of PMN suspension at the density of 2.5×10^7 cells/mL.
5. Incubate the transwells at 37°C.
6. For subsequent time points, serially transfer the monolayers at each time point into a new chemoattractant-containing well that has been prepared for the assay. Save each lower chamber to determine transmigrated cells.
7. Place inserts into a final well containing 1.0 mL of HBSS+.
8. To determined the number of PMN that have transmigrated through the tight junction but remain loosely adherent to the epithelial monolayer, centrifuge the monolayers within the well for 5 min at 50*g* and 4°C to dislodge adherent PMN.
9. Remove and discard monolayers.

3.3.3. Quantifying Migrated Cells

1. Determine the number of transmigrated PMN using a colorimetric assay for myeloperoxidase enzyme activity.

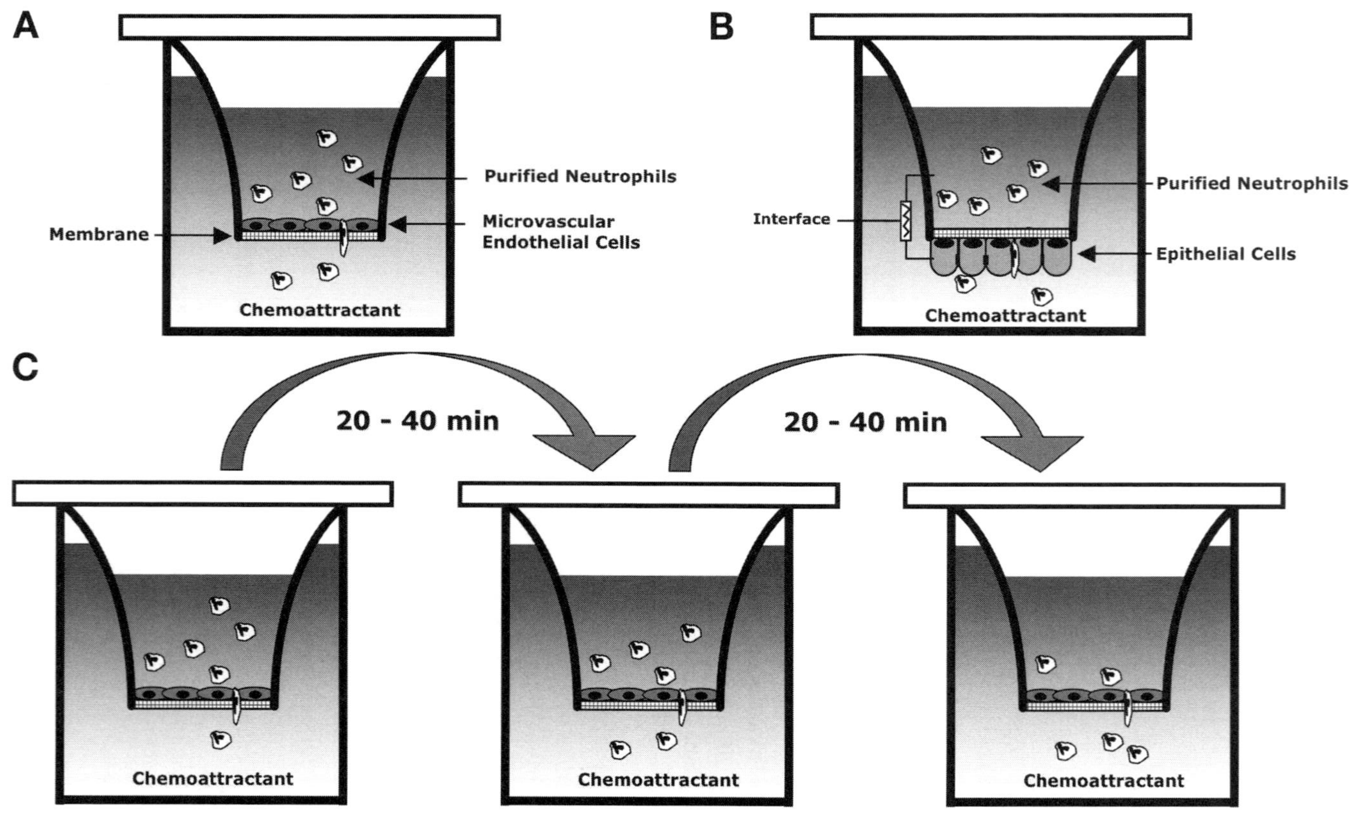

A
Purified Neutrophils
Membrane
Microvascular Endothelial Cells
Chemoattractant
B
Interface
Purified Neutrophils
Epithelial Cells
Chemoattractant
C
20 - 40 min
20 - 40 min
Chemoattractant
Chemoattractant
Chemoattractant

2. Prepare standards for the assay in volumes of 1 mL by serial dilution of the highest possible PMN concentration, assuming all applied PMN migrated through the monolayer. The range of serial dilutions should extend down to a smallest concentration of approx 2000 PMN/mL.
3. Add 50 µL of 10% Triton-X-100 to the PMN standards, bottom chambers, and wells containing dislodged cells to solubilize the cells.
4. Incubate samples on a shaker (180 rpm) at 4°C for 20 min.
5. Acidfy all PMN samples by adding 50 µL of citrate buffer per milliliter PMN sample just prior to the myeloperoxidase assay.
6. During solubilization, prepare ABTS solution.
7. Add 50 µL of 30% H_2O_2 to the ABTS solution immediately prior to the assay.
8. Transfer 75-µL aliquots from each well of solubilized PMN into wells of a flat-bottom 96-well plate.
9. Add 75 µL of ABTS/H_2O_2 solution to each well using a multichannel pipetter.
10. Incubate samples until the optical density (OD)$_{405}$ reading for the well with the highest activity is approx 2.5.
11. Analyze assay results using standard software.
12. During the analysis of kinetic assays in which the data are acquired from serial determinations of the number of PMN transmigrated across a single monolayer measured at multiple time points from multiple wells, the total number of transmigrated PMN is determined in a cumulative manner. For these experiments, the total number of transmigrated PMN is determined from the number measured within the well for that time point plus the total number of PMN transmigrated across that monolayer at all prior time points.

Fig. 3. (*Opposite page*) Models of leukocyte transmigration. In **A**, microvascular endothelial cells are plated on the upper surface of membrane permeable supports (5-µm pores). Cells are allowed to grow to confluence, and functionally assessed as a substrate for migration of human leukocytes (e.g., neutrophils). For these purposes, leukocytes are stimulated to migrate by establishment of a gradient of chemoattractant in the lower chamber. Purified human neutrophils (1×10^6) are placed in the inner well and transmigration over 90 min is determined by quantification of the detergent extractable markers (e.g., neutrophil myeloperoxidase). In **B**, epithelial cells are plated on the lower surface of membrane permeable supports (5-µm pores). Cells are allowed to grow to confluence, and can be functionally assessed (e.g., electrical determination of transepithelial resistance). In this orientation, epithelial cells can also serve as a substrate for migration of human leukocytes (e.g., neutrophils). For these purposes, leukocytes are stimulated to migrate by establishment of a gradient of chemoattractant in the lower chamber. Purified human neutrophils (1×10^6) are placed in the inner well and transmigration over 120 min is determined by quantification of the detergent extractable markers (e.g., neutrophil myeloperoxidase). **C** provides an example of how these models can be used to determine the kinetics of transmigration. For these purposes, permeable support inserts are transferred after a period of time (e.g., 20–40 min) to a well containing the equivalent concentration of chemoattractant. Cumulative determination of leukocyte markers provides an index of the kinetics of transmigration.

4. Notes

1. Alternatively, concentrated collagen solution can be prepared from rat tail tendon using methods modified from Cereijido et al. *(26)*. Rat tails are dissected to yield the shiny white tendon just below the skin. The tendon is then minced and divided into 75 mg aliquots which can be stored at −80°C. Dehydrate the 75 mg of minced tendon in 70% ethanol for 15 min then solubilize it in 100 mL of 1% (v/v) acetic acid at 4°C. Stir the mixture at 4°C overnight. The next morning, the mixture is viscous, with some undissolved materials remaining. Spin the collagen at 25,000*g* for 30 min and dialyze the spun collagen solution against at least 2 L of cold distilled water with two to three changes. The dialyzed concentrated collagen can then be diluted 100-fold to a working concentration in 60% ethanol.

2. The systematic variables influencing PMN migration in the under agarose assay have been extensively analyzed. PMN migration rates are dependent on, and thus assay consistency is critical with respect to, duration of preincubation with chemoattractant, number of PMN migrating, addition of nonmigrating cells, concentration of agarose and of gelatin, use of heat-inactivated serum, and distance between the cells and chemoattractant wells *(20)*.

3. Trypsinization of T84 cells can require incubation for anywhere from 5 to 20 min. Repeat freeze–thaw cycles will decrease the activity of trypsin-EDTA, and therefore, the time required for trypsinization can be minimized by aliquoting fresh trypsin to avoid repeat freeze–thaw cycles.

4. T84 cells generally grow better when inserts have been coated with collagen the day before plating. Type IV collagen is dissolved in 0.2% acetic acid then diluted in 60% EtOH. This process can take up to 3 d at 4°C. On the day of collagen coating, 5–150 µL (depending on insert surface diameter) of type IV collagen is used and allowed to dry for 12–18 h in a cell culture hood under ultraviolet light. Inserts do not need to be washed prior to plating.

5. When plating epithelial inverts (on the bottom surface of the support), individual inverts are kept sterile and flipped upside-down into a large Petri dish. Cells are plated and allowed to adhere to the coated support over night, then inverted into fresh media on the following day. An additional 150–200 µL of fresh media is then added to the empty upper chamber.

6. Alternative method for generating acellular, matrix-coated polyethylene supports. Supports are precoated for 30 min with 50 µL of 0.1% gelatin (derived from porcine skin; Cascade Biologics) followed by the addition of media (containing 10% newborn calf serum, Gibco/BRL) for an additional 30 min. Inserts are rinsed in HBSS+ and polymorphonuclear leukocyte transmigration assessed using similar conditions described under **Subheading 3.3.**

7. This equilibration period can also be used to preincubate either the basolateral or apical surfaces of epithelial monolayers with antibodies or other chemical agents targeted to influence the epithelial component of PMN–epithelial interactions.

8. This assay can be used for single end point measurements of transmigration and residual PMN adhesion to the apical epithelial surface. Alternatively, analysis of the kinetics of PMN transmigration can be performed by serial determinations of

transmigrated PMN at multiple intervals throughout the assay (**Fig. 3C**). If kinetic assays are desired, additional wells containing 1 mL of 10^{-6} M fMLP in HBSS+ should be added per time point for each monolayer.

References

1. Liu, Y., Shaw, S. K., Ma, S., Yang, L., Luscinskas, F. W., and Parkos, C. A. (2004) Regulation of leukocyte transmigration: cell surface interactions and signaling events. *J. Immunol.* **172,** 7–13.
2. Sherwood, E. and Toliver-Kinsky, T. (2004) Mechanisms of the inflammatory response. *Best Pract. Clin. Anaesthesiol.* **18,** 385–405.
3. Chandra, A., Enkhbaatar, P., Nakano, Y., Traber, L., and Traber, D. (2006) Sepsis: emerging role of nitric oxide and selectins. *Clinics* **61,** 71–76.
4. Jaye, D. L. and Parkos, C. A. (2000) Neutrophil migration across intestinal epithelium. *Ann. NY Acad. Sci.* **915,** 151–161.
5. Lewis, J., Lee, J., Underwood, J., Harris, A., and Lewis, C. (1999) Macrophage responses to hypoxia: relevance to disease mechanisms. *J. Leukoc. Biol.* **66,** 889–900.
6. Driscoll, K., Carter, J., Hassenbein, D., and Howard, B. (1997) Cytokines and particle-induced inflammatory cell recruitment. *Environ. Health Perspect.* **105(Suppl 5),** 1159–1164.
7. Khair, O., Davies, R., and Devalia, J. (1996) Bacterial-induced release of inflammatory mediators by bronchial epithelial cells. *Eur. Respir. J.* **9,** 1913–1922.
8. Mizgerd, J. (2002) Molecular mechanisms of neutrophil recruitment elicited by bacteria in the lungs. *Semin. Immunol.* **14,** 123–132.
9. Patel, K., Cuvelier, S., and Wiehler, S. (2002) Selectins: critical mediators of leukocyte recruitment. *Semin. Immunol.* **14,** 73–81.
10. Alon, R. and Feigelson, S. (2002) From rolling to arrest on blood vessels: leukocyte tap dancing on endothelial integrin ligands and chemokines at sub-second contacts. *Semin. Immunol.* **14,** 93–104.
11. Luscinskas, F., Ma, S., Nusrat, A., Parkos, C., and Shaw, S. (2002) Leukocyte transendothelial migration: A junctional affair. *Semin. Immunol.* **14,** 105–113.
12. Faurschou, M. and Borregaard, N. (2003) Neutrophil granules and secretory vesicles in inflammation. *Microbe. Infect.* **5,** 1317–1327.
13. Roos, D., van Bruggen, R., and Meischl, C. (2003) Oxidative killing of microbes by neutrophils. *Microbe. Infect.* **5,** 1307–1315.
14. Kobayashi, S., Voyich, J., and DeLeo, F. (2003) Regulation of the neutrophil-mediated inflammatory response to infection. *Microbe. Infect.* **5,** 1337–1344.
15. Zen, K., Babbin, B. A., Liu, Y., Whelan, J. B., Nusrat, A., and Parkos, C. A. (2004) JAM-C is a component of desmosomes and a ligand for CD11b/CD18-mediated neutrophil transepithelial migration. *Mol. Biol. Cell.* **15,** 3926–3937.
16. Louis, N. A., Hamilton, K. E., Kong, T., and Colgan, S. P. (2005) HIF-dependent induction of apical CD55 coordinates epithelial clearance of neutrophils. *FASEB J.* **19,** 950–959.

17. Liu, Y., Buhring, H. J., Zen, K., et al. (2002) Signal regulatory protein (SIRPalpha), a cellular ligand for CD47, regulates neutrophil transmigration. *J. Biol. Chem.* **277,** 10,028–10,036.
18. Ward, P. A, Cochrane, C. G., and Müller-Eberhard, H. J. (1965) The role of serum complement in chemotaxis of leukocytes *in vitro. J. Exp. Med.* **122,** 327–346.
19. Struyf, S., Gouwy, M., Dillen, C., Proost, P., Opdenakker, G., and Van Damme, J. (2005) Chemokines synergize in the recruitment of circulating neutrophils into inflamed tissue. *Eur. J. Immunol.* **35,** 1583–1591.
20. Blancquaert, A. B., Colgan, S. P., and Bruyninckx, W. J. (1989) Influence of technical parameters on the in vitro motility of equine neutrophils in the presence of streptococcal supernatant. *Vet. Immunol. Immunopathol.* **23,** 85–101.
21. Boyden, S. (1962) The chemotactic effect of mixtures of antibody antigen on polymorphonuclear leucocytes. *J. Exp. Med.* **115,** 453–466.
22. Ebrahimzadeh, P., Hogfors, C., and Braide, M. (2000) Neutrophil chemotaxis in moving gradients of fMLP. *J. Leukoc. Biol.* **67,** 51–61.
23. Parkos, C. A., Colgan, S. P., Liang, A., et al. (1996) CD 47 mediates post-adhesive events required for neutrophil migration across polarized intestinal epithelia. *J. Cell Biol.* **132,** 437–450.
24. Lawrence, D. W., Bruyninckx, W. J., Louis, N. A., et al. (2003) Antiadhesive role of apical decay-accelerating factor (CD55) in human neutrophil transmigration across mucosal epithelia. *J. Exp. Med.* **198,** 999–1010.
25. Nelson, R. D., Quie, P. G., and Simmons, R. L. (1975) Chemotaxis under agarose: a new and simple method for measuring chemotaxis and spontaneous migration of human polymorphonuclear leukocytes and monocytes. *J. Immunol.* **115,** 1650–1656.
26. Cereijido, M., Robbins, E., Dolan, W., Rotunno, C., and Sabatini, D. (1978) Polarized monolayers formed by epithelial cells on a permeable and translucent support. *J. Cell Biol.* **77,** 853–880.

V

Neutrophil Phagocytosis and Bactericidal Activity

18

Immunofluorescence and Confocal Microscopy of Neutrophils

Lee-Ann H. Allen

Summary

Neutrophils are short-lived granulocytes essential for innate host defense. We describe here methods for analysis of resting and activated cells using immunofluorescence and confocal microscopy. Procedures for stimulation of adherent and suspended cells are provided along with protocols for particle opsonization and synchronized phagocytosis. Most importantly, we describe in detail methods for rapid and efficient cell fixation and permeabilization that optimize detection of granule proteins and NADPH oxidase components. Variables that impact antigen detection (such as cell spreading, degranulation, and phagocytosis) are discussed as are methods for image acquisition and analysis.

Key Words: Neutrophil; immunofluorescence; confocal microscopy; granule; phagocytosis; serum; fibrinogen; antibody

1. Introduction

Polymorphonuclear neutrophils (PMN) are relatively small, short-lived cells characterized by a lobed nucleus and cytoplasmic granules. Biochemical analysis of subcellular fractions has been used to define the composition of gelatinase, specific, and azurophilic granules as well as secretory vesicles and to define the subcellular distribution of components of the NADPH oxidase (1,2). Although it has long been known that neutrophil granules are mobilized differentially in response to particulate and soluble stimuli, such as opsonized zymosan, phorbol myristate acetate (PMA) and formyl peptides (1), neutrophils are less commonly analyzed by fluorescence microscopy than other cell types. Consequently, events such as phagosome–granule fusion are understudied as compared with phagosome maturation in macrophages.

Several years ago, we developed methods for immunofluorescence and confocal analysis of macrophages (3–5), and we have adapted these approaches to

From: *Methods in Molecular Biology, vol. 412: Neutrophil Methods and Protocols*
Edited by: M. T. Quinn, F. R. DeLeo, and G. M. Bokoch © Humana Press Inc., Totowa, NJ

human neutrophils *(6–8)*. Herein, we describe methods of cell preparation and analysis that can be applied to resting or activated neutrophils. Our approaches address general issues, such as methods of cell fixation and permeabilization. More importantly, we address factors specific to PMN, such as cell size, adhesion, spreading, degranulation, and phagocytosis, in the context of image analysis using immunofluorescence and confocal microscopy.

2. Materials

2.1. Cell Culture

1. PMN isolated from the peripheral blood of healthy donors using dextran sedimentation and density gradient separation on Ficoll-Hypaque *(9)* (*see* Chapters 2 and 3).
2. Fresh serum to coat coverslips and also as a source of active complement factors for particle opsonization. Transfer a sample of nonheparinized venous blood (10 mL) to a sterile glass tube and incubate for 1 h at room temperature followed by 1 h on ice. Detach the clot from the tube wall using a sterile pipet tip and centrifuge at 500g for 15 min at 4°C. Transfer the serum layer to a sterile polypropylene tube and keep on ice to preserve bioactivity. Fresh serum should be used within 2 h or stored at –80°C for future use. Do not refreeze. Autologous serum is preferred, but pooled normal human serum is acceptable.
3. Heat-inactivated (HI)-fetal bovine serum (FBS): prepare by heating a 500-mL bottle of FBS (HyClone) to 56°C for 30 min with occasional swirling. Filter (0.45 µm) to remove any particulates, and store at –20°C in 50-mL aliquots.
4. Endotoxin-free HEPES-buffered RPMI-1640 medium containing L-glutamine (BioWhittaker/Cambrex). Supplement with HI-FBS (10% final concentration). Store at 4°C.
5. Endotoxin-free Hank's balanced salt solution (HBSS) containing calcium and magnesium (BioWhittaker/Cambrex) supplemented with 10 mM HEPES (pH 7.2) and 10 mM glucose. Sterile-filter (0.2 µm) and store at room temperature.
6. 35-mm tissue culture dishes (Corning).
7. Sterile 15-mL and 50-mL conical polypropylene tubes (Sarstedt).
8. Rectangular, flat-bottom aluminum pans (approx 7 × 11 inches) (from a local convenience store).
9. Refrigerated tissue culture centrifuge with swinging-bucket rotor and microplate carriers such as the Allegra 6KR (Beckman Coulter).

2.2. Particulate and Soluble Stimuli

1. Zymosan (yeast cell wall particles, MP Biomedicals): rehydrate 200 mg zymosan in 20 mL sterile phosphate-buffered saline (PBS) in a 50-mL sterile conical polypropylene tube. Vortex briefly and then sonicate for 5 min (Branson Scientific water bath sonicator). Transfer zymosan to a boiling water bath for 10 min, collect by centrifugation (400g for 10 min), decant PBS, and repeat sonication and boiling steps two more times using fresh changes of PBS. Resuspend the final zymosan pellet in

10 mL tissue culture medium (without FBS) or sterile HBSS, and store in 250-mL (5 mg) aliquots at −20°C. Prepare a working stock solution by diluting one aliquot of zymosan with three volumes of buffer or tissue culture medium (5 mg/mL final concentration). Before each use, briefly sonicate the thawed working stock to disperse any aggregates (*see* **Note 1**).

2. Complement-opsonized zymosan (COZ) particles: incubate an aliquot of the zymosan working stock with fresh human serum (50% final concentration) at 37°C for 30 min. Pellet COZ using a microfuge (2 min at ≥10,000g) and then wash twice with PBS or serum-free tissue culture medium. Keep COZ on ice and use within 2 h. Immunoglobulin (Ig)G-opsonized zymosan (IgG-Z) particles are prepared using Molecular Probes opsonizing reagent. Store IgG-Z in small aliquots at −20°C.

3. PMA (LC Laboratories): dissolve in endotoxin-free, tissue-culture-grade dimethylsulfoxide (DMSO) (Sigma-Aldrich) at a final concentration of 5 mM. Store at −80°C in small aliquots. Because PMA is rapidly inactivated in aqueous solution, thaw stocks at room temperature and use immediately after dilution into buffer or medium.

4. N-formylmethionyl-leucyl-phenylalanine (fMLP; Sigma-Aldrich): dissolve in tissue-culture-grade, endotoxin-free DMSO at a final concentration of 4 mM and stored in aliquots at −80°C.

2.3. Microscopy Supplies

1. Round glass coverslips (12 mm diameter, Fisher Scientific).
2. Fibrinogen (Sigma-Aldrich): prepare as a 0.1 mg/mL stock solution in sterile HBSS and store at −20°C.
3. Straight needle-point stainless steel forceps (Accurate Surgical and Scientific Instruments).
4. Glass Petri dishes (60 mm), store at −20°C.
5. Disposable microbeakers (10 mL capacity, Fisher Scientific).
6. Precleaned microscopy slides with frosted marking area (Fisher Scientific).
7. Lids from 24-well tissue culture cluster plates (Corning or Costar). These do not need to be sterile and can be lids saved after use in other experiments.
8. Plastic flat-bottom plastic boxes with lids (Tupperware type) large enough to accommodate one-three lids from 24-well tissue culture dishes.
9. Cardboard slide folders (Fisher Scientific).

2.4. Buffers and Other Reagents

1. Nitric acid (Fisher Scientific).
2. 70% ethanol: prepare from 95% ethanol (Fisher Scientific) using tissue-culture-grade sterile deionized water.
3. 10% netural buffered formalin solution (Sigma-Aldrich) (*see* **Note 2**).
4. Acetone (Fisher Scientific): store at room temperature and chill rapidly to −20°C prior to use (*see* **Note 3**).
5. PBS: 137 mM NaCl, 2.68 mM KCl, 8.1 mM Na$_2$HPO$_4$, and 1.47 mM KH$_2$PO$_4$ (pH 7.4). Prepare in tissue-culture-grade deionized water and store at room temperature.

6. Blocking buffer: PBS supplemented with 5 mg/mL bovine serum albumin (BSA), 0.5 mg/mL NaN$_3$ and 10% horse serum (*see* **Note 4**) (all from Sigma-Aldrich). Sterile (0.2 µm)-filter and store at 4°C.
7. Washing buffer ("PAB"): identical to blocking buffer except that it does not contain horse serum. Store at 4°C.
8. Antibodies directed against specific marker proteins of neutrophils obtained from commercial sources or prepared in your own laboratory. Antibodies used for the images shown in in this chapter are: mouse anti-human p22*phox* (mAb 44.1) (*7,8, 10,11*); rabbit anti-human p47*phox* (*6–8*); rabbit anti-human lactoferrin (*7*); mouse anti-CD66b (Biodesign); mouse anti-human CD63 (University of Iowa Developmental Studies Hybridoma Bank); and rabbit anti-human myeloperoxidase (*12*).
9. Fluorescein isothiocyanate (FITC)- or rhodamine-conjugated F(ab')$_2$ secondary antibodies (Jackson ImmunoResearch): rehydrate with sterile deionized water according to the manufacturer's directions and store at 4°C. Secondary antibodies conjugated to other fluors (such as Texas Red, AlexaFluor or cyanine dyes) can also be used (*see* **Note 5**). Alexa-conjugated secondary antibodies are available from Molecular Probes.
10. Gelvatol mounting medium (*see* **Note 6**): prepare by mixing 2.4 g of polyvinyl alcohol (Sigma-Aldrich) with 6 g of glycerol, 6 mL of deionized water, and 12 mL of 200 m*M* Tris-HCl (pH 8.0) in a beaker on a stir plate for several hours until dissolved. Heat the solution to 50°C for 10 min and then clarify by centrifugation at 5000*g* for 15 min. Decant the supernatant into a new 50-mL polypropylene tube and add 625 mg 1,4-diazobicyclo-[2.2.2]-octane (DABCO; Sigma-Aldrich). Invert the tube to dissolve DABCO. Store as 1-mL aliquots in Eppendorf tubes in a frost-free freezer at −20 or −80°C. DABCO is essential since it reduces photobleaching of fluorescence.

3. Methods

3.1. Preparation of Acid-Washed Coverslips

Acid-washing is essential because it removes manufacturing residue, oils, and other contaminants such as lipopolysaccharide (LPS) that may inadvertently activate PMN, impair cell adhesion, and/or cause high nonspecific background fluorescence after antibody staining.

1. Transfer one package of 12-mm round glass coverslips into a small glass bottle. Inside a fume hood, pour nitric acid over the coverslips, making sure that they are all submerged. Cap the bottle and incubate in the fume hood for at least 48 h.
2. Carefully decant the acid and then rinse the coverslips with 15–20 changes of sterile, deionized, tissue-culture-grade water (pour water over the coverslips, cap the bottle, rotate gently, decant the water, and repeat).
3. Rinse the coverslips with two changes of 95% ethanol to remove residual water.
4. Store coverslips in 70% ethanol at room temperature. Use after ≥6 h in ethanol.

3.2. Peparation of Coated Coverslips

Bloodstream PMN are nonadherent cells, whereas neutrophils at a site of infection have attached to, and migrated along, the extracellular matrix. Freshly isolated peripheral blood PMN will associate with glass coverslips precoated with serum proteins or purified fibrinogen (*see* **Note 7**). Adherent PMN are preferred because these cells mimic more closely cells at a site of infection and because it is easier to synchronize phagocytosis of adherent cells as compared with cells in suspension.

3.2.1. Flaming of Coverslips and Dispersal in 35-mm Dishes

1. Set up 35-mm tissue culture dishes in a rectangular metal pan. Each dish will hold three coverslips, which provide triplicate samples for each experimental condition or time point.
2. Using forceps, remove a few coverslips from their ethanol storage bottle and place on a pile of Kimwipes.
3. Fill one small microbeaker with 70% ethanol and light a Bunsen burner.
4. Working with one coverslip at a time, grasp with forceps, dunk into ethanol, and drain off excess ethanol by touching edge of the coverslip to the pile of Kimwipes.
5. Pass the coverslip through the flame and then place it into a 35-mm dish. Because the 35-mm dishes are opened only briefly, this procedure can be performed on the bench top. Failure to drain off excess ethanol will cause coverslips to shatter when passed through the flame.

3.2.2. Coating Coverslips With Serum or Fibrinogen

1. Transfer 35-mm dishes containing coverslips into a tissue culture hood.
2. Open all dishes and, using a sterile probe (such as a yellow Pipetman tip), position the coverslips in each dish such that they are not touching the side of the dish or touching one another (**Fig. 1**).
3. Pipet directly onto each coverslip 50 μL of 0.1 mg/mL fibrinogen or 50 μL of fresh serum. The coating agents spread evenly over the acid-washed glass and surface tension holds the domes of liquid in place.
4. Close all dishes, transfer back into the metal pan and place inside a 37°C tissue culture incubator for 30–60 min.
5. Rinse coverslips by flooding each dish with sterile HBSS or PBS (~3 mL). Note that failure to remove excess fibrinogen may cause coverslips to become glued to the tissue culture dish if they air-dry.

3.3. Plating PMN on Coated Coverslips

1. Dilute PMN to 10^6 cells/mL in tissue culture medium.
2. Invert tube gently to mix cells; do not vortex.
3. Place 1 mL of this suspension in each 35-mm dish containing coated coverslips. Be sure that coverslips are not overlapping one another. Push any floating coverslips to the bottom of the dish using a sterile pipet tip.

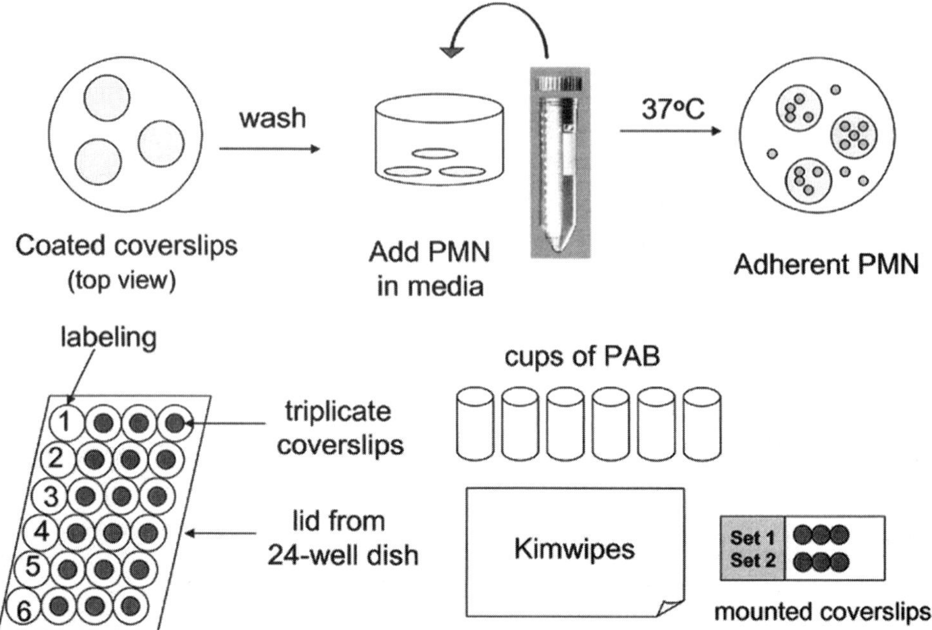

Fig. 1. Cell-plating and -staining schematic. The process of coverslip coating and neutrophil plating is shown (top row) as well as the process of cell staining and washing and coverslip mounting (bottom row).

4. Place dishes in a 37°C incubator for 1 h. Neutrophils will adhere to the coated coverslips in preference to the uncoated tissue culture dish (**Fig. 1**). For microscopy, a moderate cell density is optimal.

3.4. Stimulation of Adherent PMN

Because PMN are only loosely adherent, it is important to add stimuli to dishes carefully so as not to detach cells. For this same reason, cells are not fluid changed before addition of stimuli. Rather, particles or soluble stimuli are added to adherent neutrophils in a volume of 1 mL to achieve a total volume of 2 mL per 35-mm dish. Most stimuli will enhance PMN adhesion and spreading via their ability to trigger upregulation of β_2 integrins (*see* **Notes 7** and **8**).

3.4.1. Soluble Stimuli

PMN will be optimally activated by exposure to 200 nM PMA or 1 μM fMLP.

1. Dilute each agonist in tissue culture medium to achieve to twice the desired final concentration. Vortex.
2. Add 1 mL of either stimulus to each dish of cells.
3. Transfer to 37°C incubator for the desired amount of time (generally 0.5–5 min for fMLP and 5–30 min for PMA).

3.4.2. Synchronized Phagocytosis of Opsonized Zymosan

Centrifugation of opsonized zymosan particles onto adherent PMN at low temperature allows particle binding but not phagocytosis and rapid transfer of samples to 37°C supports synchronized ingestion.

1. Dilute COZ in cold (4°C) tissue culture medium (use 1–2 µL of 5 mg/mL COZ per dish to achieve ~2–3 phagosomes per cell).
2. Add 1 mL of cold, diluted COZ to each dish of PMN (already containing 1 mL of warm medium). Mixing warm and cold medium will reduce the overall temperature below the threshold for phagocytosis (16°C).
3. Rapidly and carefully transfer the dishes onto plate carriers in a refrigerated tissue culture centrifuge cooled to 12–15°C. Centrifuge for 2 min at approx 600*g* with maximum braking. For smaller particles such as bacteria, increase the centrifugation time to 3.5–4 min.
4. Carefully transfer dishes to 37°C incubator to allow phagocytosis or fix cells immediately (t = 0 min). If samples remained cold during centrifugation, the particles or bacteria in the 0-min samples should be cell-associated but not internalized, and forming phagosomes should be apparent within 1 min of transfer to 37°C *(6)*. For Fcγ receptor-mediated phagocytosis, samples can be cooled to 4°C instead of 15°C, but this lower temperature does not support tight binding to CD11b/CD18.

3.5. Fixation

1. Gently aspirate medium from each dish and then cover cells with 10% formalin.
2. Incubate at room temperature for 10–15 min (*see* **Note 9**). For certain applications, cells may be fixed using 4% paraformaldehyde (*see* **Note 2**).

3.6. Permeabilization

1. Fill chilled glass Petri dishes with –20°C acetone.
2. Using forceps, transfer coverslips (one at a time) into chilled acetone.
3. After 5 min, rehydrate coverslips by transferring into new tissue culture dishes filled with room-temperature PBS. Longer incubations in acetone will not harm cells. If glass dishes are not available, cells can be permeabilized using a 1:1 mixture of –20°C acetone-methanol, which is compatible with plastic (*see* **Notes 2** and **3**).

3.7. Blocking

1. Aspirate PBS from each dish and cover fixed/permeabilized cells with 2–4 mL of blocking buffer.
2. Incubate at room temperature for 1 h or overnight at 4°C.

3.8. Antibody Staining for Fluorescence Microscopy

For each antibody the optimal concentration for fluorescence microscopy must be determined empirically. As a general rule, higher concentrations of antibodies are needed for microscopy than for immunoblotting. It is also essential to determine whether the antibodies of choice exhibit nonspecific binding to opsonized zymosan, bacteria, or other particulate stimuli you may be using. In most cases, nonspecific staining can be eliminated by antibody affinity purification.

3.8.1. Primary Antibodies

1. Dilute primary antibodies in blocking buffer to the desired final concentration (*see* **Note 10**). Each coverslip requires 30 µL of antibody. For double- or triple-staining, primary antibodies can be mixed together. Prepare only what is needed for each experiment because diluted antibodies are unstable.
2. Centrifuge diluted antibodies at 9000*g* for 10 min at 4°C to pellet any aggregates. Label a 24-well dish lid and set up a pile of Kimwipes and six microbeakers of PAB (**Fig. 1**).
3. Using forceps, remove one coverslip from blocking buffer. Touch the edge of coverslip to the pile of Kimwipes to drain off liquid, rinse (swish) in one microbeaker of PAB, and drain again.
4. Dry the back (non-cell side) of the coverslip using a fresh Kimwipe and place it on the 24-well dish lid.
5. Immediately cover with 30 µL of antibody (do not let coverslips dry out). Repeat this process for other coverslips.
6. Place coverslips into a plastic box lined with moist paper towels and incubate for 1 h.

3.8.2. Washing and Secondary Antibody Staining

1. Prepare diluted secondary antibodies in blocking buffer. Appropriate dilutions will not stain cells in the absence of primary antibody. For example, we use the Jackson ImmunoResearch antibodies at 1:200 dilution. Once again, 30 µL of antibody is used per coverslip. Prepare the amount needed for each experiment. Secondary antibodies do not need to be centrifuged prior to use.
2. Lift the first coverslip off of the 24-well dish lid using forceps (grasp as close to the edge as possible) and drain primary antibody into Kimwipes.
3. Wash the coverslip sequentially in six microbeakers of PAB with thorough draining between washes.
4. After the final rinse, carefully blot dry the back of the coverslip and return it to the 24-well dish lid.
5. Cover cells with 30 µL of secondary antibody. Repeat for other coverslips.
6. Incubate for 1 h in a covered box lined with moist paper towels.

3.8.3. Nonantibody Stains

Actin filaments can be detected by staining cells with FITC-, rhodamine-, or Alexa-conjugated phalloidin. Several nucleic acid stains (available from Molecular Probes) can be used to label cell nuclei as well as microbes. Phalloidins can be added along with primary or secondary antibodies. For most DNA stains, 5–10 min incubation is sufficient. Wash samples thoroughly before proceeding (*see* **Note 5**).

3.8.4. Mounting

1. Label microscopy slides in the frosted marking area. Up to six coverslips will fit onto one slide (**Fig. 1**). It is important to place the coverslips as close as possible to the frosted marking area of the slide. Because the edges of the slide rest on the microscopy stage, it is not possible to view coverslips that are mounted is area.
2. Thaw one tube of mounting medium. Working with coverslips for one slide at a time, place six small drops (~15 µL) of gelvatol mounting medium onto the glass slide (*see* **Note 6**).
3. Wash each coverslip in six beakers of PAB as described above followed by a final rinse in deionized water. The water rinse prevents a dried salt crust from forming on mounted coverslips.
4. Blot excess water off the back of the coverslip and then invert it (cell side down) onto a bead of mounting medium (start from one edge of the coverslip and lower it slowly to avoid trapping air bubbles). Repeat for the other coverslips on this slide.
5. Gently press down on the top of each coverslip to ensure that their edges are not overlapping (this will also push out small air bubbles). If necessary, excess mounting medium can be removed using an aspirator.
6. Place slides in a cardboard slide folder; this keeps the slides flat and protected from light. Let the mounting medium set overnight at 4°C. Leftover mounting medium can be refrozen for future use. Very high humidity will prevent the mounting medium for hardening, so do not store samples in a cold room or refrigerator near dripping or condensed water.

3.9. Microscopy, Image Acquisition, and Analysis

1. Neutrophil shape, phagocytosis, and the distribution of intracellular markers can be evaluated using conventional fluorescence microscopy (*6*). The objectives on most fluorescence microscopes support phase contrast optics, which is the preferred method for analysis of cell morphology. Many confocal systems are not compatible with phase-contrast optics and the cell features that can be detected using bright field imaging or Nomarski optics are less informative. On the other hand, details regarding granule positioning (**Figs. 2–4**) and the composition of small bacterial phagosomes (*7,8*) are more readily analyzed by confocal as compared with conventional fluorescence microscopy. Advantages of confocal microscopy include the ability to zoom in on single cells and the ability to examine cells as sequential (or merged stacks) of thin optical sections. Using this approach, specific and azurophilic granules can be detected in the vicinity of COZ phagosomes prior to phagosome-granule fusion (**Fig. 4**).
2. The ability to capture several images using identical confocal and laser settings also allows different fixation and permeabilization conditions to be compared directly (**Figs. 2** and **3**), and the software packages associated with confocal microscopes allow extensive image analysis such as measurement of pixel intensity at different points within a single image (*7,13*).
3. When testing new antibodies or particulate stimuli, all samples should be carefully evaluated for specificity of intracellular staining (if the antigen distribution is known)

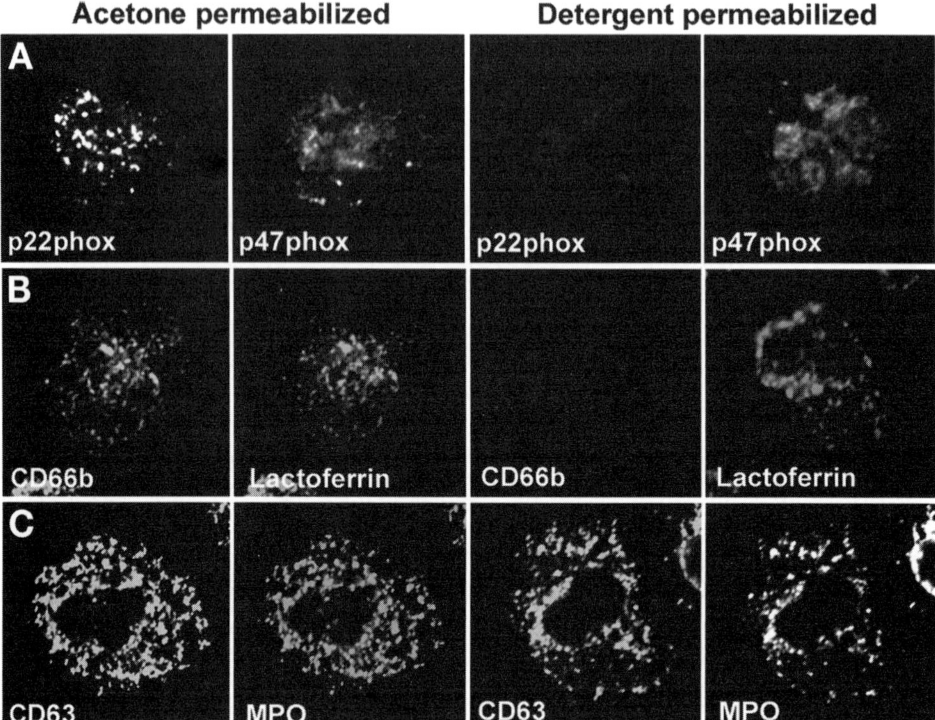

Fig. 2. Acetone permeabilization enhances detection of p22*phox* and CD66b. Neutrophils attached to serum-coated coverslips were fixed in 10% formalin for 15 min and then permeabilized using −20°C acetone or 0.1% Triton X-100 as indicated. After blocking, samples were double-stained to detect NADPH oxidase subunits p22*phox* and p47*phox* (**A**), the specific granule markers CD66b and lactoferrin (**B**), or the azurophilic granule markers CD63 and myeloperoxidase (**C**). All cells were processed in parallel and confocal images in each set were obtained using identical settings.

and absence of nonspecific staining of zymosan or bacteria. We have used the methods described here to study neutrophils from persons with chronic granulomatous disease *(6)* and to analyze PMN containing opsonized zymosan, *Neisseria meningitides*, and *Helicobacter pylori (6–8)*. The data shown in **Fig. 2** demonstrate that acetone permeabilization enhances markedly the ability to detect p22*phox* and CD66b associated with specific granules (*see* **Note 10**) and, in general, cytoplasmic granules are dispersed more evenly in acetone-permeabilized cells as compared with detergent-permeabilized neutrophils. Conversely, either permeabilization method is suitable for detection of cytosolic p47*phox*, lactoferrin inside specific granules, or the azurophillic granule markers CD63 and myeloperoxidase (MPO) (**Fig. 2**). Finally, both cell spreading and degranulation affect antigen detection (**Fig. 3**) (*see* **Note 10**).

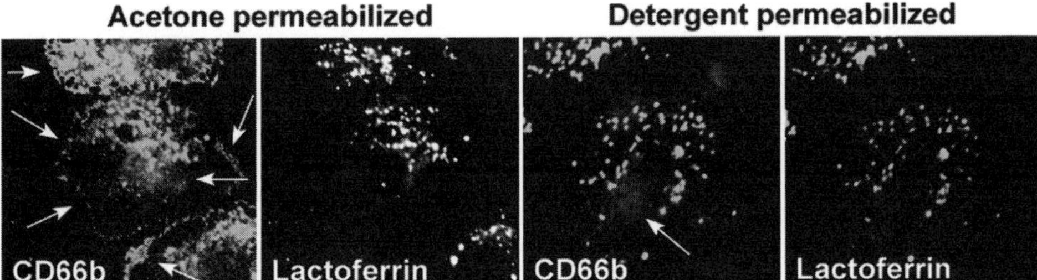

Fig. 3. Enhanced detection of CD66b in the plasma membrane of activated neutrophils. Neutrophils were stimulated with 200 n*M* phorbol myristate acetate (PMA) for 15 min and then fixed with formalin. Cells were permeabilized with acetone or Triton X-100 as indicated and then double-stained to detect CD66b and lactoferrin. CD66b delivered to the plasma membrane by PMA-stimulated degranulation is readily apparent in acetone-permeabilized cells (arrows). Higher concentrations of anti-CD66b antibody and increased confocal gain allow detection of granule-associated CD66b in detergent permeabilized cells, but these manipulations are not sufficient to detect less concentrated CD66b at the cell surface.

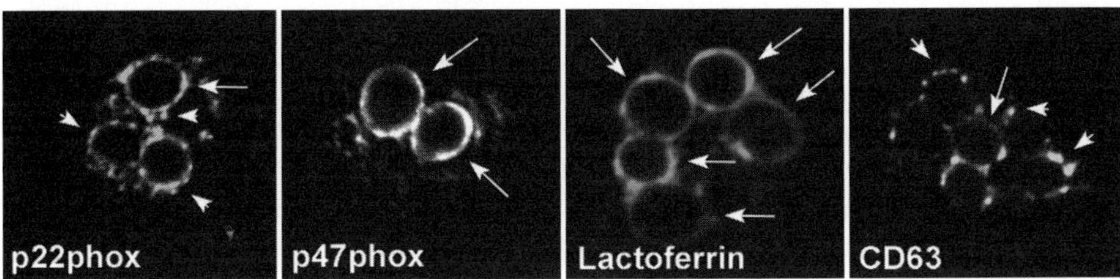

Fig. 4. Opsonized zymosan phagosomes accumulate granule markers and NADPH oxidase components. Confocal sections show accumulation of p22*phox*, p47*phox*, lactoferrin and CD63 on 15 min complement-opsonized zymosan (COZ) phagosomes (arrows). Confocal sections also demonstrate primary and secondary granule clustering near ingested COZ prior to phagosome–granule fusion (arrowheads).

4. Notes

1. Other particulate stimuli can be substituted for zymosan including live or killed microorganisms, latex beads, or opsonized sheep red blood cells. Zymosan, latex beads, and red blood cells are visible using phase-contrast, bright field, or Nomarski optics. By contrast, the small size of many bacteria makes them more difficult to detect in the absence of a direct probe (such as green fluorescent protein or a specific antibody). Most importantly, prelabeled fluorescent beads, *Escherichia coli*, or other BioParticles available from Molecular Probes should *not* be used in combination with the conventional antibody-staining approaches described here. The intense fluorescence of these particles overwhelms all other signals and appears as nonspecific fluorescence across a wide spectrum of wavelengths or channels using either conventional or confocal immunofluorescence microscopy.

2. Neutral buffered solutions of 10% formalin are equipotent to 4% paraformaldehyde. We prefer formalin because it does not need to be prepared fresh each day. However, the small amount of methanol used to stabilize formalin makes it unsuitable as a fixative when intact cell membranes are required, for example when fixing cells that have pinocytosed fluorescent dextrans or Lucifer yellow *(14)*, or when strict detection of a surface-exposed (but not intracellular) antigen or epitopes is desired *(7)*. In these instances, cells should be fixed using fresh 4% paraformaldehyde. Direct exposure of live cells to cold methanol (which can fix and permeabilize neutrophils in one step) is compatible with some (but not all) antigens.

3. Detergents such as Triton X-100 are commonly used to permeabilize fixed cells for fluorescence microscopy. However, this approach is prone to artifacts, particularly when used to detect soluble proteins *(15)*. We find that permeabilization of formalin-fixed cells with −20°C acetone (or acetone–methanol mixtures) provides superior preservation of cell morphology and enhances the sensitivity of detection of many antigens as shown here for p22*phox* and CD66b (**Figs. 2** and **3**). These data are noteworthy because the epitope on p22*phox* that is recognized by monoclonal antibody 44.1 is cytosolic *(10,11)*. As such, the limited detection conferred by 0.1% Triton-X-100 cannot be explained by differential permeabilization of granule and plasma membranes by this detergent. Similarly, we have shown previously that acetone permeabilization enhances detection of Lamp-1 on *H. pylori* phagosomes in macrophages *(13)*. On the other hand, detergent permeabilization is preferred if cells contain pinocytosed dextrans or Lucifer yellow *(14)*.

4. In our hands, the type of serum used in blocking buffer is not critical. We find that Sigma-Aldrich horse serum works well and is cost-effective (as compared with FBS or goat or donkey serum; *see also* **Note 5**).

5. For dual and triple staining, optimal results are obtained using secondary antibodies that are cross-absorbed against other species (available from Jackson Immuno-Research) and F(ab')$_2$ antibodies are preferred to whole Ig. If whole IgGs are used, it may be necessary to use blocking buffer containing serum matched to the species of secondary antibody in order to prevent nonspecific binding to Fcγ receptors. On the other hand, the choice of fluorescent conjugate is largely a matter of personal preference. For double staining, "green–red" combinations are standard. In this regard, Oregon Green and AlexaFluor488 are brighter, more photostable, and less pH-sensitive than FITC. Rhodamine and AlexaFluor546 emit strong red-orange fluorescence that is brighter than the dark red of AlexaFluor594 or Texas Red. For triple labeling, combine a far red dye such as Cy5 or AlexaFluor647 with FITC and rhodamine. Detection of ultraviolet (UV)-excitable dyes (such as 2-(4-Amidino-phenyl)-6-indolecarbamidine dihydrochloride [DAPI]) using confocal microscopy requires a special laser. DAPI and other DNA-binding dyes can be used to stain cell nuclei and ingested microbes and newer products available from Molecular Probes do not require UV excitation. Similarly, dye-conjugated phalloidins are used to detect actin filaments *(6)*.

6. We prefer the polyvinyl alcohol base of gelvatol mounting medium because it hardens rapidly and completely. The mounting medium we prepare gives consistent results,

but in our hands, this is not true for similar products obtained from commercial vendors. A common alternative to gelvatol is buffered glycerol (with or without an anti-fading agent). Because glycerol mounting solutions do not harden, coverslips must be attached to microscope slides using several layers of nail polish (which is time-consuming to apply and apt to leak).

7. We described previously the phenotype of PMN attached loosely to serum- or fibrinogen-coated coverslips *(7,8)*. Because this interaction is physiological, it is preferred to the nonspecific adhesion obtained using poly-lysine-coated surfaces. Moreover, the general "stickiness" of poly-lysine can impair phagocytosis of particulate stimuli and increase background as a result of trapping of debris. If poly-lysine is used, 0.1 mg/mL poly-D-lysine is less toxic to most cell types than poly-L-lysine. For several reasons, we also prefer coated coverslips to chamber slides (Nunc). First, coverslips are more cost-effective, in part because only six of the eight wells on each chamber slide can be used (because the wells on the far right side of the slide rest on the microscope stage). Second, neutrophils distribute evenly over coated coverslips yet bind preferentially near the edges of each chamber as a result of the meniscus of liquid in each well. Third, the binding interactions that mediate cell attachment to the proprietary slide coating are unclear and can trigger cell activation. Therefore, chamber slides should be coated with serum proteins or fibrinogen before use. Finally, 70 µL of antibody is needed for each well as compared with 30 µL for each 12-mm coverslip (despite similar surface area).

8. To stimulate PMN in suspension, dilute cells to 10^6 PMN/mL in desired medium or buffer and disperse into sterile Eppendorf tubes. Add particulate or soluble stimuli and incubate at 37°C (with tumbling). Terminate incubations by rapidly diluting each sample into 20 volumes of ice-cold PBS and keep on ice. Attach cells to acid-washed coverslips using a cytocentrifuge (such as a Shandon Cytospin). Assemble slides and filter papers as directed by the manufacturer and place a dry (uncoated) coverslip between the microscope slide and the absorbent paper with the coverslip positioned to receive the sample. Centrifuge 50,000–100,000 PMN onto each coverslip. Transfer coverslips to a tissue culture dish and continue as described for adherent cells. Do not fix and stain cells prior to cytocentrifugation. Precise synchronization of phagocytosis is not achieved using this method.

9. Both the time and temperature of cell fixation is important. Room-temperature fixation preserves cell morphology whereas chilled fixatives can cause cells to contract or detach. Cells incubated in fixative for prolonged periods of time will appear dark when using phase-contrast optics. Inadequately fixed cells will have poor organelle morphology and, because antigens are not appropriately fixed in place, antibody staining may appear less distinct or poorly localized as compared with properly fixed cells. It is also important to note that aldehyde fixation is not permanent. For this reason, samples should be stained and analyzed within 2 wk.

10. Appropriate antibody concentrations vary with changes in cell morphology and the extent of degranulation. Concentrations of antibody appropriate for activated, spread PMN that have undergone some degranulation can be too strong when applied to round, unstimulated cells. Similarly, the amount of antibody needed to detect high

concentrations of granule-associated proteins may not be sufficient to detect the same marker once it is distributed over a large phagosome or throughout the plasma membrane (**Figs. 3** and **4**). In this regard, it is important to evaluate each antibody separately and also to realize that the appropriate dilution of primary antibody may vary depending on whether the secondary antibody is conjugated to FITC or rhodamine or another fluor.

Acknowledgments

Drs. Algirdas Jesaitis (Montana State University) and William Nauseef (University of Iowa) provided some of the antibodies used for this study. This work was supported by a VA Merit Review Grant and funds from the National Institutes of Health (P01-AI44642).

References

1. Faurschou, M. and Borregaard, N. (2003) Neutrophil granules and secretory vesicles in inflammation. *Microbes Infect.* **5,** 1317–1327.
2. Nauseef, W. M. (2004) Assembly of the phagocyte NADPH oxidase. *Histochem. Cell Biol.* **122,** 277–291.
3. Allen, L.-A. H. and Aderem, A. (1995) A role for MARCKS, the α isozyme of protein kinase C and myosin I in zymosan phagocytosis by macrophages. *J. Exp. Med.* **182,** 829–840.
4. Allen, L.-A. H. and Aderem, A. (1996) Molecular definition of distinct cytoskeletal structures involved in complement- and Fc receptor-mediated phagocytosis in macrophages. *J. Exp. Med.* **184,** 627–637.
5. Allen, L.-A. H., Schlesinger, L. S., and Kang, B. (2000) Virulent strains of *Helicobacter pylori demonstrate* delayed phagocytosis and stimulate homotypic phagosome fusion in macrophages. *J. Exp. Med.* **191,** 115–127.
6. Allen, L.-A. H., DeLeo, F. R., Gallois, A., Toyoshima, S., Suzuki, K., and Nauseef, W. M. (1999) Transient association of the nicotinamide adenine dinucleotide phosphate oxidase subunits p47phox and p67phox with phagosomes in neutrophils from patients with X-linked chronic granulomatous disease. *Blood* **93,** 3521–3530.
7. Allen, L.-A. H., Beecher, B. R., Lynch, J. T., Rohner, O. V., and Wittine, L. M. (2005) *Helicobacter pylori* disrupts NADPH oxidase targeting in human neutrophils to induce extracellular superoxide release. *J. Immunol.* **174,** 3658–3667.
8. DeLeo, F. R., Allen, L.-A. H., Apicella, M., and Nauseef, W. M. (1999) NADPH oxidase activation and assembly during phagocytosis. *J. Immunol.* **163,** 6732–6740.
9. Boyum, A. (1968) Isolation of mononuclear cells and granulocytes from human blood. *J. Clin. Lab. Invest.* **21,** 77–89.
10. Burritt, J. B., Quinn, M. T., Jutila, M. A., Bond, C. W., and Jesaitis, A. J. (1995) Topological mapping of neutrophil cytochrome b epitopes with phage-display libraries. *J. Biol. Chem.* **270,** 16,974–16,980.

11. Burritt, J. B., Busse, S. C., Gizachew, D., et al. (1998) Antibody imprint of a membrane protein surface. Phagocyte flavocytochrome b. *J. Biol. Chem.* **273,** 24,847–24,852.
12. Nauseef, W. M., Root, R. K., and Malech, H. L. (1983) Biochemical and immunological anaysis of hereditary myeloperoxidase deficiency. *J. Clin. Invest.* **71,** 1297–1303.
13. Schwartz, J. T. and Allen, L.-A. H. (2006) Role of urease in megasome formation and *Helicobacter pylori* survival in macrophages. *J. Leukoc. Biol.* **79,** 1214–1225.
14. Botelho, R. J., Tapper, H., Furuya, W., Mojdami, D., and Grinstein, S. (2002) FcγR-mediated phagocytosis stimulates localized pinocytosis in human neutrophils. *J. Immmunol.* **169,** 4423–4429.
15. Melan, M. A. and Sluder, G. (1992) Redistribution and differential extraction of soluble proteins in permeabilized cultured cells: implications for immunofluorescence microscopy. *J. Cell Sci.* **101,** 731–743.

19

Assessment of Phagosome Formation and Maturation by Fluorescence Microscopy

Benjamin E. Steinberg and Sergio Grinstein

Summary

Phagocytosis of invading microorganisms by neutrophils is an important early response to infection. Here, we describe protocols designed for the quantitative study of particle internalization and of the subsequent maturation of the newly formed phagosomes using sheep red blood cells as a model target particle and neutrophils as phagocytes. Techniques to label the particles with fluorescent pH-sensitive dyes are presented, along with complement and immunoglobulin opsonization procedures. These labeling techniques, in combination with high-resolution digital imaging, allow for the quantitative assessment of phagocytosis at the single-cell level by bright field and fluorescence microscopy. Lastly, qualitative and quantitative methods of investigating the intraphagosomal pH are presented. We describe the use of a fluorescent weak base that accumulates in acidic cellular compartments and functions as a marker of phagosome acidification, as well as more quantitative phagosomal pH measurements by fluorescence ratio imaging. A brief description of the hardware and software components necessary for digital imaging is also provided.

Key Words: Acidification; ratiometric imaging; Fcγ receptor; CD11b/CD18.

1. Introduction

Neutrophils are the first cells to be recruited to the site of infection and, as such, are the body's first line of defense against invading microorganisms. On arrival, neutrophils bind and internalize the microbes by a receptor-mediated process termed phagocytosis. The ingested microorganism is trapped within a vacuole, known as the phagosome, which subsequently undergoes a maturation process whereby it fuses with intracellular granules that contain a variety of antimicrobial factors. In this way, phagocytosis and phagosome maturation play a key role in clearing infection.

From: *Methods in Molecular Biology, vol. 412: Neutrophil Methods and Protocols*
Edited by: M. T. Quinn, F. R. DeLeo, and G. M. Bokoch © Humana Press Inc., Totowa, NJ

There are two general mechanisms of particle engulfment, depending on the manner in which the particle is recognized by the phagocyte. In nonopsonic phagocytosis, receptors on the neutrophil plasma membrane directly recognize components of the microbial surface. Conversely, the microbial surface can be modified through opsonization by components of the complement system or by immunoglobulins, particularly immunoglobulin (Ig)G. The surface of some microbes is known to activate the complement system, leading to the production of C3bi, a fragment of complement component (C)3, which can subsequently be adsorbed onto the microbial surface. As the inflammatory response progresses, antibodies are produced whose Fab portion binds to the microbe. The invading microbes can thus be opsonized with C3bi, IgG, or both.

Different neutrophil receptors are engaged in opsonic and nonopsonic phagocytosis. Two main classes of opsonin receptors have been studied in detail. C3bi is recognized by the activated β2 integrin receptor CD11b/CD18 *(1)*. The exposed Fc portion of particle-bound IgG is recognized by Fc receptors that, in human neutrophils, include FcγRI (CD64), FcγRIIA (CD32), and FcγRIIIB (CD16). FcγRI is thought to function mainly after cells are primed with interferon *(2)*. Upon receptor–ligand binding, a programmed signaling cascade is initiated in the neutrophil resulting in the engulfment of the opsonized particle.

Once the particle is internalized, the phagosomal membrane and luminal contents are extensively remodeled via a series of fusion events with intracellular granules. This maturation sequence provides the phagosome with the machinery necessary to kill the engulfed particles, which includes proteases, antimicrobial peptides and proteins, and an enzymatic complex that generates reactive oxygen species. One element of the maturation process that is common to all phagocytes is a gradual acidification of the phagosomal lumen, brought about by the vacuolar type H^+-pump (V-ATPase). However, compared with phagosomes formed by macrophages, which undergo a rapid and substantial acidification, the acidification of neutrophil phagosomes is thought to be biphasic: a slight initial alkalinization is followed after a few minutes by a modest acidification *(3)*.

In this chapter, we present protocols that allow for the quantitative assessment of both phagocytosis and phagosome maturation. We consider both complement and immunoglobulin opsonization protocols of model particles, as well as techniques to label particles with fluorescent probes for the study of phagosome formation and maturation at the individual cell level. A quantitative microscopy-based analysis of phagocytosis using high-resolution digital imaging to locate cells that have engaged particles, and dual-particle labeling to discriminate adherent and internalized particles is described. Lastly, qualitative and quantitative methods to study phagosome maturation, as assessed through acidification, are presented. Microscope hardware and software setup considerations are included.

2. Materials

1. Phosphate-buffered saline (PBS): 140 mM NaCl, 5 mM KCl, 8 mM NaH$_2$PO$_4$, and 2 mM KH$_2$PO$_4$, adjusted to pH 7.4 with 1 M NaOH.
2. PBS++: PBS that has been supplemented with 0.5 mM MgCl$_2$ and 0.5 mM CaCl$_2$.
3. pH calibration solutions: 140 mM KCl, 5 µg/mL nigericin, and 10 mM of the appropriate pH buffer, adjusted to the desired pH with either 1 M KOH or 1 M HCl. The buffer should be chosen to provide optimal buffering power at the desired pH. Typical pH values used in pH calibrations for neutrophil phagosomes (with an appropriate buffer listed in parentheses) are 5.5 (2-[N-morpholino] ethanesulfonic acid [MES]), 6.0 (MES), 6.5 (MES), 7.0 (N-[2-hydroxyethyl]-piperazine-N-[2-ethane-sulfonic acid] [HEPES]), 7.5 (HEPES), and 8.0 (tris-[hydroxylmethyl] amino methanehydrochloride [Tris]). Nigericin is made up in ethanol and should be added to the calibration solution immediately before applying the solution to the cells.
4. Incubation solution: bicarbonate-free, HEPES-buffered RPMI 1640 medium.
5. Sheep red blood cells (sRBCs) and rabbit IgG and IgM fractions against sRBCs (MP Biomedicals).
6. C5-deficient serum (Sigma-Aldrich).
7. Isothiocyanate- and succinimidyl-conjugates of fluorescent probes, acidotropic dyes, and nigericin (Molecular Probes).
8. Hardware necessary for fluorescence ratio imaging: fluorescence microscope, computer-controlled shutters for both bright light and fluorescent illumination (Uniblitz D122, Vincent Associates), filter-wheel (Lambda 10-2, Sutter Instruments), high-intensity fluorescence light source (X-Cite 120, EXFO), appropriate dichroic mirrors and filters (Chroma Technology Corporation), and cooled charge coupled device (CCD) camera. The above components are available from a variety of commercial sources in addition to those listed in the parentheses.
9. Imaging software: MetaFluor®/MetaMorph® and Volocity®/Openlab® packages (Universal Imaging Corporation and Improvision, respectively). Other imaging software that can interface with shutter and filter-wheel controllers and digital cameras can similarly be used.

3. Methods

3.1. Particle Selection

1. Several different particles, including sRBCs, zymosan, latex beads, and bacteria, can be used for assays of neutrophil phagocytosis.
2. For simplicity, we consider sRBCs model particles; however, for the most part, the protocols described herein are readily adaptable to the use of any phagocytic target.

3.2. Fluorescent Labeling

Particles to be used for phagocytosis assays can be labeled with a variety of fluorophores. This allows for their direct and unambiguous identification by epifluorescence or confocal microscopy. A straightforward method of covalently labeling particles makes use of a chemical reaction between amino groups on proteins of the target particle with isothiocyanate or succinimidyl ester moieties of defined fluorophores. For example, fluorescein isothiocyanate (FITC), a pH-sensitive dye, can be conjugated to sRBCs as described below (*see* **Fig. 1C, D**). Other isothiocyanate- or succinimidyl ester-conjugates can similarly be used to label sRBCs with only minimal modifications to the protocol (*see* **Note 1**). Alternatively, fluorophore-coupled antibodies can be used to label particles.

1. Add 200 µL of a 10% suspension of sRBCs to 300 µL of PBS and spin for 10 s at 6000*g* in a microcentrifuge. Aspirate the supernatant, and resuspend the cells in 500 µL of PBS.
2. Wash the cells three times with PBS, followed by a single wash with PBS adjusted to pH 8.7. For all washes, use 500 µL of solution and the same microcentrifuge settings as in **step 1**.
3. Resuspend the sRBCs in 500 µL of PBS adjusted to pH 8.7 and containing 0.5 mg/mL of FITC. Incubate at room temperature for 120 min, rotating the sample constantly.
4. Wash the cells with PBS until little dye can be seen in the supernatant after centrifugation. A minimum of three washes is recommended.
5. Following resuspension in 500 µL of PBS, the labeled cells can now be opsonized with either C3bi or IgG (*see* **Subheading 3.3.** below).

3.3. Opsonization

Opsonization with IgG requires only the incubation of the particle with antibodies directed against it. Antibodies directed against various model particles are commercially available. The C3bi opsonization is a more complex, two-

---

Fig. 1. (*Opposite page*) Measurement of pH by ratiometric imaging. **(A)** Fluorescence excitation spectra of fluorescein isothiocyanate (FITC)-labeled sheep red blood cells (sRBCs) in solutions of varying pH. sRBCs were covalently conjugated with FITC as described in the text and suspended in solutions of varying pH (5.0 to 7.5). Uncorrected excitation spectra (shown in arbitrary fluorescence units) were recorded with a fluorescence spectrometer using a 5-nm slit width and measuring emission at 525 nm. Note that the intensity of the emitted light corresponding to excitation at 490 nm displays considerable pH sensitivity compared with that at 440 nm. The grey bars highlight the wavelength range used for pH measurements by ratiometric imaging. **(B)** The ratio of the fluorescence intensity values recorded with 490 nm and 440 nm excitation is plotted as a function of the pH of the solution. The 490 nm/440 nm ratio increases with increasing pH due primarily to the considerable pH sensitivity at the 490 nm wavelength. pH sensitivity is greatest at pH values near the pK_a of the probe, which for FITC is approx 6.3. **(C,D)** Differential interference contrast and fluorescence images, respectively, of FITC-labeled sRBCs. For the fluorescence image in **D**, cells were illuminated at 490 nm, and emitted light was captured at 535 nm.

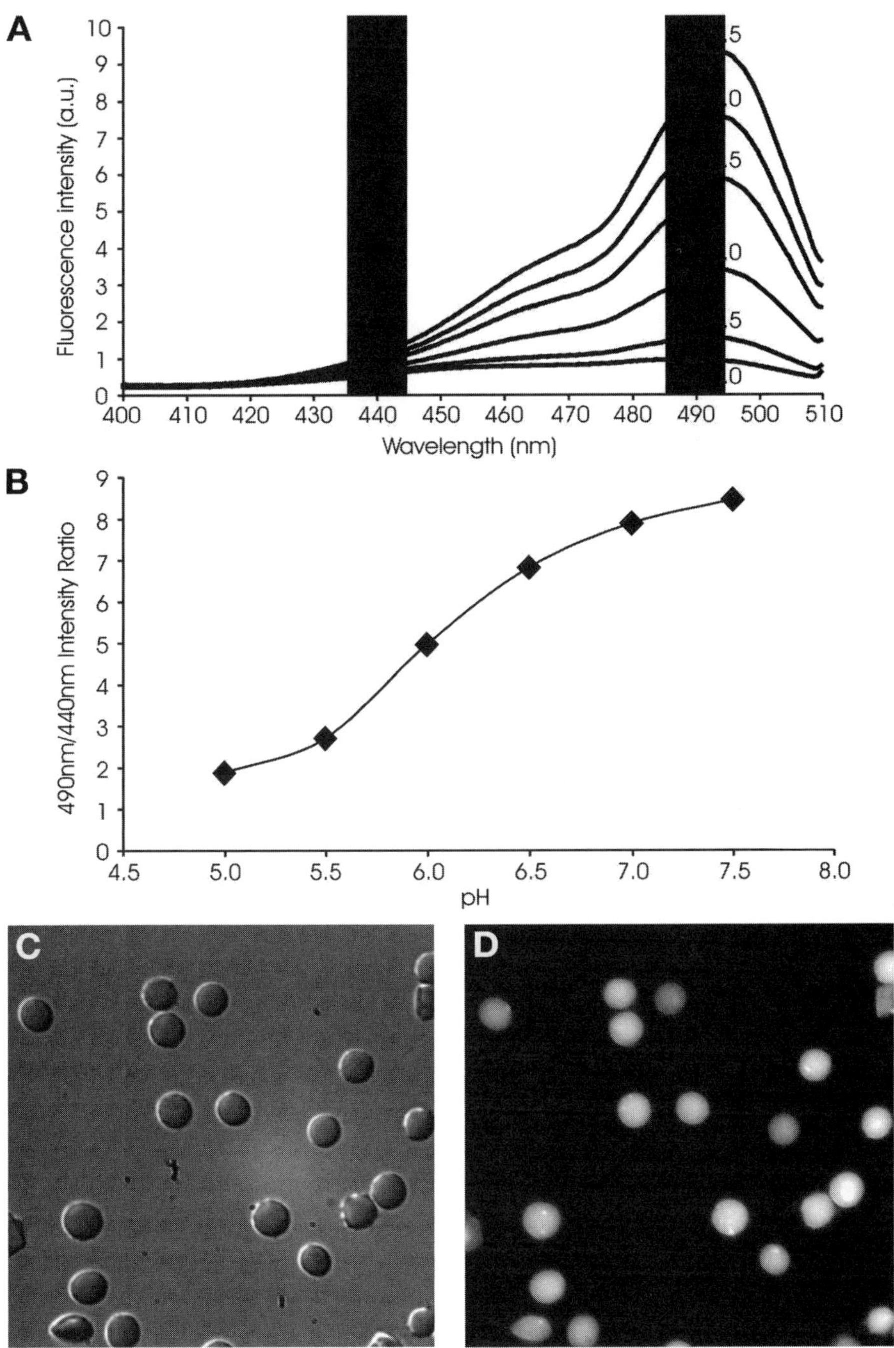

A
Fluorescence intensity (a.u.)
10
9
8
7
6
5
4
3
2
1
0
400 410 420 430 440 450 460 470 480 490 500 510
Wavelength (nm)
.5
.0
.5
.0
.5
.0
B
9
8
7
6
5
4
3
2
1
0
490nm/440nm Intensity Ratio
4.5 5.0 5.5 6.0 6.5 7.0 7.5 8.0
pH
C
D

step process whereby the sRBCs are first primed using IgM and then opsonized with C3bi in C5-deficient serum. Omission of C5 is required to avoid lysis of the sRBCs.

3.3.1. C3bi Opsonization of sRBCs

1. Add 100 µL of a 10% suspension of sRBCs to 400 µL of PBS and spin for 10 s at 6000*g* in a microcentrifuge. Aspirate the supernatant and resuspend the cells in 500 µL of PBS++. sRBCs labeled fluorescently as above can similarly be used for opsonization (*see* **Subheading 3.2.**).
2. Wash the cells once with PBS++, and resuspend in 300 µL of PBS++. For all washes, use 500 µL of solution and the same centrifugation settings outlined in **step 1**.
3. Add 40 µL of anti-sheep IgM (1.5 mg/mL stock) to the sRBC suspension and rotate for 60 min at room temperature.
4. Wash the sRBCs three times with PBS++ and resuspend in 200 µL of PBS++.
5. Add 50 µL of C5-deficient serum (see **Note 2**) and mix gently but thoroughly. It is imperative that C5-deficient serum be used, as the presence of C5 can initiate the deposition of the membrane attack complex, leading to lysis of the sRBCs.
6. Incubate at 37°C for 20 min either in a water bath or a thermoblock with gentle shaking, ensuring that the sRBCs do not settle. To accomplish this, tap the bottom of the tube every 5 min.
7. Wash once with PBS++ and resuspend in 600 µL of PBS++. The C3bi-opsonized sRBCs should be used immediately for the phagocytosis assays.

3.3.2. IgG Opsonization of sRBCs

1. Add 200 µL of a 10% suspension of sRBCs to 300 µL of PBS and spin for 10 s at 6000*g* in a microcentrifuge, aspirate the supernatant, and resuspend the cells in 500 µL of PBS. Prelabeled sRBCs can similarly be used for IgG opsonization.
2. Wash twice with PBS. For all washes, use 500 µL of solution and the same centrifugation settings outlined in **step 1**.
3. Resuspend in 200 µL of PBS and add 4 µL of rabbit anti-sheep IgG (1 mg/mL stock).
4. Incubate for 60 min at room temperature with continuous, gentle mixing.
5. Wash three times with PBS to remove excess antibody and resuspend in 600 µL of PBS (*see* **Note 3**).

3.4. Particle Presentation and Internalization

3.4.1. Particle Presentation

There are several methods for presenting particles to neutrophils depending on whether the cells are adherent or in suspension. Here, we consider neutrophils that have been allowed to sediment onto 25-mm-diameter glass coverslips that have been precoated with fibronectin or fibrinogen to mimic the extracellular matrix. Other methods of particle presentation have been described elsewhere *(4)*.

1. Place a 25-mm-diameter glass coverslip that has been precoated with 50 µg/mL fibronectin into a Leiden coverslip dish. It is important to ensure that a tight seal exists between the coverslip and the O-ring, otherwise the bathing medium will leak from the dish during imaging.
2. Add 500 µL of a neutrophil suspension (~10^6 cells/mL) into the dish and allow for the cells to attach to the matrix (approx 1–3 min).
3. Gently rinse twice with the incubation buffer to remove nonadherent neutrophils.
4. Overlay the cells with 500 µL of a suspension of labeled and opsonized sRBCs. The suspension should contain approx 50 µL of the sRBC suspension produced in either **Subheading 3.3.1.** or **3.3.2.** in 500 µL of incubation buffer.
5. Allow the adherent neutrophils sufficient time to internalize the opsonized sRBCs.

3.4.2. External and Internal Particle Discrimination

Having presented the opsonized particles to neutrophils, it is important to discriminate between those that have been actively engulfed from those that merely adhered to the cell without being internalized. It is often difficult to make this distinction visually using bright field illumination, particularly when a single cell has internalized and/or bound several particles. For this reason, two methods are presented to unambiguously classify a particle as having been internalized or not.

3.4.2.1. Hypotonic Lysis

In contrast to sRBCs, neutrophils have the capacity to regulate their cell volume in response to an acute hypotonic stress. By briefly exposing the preparation to a hypotonic solution, any sRBCs that remain adherent to the surface of the neutrophil will swell and lyse. However, those sRBCs that were fully internalized by the neutrophil are protected from lysis and retain their hemoglobin, facilitating identification. Although quick and effective, this method is not always well suited for use with labeled sRBCs (*see* **Note 4**).

1. Following the phagocytosis protocol, aspirate the bathing solution and add sterile water to the cells for 20 s.
2. Aspirate the water and replace with PBS. Visualize the preparation by phase or differential interference contrast microscopy. Any external sRBCs will have lysed and, although the ghosts may remain adherent to the neutrophils, they are no longer visible. Only internalized sRBCs associated with the neutrophils are visible and can be readily scored.

3.4.2.2. Dual-Labeling Discrimination

Internalized and external sRBCs can be differentially labeled with an impermeant fluorophore in order to discriminate between the two. External particles are labeled by the addition of a secondary antibody conjugated to a fluorophore of a distinct color from that of the dye used for direct sRBC labeling (*see* **Subheading 3.2.**). This secondary antibody is directed against the IgG used to opsonize the sRBCs. In the case of C3bi-opsonized sRBCs, an anti-IgM secondary must be used. A general protocol is provided.

1. Following phagocytosis, the cells are washed with cold PBS (4°C) and placed on ice to arrest any further phagocytosis.
2. Add the fluorophore-conjugated secondary antibody (1:1000 of 0.6 mg/mL stock) and incubate on ice for 5–10 min. Wash away unbound secondary antibody using cold PBS.

3.4.3. Determination of Phagocytic Index

Following particle phagocytosis and labeling of external particles, the neutrophils are visualized by bright field and/or fluorescence microscopy using appropriate filter sets. Quantification can be performed immediately or digital images can be acquired for subsequent analysis.

1. Set the preparation on the microscope stage. Use of 63× or 100× objectives is recommended to clearly resolve the internalized particles.
2. Acquire images of multiple, randomly chosen fields of view. If using hypotonic lysis, where external labeling is obviated, only bright field images need to be acquired. When using dual fluorescent labeling, separately acquire fluorescence images in the channels corresponding to total and external sRBCs, along with a bright field image to visualize the neutrophils themselves.
3. For each field, count the total number of sRBCs, the total number of external sRBCs, and the total number of neutrophils.
4. Compile the data and compute the phagocytic index (PI).
5. Phagocytic efficiency is usually expressed as the PI, which is defined as the number of particles ingested per neutrophil (or other phagocyte). The PI is calculated as:

$$PI = (\text{Total sRBCs} - \text{external sRBCs}) / \text{total number of neutrophils} \qquad (1)$$

6. In a single field or image, the number of internalized particles equals the number of external particles subtracted from the total number of particles. When using hypotonic lysis, the number of external particles is zero. Some investigators normalize the PI per 100 phagocytes.

3.5. Phagosome Maturation and Acidification

Phagosomal acidification can be readily monitored in a qualitative fashion using an acidotropic dye (*see* **Notes 5** and **6**). The pH of the phagosome can be measured more precisely by recording the emission of a fluorescent pH indicator attached to the particle. However, for a probe monitored at single excitation and emission wavelengths, changes in fluorescence can be caused by factors other than the pH. For example, photobleaching or dye leakage can result in a pH-independent decrease in fluorescence. Similarly, changes in focal plane during the course of the experiment, caused by neutrophil motility or by relocalization of phagosomes within the cell, can alter fluorescence in either direction, regardless of pH. Ratiometric imaging can overcome such artifacts. This technique makes use of a fluorescent dye that is differentially sensitive to pH at two distinct wavelengths. One of the wavelengths is highly sensitive to pH, whereas the other is relatively insensitive (or, ideally, sensitive in the opposite direction) to pH. For example, with FITC, excitation at 490 nm is highly pH-sensitive in comparison with excitation at 440 nm (*see* **Fig. 1A**). A ratio can be computed by recording fluorescence changes at the two wavelengths (*see* **Fig. 1B**), which normalizes the pH measurement with respect to the concentration of the dye within the phagosome. Moreover, with this method, an *in situ* calibration can be constructed to translate the measured ratio to pH by equilibrating the intraphagosomal pH with the extracellular pH using the K^+/H^+ ionophore nigericin. For a more detailed description of ratiometric imaging, please refer to **ref. 5** and other texts on fluorescence microscopy.

3.5.1. Phagosome pH Measurement

To measure phagosomal pH, an appropriate probe must be used. A variety of ratiometric, pH-sensitive fluorophores are commercially available. A suitable dye should be readily targeted to the phagosome and must have a pK_a near the expected pH of the compartment of interest (*see* **Note 7**). Here, we describe a protocol using FITC as our pH indicator (*see* **Note 8**), which we load into the phagosome by covalent conjugation to sRBCs (*see* **Subheading 3.2.**). The protocol considers single-cell analysis, but is easily adapted to population-based measurements (*see* **Note 9**).

1. Place a 25-mm-diameter glass coverslip that has been precoated with fibronectin into a Leiden dish. It is important to ensure that a tight seal exists between the coverslip and the O-ring to prevent the bathing medium from leaking during imaging.
2. Add 500 µL of a neutrophil suspension (~10^6 cells/mL) into the dish and allow for the cells to attach to the matrix.
3. Wash twice with incubation buffer to remove nonadherent neutrophils.
4. Add 1 mL of PBS to the Leiden coverslip dish and place it in its chamber holder maintained at 37°C above the appropriate objective. Either 63× or 100× objectives are recommended to clearly resolve the internalized particles. Find a field containing several adherent neutrophils.
5. Overlay the cells with 500 µL of a suspension of FITC-labeled and opsonized sRBCs.
6. Monitor the cells using bright field video microscopy until one or more sRBCs are seen to bind to a neutrophil.
7. As the particle is internalized and for the duration of the assay, acquire fluorescence images by alternately illuminating the field with 490 nm and 440 nm wavelength light.

3.5.2. In Situ *Calibration*

1. Perform an *in situ* calibration after the desired maturation time course.
2. Wash the cells once with the initial calibration solution, and overlay the cells with 1 mL of the same calibration solution containing nigericin.
3. Make sure that the cells of interest remain in the field of view and in focus during and after all solution changes.
4. Allow 3–5 min for equilibration of the phagosomal pH with that of the calibration solution and then acquire a series of images. Usually, three to five images are sufficient. These images will be used to calculate the 490 nm/440 nm fluorescence ratio value corresponding to the pH of the first calibration solution.
5. Repeat **steps 1–4** for calibration solutions of all the desired pH values.

3.5.3. *Image Analysis*

1. Using image analysis software, define a region of interest (ROI) covering the internalized particle and record the average fluorescence intensity of that ROI in both the 490 nm and 440 nm channels (*see* **Note 10**).
2. Obtain background readings at both wavelengths from a comparable size ROI in a region outside the cells and subtract this value from the phagosomal readings.
3. From the background-corrected values, compute the 490 nm/440 nm ratio.
4. Proceed through the time course and calibration, ensuring that the ROI is aligned over the phagosome in case movement has occurred.
5. Construct a calibration curve by plotting the background-corrected 490 nm/440 nm ratio as a function of pH.
6. To determine the pH of the phagosome throughout the time course, interpolate the pH from the calibration curve using the measured ratios.

4. Notes

1. When labeling sRBCs with a fluorescent compound, it is important to consider the solvent used to dissolve the fluorophore. Several succinimidyl ester conjugates have poor water solubility and must be made up in either dimethylsulfoxide (DMSO) or *N,N*-dimethylformamide (DMF). If added to suspensions of sRBCs at high concentration, these solvents can affect the cell membranes and cause cell lysis. Accordingly, when using fluorophores dissolved in such solvents, the conjugation step should be carried out in a dilute suspension of both sRBCs and fluorophore solvent. For example, 50 µL of 10 mg/mL of a succinimidyl ester conjugate made up in DMF can be added to 20 mL of a sRBC suspension to minimize red cell lysis.
2. The C5-deficient deficient serum should be aliquoted and kept at −80°C. Thaw the serum by briefly immersing and gently swirling the tube in a 37°C water bath without allowing it to warm up. This should be done immediately before adding it to the sRBCs.
3. In contrast to C3bi-opsonized sRBCs, which should not be stored for extended periods of time, sRBCs opsonized with IgG can be stored rotating at 4°C for several hours, even days.

4. Hypotonic lysis of fluorescently labeled sRBCs can sometimes prove problematic as the labeled protein from lysed sRBCs can become adherent to, and internalized by, surrounding neutrophils. Moreover, the membranes of any bound, externally adherent sRBCs that were not internalized can remain associated to the neutrophil surface. This undesired neutrophil-associated fluorescence might interfere with particle discrimination. It is also cosmetically undesirable for image presentation.

5. Fluorescent acidotropic compounds are weak bases that selectively accumulate in cellular compartments with low pH. These dyes permeate across cell membranes becoming protonated and retained within acidic compartments. Acidotropic dyes of assorted fluorescent colors are commercially available through Molecular Probes. Their LysoTracker® probes offer a simple, albeit qualitative, assessment of phagosomal pH, and can be used as per the manufacturer's instructions.

6. Two important points should be considered when using acidotropic dyes in neutrophils. First, the accumulation of the dye is not always readily reversible, even when the cellular compartment is artificially alkalinized by the addition of a weakly basic, membrane-permeant compound or of ionophores. Second, the neutrophil phagosome does not undergo a substantial acidification during its maturation. As a result, certain acidotropic dyes may not accumulate visibly within the phagosome. Failure of neutrophil phagosomes to acidify in the early stages has been attributed to proton consumption by products of the NADPH oxidase. When NADPH oxidase inhibitors such as diphenylene iodonium (DPI) or apocynin are present, a significant acidification of the phagosomal lumen can be observed *(3)*. In such cases, the acidotropic dye should accumulate within the phagosome. Although acidotropic dyes offer a simple, qualitative assessment of intraphagosomal pH, the most accurate method of evaluating phagosome acidification remains the direct measurement of the organellar pH by fluorescence ratiometric imaging.

7. pH indicators with different pKa and spectral properties can be used to measure the pH within the phagosome. Commonly used dyes (with their pKa values in parenthesis) include: FITC (6.3), Oregon Green 514 (4.8) and SNAFL-2 (7.7). Combinations of dyes can also be used to adjust the composite pKa nearer to the desired range.

8. Fluorescein derivatives, such as FITC and Oregon Green 514, are susceptible to chlorination by cellular peroxidases. Therefore, an active myeloperoxidase can severely impact on pH measurements employing a fluorescein derivative. To circumvent this artifact, it is imperative that neutrophil myeloperoxidase be inhibited immediately before and during phagocytosis. One mode of inhibition is through incubation with medium supplemented with 5 m*M* NaN$_3$ *(3)*. Note that although NaN$_3$ is an effective inhibitor of the respiratory chain, neutrophils rely mostly on anaerobic sources for energy. As a result, the NaN$_3$ should not grossly affect neutrophil viability and function.

9. The hardware required for fluorescence ratio imaging is configured around a fluorescence microscope modified to allow concomitant differential interference contrast video microscopy with fluorescence digital imaging. The general components include: (1) computer-controlled shutters for the halogen illumination and for the

fluorescence light source, which are required to toggle between bright field and fluorescence illumination, (2) a filter wheel, used to rapidly alternate between excitation at 490 nm and 440 nm, which allows for the acquisition of images that approximate a single time point, (3) the 490 nm and 440 nm light, which is reflected to the sample using a 510 nm dichroic mirror that, in turn, transmits the emitted light towards a 535 nm emission filter, and (4) a cooled CCD camera with which the filtered light is then captured.

10. Ratiometric imaging requires software that is able to control image acquisition, shutter settings, and excitation and emission filter selection. A variety of software packages are commercially available that accommodate these requirements.

Acknowledgments

B. E. S. is supported by a graduate studentship from the McLaughlin Center for Molecular Medicine. S. G. is the current holder of the Pitblado Chair in Cell Biology. Research in the authors' laboratory is supported by the Heart and Stroke Foundation of Canada and by the Canadian Institutes of Health Research.

References

1. Ehlers, M. R. (2000) CR3: a general purpose adhesion-recognition receptor essential for innate immunity *Microbes Infect.* **2,** 289–294.
2. Lee, W. L., Harrison, R. E., and Grinstein, S. (2003) Phagocytosis by neutrophils *Microbes Infect.* **5,** 1299–1306.
3. Jankowski, A., Scott, C. C., and Grinstein, S. (2002) Determinants of the phagosomal pH in neutrophils *J. Biol. Chem.* **277,** 6059–6066.
4. Morris, M. R., Dewitt, S., Laffafian, I., and Hallett, M. B. (2003) Phagocytosis by inflammatory phagocytes: experimental strategies for stimulation and quantification *Methods Mol. Biol.* **225,** 35–46.
5. Demaurex, N., Romanek, R., Rotstein, O., and Grinstein, S. (1998) Measurement of cytosolic pH in single cells by dual excitation fluorescence imaging. Simultaneous visualization using differential interference contrast optics, in *Cell Biology: A Laboratory Handbook* (Celis, J., ed.). Academic, San Diego, pp. 380–386.

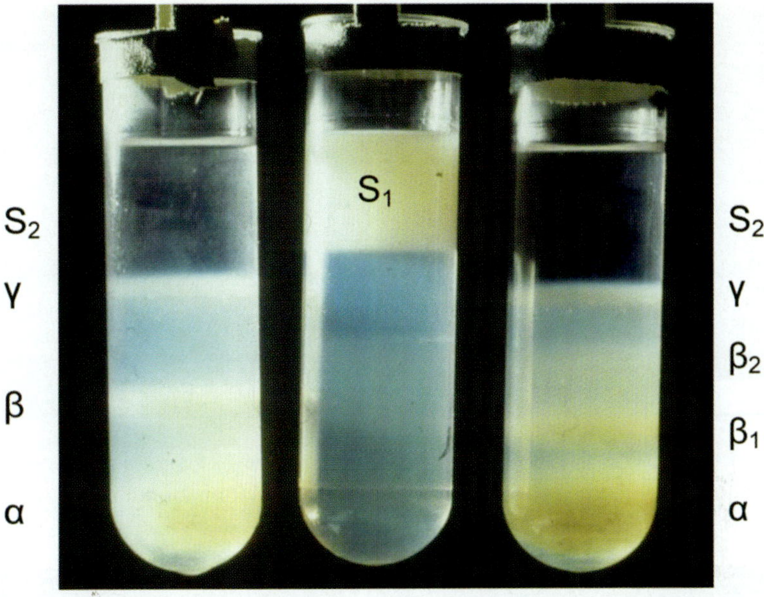

Color Plate 1, Ch. 4, Fig. 5 (*see* full caption on p. 48 and discussion in Chapter 4.) Photograph of the three-layer Percoll density gradient before and after centrifugation.

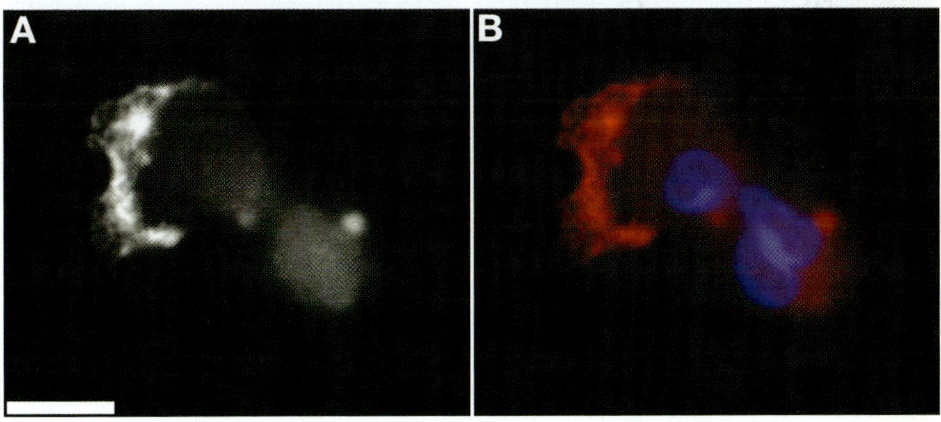

Color Plate 2, Ch. 14, Fig. 1 (*see* full caption on p. 220 and discussion in Chapter 14.) Immunofluorescent staining for actin in a primary human neutrophil.

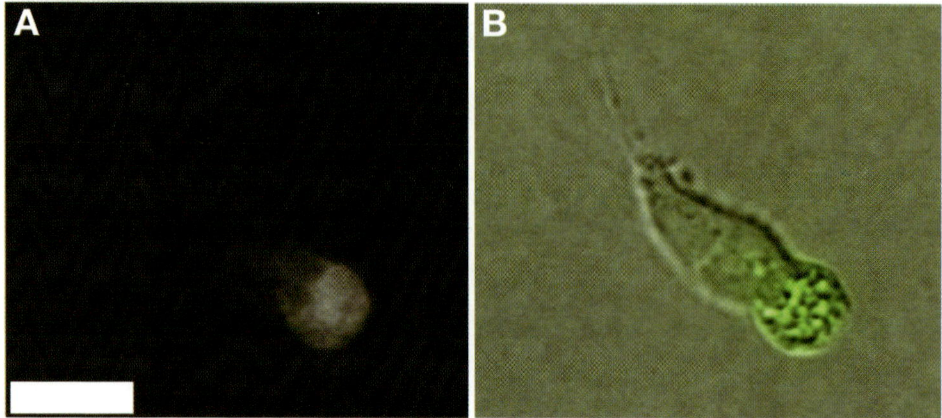

Color Plate 3, Ch. 14, Fig. 2 (*see* full caption on p. 221 and discussion in Chapter 14.) Immunofluorescent staining for G_M3 in a primary human neutrophil.

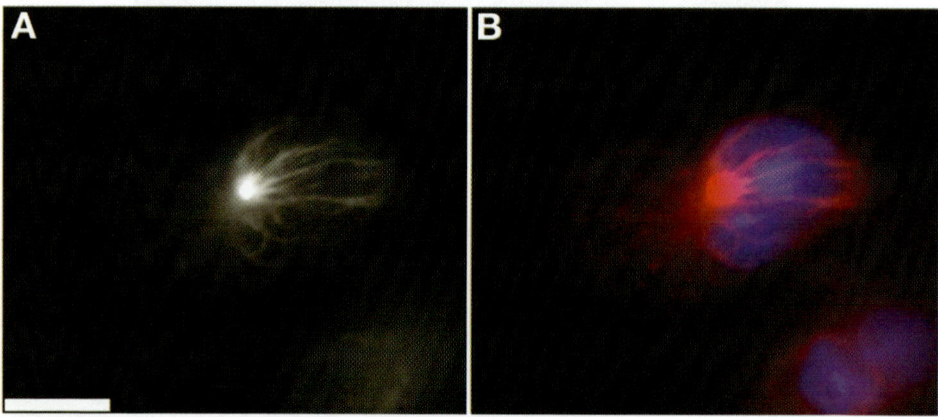

Color Plate 4, Ch. 14, Fig. 3 (*see* full caption on p. 222 and discussion in Chapter 14.) Immunofluorescent staining for tubulin in a primary human neutro-

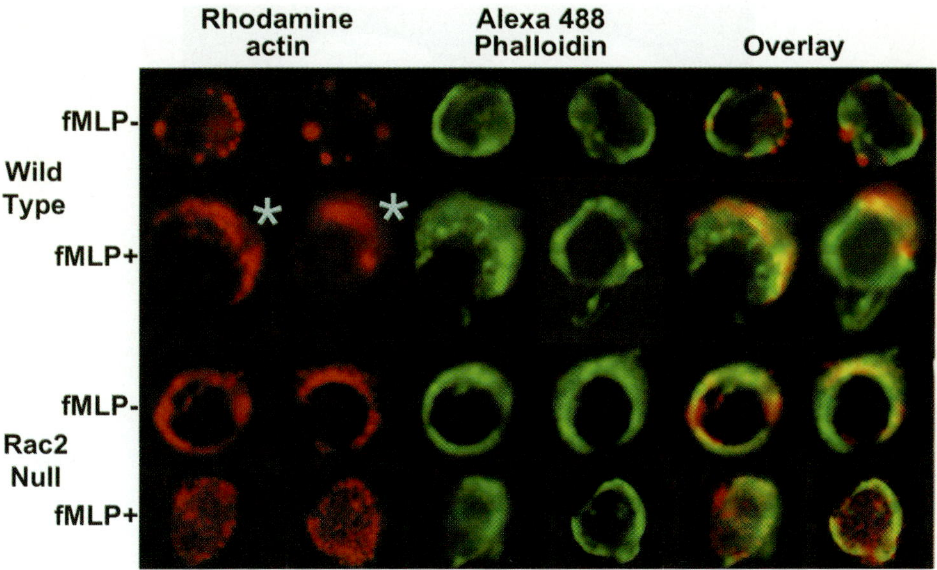

Color Plate 5, Ch. 15, Fig. 2 (*see* full caption on p. 236 and discussion in Chapter 15.) Localizing free-actin barbed ends generated after fLMP stimulation in neutrophils.

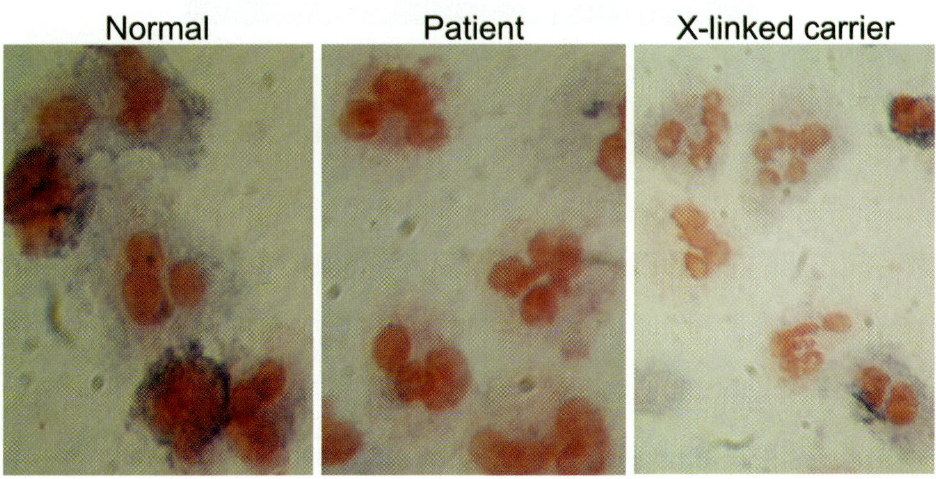

Color Plate 6, Ch. 31, Fig. 1 (*see* full caption on p. 509 and discussion in Chapter 32.) Nitroblue tetrazolium (NBT) assay.

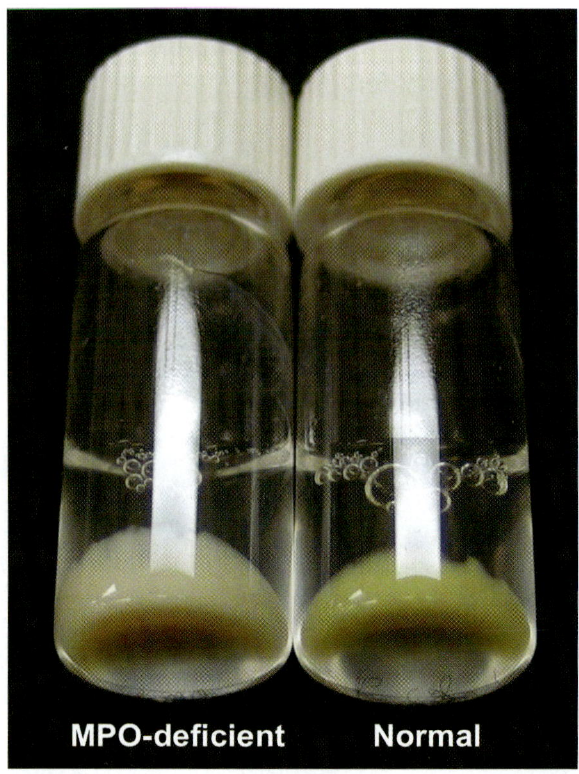

Color Plate 7, Ch. 32, Fig. 1 (*see* full caption on p. 526 and discussion in Chapter 32.) Comparison of polymorphonuclear neutrophils (PMN) from PMN from healthy subjects (Normal) and those with myeloperoxidase (MPO) deficiency (MPO-deficient).

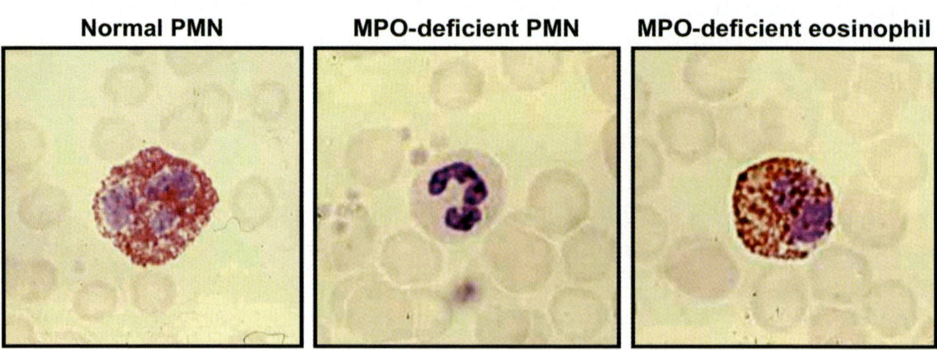

Color Plate 8, Ch. 32, Fig. 2 (*see* full caption on p. 530 and discussion in Chapter 32.) Peroxidase activity in polymorphonuclear neutrophils (PMN) from a healthy subject or an individual with myeloperoxidase deficiency.

20

Analysis of Neutrophil Membrane Traffic During Phagocytosis

Per Lönnbro, Pontus Nordenfelt, and Hans Tapper

Summary

In this chapter, we describe methods to study membrane traffic during phagosome formation and maturation. Although it is convenient to define events as occurring either before (i.e., during formation) or after (during maturation) the creation of a sealed phagosome, it might not be correct to assume that this is reflected by a sudden change in the membrane traffic events involved. Nevertheless, formation events are studied by approaches different from those of maturation events. Before closure of a phagosome, methods relying on immunochemistry and fluorescent markers are employed. Once phagosomes have formed, these can be isolated and additional methods can be used for their characterization. Here, we describe (1) methods to study membrane traffic during phagosome formation and (2) magnetic purification of bacteria-containing phagosomes for maturation studies.

Key Words: Phagocytic cell; phagosome; phagocytosis; membrane traffic; secretion; pinocytosis; bacteria; phagocytosis.

1. Introduction

Phagocytosis of microorganisms by neutrophils is accompanied by intense membrane traffic that, at least in the case of azurophilic granule secretion, is restricted to the cellular region closest to the phagocytic prey. Even before closure of a forming phagosome, localized secretion of azurophilic granules and localized pinosome formation is triggered *(1–3)* (*see* **Fig. 1**). These and other secretory and pinocytic processes must be in balance during the formation and the subsequent maturation of a phagosome, allowing both its content and the properties of its membrane to change with time. An important parameter of a secretory response is its localization. During phagosome formation in the human neutrophil, azurophilic granules fuse with the plasma membrane preferentially in the vicinity of nascent phagosomes. Other, more easily triggered granular

From: *Methods in Molecular Biology, vol. 412: Neutrophil Methods and Protocols*
Edited by: M. T. Quinn, F. R. DeLeo, and G. M. Bokoch © Humana Press Inc., Totowa, NJ

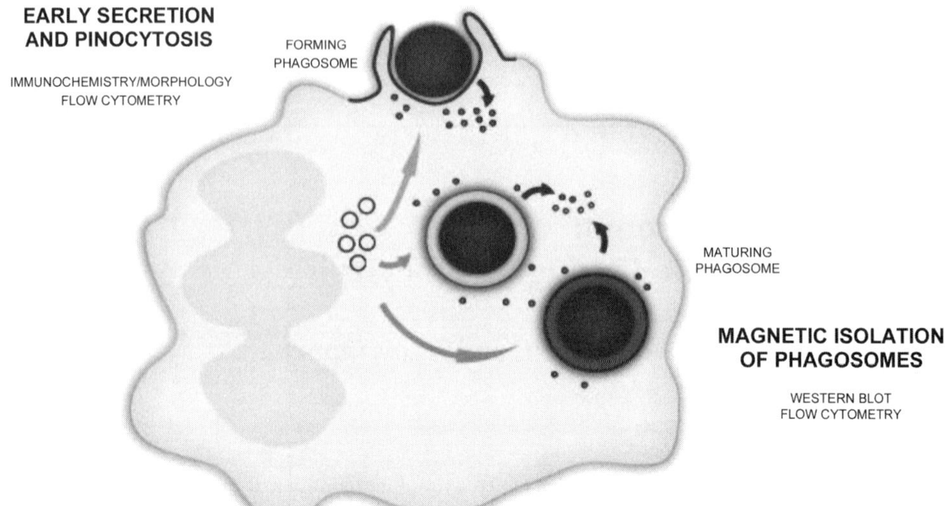

Fig. 1. Membrane traffic during phagosome formation and maturation. The initial contact between a phagocyte and its prey triggers early membrane remodeling processes that continue, but progressively change, as a phagosome pinches off from the plasma membrane and matures. Techniques used to study these processes are indicated.

compartments are less locally restricted and are secreted also towards other regions of the cell *(1,2)*.

Immunochemical detection of granule membrane markers is a sensitive method that can be used both for studies of localization of a secretory response by immunofluorescence or electron microscopy and for quantitative studies by flow cytometry. The procedure measures the insertion of granule membrane proteins (such as CD63/LAMP-3 [azurophilic granules] or CD66b [specific granules]) into the plasmalemma *(2,4)*. The epitopes of marker proteins are localized in the lumen of the granules and, in nonpermeabilized cells, are accessible to staining by antibodies only after exposure at the cell surface as a result of exocytosis (*see* **Note 1**).

In the human neutrophil, secretion towards a site of phagosome formation is accompanied by localized activation of pinocytosis *(3)* (*see* **Fig. 2**). This can be studied using a fluid-phase marker, such as Lucifer Yellow or using amphiphilic fluorescent dyes, such as FM 1-43. The latter membrane probe is unable to enter intact cells, and fluoresces brightly only after partitioning into the outer monolayer of the surface membrane *(5)*. Both secretory and pinocytic processes continue at the phagosome membrane during the maturation stage, after detachment of the phagosome from the plasma membrane. The methods used to study forming phagosomes can therefore be applied also to studies of maturing phago-

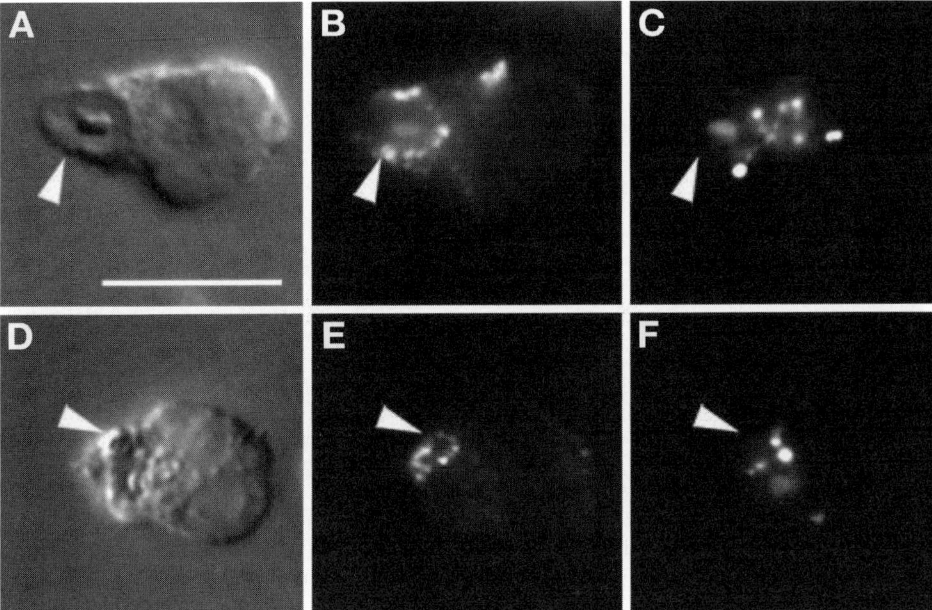

Fig. 2. Azurophilic granule secretion and pinosome formation. Secretory events are localized in ATRA-differentiated HL-60 cells after a 1-min interaction with immunoglobulin G-opsonized prey (arrowheads: zymosan particle, **A–C**; BMJ71 strain *Streptococcus pyogenes* bacterium, **D–F**). Azurophilic granule secretion is indicated by immunofluorescent staining of plasma membrane-located CD63 (**B,E**) and pinosome formation by internalization of Lucifer Yellow (**C,F**). Phagocytosis was inhibited by cytochalasin D. Bar = 10 µm.

somes, with slight protocol modifications (e.g., using permeabilized cells and/or longer chase periods). Phagosome maturation can also be studied by analyzing the composition of purified phagosomes, as described below.

Phagosome research can be performed using latex beads as model prey *(6)*. However, for studies of pathogen-mediated modulation of phagosome maturation, real bacteria must be used as prey. Isolation of bacterial phagosomes can be achieved by employing a series of ultracentrifugation steps using combinations of various density gradients *(7)*, but is dependent on the density, size, and shape of the bacterial species. An alternative approach presented here is to sparsely coat the bacteria with small superparamagnetic particles. Phagosomes containing such magnetically susceptible bacteria can then be purified by magnetic separation.

Some pathogenic bacteria, such as *Streptococcus pyogenes*, can interfere with the ability of a phagocytic cell to remodel early phagosomes into bactericidal compartments *(8–10)*. This occurs by modulation of fusion/fission events at

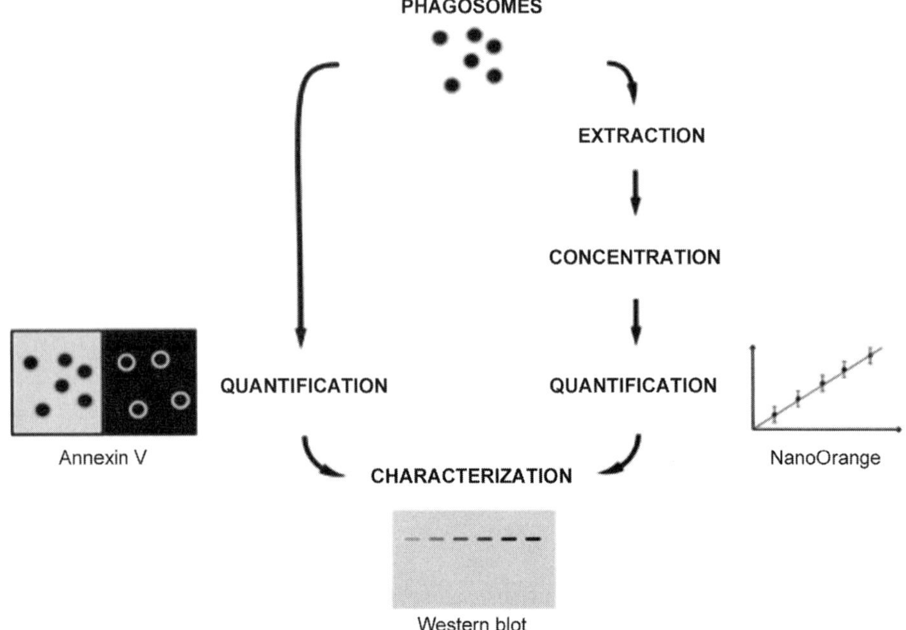

Fig. 3. Schematic view of methods used for quantification and characterization of isolated phagosomes. The yield and purity of phagosomes can be determined by visual inspection after staining with fluorescently labeled Annexin V. In parallel, protein content of phagosomes can be quantified using the NanoOrange method.

the phagosome membrane. The protocol described here for the preparation of superparamagnetic prey was developed for *Streptococcus pyogenes* bacteria, but can be adapted to other microorganisms. The technique requires only the presence of free amino groups on the surface of the bacteria. These groups are then covalently cross-linked to the amino groups on the superparamagnetic particles by means of glutaraldehyde-derived carbon bridges (*see* **Note 2**). Further, the protocol describes the disruption of the cells (*see also* **Note 3**), the magnetic purification step (*see also* **Note 4**), and a method for visual confirmation of a successful experiment (the Annexin V-staining procedure, *see also* **Note 5** and **Figs. 3** and **4**). Finally, we describe methods for quantitation of the yield and for the analysis of phagosome maturation by Western blot.

2. Materials

2.1. Secretion to the Plasma Membrane

1. Na-medium: 5.6 mM D(+)-glucose, 127 mM NaCl, 5.4 mM KCl, 1.8 mM CaCl$_2$, 0.8 mM MgSO$_4$, 1.2 mM KH$_2$PO$_4$, 10 mM 2-[4-(2-hydroxyethyl)-1-piperazinyl]ethane-sulfonic acid (HEPES), pH 7.3.

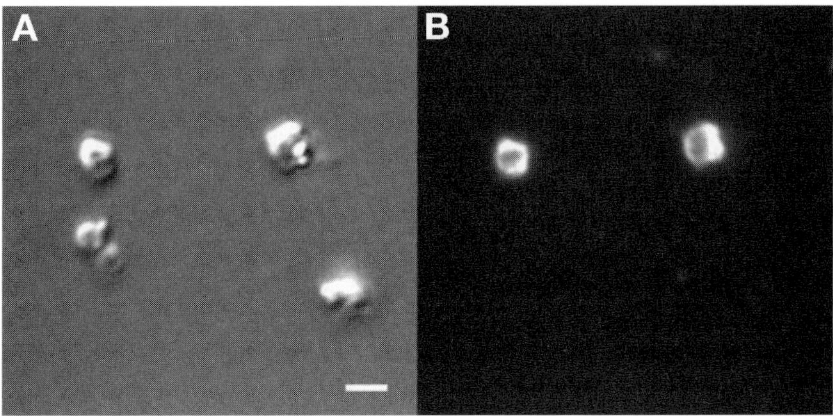

Fig. 4. Annexin V-staining of phagosome membrane. Yield and purity of magnetically retrieved *Streptococcus pyogenes*-containing phagosomes can be determined by visual inspection. After staining with fluorescent annexin V, two of the particles observed in **A** can be identified as phagosomes (**B**). Bar = 1 μm.

2. Blocking buffer: phosphate-buffered saline (PBS; 2.7 mM KCl, 137 mM NaCl, 1.5 mM KH$_2$PO$_4$, 8.1 mM Na$_2$HPO$_4$, pH 7.3) containing 0.1% bovine serum albumin (BSA) and 5% goat serum (*see* **Note 6**).
3. Antibodies recognizing granule membrane proteins that are exposed at the external aspect of the plasma membrane after exocytosis, such as Abs against CD63 (azurophilic granules), CD66b (specific granules), and CD35 (secretory vesicles).
4. Alexa-conjugated secondary antibodies (Molecular Probes) (*see* **Note 7**).
5. Paraformaldehyde (PFA): 10% stock solution (Becton Dickinson).
6. Permeabilization buffer: blocking buffer supplemented with 0.2% Tween-20 and 0.02% Triton X-100.
7. Poly-L-lysine coated glass coverslips (poly-L-lysine hydrobromide, average molecular mass 99.5 kDa, Sigma) (*see* **Note 8**) and ProLong Antifade Kit (Molecular Probes).

2.2. Pinosome Formation

1. Na-medium, *see* **Subheading 2.1., step 1**.
2. Lucifer Yellow or FM 1-43FX (Molecular Probes).
3. PBS: 2.7 mM KCl, 137 mM NaCl, 1.5 mM KH$_2$PO$_4$, 8.1 mM Na$_2$HPO$_4$, pH 7.3.
4. PFA, *see* **Subheading 2.1., step 5**.
5. Poly-L-lysine coated glass coverslips and ProLong Antifade Kit, *see* **Subheading 2.1., step 7**.

2.3. Magnetic Purification Protocol

2.3.1. Preparation of Prey

1. PBS, *see* **Subheading 2.2., step 3**.
2. Ultrasonic water bath, such as the XB2 (Grant Instruments).

3. LIVE/DEAD BacLight L-7007 bacterial viability kit (Molecular Probes).
4. Heat block, such as Thermomixer Compact (Eppendorf).
5. Syringe filters, 0.2 μm, such as Filtropur (Sarstedt).
6. BioMag BM546 superparamagnetic particles (Bangs Laboratories, Fishers, IN) (*see* **Note 9**).
7. Magnetic rack, such as the Dynal MPC-M (Dynal).
8. Coupling buffer: 0.01 *M* pyridine, pH 6.0. Shelf-life: several years. Cold storage is preferable.
9. Cross-linking solution: 5% glutaraldehyde (EM-grade aqueous stock solution, 25%) and 0.05% Tween 20 in coupling buffer. Mix 1 mL in an Eppendorf tube just before use.
10. Rotator, capable of 7–18 rpm, such as the LD-79 (Labinco).
11. Bürker chamber.
12. Quenching solution: 1 *M* glycine, 1% BSA, pH 8.0. Mix just before use.
13. PBS with 0.05% BSA. Add the BSA just before use.

2.3.2. Nitrogen Cavitation

1. Cell disruption bomb (Parr Instrument Company, Moline, IL).
2. Cavitation buffer: protease inhibitor cocktail (Complete Mini, Roche Diagnostics; one tablet per 2 mL of solution [for ~5·10^8 cells]; includes 5 m*M* disodium ethylenediamine tetraacetate [Na_2EDTA]), 7 m*M* $MgCl_2$, 0.25 *M* sucrose, 10 m*M* HEPES, pH 7.3. Buffered sucrose solution is kept for long-term storage at 4–8°C and mixed with the inhibitor cocktail on the day of the experiment. To each 1-mL aliquot, add 2 U/10^6 cells of Benzonase endonuclease (purity ≤ 90%, ≤ 25 U/μL, Merck) (*see* **Note 10**).

2.3.3. Magnetic Separation

1. PickPen magnetic probe (Bio-Nobile), for maximal purity, or a magnetic rack, such as the Dynal MPC-M (Dynal).
2. Wash buffer: same as cavitation buffer (*see* **Subheading 2.3.2.**), but with one tablet Complete Mini per 10 mL of solution, 3 m*M* $MgCl_2$, and 190 U/mL Benzonase.
3. 1-mL syringe with 27-G needle.
4. 96-well microtiter plates.

2.4. The Annexin V Method

1. Bürker chamber (for total count, using 20× objective) and regular microscope slide and coverslip (for identification of phagosomes, using 100× objective).
2. Annexin V conjugated to a fluorophore and a Ca^{2+}-containing buffer, such as found in the Vybrant Apoptosis Assay Kits (Molecular Probes).

2.5. Quantification of Phagosomal Protein Using Microcon and NanoOrange

1. Protease inhibitor solution: PBS with protease inhibitors (one tablet of Complete Mini per 10 mL, Roche Diagnostics).
2. Extraction buffer: protease inhibitor solution containing 0.1% Triton X-100.

3. Microcon YM-3 Centrifugal Filter Unit (Millipore).
4. NanoOrange protein quantification kit (Invitrogen).
    a. Prepare protein quantitation diluent by adding 1 mL of Component B to 9 mL of distilled H_2O.
    b. Prepare NanoOrange working solution (NOWS) by diluting component A 500 times in quantitation diluent (add 20 μL of Component A to 10 mL of quantitation diluent). Use within a few hours. Protect from light.
    c. Prepare a 10 μg/mL solution of BSA by diluting Component C 200 times with NOWS (add 10 μL of Component C to 1990 μl of NOWS).
5. Fluorescence plate reader, such as the Fluostar Optima (BMG Labtech).

2.6. Sodium Dodecyl Sulfate-Polyacrylamide Gel Electrophoresis

1. NuPAGE Novex precast Bis Tris 4–12% gradient gels (Invitrogen). Shelf life of 12 mo at 4–25°C.
2. Running buffer, NuPAGE MOPS sodium dodecyl sulfate (SDS) (20X): 50 m*M* MOPS, 5 m*M* Tris base, 0.1% SDS, 1 m*M* EDTA, pH 7.7. Stable for 6 mo at 4°C.
3. Antioxidant, NuPAGE Antioxidant.
4. Sample buffer, NuPAGE LDS Sample Buffer (4X).
5. Prestained molecular mass markers: Kaleidoscope markers (Bio-Rad).

2.7. Western Blot

1. Transfer buffer, NuPAGE Transfer Buffer (20X): 25 m*M* Bicine, 25 m*M* Bis-Tris, 1 m*M* EDTA, pH 7.2. Stable for 6 mo at 4°C.
2. Transfer buffer for semi-dry blotting: 10.0 mL NuPAGE Transfer Buffer (20X), 0.1 mL NuPAGE Antioxidant for reduced samples, 10.0 mL methanol and 79.9 mL deionized water.
3. Polyvinylidene difluoride (PVDF) membrane (Millipore).
4. PBS with Tween (PBST): 1000 mL PBS with 0.05% Tween-20.
5. Blocking buffer: PBST with 5% (w/v) skim milk.
6. Antibody buffer: PBST with 1% (w/v) skim milk.
7. Enhanced chemiluminescence (ECL) reagent: Pierce SuperSignal West Dura Extended Duration Substrate (Pierce Biotechnology).
8. Primary antibodies used in **Fig. 5** are: anti-cathepsin D, 1:5000, rabbit polyclonal (sc-10725, Santa Cruz Biotech); anti-CD63, 1:2500, rabbit polyclonal (sc-15363, Santa Cruz Biotech); anti-LAMP-1, 1:5000, mouse monoclonal (L76620, Transduction Laboratories); anti-myeloperoxidase, 1:5000, rabbit polyclonal *(11)*.
9. Secondary antibodies: anti-mouse, 1:200 000, horseradish peroxidase (HRP)-conjugated (Pierce Biotechnology); anti-rabbit, 1:200 000, HRP-conjugated (Pierce Biotechnology).

3. Methods

3.1. Secretion to the Plasma Membrane

1. For a synchronized presentation, mix cells and phagocytic prey (final volume 0.2–0.5 mL) in a 1.7-mL Eppendorf tube (*see* **Note 11**).

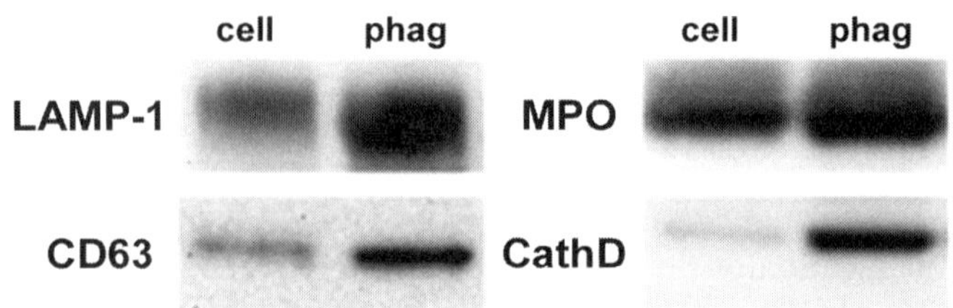

Fig. 5. Western blots of cell and phagosome lysates. Extracted material from 10^5 differentiated HL-60 cells and 6 µg of phagosome protein (from 1-min phagosomes containing magnetic *Streptococcus pyogenes* BMJ71 bacteria) was loaded onto 4–12% gradient gels. Strong bands representing azurophilic granule content markers (myeloperoxidase and cathepsin D) and the membrane markers LAMP-1 and CD63 (low-molecular-mass form) are observed for the phagosome lysates.

2. Heat the tube in a 37°C water bath for approx 1 min and then briefly pellet the mixture (20 s at 12 000g, with time 0 s defined as when centrifuge reaches target speed.
3. Transfer the tube to a 37°C water bath, and at time 1 min resuspend the pellet using an automatic pipet.
4. If necessary, reduce the number of free bacteria by centrifuging (5 min, 200g, 4°C) in a swinging-bucket rotor. Thereafter, optionally perform a chase period (in 1 mL new medium at 37°C) to allow continued phagosome maturation. Halt incubation by placing samples on ice.
5. Following the presentation/phagocytosis steps, centrifuge each sample at 4°C (30 s, 12,000g).
6. Remove the presentation medium and replace with ice-cold blocking buffer (PBS containing 0.1% bovine serum albumin and 5% goat serum (*see* **Note 12**).
7. Incubate for 10 min on ice.
8. Following the blocking step, centrifuge as above and replace supernatant with blocking buffer containing the primary antibodies (anti-CD63 1:600, anti-CD66b 1:1000), and incubate for 15 min on ice (*see* **Note 13**).
9. Wash the cells twice in ice-cold blocking buffer (*see* **Note 12**).
10. Add appropriate Alexa-labeled secondary antibodies in blocking buffer at a dilution of 1:600 and incubate for 15 min on ice.
11. Wash the cells twice in ice-cold blocking buffer.
12. Analyze secretion by flow cytometry and microscopy.
    a. For flow cytometric analysis, gate the neutrophils based on forward and side scatter and collect fluorescence data of at least 10,000 events. To allow selective analysis of cells interacting with prey, use particles labeled prior to experimentation with a fluorophore compatible with the flow cytometer laser, such as ethidium bromide.
    b. For microscopic analysis, fix the samples using 1% PFA for 15 min at 4°C, followed by a 45-min incubation at room temperature.

3.2. Secretion to Phagosome

1. Perform the synchronized presentation described above through **Subheading 3.1., step 5**.
2. Fix the samples using 1% PFA, for 15 min at 4°C followed by 45 min incubation at room temperature.
3. Permeabilize the cells using blocking buffer supplemented with 0.2% Tween-20 and 0.02% Triton X-100 for 20 min at room temperature.
4. Incubate permeabilized cells with antibodies against the prey or intracellular structures of interest (3 h at room temperature in blocking buffer containing Tween-20 and Triton). For example, localization of a secretory response relative to internalized streptococci requires incubation of the samples with a rabbit antistreptococcal antibody, diluted 1:1000.
5. Wash cells twice using blocking buffer containing Tween-20 and Triton X-100.
6. Incubate the cells at room temperature for 60 min with an appropriate secondary antibody.
7. Wash the cells twice in ice-cold blocking buffer and once using ice-cold PBS.
8. Allow cells to sediment onto poly-L-lysine-coated cover slips for 30 min, and mount these onto glass slides using ProLong Gold anti-fade kit or equivalent.

3.3. Pinosome Formation

1. Perform the synchronized presentation described above in the presence of either Lucifer Yellow (1 mg/mL) or FM 1-43FX (10 μM). Place samples on ice (*see* **Note 14**).
2. Wash the cells three times in ice-cold PBS.
3. Optional: use a chase period at 37°C to allow continued phagosome maturation. Halt incubation by putting sample on ice.
4. Quantify pinosome formation by microscopic, flow cytometric, or fluorometric analysis.
   a. For microscopic analysis, fix the samples using 1–4% PFA for 15 min at 4°C, followed by a 45-min incubation at room temperature (*see* **Note 15**). Attach cells to poly-L-lysine coated cover slips and mount as described above.
   b. Analyze by flow cytometry to determine the amount of internalized pinosome marker.
   c. The amount of internalized Lucifer Yellow can also be quantified in a spectrofluorometer. Resuspend cells in Na-medium containing 0.1% Triton X-100. Remove cell components not solubilized by Triton with centrifugation, and analyze supernatants with a spectrofluorometer using $\lambda_{ex} = 410$ nm and $\lambda_{em} = 520$ nm.

3.4. Magnetic Purification Protocol

3.4.1. Preparation of Prey

1. All solutions, tubes, and pipet tips should be sterile.
2. All steps are performed at room temperature, unless otherwise stated. Pyridine- and glutaraldehyde-containing solutions should be handled in a ventilated hood.

3. Wash the streptococci in 4×10 mL of PBS buffer (screw-cap glass tubes). Resuspend in 1 mL of PBS in 1.7 mL Eppendorf tubes.

4. Sonicate for 3×30 s in an ultrasonic water bath to break up bacterial aggregates and large chains.

5. Estimate the bacterial concentration (usually in the order of $1 \cdot 10^9$/mL) by optical density (OD) measurements.

6. Assess the viability of the bacteria with an inverted fluorescence microscope, using the LIVE/DEAD BacLight L-7007 bacterial viability kit. If the viability is higher than 90%, the bacteria are acceptable for further use.

7. Heat-kill the bacteria at 80°C for 5 min in a heat block (agitation at 800 rpm); then quickly cool them down by immersing the sample tubes in ice water.

8. Determine the concentration of the BioMag BM546 particle stock suspension (~$8 \cdot 10^9$ particles/mL) by counting the particles in a Bürker chamber. Sediment for 5 min before counting.

9. Dilute 50 µL of the BioMag BM546 stock suspension with H_2O to 1000 µL in a 1.7-mL Eppendorf tube (*see* **Note 16**).

10. Centrifuge ($1000g$ for 30 s), collect 900 µL of the supernatant, and discard the rest. This supernatant contains particles small enough to be unobtrusive when conjugated to the surface of streptococci.

11. Place the tube containing the supernatant onto a magnetic rack, and allow the BioMag particles to migrate to the magnetic wall perpendicular to gravity for 10 min.

12. Exchange the solution for 1 mL of coupling buffer, using a pipet. Repeat this step three times. To reduce particle losses, always leave approximately 40 µL of liquid at the bottom of the tube.

13. Resuspend particles in 1 mL of cross-linking solution, and rotate the mixture for three hours at 18 rpm.

14. Wash four times in coupling buffer, separating the particles from the solution as described in **steps 11** and **12**.

15. Centrifuge the desired amount of heat-killed bacteria (6 min, $12,000g$), and resuspend them in an appropriate amount of coupling buffer (*see* **Note 17**). When mixing the streptococci with the pretreated magnetite particles the nominal bacteria: particle ratio should be 10:1 (*see* **Note 17**).

16. Sonicate the mixture for 30 s and place the tube on a rotator (11 rpm) for 1–3 h at room temperature. Transfer the rotator to a cold room and incubate overnight.

17. Separate the suspension for 10 min on the magnetic rack.

18. Resuspend the superparamagnetic bacteria in quenching solution, briefly sonicate, and incubate on a rotator (18 rpm) for 30 min.

19. Briefly sonicate the superparamagnetic bacteria, and wash them in 4×1 mL of PBS (containing 0.05% BSA), using the magnetic rack.

20. Resuspend the final preparation in 1 mL of PBS/BSA and store at 4°C, for up to 1 mo after preparation. Aliquots taken from this stock suspension should be briefly sonicated before use. The final concentration of the superparamagnetic bacteria can be determined by counting in a Bürker chamber (usually at dilution 1:100; allow 5 min sedimentation).

3.4.2. Nitrogen Cavitation

1. Connect a N_2 gas container to a cell disruption bomb that has a thin 4-inch latex tube attached to its elution valve (*see* **Note 18**). Keep the cell disruption bomb, cell samples and solutions on ice during the cavitation procedure.
2. Perform phagocytosis procedure, as described above (*see* **Note 19**).
3. Centrifuge the cells (5 min, 200g, 4°C) in a swinging-bucket rotor and resuspend the pellet in 1 mL of cavitation buffer.
4. Transfer cells to the sample compartment of the cell disruption bomb.
5. Increase the N_2 partial pressure to 1.0–1.4 MPa (150–200 psi), and maintain this level for 20 min. Wear safety goggles.
6. Release the pressure by gently opening the valve. Caution: exercise extreme care, to avoid loss of material. The eluate is released through the latex tube and collected in a 50-mL polypropylene conical tube (*see* **Note 20**).
7. Rinse the bomb cavity and latex tube with deionized water between samples.
8. Remove remaining intact cells and nuclei by centrifugation (5 min, 200g, 4°C) in a swinging-bucket rotor, and use the supernatant for the magnetic separation procedure.

3.4.3. Magnetic Separation

1. Perform the magnetic separation preferably on ice.
2. Magnetic probe alternative (high purity):
   a. Aspirate the cavitate supernatant four times, using a 1-mL syringe with a 27-G needle (*see* **Note 21**).
   b. Transfer $4 \times 250\,\mu L$ of each 1-mL sample to the wells of a 96-well microtiter plate.
   c. Attach a dedicated silicon tip to the PickPen and extend the magnetic end of the PickPen into the tip.
   d. Dip the PickPen tip into the suspension, using a gentle circular motion and then keep it immersed for up to 2 min.
   e. Withdraw the PickPen and dip it into $250\,\mu L$ of isotonic protease-inhibitor buffer that was placed in another well.
   f. Release the phagosomes by retracting the PickPen magnet and stirring the silicon tip for a while. We recommend that **steps d–f** be repeated until no material is visible on the tip (usually twice for transfers from the original suspension, and once for transfers from each consecutive wash bath).
3. Magnetic rack alternative:
   a. Place the cavitated 1-mL samples onto the magnetic rack and allow 5–10 min separation.
   b. Aspirate approx $200\,\mu L$ of the liquid from the bottom of the tube, thereby removing any gravitationally sedimented material.
   c. Remove the rest of the buffer while keeping the tip of the pipet just below the gradually receding liquid surface (*see* **Note 22**).
   d. Leave approx $40\,\mu L$ of buffer at the bottom of the tube.
   e. Perform the above wash procedure three times, using 3×1 mL of isotonic protease-inhibitor buffer.

3.5. Purity Assessment and Quantification

3.5.1. Estimating the Number of Phagosomes—The Annexin V Method

1. Set aside 25 µL from each purified phagosome sample.
2. Transfer approx 10 µL to a Bürker chamber, allow 5 min sedimentation, and count the bacteria in a microscope (20× objective). This will give the concentration of all bacteria in the sample, including those surrounded by phagosome membranes.
3. Mix another 10 µL of sample with, e.g., 5 µL of 5X Ca^{2+}-containing buffer and 1 µL of annexin V–Alexa Fluor 488 solution (*see* **Note 23**).
4. Incubate sample in the dark at room temperature for 15 min.
5. Apply the mixture to a microscope slide, and overlay with a coverslip.
6. Use a fluorescence microscope (100× objective) to determine the ratio of the number of bacteria surrounded by unbroken fluorescent rings (i.e., intact phagosomes) to the total number of bacteria (*see* **Note 24**).
7. Calculate the phagosome concentration in the original sample by multiplying the concentration of bacteria in the original sample with the phagosome:bacteria ratio.

3.5.2. Quantification of Phagosomal Protein Content Using Microcon and NanoOrange

1. These instructions assume the use of Microcon and NanoOrange. It is also assumed that phagosomes have been isolated using the described purification protocol.
2. Centrifuge the phagosome-containing samples (10 min, 20,000*g*, 4°C).
3. Remove the supernatant, and resuspend the pellet in 50 µL of ice-cold extraction buffer (*see* **Note 25**).
4. Incubate samples for 20 min on ice.
5. Add 450 µL of protease inhibitor solution to the samples (*see* **Note 26**).
6. Centrifuge the samples (10 min, 20,000*g*, 4°C).
7. Aspirate the supernatants and proceed to Microcon concentration (*see* **Note 27**).
8. Assemble Microcon sample containers with permeate tubes and add the samples to the containers (*see* **Note 28**).
9. Centrifuge at 14,000*g* for 140 min at 4°C. This will generate approx 50 µL of retentate from each sample.
10. Remove the sample containers, put them inside new tubes, and then invert them in order to collect the retentate.
11. Centrifuge at 1000*g* for 3 min to collect the concentrated sample.
12. Prepare samples in triplicate for the standard curve (range 1–10 µg/mL). Add 5 µL of PBS containing 0.01% Triton X-100 (*see* **Note 29**).
13. Dilute phagosomal samples 70 times with NOWS (add 5 µL sample to 345 µL of NOWS).
14. Incubate samples at 95°C for 10 min; keep protected from light.
15. Let samples cool at room temperature for 20 min; keep protected from light.
16. Transfer 300 µL from each sample to an enzyme-linked immunosorbent assay (ELISA) plate.

17. Measure the fluorescence at excitation 485 nm and emission 590 nm.
18. Subtract the blank value (mean of six blank samples) and make a standard curve (*see* **Note 30**).
19. Subtract the value of the blank from the sample readings and use standard curve to determine protein content.

3.6. Preparation of Samples for Gel Electrophoresis

1. For use as control in the Western blot, prepare a cell lysate of the cell type used for the phagosome preparation: wash cells three times with PBS, centrifuging between each wash (3 min, 2000g, 4°C).
2. Resuspend pellet in ice-cold extraction buffer, and incubate for 20 min on ice.
3. Centrifuge samples (10 min, 12,000g, 4°C), aspirate supernatants, and discard pellets.
4. Prepare cell and phagosome lysates for reduction by adding 6.25 µL of sample buffer and 2.5 µL of reducing agent (*see* **Note 31**).
5. Add H_2O to a final volume of 25 µL before incubating the samples at 70°C for 10 min.

3.7. SDS-PAGE

1. These instructions assume the use of an XCell SureLock Mini-Cell system (Invitrogen) with NuPAGE gels (4–12% gradient, 8 × 8 cm, 10 1.0-mm wells).
2. Remove the gel cassette from the pouch and rinse with deionized water. Peel off the tape from the bottom of the cassette.
3. With one smooth motion, pull out the comb from the cassette and rinse the wells with running buffer three times.
4. Place the gels in the Mini-Cell with the sample wells facing inwards. If using only one gel, insert the provided plastic substitute. Lock with the supplied wedge.
5. Before filling the system with running buffer, add 500 µL of antioxidant to 200 mL running buffer for use in the inner chamber.
6. Check the inner chamber for leaks with running buffer before filling it until it covers the wells.
7. Load the wells with 25 µL of reduced sample (as described earlier) or 10 µL of pre-stained molecular mass marker. Empty wells are loaded with sample buffer diluted in distilled H_2O.
8. Fill the outer chamber with 600 mL of running buffer. It is possible to re-use the buffer for the outer chamber several times.
9. Run the gels at 200 V for 50 min.
10. After electrophoresis, use the provided gel knife to open the cassette. Place the cassette with the well side up, insert the knife and push upwards. Repeat for all sides and discard the upper plate, leaving the gel on the bottom one. Proceed to Western blot protocol.

3.8. Western Blot

1. These instructions assume the use of the semi-dry unit from Bio-Rad and that the gel electrophoresis has been executed as described in the previous section.

2. Cut two PVDF membranes (*see* **Note 32**) and two filter papers to the size of the gel (8 × 8 cm). Activate the PVDF membranes by placing them in methanol for 10 s and then rinsing them with deionized water. Place the membranes and the filter papers in 100 mL transfer buffer and soak for 10 min.

3. Prepare a sandwich on the semi-dry unit by placing one thick filter paper on both sides of two PVDF membranes.

4. Cut off the wells from the gel and then remove it from the cassette by pushing the gel knife through the hole on the backside of the cassette, holding the gel over the filter-membrane stack. Cut away the lip of the gel and place the presoaked filter paper on top (*see* **Note 33**).

5. Close the semi-dry unit and transfer the proteins at 15 V for 45 min (*see* **Note 34**).

6. Once the transfer is complete, put the membranes in heat-sealable plastic bags and incubate overnight with blocking buffer at 4°C.

7. Discard blocking buffer, and incubate membrane with primary antibody in buffer for 2 h.

8. Wash membrane three times with PBST (5, 10, and 15 min).

9. Incubate with secondary antibody (matching species) diluted in buffer for 1 h.

10. Wash three times with PBST (5, 10, and 15 min).

11. Mix the two components of the ECL reagent. Incubate 5 min with ECL reagent (*see* **Note 35**).

12. Pour off the solution, and remove excess fluid by holding the membrane in one corner with gloved hands and letting the other corner touch tissue paper. Place the membrane in a new bag, exercising caution not to smear the surface.

13. Expose in camera (usually between 10 s and 10 min).

4. Notes

1. One caveat of the method is that steady-state distributions of membrane markers are measured, reflecting a balance between secretory and pinocytic activities. Analysis of secretion by measuring the release of a soluble content marker may be less confounded by pinocytic reuptake.

2. Because of the nature of the procedure, live bacteria can not be used. However, heat-killed *S. pyogenes* bacteria have been shown to affect phagosomal maturation in a manner similar to, albeit somewhat less pronounced than, that of live bacteria *(9,10)*.

3. There are a number of different methods available for disruption of cells, notably mechanical homogenization, sonication, and nitrogen cavitation. The latter, which is the most gentle, relies on the pressurized dissolution of inert gas molecules inside cells followed by the emergence of cell-disrupting microscopic bubbles once the pressure is released.

4. The superparamagnetic phagosomes can be purified by using either a stationary magnet (magnetic rack) or a movable one (magnetic probe). In the former case, a minimum of three washes is required. In the latter case, one transfer to a fresh wash bath is sufficient to obtain relatively pure phagosomes.

5. One way of estimating the amount of phagosome membrane in the magnetically purified samples is to determine the number of phagosomes. In order to distinguish phagosomes from bacteria not surrounded by a phagosome membrane a suitable marker is required. Annexin V conjugated to a fluorophore is used here because of its strong affinity for phosphatidyl serine, a component of the inner face of the cytoplasmic membrane and of the outer face of the phagosome membrane. The fluorescent phagosomes can be detected either in a fluorescence microscope or in a flow cytometer.

6. Preferably, use serum from the same species as that in which secondary antibody was raised.

7. The use of F(ab')$_2$ fragments as secondary antibody is recommended in order to avoid binding to Fc-receptors. Antibody dilutions must be tested empirically.

8. Coating of coverslips: wash glass coverslips with methanol and overlay with 0.25 mL poly-L-lysine (0.2 mg/mL in water). Evaporate the added fluid at 50–65°C, and then wash twice with distilled H_2O.

9. BioMag BM546 is an aqueous suspension of superparamagnetic particles composed of >90% magnetite (Fe_3O_4) and an inert silane coating containing free amino groups. The size range of the particles is wide.

10. Benzonase is an endonuclease able to digest DNA and RNA even at temperatures as low as 4°C. Because its activity requires the presence of Mg^{2+}, 3 mM $MgCl_2$ is added to the N_2 cavitation buffer. This amount was determined empirically. Use of Benzonase is strongly recommended during N_2 cavitation and subsequent washes, because otherwise, some aggregation of phagosomes with cellular debris and DNA may be observed, leading to decreased yield of free phagosomes.

11. Use a particle:cell ratio that does not exceed the phagocytic capability of the cells.

12. Strict control of temperature requires use of a thermostated centrifuge set to 4°C.

13. Optionally, PFA can be used to fix the cells prior to antibody staining, but in this case control experiments must be performed to demonstrate inaccessibility of antibodies to intracellular structures. Antibody dilutions have to be tested empirically.

14. Inevitably, some sealed phagosomes containing pinosome marker will form during the presentation protocol. To minimize phagocytosis, shorten the presentation step to 30 s, or preincubate the cells with 10 µM cytochalasin D (prevents actin polymerization) for 20 min, and perform the presentation step in the continued presence of this inhibitor.

15. Lucifer Yellow contains a carbohydrazide (CH) group that allows it to be covalently linked to surrounding biomolecules during aldehyde fixation. To allow identification in the microscope, the bacteria can be prelabeled with a fluorophore, such as a suitable Alexa dye.

16. Do not use PBS, which will cause some aggregation of the magnetite particles.

17. This value was empirically arrived at for the *S. pyogenes*/BioMag BM546 system by microscopic analysis (100× objective). It may differ for other species of pathogens. The particle coverage of the bacterial cell wall should be very sparse, so as not to hinder the interaction of the bacterial surface with phagocytic cells. For example, if the concentration of the magnetite particles is $0.4 \cdot 10^9$/mL ($8 \cdot 10^9$/mL stock

× 50 µL) and the available amount of bacteria is $1 \cdot 10^9$, mix 250 µL of the particle suspension (i.e., $0.1 \cdot 10^9$ particles) with the $1 \cdot 10^9$ bacteria suspended in 750 µL coupling buffer in an Eppendorf tube. Note that the concentration of magnetite particles is based on counting the particles in the original stock suspension, not the postcentrifugal supernatant of smaller particles. Because the empirically-tested coupling ratio is based on these precentrifugal figures, this is no problem.

18. Before performing the first nitrogen cavitation, it is advantageous to flush some buffer through the elution valve of the bomb.

19. Isolating phagosomes to a high degree of purity requires a considerably larger amount of cells and bacteria (~100 times more) than when preparing samples for immunofluorescence. Increasing the concentration of cells and bacteria in a single phagocytosis presentation could result in a decrease in phagocytosis efficiency. Therefore, we recommend performing multiple presentations using a multichannel pipet, and then pooling the aliquots before transfer to the cavitation buffer.

20. To minimize potential losses it is advisable to stretch some parafilm over the opening of the 50-mL tube.

21. By forcing the cavitated solution through a very narrow passage (the needle), shear stress is applied. This is done in order to homogenize the solution and thus minimize the number of phagosomes stuck in cellular debris.

22. For maximal retrieval of the superaparamagnetic phagosomes, exchange of the cavitation buffer requires careful handling. At this stage, it is advantageous to increase the magnetic field by mounting an additional magnet (such as a BioMag Separator magnetic plate) onto the back of the magnetic rack.

23. Because the binding of annexin V is Ca^{2+}-dependent, it is important to replace part of the calcium-free isotonic protease-inhibitor buffer with a concentrated Ca^{2+}-containing buffer. This is preferably accomplished by mixing, as described, instead of centrifugation and removal of supernatant. Note also that no fixation and permeabilization is required, because the target of annexin V (i.e., phosphatidyl serine) is found at the exterior of the phagosomal membrane.

24. Counting phagosomes in the microscope is very time-consuming, but it is the safest way to identify true phagosomal rings. When counting phagosomes, do not include bacteria with incomplete fluorescent rings or bacteria surrounded by weak punctate staining, as the latter could represent unspecific binding of annexin V to magnetite particles.

25. Make sure that the incubation period in extraction buffer is identical for all samples.

26. Because of interference of detergents with the assay *(12)*, a buffer change is performed using Microcon centrifugation filter units (after dilution of the samples with protease inhibitor solution to lower the Triton concentration). This strategy allows accurate analysis of the low protein content in the phagosomal samples.

27. For control experiments, resuspend the bacterial pellets in 50 µL of protease inhibitor solution and keep at 4°C.

28. For use as controls in the NanoOrange assay, it is advisable to include samples containing only extraction buffer.

29. Prepare a 10 µg/mL stock solution of BSA in NOWS. Put 345, 210, 105, and 35 µL of this solution in Eppendorf tubes. Add NOWS so that all tubes contain a total volume of 345 µL. Include six tubes containing only NOWS to use as blanks. Finally, add 5 µL of PBS with 0.01% Triton X-100 (to simulate the sample). Use these samples to generate standard curve and background controls (1, 3, 6, and 10 µg/mL, plus blanks).

30. As mentioned earlier, this assay is detergent-sensitive and will only give reliable results if the background caused by the Triton X-100 can be accurately measured, hence the six blank samples. The upper limit for Triton X-100 is 0.001%, according to the manufacturer, and the end concentration in our assay is 0.00014% (50 µL Triton 0.1% + 450 µL PBS = 0.01% Triton; 5 µL sample with 0.01% Triton + 345 µL NOWS = 0.00014% Triton X-100).

31. Add the reducing agent immediately prior to electrophoresis to obtain the best results.

32. When testing new material, it is advisable to use two membranes in case the transfer times for the proteins are unknown. By utilizing prestained markers it is possible to follow the transfer and adjust the conditions for optimal results.

33. Rolling the assembled sandwich with a test tube will improve transfer by removing bubbles.

34. The transfer time has to be adjusted depending on the size of the target protein. Approximately 45 min is suitable for medium-sized proteins. For large proteins, it is also advisable to add SDS and/or incubate the PVDF membrane without methanol in the transfer buffer.

35. Using a pair of tweezers, place the membrane onto a flat clean surface. Apply approx 2.5 mL of reagent with a pipet. The surface tension will keep the solution in place.

Acknowledgments

This work was supported by the Swedish Research Council (grant nr. 2002-6479), the Crafoord Foundation, the Greta and Johan Kock Foundation, the Kungliga Fysiografiska Sällskapet, the Thelma Zoéga Foundation, and the Alfred Österlund Foundation.

References

1. Nauclér, C., Sundler, R., Grinstein, S., and Tapper, H. (2002) Signaling to localized degranulation in neutrophils adherent to immune complexes. *J. Leukoc. Biol.* **71**, 701–710.

2. Tapper, H., Furuya, W., and Grinstein, S. (2002) Localized exocytosis of primary (lysosomal) granules during phagocytosis: role of Ca^{2+}-dependent tyrosine phosphorylation and microtubules. *J. Immunol.* **168**, 5284–5296.

3. Botelho, R. J., Tapper, H., Furuya, W., Mojdami, D., and Grinstein, S. (2002) FcγR-mediated phagocytosis stimulates localized pinocytosis in human neutrophils. *J. Immunol.* **169**, 4423–4429.

4. Niessen, H. W. M. and Verhoeven, A. J. (1992) Differential up-regulation of specific and azurophilic granule membrane markers in electropermeabilized neutrophils. *Cell. Signalling* **4,** 501–509.
5. Smith, C. B. and Betz, W. J. (1996) Simultaneous independent measurement of endocytosis and exocytosis. *Nature* **380,** 531–534.
6. Desjardins, M. and Griffiths, G. (2003) Phagocytosis: latex leads the way. *Curr. Opin. Cell Biol.* **15,** 498–503.
7. Lührmann, A. and Haas, A. (2001) A method to purify bacteria-containing phagosomes from infected macrophages. *Methods Cell Sci.* **22,** 329–341.
8. Staali, L., Mörgelin, M., Björck, L., and Tapper, H. (2003) *Streptococcus pyogenes* bacteria expressing M and M-like surface proteins are phagocytosed but survive inside human neutrophils. *Cell. Microbiol.* **5,** 253–265.
9. Staali, L., Bauer, S., Mörgelin, M., Björck, L., and Tapper, H. (2006) *Streptococcus pyogenes* bacteria modulate membrane traffic in human neutrophils and selectively inhibit azurophilic granule fusion with phagosomes. *Cell. Microbiol.* **8,** 690–703.
10. Bauer, S. and Tapper, H. (2004) Membrane retrieval in neutrophils during phagocytosis: inhibition by M protein-expressing *S. pyogenes* bacteria. *J. Leukoc. Biol.* **76,** 1–9.
11. Olsson, I., Olofsson, T., and Odeberg, H. (1972) Myeloperoxidase-mediated iodination in granulocytes. *Scand. J. Haematol.* **9,** 483–91.
12. Jones, L. J., Haugland, R. P., and Singer, V. L. (2003) Development and characterization of the NanoOrange protein quantitation assay: a fluorescence-based assay of proteins in solution. *Biotechniques* **34,** 850–854, 856, 858 *passim.*

21

Analysis of Neutrophil Bactericidal Activity

Jessie N. Green, Christine C. Winterbourn, and Mark B. Hampton

Summary

The primary function of neutrophils is to engulf and destroy invading pathogens. If the bactericidal capacity of neutrophils is defective, an individual may suffer from enhanced susceptibility to potentially fatal microbial infection. To identify such defects, and to investigate the mechanisms used to kill bacteria, the bactericidal activity of neutrophils must be accurately quantified. This chapter provides details of a comprehensive microbiological technique that quantifies neutrophil bactericidal activity by measuring the loss of viability of ingested bacteria over time. Two variations of this technique are presented: a simple "one-step" protocol providing a composite measure of phagocytosis and killing, and a more advanced "two-step" protocol that allows calculation of separate rate constants for both of these processes.

Key Words: Neutrophil; bacteria; killing; bactericidal activity; *Staphylococcus aureus*; method.

1. Introduction

Neutrophils are the immune system's key defenders against bacterial infection; their primary function being to ingest and destroy invading pathogens. They achieve this by engulfing the pathogen into an intracellular compartment called the phagosome, and then subjecting it to an array of both oxygen-independent and oxygen-dependent killing mechanisms. These include the release of antimicrobial and proteolytic peptides into the phagosomal space, along with the production of reactive oxygen species by the NADPH oxidase complex that assembles in the membrane. The relative importance of these pathways in bacterial killing remains controversial. In particular, it is not clear whether the oxidative mechanisms act on bacteria directly, or whether they facilitate the action of the nonoxidative factors *(1–3)*. A key tool in unravelling the mechanisms involved in bacterial killing by neutrophils is accurate quantification of bactericidal activity.

From: *Methods in Molecular Biology, vol. 412: Neutrophil Methods and Protocols*
Edited by: M. T. Quinn, F. R. DeLeo, and G. M. Bokoch © Humana Press Inc., Totowa, NJ

Bactericidal activity is quantified by measuring the loss in viability of bacteria co-cultured with neutrophils. Methods include measuring the ability of bacteria to form colonies on nutrient agar, or to incorporate [³H]-thymidine into newly synthesized DNA. Alternatively, fluorescent dyes such as acridine orange can distinguish viable from dead bacteria by intercalating into the less structurally organized DNA of the dead bacteria, resulting in a shift in emission peak. The advantages and disadvantages of these various methods have been reviewed elsewhere *(4)*.

This chapter provides details of a versatile microbiological technique, requiring only standard laboratory equipment, in which neutrophils and bacteria are mixed and the loss of bacterial viability over time is measured by plating of diluted samples and colony counting. It is an adaptation of our previously published methods *(4,5)* and includes an improved neutrophil lysis step *(6)*. Two variations of the procedure are described, as illustrated in **Fig. 1**. In the one-step" protocol, neutrophils are not separated from uningested bacteria. Therefore, a composite measure of both phagocytosis and killing is obtained. The "two-step" protocol incorporates a differential centrifugation step *(7,8)*, which separates uningested (or extracellular) bacteria from those ingested by the neutrophils (intracellular bacteria). The number of viable extracellular and intracellular bacteria is measured at various times, and a kinetic analysis is then undertaken, allowing separate rate constants to be calculated for both phagocytosis and killing. The "one-step" protocol has the advantage of requiring significantly less manual sample processing and can be used when it is not critical to distinguish which function (phagocytosis or killing) is affected. It is also useful for screening numerous samples or conditions. If differences are detected, samples can then be investigated in more depth using the two-step protocol. The two-step protocol is preferable for investigating defects specific to killing, and for the study of bactericidal mechanisms.

2. Materials

1. Tryptic Soy Broth: prepare according to the manufacturers directions and autoclave before use.
2. Phosphate-buffered saline (PBS): to prepare a 10 X stock, dissolve 2 g of KH_2PO_4 (18 mM), 11.4 g of Na_2HPO_4 (100 mM), and 2.4 g of KCl (27 mM) in 900 mL of H_2O. Adjust to pH 7.4 with HCl if necessary and add 80 g of NaCl (1.37 M). Bring the final volume to 1 L. Store at room temperature. Prepare a working solution by dilution of one part with nine parts water, and autoclave before use (*see* **Note 1**).
3. Hank's balanced salt solution (HBSS): 1 mM $CaCl_2$, 0.5 mM $MgCl_2$, and 1 mg/mL glucose in PBS. Filter sterilize before storage for up to 1 mo at 4°C.
4. Autologous human serum: to prepare approx 1 mL of serum, reserve 2 mL of blood (without anticoagulant) from the donor whose neutrophils are to be used in the experiment. Allow blood to clot in a glass tube at room temperature for approx 1 h.

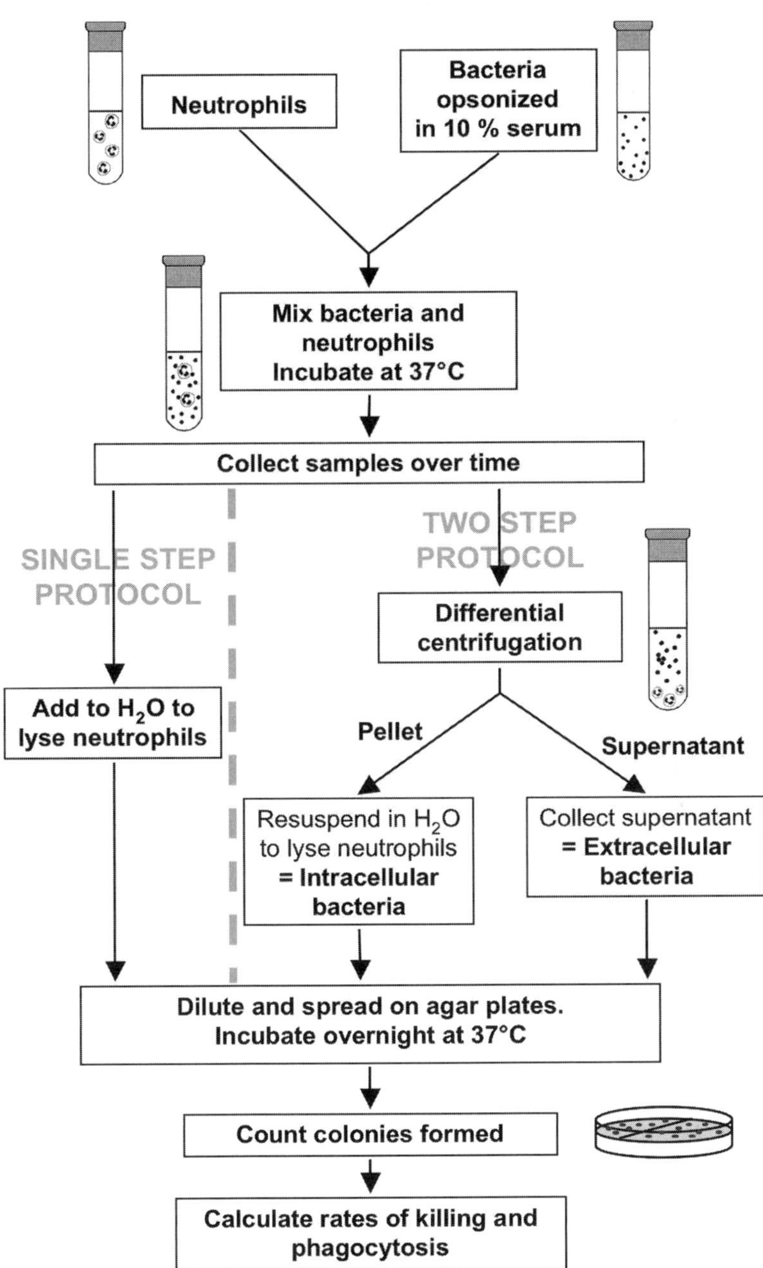

Fig. 1. Outline of the presented methods for measuring bactericidal activity of neutrophils.

Intermittently "ring" the clot by running a Pasteur pipet tip around the inside of the tube. Separate the serum from the clot by centrifugation at 3000g for 4 min, and transfer the serum to a clean tube.

5. pH 11 H_2O: Bring sterile H_2O to pH 11.0 by adding NaOH immediately prior to use. If pH decreases during the neutrophil lysis step, then a low concentration (5 mM) of an appropriate salt, e.g., sodium carbonate, should be used to enhance buffering capacity.
6. Columbia sheep blood agar plates: these are suitable for the fastidious growth requirements of *Staphylococcus aureus*.

3. Methods

These protocols have been optimized for use with the target bacteria, *S. aureus*. Other target micro-organisms can be used; however, sampling times and dilution factors may need adapting.

3.1. Neutrophil Preparation

1. Isolate neutrophils from whole blood as described in detail elsewhere in this volume (*see* **Note 2**). Ten milliliters of blood will typically yield sufficient neutrophils to test at least two conditions.
2. Resuspend neutrophils to a final concentration of 1 × 10^7 cells/mL in HBSS. Keep at room temperature and use within 1 h.

3.2. Preparation of Bacteria

1. Inoculate 5 mL of sterile tryptic soy broth with a single colony of *S. aureus* grown on sheep blood agar. Grow in a shaking incubator overnight at 37°C.
2. Take a 1-mL sample of the overnight culture and centrifuge at 1000g for 4 min to pellet the bacteria. Wash the pellet by resuspending in PBS, then centrifuge again and repeat the wash. Resuspend the final pellet in 2 mL of HBSS.
3. Calculate the concentration of bacteria in the sample by measuring the optical density at 550 nm and relating to a previously established standard curve of optical density vs colony-forming units (CFU).
4. Opsonize the bacteria by suspending 1 × 10^8 CFU in 1 mL of HBSS containing 10% autologous serum in a glass tube (*see* **Note 3**). Rotate end-over-end (6 rpm) for 20 min at 37°C.
5. Use opsonised bacteria immediately in either the one-step (**Subheading 3.3.**) or two-step (**Subheading 3.4.**) protocol.

3.3. One-Step Assay of Neutrophil Bactericidal Activity

1. Prepare two tubes—an experimental and a control—for each condition tested.
2. For the experimental tube, mix 300 μL of freshly prepared neutrophils (at 1 × 10^7 cells/mL) with 30 μL of autologous serum, and incubate at 37°C for 10 min to prewarm. Add 300 μL of freshly opsonized bacteria to start the reaction. The final ratio of bacteria to neutrophils is 10:1 and the serum concentration is approx 10% (*see* **Note 4**).

Table 1
Dilution Guidelines for the One-Step Protocol

	Sample time			
	0 min	5 min	10 min	20 min
Dilutions for One Step Protocol				
Approx. undiluted concentration /mL[a]	5×10^7	4×10^7	3×10^7	2×10^7
Dilution due to sample processing[b]	1 in 50	1 in 50	1 in 50	1 in 50
Dilution incurred by plating 20 µL[b]	1 in 50	1 in 50	1 in 50	1 in 50
Additional dilution required to get approx 50 colonies per 20 µL[c]	1 in 400	1 in 320	1 in 240	1 in 160
Total dilution factor	1×10^6	8×10^5	6×10^5	4×10^5

[a]From experience, these are the approximate bacterial concentrations we have come to expect at time of sampling. These values are likely to vary under different conditions. If the numbers of bacteria being detected are too high (>100) or low (<20) to make accurate colony counts, the additional dilutions should be adjusted accordingly.

[b]These are dilutions that will be automatically incurred though following the protocol as written.

[c]These are the further dilutions called for in the final step of either the one-step or two-step protocol.

3. Prepare an identical "control" tube, replacing the neutrophils with 300 µL of HBSS. The control allows direct measurement of the starting number of bacteria, their growth over the time course of the experiment, any negative impact of opsonization and dilution in water at pH 11.0, and the possible effects of experimental drugs on bacterial growth.

4. Incubate the tubes at 37°C with end-over-end rotation (6 rpm) (*see* **Note 5**). Take 50 µL samples from the experimental tube at 5, 10, and 20 min, and from the control at 0 and 20 min (*see* **Note 6**) and dilute into 2.45 mL of H_2O, pH 11.0 (*see* **Note 7**).

5. Allow the neutrophils to lyse by standing at room temperature for 5–10 min, then vortex vigorously for approx 5 s to disperse the bacteria.

6. Dilute each sample further in water at pH 11.0 to give a bacterial concentration of approx 2500–5000 CFU per mL (*see* **Table 1**). Spread 20 µL on half an agar plate, giving approx 50–100 CFU per half-plate. Plate four half-plates per sample. Incubate the plates overnight at 37°C, and count the number of colonies formed.

3.4. Two-Step Assay of Neutrophil Bactericidal Activity

1. Carry out **Subheading 3.3.**, steps **1–3**.

2. Incubate the tubes at 37°C with end-over-end rotation (6 rpm) (*see* **Note 5**). Take 50 µL samples from the experimental tube at 5, 10, and 20 min and from the control at 0 and 20 min (*see* **Note 6**).
3. Dilute samples into 950 µL of ice-cold PBS to halt neutrophil activity.
4. Centrifuge each sample at 100*g* for 5 min at 4°C (*see* **Note 8**).
5. Collect the supernatant, taking care not to disturb pellet (*see* **Note 9**), and wash the pellet twice more with 1 mL of ice-cold PBS, pooling the supernatants. The pooled supernatant contains the bacteria that have not been phagocytosed (*extracellular* bacteria), whereas the bacteria that have been phagocytosed are in the neutrophil pellet (*intracellular* bacteria).
6. Resuspend the pellet in 2.5 mL of H_2O, pH 11.0. Stand at room temperature for 5–10 min, then vortex thoroughly for approx 5 s.
7. Dilute each sample further in water at pH 11.0 to give a bacterial concentration of approx 2500–5000 CFU per mL (*see* **Table 2**). Spread 20 µL on half an agar plate, giving approx 50–100 CFU per half-plate. Plate four half-plates per sample. Incubate the plates overnight at 37°C, and count the number of colonies formed.

3.5. Data Analysis for One-Step Assay

1. Convert colony counts back to the original bacterial concentrations by multiplying by the appropriate dilution factor (*see* **Table 1**).
2. Plot raw values against time (as illustrated in **Fig. 2**), or converted to percentage killing or percentage survival relative to the initial number of starting bacteria. Expression of data as percentages normalizes any variation in the initial concentration of bacteria and enables experiments on different days to be compared. Different experiments can also be compared by obtaining the slope of a semi-log plot of the data. The slope provides a single rate which is a composite measure phagocytosis and killing.
3. **Figure 2** shows data obtained from a one-step experiment using neutrophils from a healthy donor and the same neutrophils treated with diphenyleneiodonium (DPI), a general inhibitor of flavoproteins including NADPH oxidase. It is immediately obvious that there is significant loss in bacteria viability when incubated with healthy neutrophils, but that DPI blocks the majority of killing.

3.6. Data Analysis for Two-Step Assay

1. Using data obtained with the two-step protocol, killing can be quantified using a kinetic analysis. We have shown previously that both phagocytosis and killing approximate first order processes *(5)*. On this basis, rate constants for the processes can be calculated.
2. The two-step protocol measures the changes in both viable extracellular and intracellular bacteria over time. The number of extracellular bacteria (*A*) decreases with time, while the number of viable intracellular bacteria (*B*) initially increases but then decreases as the bacteria are killed (*C*). These two events can be represented as occurring in series.

Table 2
Dilution Guidelines for the Two-Step Protocol

	Sample							
	Control		Extracellular			Intracellular		
	0 min	20 min	5 min	10 min	20 min	5 min	10 min	20 min
Dilutions for Two Step Protocol								
Approx. undiluted concentration /mL[a]	5×10^7	5×10^7	2×10^7	1.5×10^7	5×10^6	1×10^7	1×10^7	5×10^6
Dilution due to sample processing[b]	1 in 50	1 in 50	1 in 60	1 in 60	1 in 60	1 in 50	1 in 50	1 in 50
Dilution incurred by plating 20 µL[b]	1 in 50	1 in 50	1 in 50	1 in 50	1 in 50	1 in 50	1 in 50	1 in 50
Additional dilution required to get approx 50 bacteria per 20 µL[c]	1 in 400	1 in 400	1 in 133	1 in 100	1 in 33	1 in 80	1 in 80	1 in 40
Total dilution factor	1×10^6	1×10^6	4×10^5	3×10^5	1×10^5	2×10^5	2×10^5	1×10^5

[a]From experience, these are the approximate bacterial concentrations we have come to expect at time of sampling. These values are likely to vary under different conditions. If the numbers of bacteria being detected are too high (>100) or low (<20) to make accurate colony counts, the additional dilutions should be adjusted accordingly.

[b]These are dilutions that will be automatically incurred though following the protocol as written.

[c]These are the further dilutions called for in the final step of either the one-step or two-step protocol.

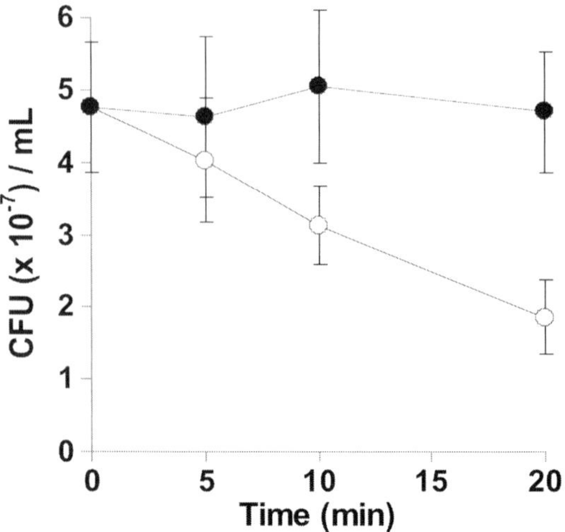

Fig. 2. Analysis of killing of *Staphylococcus aureus* by neutrophils in the presence or absence of diphenylene iodonium (DPI) using the one-step protocol. **(A)** *S. aureus* was incubated with normal neutrophils (open circles) or neutrophils treated with 10 µ*M* DPI (solid circles) and samples taken over time. Colony counts of surviving bacteria were multiplied by the appropriate dilution factors to give a bacterial concentration. Each point is the mean ± SD of four colony counts from a single experiment. Bacterial killing appears to be completely inhibited, illustrating the importance of reactive oxygen species production by NADPH oxidase for neutrophil bactericidal activity towards *S. aureus*.

$$A \xrightarrow{k_p} B \xrightarrow{k_k} C$$

By obtaining values for *A* and *B* using the two-step protocol, rate constants for phagocytosis (k_p) and killing (k_k) can be calculated.

3. Phagocytosis and killing are represented by equations 1 and 2, which can be integrated to give equations 3 and 4, where A_0 = the initial number of bacteria added to the system and t = time.

$$\frac{d[A]}{dt} = k_p[A] \tag{1}$$

$$\frac{d[B]}{dt} = k_p[A] - k_k[B] \tag{2}$$

$$A = A_0 e^{-k_p t} \tag{3}$$

$$B = \frac{A_0 k_p}{k_k - k_p} (e^{-k_p t} - e^{-k_k t}) \tag{4}$$

Solving for k_p from **eq. 3** involves obtaining the slope of a semi-log plot of *A* with time. Knowing k_p and A_0, **eq. 4** can be solved numerically for k_k. The k_k values for each timepoint are then averaged, giving an overall k_k value. k_p and k_k can also be converted to half-lives of phagocytosis and killing using the equation $t_{1/2} = \ln2 / k$.

4. To execute these mathematical analyses we have developed an Excel file which can be downloaded from http://www.chmeds.ac.nz/research/freerad/assays.htm, along with the instructions for its use. The calculations executed by this file are illustrated in **Figs. 3** and **4,** which show the graphical outputs obtained after each step of the calculation is completed. Data for these analyses were obtained in a two-step assay using neutrophils from a healthy donor (**Fig. 3**) and the same neutrophils treated with the NADPH oxidase inhibitor, DPI (**Fig. 4**). Comparison of the final graphs in these figures (**Figs. 3E** and **4E**) illustrates that although DPI has little effect on phagocytosis, it inhibits the killing of ingested bacteria almost completely. This effect is further illustrated in **Fig. 5** which summarizes rate constant data from four independent experiments similar to those described in **Figs. 3** and **4.**

4. Notes

1. Reagents and equipment should be sterilized before use for all the procedures described in this chapter. Subsequent handling can be undertaken on the bench because contamination will be very minor in comparison to the numbers of experimental bacteria present. However, when plating samples, an aseptic technique should be used to avoid contamination of the plate with airborne micro-organisms.

2. We use Ficoll-Hypaque density-gradient separation and dextran sedimentation, followed by hypotonic lysis of erythrocytes. This is gentler than ammonium chloride lysis and helps to preserve neutrophil function.

3. Glass tubes should be used at all stages throughout the protocols. Plastic tubes should be avoided as some bacteria (including *S. aureus*) have a tendency to stick to the plastic leading to inconsistent results between experiments. For steps requiring the tube to be rotated, we use small blood collection-type glass test tubes with tight-fitting rubber bungs.

4. The method can be used with other bacteria: neutrophil ratios, giving comparable results. If the ratio is changed, dilutions must be adjusted accordingly.

5. Continuous, slow, end-over-end rotation of the tubes is very important to prevent the neutrophils from sedimenting and clumping, and to ensure continual contact between neutrophils and uningested bacteria. The rotation speed must not be too vigorous or neutrophil function is disrupted.

6. The time points at which viable bacteria are measured should be selected to coincide with the period when most of the phagocytosis and killing occurs. For *S. aureus* at 10:1, sampling at 5, 10, and 20 min is appropriate. Samples taken at longer times generally show little further decrease in bacterial numbers as the rate of killing begins to decrease after 20–30 min. The killing mechanisms at these later times may differ from those functioning during the initial rapid killing period and the contribution of processes important for killing the bulk of the bacteria may be obscured. If using other bacterial species or ratios, it may be necessary to adjust the sampling times.

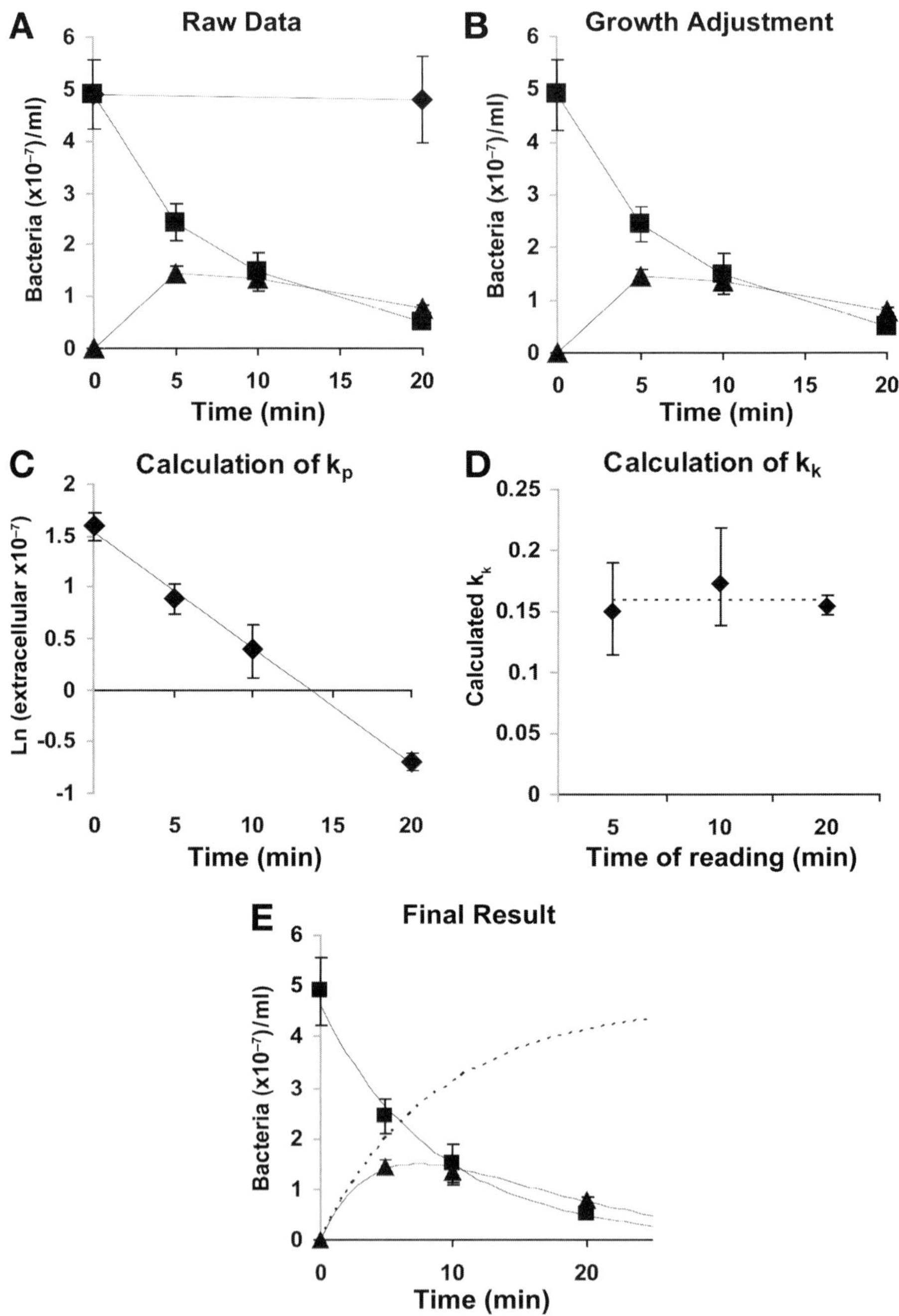
A
Raw Data
Bacteria (x10⁻⁷)/ml
Time (min)
B
Growth Adjustment
Bacteria (x10⁻⁷)/ml
Time (min)
C
Calculation of k_p
Ln (extracellular x10⁻⁷)
Time (min)
D
Calculation of k_k
Calculated k_k
Time of reading (min)
E
Final Result
Bacteria (x10⁻⁷)/ml
Time (min)

7. Dilution into H_2O, pH 11.0 results in osmotic lysis of the neutrophils without affecting the viability of *S. aureus*. When water at neutral pH is used, there is incomplete lysis of the neutrophils and ineffective dispersal of bacteria associated with the cell debris *(6)*. This results in an overestimation of bactericidal activity due to several bacteria only producing a single colony. To overcome this problem, H_2O at pH 11.0 is used *(6,9)*. Decleva et al. *(6)* found this to be much more effective than other lysis methods, including dilution into neutral pH H_2O or into detergent-containing buffer. We and others have previously used 0.1% saponin in PBS *(5,10)*. However, recent experiments in our laboratory have shown that this also gives ineffective dispersal and confirmed that osmotic lysis in H_2O at pH 11.0 is more effective.

Another potential problem recently brought to our attention is the precipitation of neutrophil DNA during lysis. This is observed as a stringy precipitate following vortexing of the lysed sample. Bacteria can become trapped in the precipitate, hindering their dispersal and, as described above, cause bactericidal activity to be overestimated. Use of H_2O at pH 11.0 for all subsequent dilution steps will assist dispersion. However, if significant bacterial clumping is problematic, then after the lysis step, bacteria can be pelleted and resuspended in HBSS containing DNAase prior to subsequent dilution (H. Rosen, University of Washington, personal communication).

8. This is a low-speed centrifugation to sediment the neutrophils and their ingested bacteria, but not extracellular bacteria. A small, diffuse white pellet of neutrophils should be seen in the experimental tube and no bacterial pellet in the control tube. If this is not the case, the differential centrifugation speed should be adjusted.

Fig. 3. (*Opposite page*) Calculation of rates of phagocytosis (k_p) and killing (k_k) of *S. aureus* by neutrophils from a healthy donor using two-step protocol results. (**A**) Colony counts of control (diamonds), extracellular (squares), and intracellular (triangles) bacteria were obtained using the two-step protocol with neutrophils from a healthy donor. These have been converted back to bacterial concentrations using dilution factors in **Table 2**, and plotted. Each point is the mean ± SD of four colony counts. (**B**) Bacterial numbers from **A** were adjusted at each time according to the growth of control bacteria. (**C**) k_p was determined from a semi-log plot of extracellular bacteria with time, where the slope of the regression line = k_p. In this example $k_p = 0.11$ min^{-1}, $t_{1/2} = 6.2$ min. (**D**) Having obtained k_p, the rates of killing (k_k) for each sampling time could be calculated. The dotted line is the mean k_k. In this example $k_k = 0.16$ min^{-1}, $t_{1/2} = 4.4$ min. The error bars in this graph reflect the error in the intracellular colony counts plus the degree of phagocytosis at that particular time. If limited phagocytosis has occurred, a small variation in the intracellular colony counts will have a much larger effect on the calculated k_k value than if a lot of bacteria have been phagocytosed. If the error observed in calculating k_k at 5 min is consistently high then it would be necessary to make the sampling times later. (**E**) Theoretical curves of extracellular and intracellular bacteria were generated using the calculated rate constants, and the experimentally obtained values from **B** plotted for comparison. The dotted line represents the expected number of viable intracellular bacteria if none were killed.

   Green, Winterbourn, and Hampton

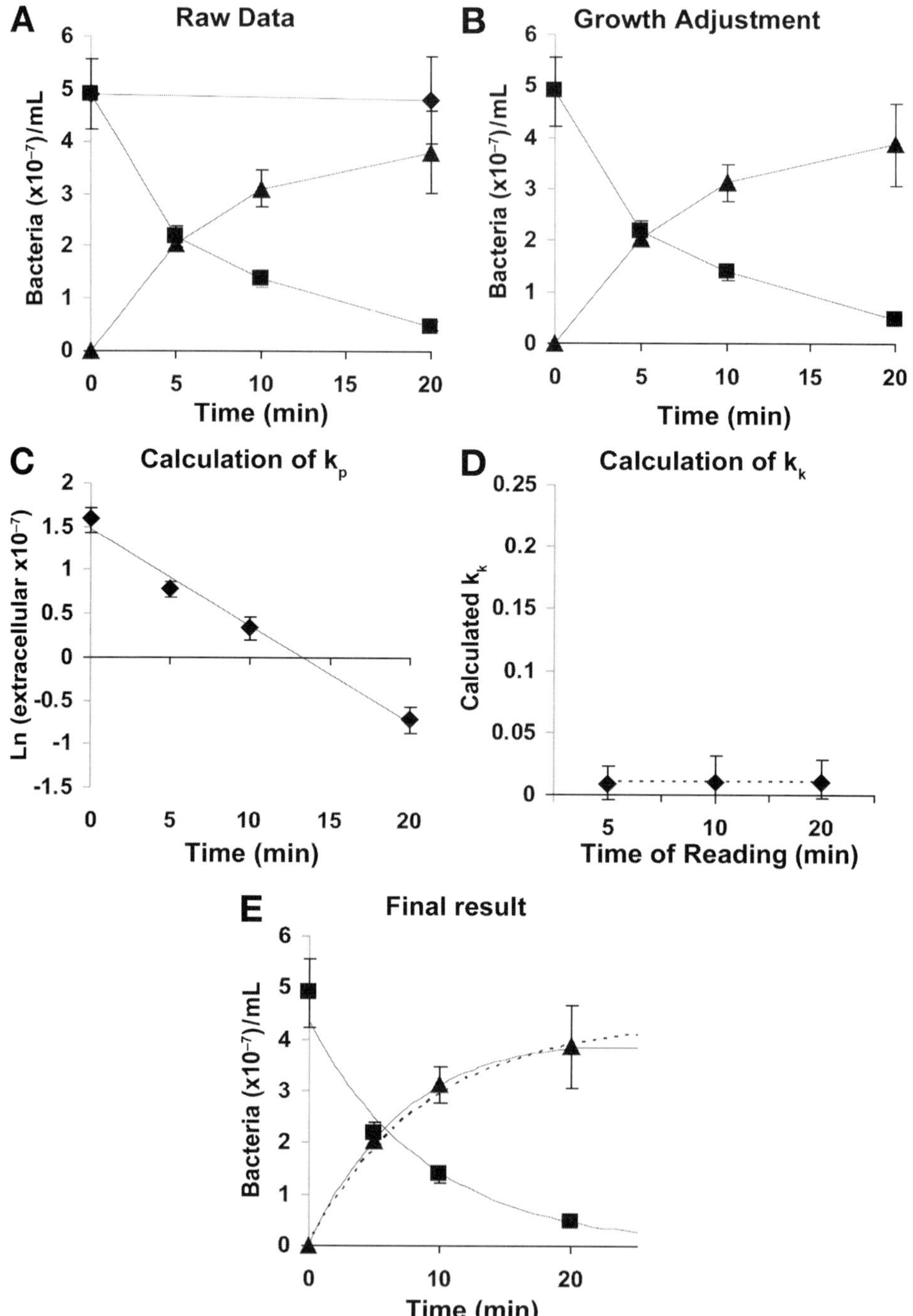

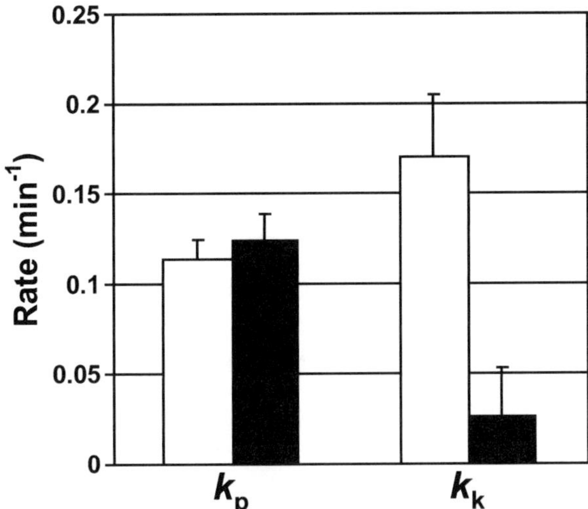

Fig. 5. Comparison of rate constants for phagocytosis (k_p) and killing (k_k) in the presence and absence of DPI. The two-step protocol was undertaken using neutrophils from a healthy donor (open bars) or the same neutrophils treated with 10 µM diphenylene iodonium (DPI) (closed bars), and k_p and k_k calculated. Each bar represents the mean and SD of four independent experiments. There is no significant difference between the k_p of normal and DPI-treated neutrophils. However, k_k is decreased by approx 84 % in the presence of DPI.

9. As the centrifugation speed is low, the pellet does not adhere strongly to the tube wall and care must be taken to avoid disturbing it as the supernatant is removed. A volume of approx 100–150 µL should be left in the tube after the first two spins, then as much possible removed after the final spin. The residual volume after the final spin should be taken into account when resuspending to the final volume of 2.5 mL.

Acknowledgments

The authors would like to thank Michael Bullock for development of the Excel data analysis file.

Fig. 4. (*Opposite page*) Calculation of rates of phagocytosis (k_p) and killing (k_k) of *Staphylococcus aureus* by neutrophils treated with diphenylene iodonium (DPI) using two-step protocol results. An experiment identical to that presented in **Fig. 3** was undertaken, except that the reaction was carried out in the presence of 10 µM DPI. Refer to the legend of **Fig. 3** for figure descriptions. In this example $k_p = 0.11$ min⁻¹, $t_{1/2} = 6.2$ min and $k_k = 0.01$ min⁻¹, $t_{1/2} = 69$ min.

References

1. Hampton, M. B., Kettle, A. J., and Winterbourn, C. C. (1998) Inside the neutrophil phagosome: oxidants, myeloperoxidase, and bacterial killing. *Blood* **92,** 3007–3017.
2. Reeves, E. P., Lu, H., Jacobs, H. L., et al. (2002) Killing activity of neutrophils is mediated through activation of proteases by K+ flux. *Nature* **416,** 291–297.
3. Roos, D. and Winterbourn, C. C. (2002) Immunology–lethal weapons. *Science* **296,** 669–671.
4. Hampton, M. B. and Winterbourn, C. C. (1999) Methods for quantifying phagocytosis and bacterial killing by human neutrophils. *J. Immunol. Meth.* **232,** 15–22.
5. Hampton, M. B., Vissers, M. C. M., and Winterbourn, C. C. (1994) A single assay for measuring the rates of phagocytosis and bacterial killing by neutrophils. *J. Leukoc. Biol.* **55,** 147–152.
6. Decleva, E., Menegazzi, R., Busetto, S., Patriarca, P., and Dri, P. (2006) Common methodology is inadequate for studies on the microbicidal activity of neutrophils. *J. Leukoc. Biol.* **79,** 87–94.
7. Cohn, Z. A. and Morse, S. I. (1959) Interactions between rabbit polymorphonuclear leucocytes and Staphylococci. *J. Exp. Med.* **110,** 419–443.
8. Leijh, P. C. J., van Furth, R., and Van Zwet, T. L. (1986) *In vitro* determination of phagocytosis and intracellular killing by polymorphonuclear and mononuclear phagocytes, in *Handbook of Experimental Immunology, Vol. 2.* Blackwell Scientific Publishers, Oxford, England, pp. 46.41–46.21.
9. Gargan, R. A., Brumfitt, W., and Hamilton-Miller, J. M. T. (1989) Failure of water to lyse polymorphonuclear neutrophils completely: role of pH and implications for assessment of bacterial killing. *J. Immunol. Meth.* **124,** 289–291.
10. Hamers, M. N., Bot, A. A. M., Weening, R. S., Sips, H. J., and Roos, D. (1984) Kinetics and mechanism of the bactericidal action of human neutrophils against *Escherichia coli. Blood* **64,** 635–641.

22

A Skin Chamber Technique as a Human Model for Studies of Aseptic Inflammatory Reactions

Per Follin and Claes Dahlgren

Summary

Using a combination of induced skin blistering and collection chambers permits dynamic studies of the aseptic inflammatory reaction in humans. Blisters filled with interstitial fluid can be generated by applying negative pressure to normal skin for up to 2 h. The blisters are subsequently denuded to form superficial "skin windows" that are well defined with regard to area and depth. The denuded areas are covered with a separate collection chamber filled with a suitable medium and left for 18–24 h. During this period, neutrophils and inflammatory agents accumulate in the chamber medium, and sequential events in the inflammatory process can be studied by repeated sampling. Inactive medium or isolated peripheral blood cells from the same individual can be used as controls for both cellular functions and the pro-/anti-inflammatory mediators that are generated or released.

Key Words: Aseptic; inflammation; leukocytes; cytokines; neutrophil migration in vivo; suction blister; skin chamber; exudate cells; exocytose; and chemoattractant.

1. Introduction

In response to an acute local inflammation, circulating neutrophils are recruited from the blood stream and accumulate at the affected site. This dynamic process, referred to as exudation, involves several neutrophil functions that are all essential for an intact and normally operating immune system. Several well characterized methods for studying neutrophil migration in vitro have been developed, but the mobility observed in these tests does not necessarily reflect the situation in vivo. The in vitro assays do not take into consideration changes in local blood flow and vascular permeability, interaction between leukocytes and capillary endothelium, tissue architecture, and the effect of endogenous mediators on migrating cells. Accordingly, some types of studies of the complexity of the exudation process are preferentially, or by necessity, performed in vivo, and,

From: *Methods in Molecular Biology, vol. 412: Neutrophil Methods and Protocols*
Edited by: M. T. Quinn, F. R. DeLeo, and G. M. Bokoch © Humana Press Inc., Totowa, NJ

as aspects such as species differences, it may be important to determine the response to an inflammatory challenge in human subjects who give informed consent. Various techniques to create a suitable "skin window" have been developed over the years, but those methods have generally entailed poor reproducibility and traumatization of the skin.

Introduction of a skin blistering procedure in the 1960s provided a standardized approach with minimized trauma. When a comparatively low negative pressure is applied to normal skin, the epidermis detaches from the underlying dermis *(1)*. The physical force simultaneously induces a loosening of the deeper epidermal structures, accumulation of fluids, and consistent formation of a cleft between the basal membrane and the basal lamina *(2–4)*. In that way, dermal papillae are exposed on the blister floor, without damaging their capillaries or causing bleeding. Blister formation *per se* triggers a transient inflammatory response, but the volume of the blister fluid is small and relatively few cells are recruited. To increase the number of experiments that can be performed with the blister material, several modifications have been made. We describe here methods for combining the blister technique with separate collection chambers, which allows improved sequential analysis of both soluble components and cellular contents at the inflammatory site. Selection of proper control cells and technical problems/risks are included.

2. Materials

1. Skin blistering suction chamber: blistering devices are primarily designed for dermatological research and are commercially available (Electronic Diversities, Finksburg, MD, http://www.electdiv.com), but they can also be made to order by local manufacturers. The type of acrylic suction chamber described here (**Fig. 1**) was custom-made at the hospital repair shop, and it is 40 mm in diameter and 18 mm high. The bottom surface of the chamber is flat, and it has three holes that have a diameter of 5 mm and are situated 8 mm apart in the shape of a triangle. These holes are connected by tubing to the vacuum pump.
2. Collection chamber: we currently use a custom-made acrylic collection chamber that has the same cylindrical shape and outer dimension as the suction chamber, but the collection holes (7 mm in diameter) are somewhat larger than the lesions (**Fig. 2**). Each collection chamber has a rubber gasket fixed in a groove on its lower surface to improve contact with the surface of the skin and to prevent leakage (*see* **Note 1**).
3. Vacuum pump: we currently use a portable handheld Dermovac pump (Ventipress Oy, Lappeenranta, Finland), which is made of Plexiglas and equipped with an aneroid gauge, a spring-operated air lock, a pressure regulating valve, and an outlet to the tubing and adjacent suction chamber(s). This pump has two advantages: it is very easy to operate, and it allows the subject to move freely (*see* **Note 2**).
4. Autologous serum (obtained by vein puncture).

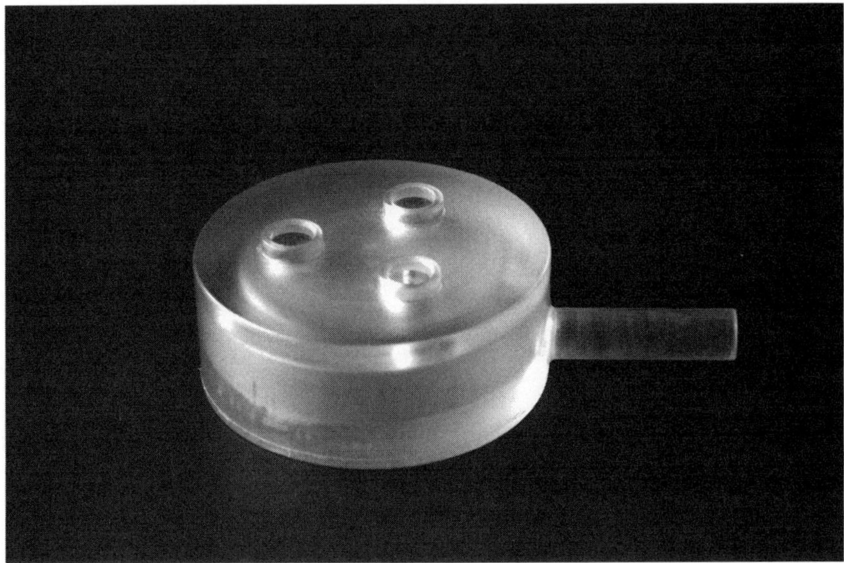

Fig. 1. Acrylic skin blistering suction chamber.

5. Autologous plasma (anticoagulated with citrate).
6. Hypoallergenic surgical tape (Millipore).
7. Sterilized surgical scissors and tweezers.
8. Sterilized skin chambers (**Fig. 2**).
9. Phosphate-buffered saline (PBS): 136 mM NaCl, 2.7 mM KCl, 7.7 mM Na$_2$HPO$_4$, and 1.5 mM KH$_2$PO$_4$, pH 7.3.
10. Krebs Ringer glucose buffer (KRG): 120 mM NaCl, 5 mM KCl, 1.7 mM KH$_2$PO$_4$, 8.3 mM Na$_2$HPO$_4$, 10 mM glucose, 1 mM CaCl$_2$, and 1.5 mM MgCl$_2$, pH 7.3.
11. Peripheral blood cells (i.e., controls) are isolated according to the technique described by Bøyum and are stored on melting ice pending further experiments.

3. Methods

3.1. Selection of Study Subjects

1. According to the WMA Declaration of Helsinki, all experiments performed on human subjects must be approved by an ethics committee. After receiving such approval, patients or healthy volunteers of appropriate age and sex can be approached and properly informed.
2. Once they have given informed consent, the experiments can be initiated.

3.2. Skin Blistering

1. Connect the suction chamber to the vacuum pump before mounting the chamber on subject's skin (*see* **Note 2**).

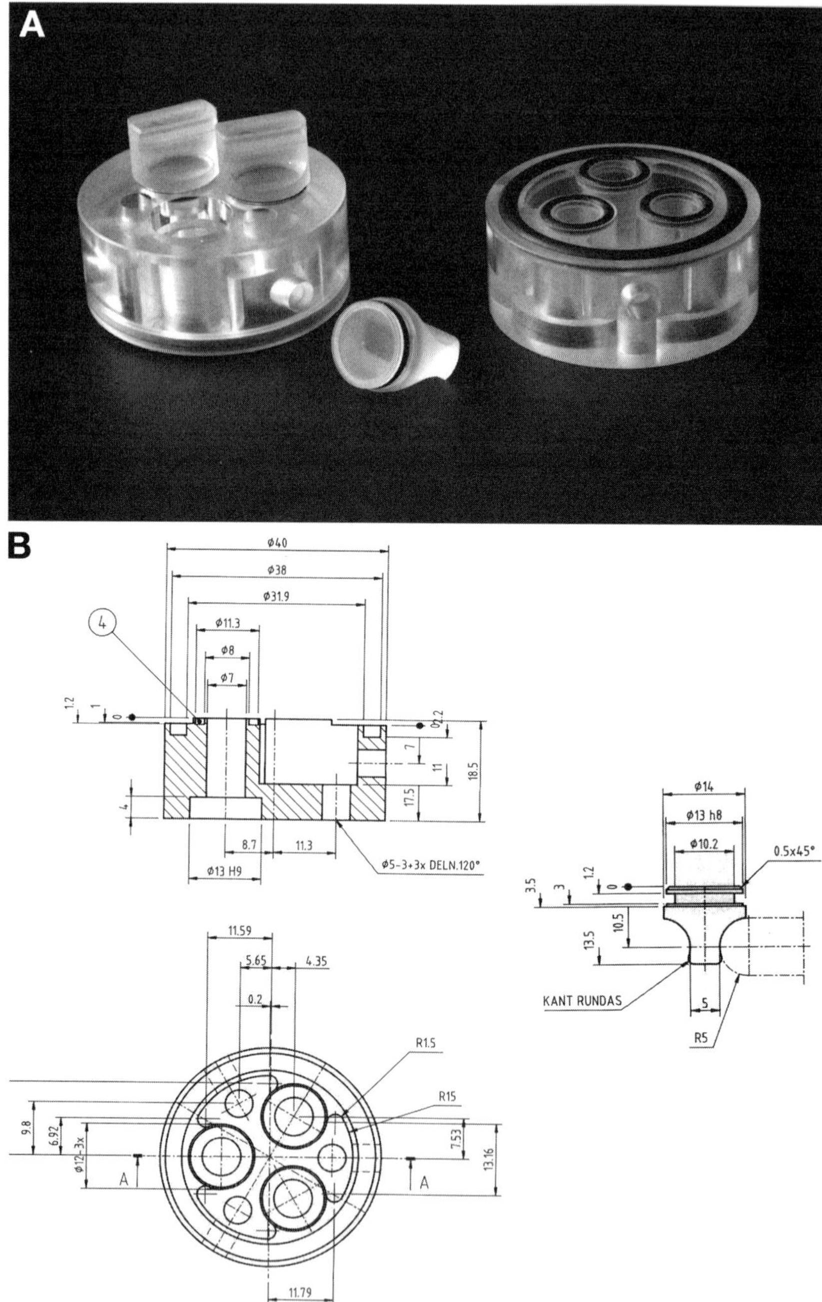

Fig. 2. Three-compartment acrylic collection chambers. (**A**) Each compartment is sealed at the upper end with a removable stopper and at the lower surface (facing the skin) with a rubber gasket. (**B**) Technical drawing of the collection chamber shown in **A**.

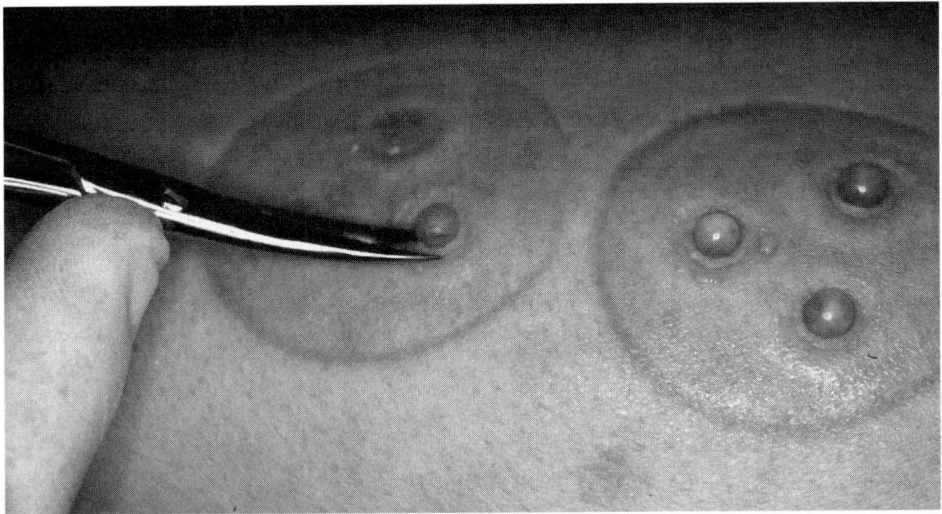

Fig. 3. Denudation of skin blisters using sterilized scissors and tweezers.

2. Select desired location to form blister. We prefer to induce skin blistering on the volar side of the nondominant forearm, which, in most normal-sized subjects, is large enough to fit a pair of suction chambers, if required (*see* **Note 3**).
3. To cause the chamber to adhere to the skin, manually apply light pressure on the chamber while the air is being pumped out. As the pressure is lowered, the suction chamber will stick to the skin, and the manual pressure can be decreased.
4. After the chamber is fixed, secure it more firmly with hypoallergenic surgical tape, and cover it with an elastic bandage (*see* **Note 4**).
5. Skin blisters can be induced within 2 h by applying a negative pressure of 200 mmHg (~27,000 Pascal or 0.27 Bar) in the suction chamber. The blister formation is painless, but the subject may experience a transient, local itching sensation during the epidermal detachment.
6. The time required to create blisters is inversely correlated with the vacuum pressure used (from approx 100 to 700 mm Hg) *(5)*. The pressure should be adjusted so that blistering occurs within a reasonable amount of time and without any barotraumas, such as capillary bleeding, bruising, petechia, edema, or simply blister rupture (*see* **Note 5**).

3.3. Unroofing Blisters

1. After approx 2 h of blister formation, the negative pressure is removed, and the suction chamber is detached
2. The roofs of the blisters are carefully removed with sterilized scissors and tweezers (**Fig. 3**). This leaves three uniform lesions (skin windows), each covering an area of approx 0.2 cm^2.

3.4. Application and Filling of Skin (Collection) Chambers

1. Place a collection chamber with holes that match the unroofed blisters (**Fig. 2**) over the skin lesions. Such chambers must be adequately sterilized between experiments to avoid any possible transmission of blood-borne infections from one subject to another (*see* **Note 6**).
2. Fill each compartment in the collection chamber with 0.5 mL of a medium designed to attract neutrophils. Autologous serum is often used as the attractant medium because it offers chemotactic potency and good cellular yield, but it may be overly potent in certain experimental settings (*see* **Note 7**).
3. The choice of attracting medium has a major impact on the accumulation of cells in the collection chamber (*see* **Note 8**).
4. Seal the compartments with tightly fitting stoppers that can easily be removed for sampling and/or addition of new medium during the experiment.
5. Secure the sealed collection chamber to the forearm with hypoallergenic surgical tape and an elastic bandage, and leave it in place for 18–24 h.
6. During the incubation period, neutrophils migrate into the chamber medium. The number of cells recovered in the chamber is correlated with both the area of the skin lesion *(6,7)* and the incubation time *(7–9)*. A larger skin lesion exposes a larger part of the dermal capillary bed and consequently offers a greater yield of exudate cells (*see* **Note 9**).
7. Sampling or refilling of the collection chamber can be performed during an experiment with the device still in place. Carefully remove the stoppers in the three compartments one by one to gain access to the chamber (*see* **Note 10**).

3.5. Collecting Neutrophil Exudates

1. Remove the skin chamber from the subject's arm after the desired incubation period. To prevent leakage during detachment of the device, apply manual pressure when the surgical tape is cut. Rotate the arm so the chamber is turned upside-down before it is finally detached from the skin.
2. Aspirate the medium containing the exudate neutrophils from the "bottom" of the chamber.
3. Rinse each chamber compartment twice with 0.5 mL of PBS, collect the rinse solution, and combine with the aspirate from **step 2**.
4. Nonexudate cells that are to serve as controls should be isolated and prepared under suitable conditions (*see* **Note 11**).

3.6. Analysis of Exudate Cell Number and Viability

1. Cell number can be evaluatecd by performing standard differential cell counts using Wright's stain (*see* **Note 12**).
2. Cell viability can be evaluated by microscopic analysis of cells suspended in trypan blue solution (*see* **Note 13**).

3.7. Specific Considerations, Technical Problems, and Risks

1. Risk for bacterial infection: viable bacteria are rarely found in the chamber medium *(7,8,10,11)*, and there are no reports of clinical infections. Therefore, it can be

concluded that the phagocytes that accumulate in the chamber can efficiently accomplish their antimicrobial mission, resulting in a negligible risk of bacterial infection.

2. Chamber leakage: there may be some leakage from the collection compartments, primarily as a result of improper fixation of the chamber or loose stoppers.
3. Local hypersensitivity: removal of the exudation chamber exposes the denuded area with bare nerve endings, which is sensitive to touch, just like any other superficial abrasion. The denuded area normally heals within a few days without scar formation.
4. Hyperpigmentation can occasionally occur, but generally vanishes within a year, whereas dyspigmentation is rarely seen after skin blistering. To reduce the risk of dyspigmentation, it may be preferable to induce very small blisters (approx 0.07 cm^2) *(7)*.

3.8. Analysis of Exudate Cell Function

Studies of human neutrophils isolated from an aseptic inflammation have revealed that, during exudation, neutrophils are functionally modulated in several ways as compared to such cells isolated in parallel from peripheral blood. A variety of functional analyses can be performed on human neutrophils isolated from skin chambers, including analysis of neutrophil granule mobilization (*see* **Note 14**), priming (*see* **Note 15**), respiratory burst activity (*see* **Note 16**), and down-upregulation of cellsurface receptors (*see* **Note 17**). The methods for these assays are described in elsewhere in this volume.

3.9. Analysis of Inflammatory Mediators in Exudate Fluid

Development of an aseptic inflammation in tissues is accompanied by accumulation of both inflammatory cells and inflammatory mediators, thus it is also possible to evaluate the kinetics and concentrations of endogenous chemotactic factors in the skin chamber fluid, such as complement factor (C)5a (*see* **Note 18**) and interleukin (IL)-8 (*see* **Note 19**), as well as the accumulation of granule-derived constituents (*see* **Note 20**).

4. Notes

1. It is possible that the experimental results can be influenced by the design of the collection chamber and/or the material of which it is made. Over the years, several such chambers have been developed in many different materials, including plastics *(12)*, modified bottle caps (also plastics) *(11)*, epoxy (Araldite M, Ciba-Geigy) *(6)*, plexiglas *(13)*, acrylic *(7,14–16)*, polycarbonate (Neuro Probe, Inc., Gaithersburg, MD) *(9,17)*, siliconized rubber (ordinary rubber stoppers for infusion bottles) *(8,18)*, siliconized glass *(19)*, and undefined metals *(20)*. However, no studies have been published that have attempted to compare collection chambers that have different designs or are made of different materials, and hence there are no sound scientific findings on which to base a choice in this context.

2. Other suitable pumps are also available, some of which can be equipped with heating (Electronic Diversities, Finksburg, MD, http://www.electdiv.com; Lincoln Customer Service, St. Louis, MO, http://www.mityvac.com.
3. The chambers can also be mounted on almost any part of the body, including the abdomen *(10,21)*. The time required for blisters to appear and the yield of exudate cells may vary between locations *(10)*.
4. In some cases (e.g., when a subject has very dry skin), it is necessary to tighten the attachment by applying a thin layer of petroleum jelly between the chamber and the skin.
5. Besides the applied negative pressure, blister formation is influenced by several other factors that might require consideration. There is obviously variation between individual subjects as well as between different parts of the body. The blistering time is also affected by the temperature of the skin *(22)*, thus it may be advised by some authors *(8,9,17)*, albeit not necessary, to use external heating.
6. Autoclaving is probably the most appropriate method, provided the material used to manufacture the collection chambers can withstand high temperatures and damp pressure. However, that is not always the case, which is why gas sterilization (ethylene oxide) may be preferable. The device should also be cleaned in a way that minimizes the risk of remaining traces of detergents, organic solvents, or antiseptics.
7. An investigation in which control neutrophils were incubated in serum has revealed substantial upregulation of adhesion molecules (MAC-1), which indicates that serum can induce exocytosis of easily mobilized intracellular components (gelatinase granules and secretory vesicles), whereas plasma does not *(16)*. Autologous ascitic fluid has also been used as a medium in a study of neutrophil function in relation to liver diseases *(23)*.
8. Autologous serum contains components generated by the coagulation and the complement system, which directly or indirectly attract inflammatory cells *(7–11, 13,14,17)*. Wandall *(6)* compared the efficacy of undiluted and diluted serum and concluded that the number of cells accumulating in the skin chambers is directly proportional to the degree of dilution. The attractants in serum are not inactivated by heating (56°C for 30–60 min) *(6,17,24)*, as long as the temperature does not get too high *(12)*. Compared to serum, plasma is a poor attractant *(25)*, and it may also be less suitable, because prolonged incubation (≥18 h) in plasma can result in coagulation and thereby loss of cells that are trapped in the clotted material (unpublished observation). Nevertheless, plasma can be used as collecting medium without pronounced reduction in neutrophil yield, if the chambers are first filled with serum for a certain length of time *(16)*. As an alternative, some investigators have chosen to use buffered isotonic saline in the collection chamber *(19,20)*.
9. Studies of the kinetics have shown that there is usually a lag phase of 2–4 h before the number of neutrophils increases in the collection chamber. It has been suggested that this early phase is related to events that take place in the endothelium rather than in the chamber medium *(12)*. The latency period is followed by a rapid rise in neutrophils to almost double the number within 18–24 h *(7–9)*. It is pos-

sible that some cells harvested after 24 h have been in the chamber fluid for up to 20 h, although the in-chamber time will be much shorter for a majority of the cells.

10. Longitudinal sampling can be performed to study the dynamics of accumulation of cells and inflammatory mediators *(17)*, and the medium can be changed according to the type of chemotactic activation that is required in the experimental setting. For example, after an initial 18 h period of exposure to serum used as the attractant to induce extravasation, the chamber fluid can be replaced with plasma for the final 6 h, which enables collection of migrating cells in a medium that causes less activation *(16)*.

11. To allow direct comparison with exudate cells, it is necessary to expose control cells isolated from peripheral blood to the same conditions as the exudate cells, with the exception of the extravasation process. This is not completely possible, because the contents of the chamber change dynamically during an experiment, and the exudate cells enter the chamber at different time points. Depending on the experimental setup used and the precise objective of a study, it will be necessary to select an appropriate chamber attractant (*see* **Note 8**) and adequately matched control cells, taking into account both the advantages and disadvantages of the options available.

12. Shortly after blister formation is induced (i.e., within 2 h), a small number of undefined cells are present in the blister fluid *(17)*. Studies of the longitudinal changes in differential cell count in the contents of the skin chamber compartments revealed that mononuclear cells appear during the early phase (first 8 h), but these cells are soon (within 10–24 h) outnumbered by the predominate polymorphonuclear cells (representing about 90–98% of all cells detected) *(7,8,17)*.

13. A large fraction (92–98%) of neutrophils that accumulate in skin chambers 10–24 h after application are resistant to trypan blue staining, which indicates preserved viability *(8,11)*. A small proportion of the neutrophils (<1%) collected after 6 h exhibit cytoplasmic vacuolization, and the proportion is gradually increased to around 5% after 24 h *(8)*. The observed vacuolization may indicate the induction of an apoptotic rather than a necrotic process, but proper studies of apoptosis in exudate cells have not yet been performed to clarify that issue. The rather crude viability tests suggest that most cells found in the chamber medium are viable, and that assumption is supported by the reported absence of lactate dehydrogenase (LDH) in the chamber fluid *(26)*. The composition of the medium is also highly important for neutrophil viability, as illustrated by a study in which viability (determined by trypan blue exclusion) decreased to 50–80% when saline medium was used *(11)*.

14. In the course of neutrophil migration, subcellular granules are released. Sequential mobilization of granules and vesicles has been demonstrated in vitro, which indicates a hierarchy among these intracellular components. Investigations of the release of specific and azurophilic granules have shown that these elements are also released in a particular order in vivo *(15,18,26,27)*. Nevertheless, a somewhat more complex approach is needed to study the release in vivo of other; more easily mobilized constituents, such as gelatinase granules and the even more easily acti-

vated secretory vesicles. The autologous serum in skin chambers must be replaced with autologous plasma during the last 6 h of incubation because the inflammatory mediators present in serum can mobilize granule/vesicle subsets independent of the extravasation process. Analysis of cell-surface exposed granule markers by flow cytometry (fluorescence-activated cell sorting [FACS]) can be combined with investigation of granule matrix components found in both exudate and peripheral blood cells, as well as in the collection medium. We have conducted a study using this experimental strategy *(16)* and found that the proportions of granules released by neutrophils obtained through exudation in vivo occurred in reverse order compared to the formation seen during differentiation. More precisely, all secretory vesicles and around 40% of the gelatinase granules, 20% of the specific granules, and 10% of the azurophilic granules were released. Furthermore, we found that the cell-surface glycoprotein L-selectin (which mediates neutrophil rolling along the endothelial cells prior to extravasation) is completely shed on cells after exudation in vivo, an observation that has also been made by other researchers *(28–30)*. Such casting off of L-selectin differs between healthy individuals and patients undergoing dialysis as a result of renal disease, but the mechanism underlying this dissimilarity has not yet been identified *(29,30)*. Experiments focused on adhesion receptor subunits have demonstrated that extravasation into skin chambers upregulates the α2β1 integrin (VLA-2) *(31)*, and that finding supports a role for this receptor in neutrophil migration in interstitial tissues *(31,32)*.

15. Exposing neutrophils in vitro to a low, nonactivating concentration of an agonist prior to stimulation with a heterologous agent can induce an increased metabolic response, a phenomenon that is often referred to as priming. Cellular reactions to chemoattractants can also be downregulated by desensitization *(33)*. Parallel quantification of the respiratory burst response in neutrophils from exudate and peripheral blood has shown that the exudate cells are metabolically primed for activation by the chemotactic peptides fMLF and WKYMVM, which operate through the formyl peptide receptor (FPR) and the closely related FPRL1, respectively *(14,15,18,27, 34–37)*. This primed response is agonist-specific, as illustrated by the metabolic reaction to phorbol myristate acetate (PMA), an agent that bypasses cell-surface receptors and directly activates protein kinase C (PKC) *(34,38)*. The responses to PMA are of equal magnitude in both types of neutrophils, even though the lag phase, which precedes the membrane depolarization associated with cell activation, is shorter in the exudate cells. The latter phenomenon may be due to translocation (but not activation) of cytosolic PKC to the plasma membrane *(38)*.

16. Granule membranes contain a number of different receptors, including the opsonin receptor CR3 and the chemoattractant receptors FPR and FPRL1 *(39)*. Therefore, granule mobilization is accompanied by exposure of new receptors originating from these stores. It is an attractive idea that receptor mobilization through fusion of granule membranes with the plasma membrane may be a mechanism behind the primed response in exudate cells. Indeed, it has been observed that increased exposure of CR3 and FPR on the surface of exudate neutrophils is associated with greater responsiveness to ligands that bind to these receptors *(27,34)*.

17. Chemotactic peptides IL-8 and C5a are poor activators of exudate neutrophils, suggesting the cells are desensitized to these agonists during extravasation *(36)*. Because both IL-8 and C5a can be detected in skin chamber medium, this deactivation is probably due to homologous receptor desensitization *(36)*. Exudate neutrophils, but not neutrophils isolated from peripheral blood, are activated by the β-galactoside-binding proinflammatory lectins galectin-1 and -3. Blood neutrophils also become responsive to galectins, if they are primed with known secretagogues, and it has been proposed that the mechanism behind this effect is receptor upregulation by degranulation, with CEACAM being a candidate receptor *(40–42)*. The switch from galectin insensitivity to sensitivity during the extravasation process supports the concept that receptor upregulation by granule mobilization is an important event that enhances the responsiveness of (primes) exudate cells.

18. The potency of endogenous inflammatory mediators derived from C5a during activation of the complement system is well known from experiments performed both in vitro and in vivo *(43)*. Serum contains only small amounts of C5a, but analyses have shown that substantial activation occurs in the skin chamber medium, detected as a significant increase in the amount of $C5a/C5a_{des\ Arg}$ during the process of extravasation *(17,36)*. A study of the kinetics by longitudinal sampling of aliquots of chamber medium has suggested that the main chemotactic effect of C5a is exerted during the initial phase of the exudative process *(17)*. Scheja and Forsgren *(43)* analyzed the arachidonic-acid-derived attractant LTB_4 by high-performance liquid chromatography (HPLC) but could not detect that substance in skin chamber medium after 24 h of incubation. However, those investigators did find a small amount of the dicarboxylic acid metabolite (20-COOH LTB_4). By comparison, Kuhns et al. *(28)* used a very sensitive assay system and, during 24 h of incubation, found a gradual accumulation of LTB_4 but did not detect leukotriene/s in early blister fluid.

19. Early studies in vitro demonstrated the chemotactic potency of tissue-derived IL-8. The role of this peptide as an important chemoattractant during in vivo exudation was suggested when it was found in significant amounts in human skin chamber medium *(36)*. Longitudinal measurements have shown that production of IL-8 is most pronounced after 8–12 h of skin chamber application, although small amounts can be found in the chamber medium after only a few hours *(17)*. Several different cell lines, including fibroblasts, endothelial cells, epithelial cells, monocytes, and neutrophils *(44)*, can generate IL-8 upon appropriate stimulation. Larger amounts of IL-8 are found in exudate neutrophils than in neutrophils isolated from peripheral blood, indicating that exudation leads to increased production of and/or association with this cytokine (bound either to the cell surface or to subcellular compartments). The amount of cell-coupled IL-8 also rises during prolonged incubation of exudate neutrophils in serum *(24)*, which suggests that these leukocytes have the capacity to synthesize IL8.

20. In addition to endogenous chemotactic factors, substantial amounts of various neutrophil granule constituents are released and can be detected in the skin chamber medium *(16)*. These factors are no doubt involved in killing contaminating

microbes and initiating processes included in tissue repair. In support of that assumption, researchers studying gene expression in exudate neutrophils have observed upregulation of several genes grouped in a pattern that suggests that exudation can promote wound healing *(45)*.

References

1. Kiistala, U. and Mustakallio, K. K. (1964) In-vivo separation of epidermis by production of suction blisters. *Lancet* **1**, 1444–4445.
2. Bork, K. (1978) Physical forces in blister formation. The role of colloid osmotic pressure and of total osmolarity in fluid migration into rising blister. *J. Invest. Derm.* **71**, 209–212.
3. Kiistala, U. and Mustakallio, K. K. (1967) Dermo-epidermal separation with suction. Electron microscopic and histochemical study of cellular dynamics of initial events of blistering on human skin. *J. Invest. Derm.* **48**, 466–477.
4. Hunter, J., McVittie, E., and Comaish, J. S. (1974) Light and electron microscopic studies of physical injury to the skin. I. Suction *Br. J. Dermatol.* **90**, 481–490.
5. Lowe, B. L. and van der Leuen, J. C. (1968) Suction blisters and dermal-epidermal adherence. *J. Invest. Dermatol.* **50**, 308–314.
6. Wandall, J. H. (1980) Leukocyte mobilization to skin lesions. *Acta. Pathol. Microbiol. Immunol. Scand. (C)* **88**, 255–261.
7. Scheja, A. and Forsgren, A. (1985) A skin chamber technique for leukocyte migration studies; description and reproducibility. *Acta. Pathol. Microbiol. Immunol. Scand. (C)* **93**, 25–30.
8. Hellum, K. B. and Solberg, C. O. (1977) Human leukocyte migration: studies with an improved skin chamber technique. *Acta. Pathol. Microbiol. Immunol. Scand. (C)* **85**, 413–423.
9. Zimmerli, W. and Gallin, J. I. (1987) Monocytes accumulate on Rebuck skin window coverslips but not in skin chamber fluid. A comparative evaluation of two in vivo migration models. *J. Immunol. Meth.* **96**, 11–17.
10. Koivuranta-Vaara, P. (1985) Neutrophil migration in vivo: analysis of a skin window technique. *J. Immunol. Meth.* **79**, 71–78.
11. Senn, H., Holland, J. F., and Banerjee, T. (1969) Kinetic and comparative studies on localized leukocyte mobilization in normal man. *J. Lab. Clin. Med.* **74**, 742–756.
12. Gowland, E. (1964) Studies on the emigration of polymorphonuclear leukocytes from skin lesions in man. *J. Pathol. Bacteriol.* **87**, 347–352.
13. Mass, M. F., Dean, P. B., Weston, W. L., and Humbert, J. R. (1975) Leukocyte migration in vivo: a new method of study. *J. Lab. Clin. Med.* **86**, 1040–1046.
14. Scheja, A. and Forsgren, A. (1985) Functional properties of polymorphonuclear leukocyter accumulated in a skin chamber. *Acta. Pathol. Microbiol. Immunol. Scand. (C)* **93**, 31–36.
15. Follin, P., Briheim, G., and Dahlgren, C. (1991) Mechanisms in neutrophil priming: characterization of the oxidative response induced by formylmethionyl-leucyl-phenylalanine in human exudated cells. *Scand. J. Immunol.* **34**, 317–322.

16. Sengeløv, H., Follin, P., Kjeldsen, L., Lollike, K., Dahlgren, C., and Borregaard, N. (1995) Mobilization of granules and secretory vesicles during in vivo exudation in human neutrophils. *J. Immunol.* **154,** 4157–4165.
17. Kuhns, D. B., DeCarlo, E., Hawk, D. M., and Gallin, J. I. (1992) Dynamics of the cellular and humoral components of the inflammatory response elicited in skin blisters in human. *J. Clin. Invest.* **89,** 1734–1740.
18. Wright, D.G. and Gallin, J.I. (1979) Secretory responses of human neutrophils: exocytosis of specific (secondary) granules by human neutrophils during adherence in vitro and during exudation in vivo. *J. Immunol.* **123,** 285–294.
19. Perillie, P. E. and Finch, S. C. (1964) Quantitative studies of the lokal exudative cellular reaction in acute leukemia. *J. Clin. Invest.* **43,** 425–430.
20. Southam, C. M. and Levin, A. G. (1966) A quantitative Rebuck technique. *Blood* **27,** 734–738.
21. Kiistala, U. (1968) Suction blistering device for separation of viable epidermis from dermis. *J. Invest. Derm.* **50,** 129–137.
22. Kiistala, U. (1972) Dermal-epidermal separation. II. External factors in suction blister formation with special reference to the effect of temperature. *Ann. Clin. Res.* **4,** 236–246.
23. Fiuza, C., Salcedo, M., Clemente, G., and Tellado, J. M. (2000) In vivo neutrophil dysfunction in cirrhotic patients with advanced liver disease. *J. Infect. Dis.* **182,** 526–533.
24. Kuhns, D. B. and Gallin, J. I. (1995) Increased cell-associated IL-8 in human exudative and A23187-treated peripheral blood neutrophil. *J. Immunol.* **154,** 6556–6562.
25. Elmgreen, J. (1985) Neutrophil recruitment in skin window chambers–activation by complement. *Acta Pathol. Microbiol. Immunol. Scand. (C)* **93,** 139–142.
26. Scheja, A., Forsgren, A., and Ohlsson, K. (1986) Kinetics of enzymes released from polymorphonucler leucocytes in a skin chamber. *Clin. Exp. Rheum.* **4,** 37–41.
27. Zimmerli, W., Seligmann, B., and Gallin, J. I. (1986) Exudation primes human and guinea pig neutrophils for subsequent responsiveness to chemotactic peptide *N*-formylmethionylleucylphenylalanine and increases complement componen C3bi receptor expression. *J. Clin. Invest.* **77,** 925–933.
28. Kuhns, D. B., Long Priel, D. A., and Gallin, J. I. (1995) Loss of L-selectin (CD62L) on human neutrophils following exudation *in vivo. Cell. Immunol.* **164,** 306–310.
29. Dadfar, E., Lundahl, J., and Jacobson, S. H. (2004) Granulocyte extravasation and recruitment to sites of interstitial inflammation in patients with renal failure. *Am. J. Nephrol.* **24,** 330–339.
30. Dadfar, E., Lundahl, J., Fernvik, E., Nopp, A., Hylander, B., and Jacobson, S. H. (2004) Leukocyte CD11b and CD64L expression in response to interstitial inflammation in CAPD patients. *Perit. Dial. Int.* **24,** 28–36.
31. Werr, J., Johansson, J., Eriksson, E. E., Hedqvist, P., Ruoslahti, E., and Lindblom, L. (2000) Integrin $\alpha2\beta1$ (VLA-2) is a principal receptor used by neutrophils for locomotion in extravascular tissue. *Blood* **95,** 1804–1809.

32. Lindblom, L. and Werr, J. (2002) Integrin-dependent neutrophil migration in extravascular tissue. *Semin. Immunol.* **14,** 115–121.

33. Fu, H., Karlsson, J., Bylund, J., Movitz, C., Karlsson, A., and Dahlgren, C. (2006) Ligand recognition and activation of formyl peptide receptors in neutrophils. *J. Leukoc. Biol.* **79,** 247–256.

34. Briheim, G., Coble, B., Stendahl, O., and Dahlgren, C. (1988) Exudated polymorphonuclear leukocytes isolated from skin chambers are primed for enhanced response to subsequent stimulation with chemoattractant f-Met-Leu-Phe and C3-opsonized yeast particles. *Inflammation* **12,** 141–152.

35. Follin, P. and Dahlgren, C. (1990) Altered O_2^-/H_2O_2 production ratio by in vitro and in vivo primed human neutrophils. *Biochem. Biophys. Res. Commun.* **167,** 970–976.

36. Follin, P. Wymann, M. P., Dewald, B., Ceska, M., and Dahlgren, C. (1991b) Human neutrophil migration into skin chambers is associated with production of NAP-1/IL-8 and C5a. *Eur. J. Haematol.* **47,** 71–76.

37. Dahlgren, C., Christophe, T., Boulay, F., Madianos, P. N., Rabiet, M. J., and Karlsson, A. (2000) The synthetic chemoattractant Tsp-Lys-Tyr-Met-Val-DMet activates neutrophils preferentially through the lipoxin A_4 receptor. *Blood* **95,** 1810–1818.

38. Zimmerli, W., Seligmann, B. E., and Gallin, J. I. (1987) Neutrophils are hyperpolarized after exudation and show an increased depolarization response to formyl-peptide but not to phorbol myristate acetate. *Eur. J. Clin. Invest.* **17,** 435–441.

39. Bylund, J., Karlsson, A., Boulay, F., and Dahlgren, C. (2002) Lipopolysaccharide-induced granule mobilization and priming of the neutrophil response to *Helicobacter pylori* peptide Hp(2-20), which activates formyl peptide receptor-like 1. *Infect. Immun.* **70,** 2908–2914.

40. Karlsson, A., Follin, P., Leffler, H., and Dahlgren, C. (1998) Galectin-3 activates the NADPH-oxidase in exudated but not peripheral blood neutrophils. *Blood* **91,** 3430–3438.

41. Almkvist, J., Dahlgren, C., Leffler, H., and Karlsson, A. (2002) Activation of the neutrophil nicotinamide adenine dinucleotide phosphate oxidase by Galectin-1. *J. Immunol.* **168,** 4034–4041.

42. Feuk-Lagerstedt, E., Samuelsson, M., Mosgoeller, W., et al. (2002) The presence of stomatin in detergent-insoluble domains of neutrophil granule membranes. *J. Leukoc. Biol.* **72,** 970–977.

43. Scheja, A. and Forsgren, A. (1986) Effects of LTB_4 and its isomers on human leucocyte migration into skin chambers. *Clin. Exp. Rheum.* **4,** 323–329.

44. Pellmé, S., Mörgelin, M., Tapper, H., Mellqvist, U-H., Dahlgren, C., and Karlsson, A. (2006) Lokalization of human neutrophil interleukin-8 (CXCL-8) to organell(s) distinct from the classical granules and secretory vesicles. *J. Leukoc. Biol.* **79,** 1–10.

45. Tielgaard-Mörch, K., Knudsen, S., Follin, P., and Borregaard, N. (2004) The transcriptional activation program of human neutrophils in skin lesions support their important role in wound healing. *J. Immunol.* **172,** 7684–7693.

VI

NADPH Oxidase
and Production of Reactive Oxygen Species

23

Measurement of Respiratory Burst Products Generated by Professional Phagocytes

Claes Dahlgren, Anna Karlsson, and Johan Bylund

Summary

Activation of professional phagocytes, potent microbial killers of our innate immune system, is associated with an increase in cellular consumption of molecular oxygen (O_2). The burst of O_2 consumption is utilized by an NADPH-oxidase to generate highly-reactive oxygen species (ROS) starting with one and two electron reductions to generate superoxide anion (O_2^-) and hydrogen peroxide (H_2O_2), respectively. ROS are strongly bactericidal but may also cause tissue destruction and induce apoptosis in other immune competent cells of both the innate and the adaptive immune systems. Thus, the development of basic techniques to measure/quantify ROS generation/release by phagocytes during activation of the respiratory burst is of great importance, and a large number of techniques have been used for this purpose. Three of these techniques, chemiluminescence amplified by luminol/isoluminol, the absorbance change following reduction of cytochrome c, and the fluorescence increase upon oxidation of p-hydroxyphenylacetate, are described in detail in this chapter. These techniques can be valuable tools in research spanning from basic phagocyte biology to more clinically-oriented research on innate immune mechanisms and inflammation.

Key Words: Reactive oxygen species; superoxide; hydrogen peroxide; myeloperoxide; intracellular NADPH-oxidase activity; plasma membrane NADPH-oxidase activity; subcellular granules; chemiluminescence; cytochrome c reduction; PHPA oxidation.

1. Introduction

Professional phagocytes of the innate immune system increase their consumption of O_2 during phagocytosis of microbial intruders or upon interaction with inflammatory mediators. The electron transport system responsible for O_2 consumption, the NADPH-oxidase, is essential for protection against invading pathogens, as demonstrated by the susceptibility of persons with a nonfunctional oxidase (e.g., suffering from chronic granulomatous disease [CGD]) to bacterial and fungal infections *(1,2)* (*see* **Note 1**).

From: *Methods in Molecular Biology, vol. 412: Neutrophil Methods and Protocols*
Edited by: M. T. Quinn, F. R. DeLeo, and G. M. Bokoch © Humana Press Inc., Totowa, NJ

In resting phagocytes of healthy individuals, the cytosolic components of the oxidase are separated from the membrane components. During phagocyte activation, the cytosolic and membrane components assemble to form a functional multi-component electron-transfer system, which catalyzes the reduction of molecular O_2 at the expense of cytosolic NADPH *(3)*. Electrons are ferried from the cytosol across a membrane and delivered to O_2 present in an intracellular compartment (e.g., a phagosome) or extracellularly *(3,4)*, although the intracellular signals responsible for assembly of the active oxidase are in many respects unknown. The primary product of the NADPH oxidase, superoxide anion (O_2^-), and its initial metabolite, hydrogen peroxide (H_2O_2), are not sufficiently reactive to account for the bactericidal effects. Indeed, these ROS are metabolized into other reactive oxygen metabolites that are strongly antimicrobial (*see* **Note 2**). On the other hand, the precise role of ROS in microbial killing is a still matter of debate *(2,5,6)*. Thus, it is of great importance to have access to adequate techniques for analysis of the generation/release of phagocyte respiratory burst products, not only in the extracellular milieu, but also inside these cells. We describe here some basic, easy to perform techniques that are frequently used to follow cellular production of O_2^- or H_2O_2.

2. Materials

1. 5 *M* NaOH stock solution.
2. 5-amino-2,3-dihydro-1,4-phtalazinedione (Luminol) solution: prepare stock solution by dissolving luminol powder (Sigma) to 50 m*M* in freshly-prepared 0.1 *M* NaOH (from the 5-*M* stock). Luminol stock solution can be stored at room temperature, protected from light, for several months. Prepare 500 µ*M* luminol working solution fresh on the day of the experiment.
3. 6-amino-2,3-dihydro-1,4-phtalazinedione (Isoluminol) solution: prepare stock solution by dissolving isoluminol powder (Sigma) to 50 m*M* in freshly prepared 0.1 *M* NaOH (from the 5-*M* stock). Isoluminol stock solution can be stored at room temperature, protected from light, for several months. Prepare 500 µ*M* isoluminol working solution fresh on the day of the experiment.
4. Horseradish peroxidase (HRP) solution: dissolve HRP powder (Boehringer-Mannheim) in physiological saline to 80 U/mL. Store in small aliquots at −20°C.
5. p-hydroxyphenylacetate (PHPA) solution: dissolve PHPA powder (Sigma) in physiological saline to 10 mg/mL. Store in small aliquots at −70°C.
6. Superoxide dismutase (SOD) solution: dissolve SOD powder (Boehringer-Mannheim) in physiological saline to 5000 U/mL. Store in small aliquots at −70°C.
7. Catalase solution: wash particulate catalase (Boehringer-Mannheim) twice by suspending in H_2O and centrifugation at 15,000g for 10 min. Dissolve the final pellet in physiological saline to 200,000 U/mL. Store in small aliquots at −70°C.
8. Sodium azide (NaN_3) solution: dissolve NaN_3 (Sigma) in KRG. NaN_3 solution can be stored at 4°C for several months.

9. Cytochrome c (cytC) solution: dissolve cytC powder (Boehringer-Mannheim) in KRG at 15 mg/mL. Prepare solution fresh on the day of the experiment.
10. Fresh blood or cells in culture
11. Equipment for cell separation
12. Chemiluminometer, preferentially equipped with a temperature-controlled sample holder.
13. Photometer, preferentially equipped with a temperate cuvet holder.
14. Fluorometer, preferentially with a temperate cuvet holder.
15. Neutrophils purified from peripheral blood by standard techniques (or any cell that can generate ROS). Store the cells on melting ice until use.
16. A balanced-salt solution, such as Krebs-Ringer phosphate buffer: 120 mM NaCl, 5 mM KCl, 1.7 mM KH$_2$PO$_4$, 8.3 mM Na$_2$HPO$_4$, 10 mM glucose, 1 mM CaCl$_2$, and 1.5 mM MgCl$_2$, pH 7.3.

3. Methods

A large number of techniques have been developed over the years to measure the cellular production of different ROS (**Table 1**). Ideally, techniques for measuring cellular production of ROS should: (1) be specific for a particular oxygen metabolite, (2) be sensitive, (3) not interfere with cellular function, (4) distinguish the localization of the ROS production, (5) not require specialized laboratory equipment that is expensive or complicated to handle, and (6) be easy to standardize. No single technique has hitherto been found that satisfy all of these criteria (*see* **Note 3**). The drawbacks of the various techniques differ, and more than one technique usually must be included in the methodological repertoire of oxygen radical scientists. Three of the techniques that we have used to characterize respiratory burst activity in human phagocytes are described here in detail, and these methods serve as valuable tools in both basic and clinically-oriented research involving phagocyte function.

3.1. Detection of Extracellular ROS

It is generally assumed that the NADPH-oxidase is assembled and activated either in the plasma membrane or in the membranes of internalized phagosomes *(7,8)*. Reactive oxygen species (ROS) generated will then either be released from the cells (activation in the plasma membrane) or be retained inside the phagocyte (activation in the granule phagosomal membrane). Thus, it is important to measure both intra- and extracellular ROS production by phagocytes treated with various activating stimuli.

3.1.1. Extracellular O$_2^-$ Detection by Isoluminol-amplified Chemiluminescence

There are several dyes that after being excited by ROS release energy in the form of light (i.e., chemiluminescence). Among these dyes, the membrane-

Table 1
Techniques Used for Measuring Cellular Production of Different Reactive Oxygen Species

Technique	Measuring principle	Cellular localization	Comment
Superoxide anion			
Photometry	SOD-inhibitable reduction of cytochrome c	extracellular	Easy to follow kinetics of the response, provided that the stimulus is non-particulate; H_2O_2 may interfere with the assay; low sensitivity.
Luminometry	Peroxidase-dependent isoluminol-amplified chemiluminescence	extracellular	High sensitivity; easy to follow kinetics of the response; detects O_2^- despite the requirement for a peroxidase.
	Lucigenin-amplified chemiluminescence	extracellular	High sensitivity, but less than the isoluminol system; easy to follow kinetics of the response.
Precipitation reaction	NBT reduction	intracellular	Simple to count the number of positive cells microscopically, but laborious to make quantitative; should include SOD and catalase to remove released metabolites.
Hydrogen peroxide			
Fluorometry	Peroxidase-dependent oxidation of PHPA ± azide	entracellular intracellular	Fluorescence increases, making kinetics easy to follow; SOD is required for conversion of O_2^- to H_2O_2; low sensitivity. NaN_3 inactivates MPO and catalase, thus allowing H_2O_2 generated intracellularly to leak out and be detected extracellularly.
	Peroxidase-dependent oxidation of scopoletin ± azide	extracellular intracellular	Fluorescence decreases, making kinetics more difficult to follow; SOD is required for conversion of O_2^- to H_2O_2; higher sensitivity than the PHPA system. See above regarding NaN_3.

Nonidentified oxygen radical

Precipitation reaction	DAB oxidation	intracellular	Simple to count the number of positive cells microscopically, laborious to make quantitative; should include SOD and catalase to remove released metabolites.
Fluorometry	Oxidation of 2,7-dichlorofluorescein or dihydrorhodamine	intracellular	Many different oxidants can change the fluorescence of these substrates, making it difficult to use the technique quantitatively; difficult to follow kinetics.
Luminometry	Peroxidase-dependent luminol-amplified chemiluminescence	intracellular	Possibly detects O_2^-; high sensitivity; easy to follow kinetics; dependent on extracellular peroxidase; should include SOD and catalase to remove extracellularly produced metabolites.

SOD, superoxide dismutase; NBT, p-Nitroblue tetrazolium; PHPA, p-hydroxyphenylacetate; MPO, myeloperoxidase; DAB, 3',3'-Diaminobenzidine.

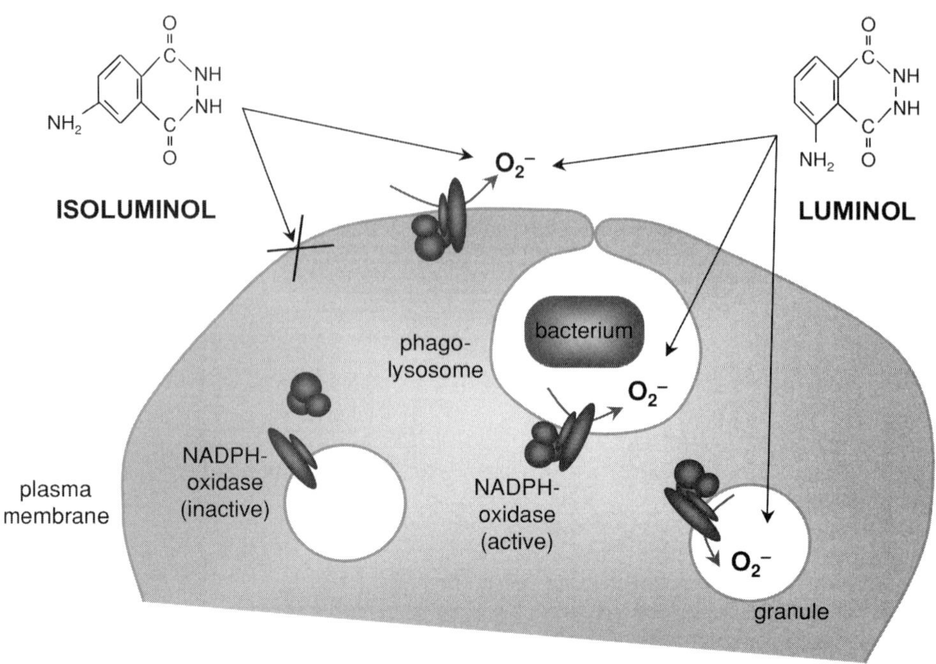

Fig. 1. Molecular structure and membrane permeability of luminol and isoluminol.

permeable luminol (**Fig. 1**) is the most intensively investigated and most frequently used in the free radical research field (*see* **Note 4**). Luminol is an activity amplifier that decreases the detection limit substantially, thus making the technique very sensitive (*see* **Note 5**). Changing the position of the amino group in the phthalate ring of luminol does not change the ability of the molecule to detect ROS. However, moving the amino group away from the first carbon atom in the aromatic ring makes the molecule more hydrophilic and less able to transverse biological membranes. Thus, isoluminol (**Fig. 1**) can be used to exclusively measure ROS *released* from activate cells *(9,10)*. Although this method provides a relative measure of ROS, it does not quantify actual moles of O_2^- present in the sample.

1. Add 600 µL of Krebs Ringer glucose buffer (KRG) to Eppendorf tubes or luminometer tubes for each sample (*see* **Note 6**).
2. Add 100 µL of fresh 500 µ*M* isoluminol solution.
3. Add 100 µL of HRP solution (*see* **Note 7**).
4. Add 100 µL of cells in KRG (10^2–10^6 cells, depending on the activity of the cells and the sensitivity of the luminometer) (*see* **Note 8**).
5. Equilibrate samples to the desired temperature.

6. Add 100 μL of a stimulus dissolved in KRG to activate the cells (*see* **Note 9**).
7. Measure chemiluminescence with an ordinary β-counter or one of the many luminometers present on the market (preferentially equipped with a thermostated sample holder), including those adapted to microtiter plates (*see* **Note 10**).

3.1.2. Extracellular O_2^- Detection by Reduction of Cytochrome c

ROS can reduce a number of different substrates, and techniques that exploit the absorbance change of chromogenic substrates are commonly used to measure ROS production. Here, we describe a simple and highly-reproducible technique that relies on the reduction of the membrane-impermeable substrate cytC by O_2^-. The reduction of cytC can be detected by the change in absorbance at 550 nm with a spectrophotometer. As there is a one-to-one molar stoichiometry between O_2^- produced and cytochrome c molecules reduced, the actual amount of O_2^- produced can be quantified with this technique.

1. Prepare two cuvets for each sample (i.e., a sample and a reference cuvet) (*see* **Note 6**).
2. Add 700 μL of KRG to the sample cuvet and 690 μL of KRG to the reference cuvet.
3. Add 100 μL of cytC solution to both cuvets.
4. Add 10 μL of SOD solution to the reference cuvet only (*see* **Note 11**).
5. Add 100 μL of cells ($\geq 5 \times 10^5$ cells) in KRG to both cuvets.
6. Equilibrate both samples to the desired temperature in a thermostated cuvet holder or water bath.
7. Add 100 μL of a stimulus dissolved in KRG to activate the cells (*see* **Note 9**).
8. Continuously monitor the change is absorbance at 550 nm in a spectrophotometer (preferentially equipped with a thermostated sample holder) (*see* **Note 10**).
9. Determine the change in absorbance (ΔOD_{550}) for each sample by subtracting the absorbance of the reference cuvet containing SOD from that of the sample cuvet for each sample for each time point.
10. Calculate the molar amount of O_2^- generated per unit time and volume using the Beer-Lambert law with an extinction coefficient (ε) of 21.1 mM^{-1}cm^{-1} for reduced cytC at 550 nm. For a standard 1 mL assay containing 10^6 cells in a cuvet with a 1-cm light path, the amount of O_2^- generated can easily be calculated using **eq. 1** (*see* **Notes 12** and **13**):

$$\Delta OD_{550} \times 47.4 = \text{nmoles } O_2^-/10^6 \text{ cells/time unit} \qquad (1)$$

11. When a particulate prey (e.g., bacteria) is used to activate the cells, a discontinuous assay system has to be used in which the samples are centrifuged (to remove the bacteria that could otherwise scatter the light) and O_2^- production is determined from the sample and reference supernatants (*see* **Note 14**).

3.1.3. Extracellular H_2O_2 Detection by PHPA Fluorescence

Membrane-impermeable PHPA is oxidized by H_2O_2 in the presence of HRP (which also does not penetrate intact phagocytes), resulting in an increase in fluorescence. Production of H_2O_2 can subsequently be followed continuously using a fluorometer (preferentially one with a thermostated cuvette holder), including microtiter plate fluorometers.

1. Add 550 µL of KRG to Eppendorf tubes for each sample (*see* **Note 6**).
2. Add 100 µL of PHPA solution.
3. Add 50 µL of HRP solution.
4. Add 100 µL of SOD solution (*see* **Note 15**).
5. Add 100 µL of cells ($\geq 5 \times 10^5$ cells) in KRG.
6. Equilibrate samples to the desired temperature.
7. Add 100 µL of a stimulus dissolved in KRG to activate the cells (*see* **Note 9**).
8. Measure change in fluorescence emission at 400 nm with an excitation wavelength of 317 nm.
9. Calculate the amount of H_2O_2 produced from a standard curve of the PHPA system, calibrated with different concentrations of H_2O_2 added to nonactivated cell samples (*see* **Note 16**).

3.2. Detection of Intracellular ROS

It is clear that an intracellular activation of the oxidase can be achieved also in the absence of phagocytosis *(11)*, indicating that the NADPH oxidase can be assembled and activated also in a flavocytochrome b-containing intracellular membrane distinct from the phagosome (*see* **Note 17**). Here we describe methods for measuring intracellular ROS production by phagocytes treated with various activating stimuli.

3.2.1. Intracellular ROS Detection
by Luminol-Amplified Chemiluminescence

The membrane-permeable dye, luminol, is excited by phagocyte-generated ROS, resulting in chemiluminescence. By adding the membrane-impermeable enzymes, SOD and catalase, to reaction mixtures to remove secreted O_2^- and H_2O_2, ROS generated specifically in intracellular compartments can be measured.

1. Add 680 µL of KRG to Eppendorf tubes or luminometer tubes for each sample (*see* **Note 6**).
2. Add 100 µL of fresh 500 µ*M* luminol solution.
3. Add 10 µL of SOD solution.
4. Add 10 µL of catalase solution.
5. Add 100 µL of cells in KRG (10^2–10^6 cells, depending on the activity of the cells and the sensitivity of the luminometer) (*see* **Note 8**).
6. Equilibrate samples to the desired temperature.
7. Add 100 µL of a stimulus dissolved in KRG to activate the cells (*see* **Note 9**).
8. Measure chemiluminescence with a β-counter or luminometer (preferentially equipped with a thermostated sample holder), including those adapted to microtiter plates (*see* **Notes 10** and **18**).

3.2.2. Intracellular H_2O_2 Detection by PHPA Fluorescence

As mentioned above, some of ROS generated in neutrophils will be retained inside the phagocyte (*see* **Note 17**). O_2^- generated inside an intracellular compartment cannot normally be detected extracellularly, simply because neutrophils lack the anion channels needed for O_2^- to pass biological membranes *(12)*. Similarly, H_2O_2 generated in an intracellular compartment cannot be measured extracellularly, but for different reasons. H_2O_2 can pass biological membranes but is rapidly consumed by endogenous peroxidases and catalase on its way from intracellular compartments to the plasma membrane. However, if endogenous myeloperoxidase (MPO) and catalase are inhibited (e.g., with NaN_3), then the PHPA technique described above can also be used to determine intracellular production of H_2O_2 *(13,14)*.

1. Prepare two Eppendorf tubes for each sample (i.e., a sample and a reference tube) (*see* **Note 6**).
2. Add 550 µL of KRG to the sample tube and 600 µL of KRG to the reference tube.
3. Add 100 µL of PHPA solution.
4. Add 50 µL of HRP solution.
5. Add 100 µL of NaN_3 to the reference sample only (*see* **Note 19**).
6. Add 100 µL of cells ($\geq 5 \times 10^5$ cells) in KRG.
7. Equilibrate samples to the desired temperature.
8. Add 100 µL of a stimulus dissolved in KRG to activate the cells (*see* **Note 9**).
9. Measure change in fluorescence emission at 400 nm with an excitation wavelength of 317 nm.
10. Calculate intracellular H_2O_2 generation by subtracting the fluorescence of samples containing NaN_3 (extracellular H_2O_2) from that of samples without NaN_3 (total H_2O_2).
11. Calculate the amount of H_2O_2 produced from a standard curve of the PHPA system, calibrated with different concentrations of H_2O_2 added to nonactivated cell samples (*see* **Note 16**).

4. Notes

1. While the primary role of phagocyte-derived ROS is in defense against invading microorganisms, these oxidants can also damage "innocent bystanders," resulting in the tissue destruction associated with many inflammatory diseases. The fact that ROS generated by phagocytes can function as regulators of other immune reactive cells *(15–17)* and apoptotic/necrotic processes *(18,19)* further demonstrates the impor-tance of these molecules in immune regulation during health and disease *(20)*.
2. Subsequent to receptor activation in neutrophils (or sometimes through receptor independent activation of the cells), O_2 undergoes a one- or two-electron reduction to form O_2^- or H_2O_2. The electron donor, NADPH, is formed by the oxidation of glucose in the hexose monophosphate shunt. The ROS may be further reduced by a number of different cellular protection systems, including SOD (catalyzes the reduction of O_2^- to H_2O_2), catalase (catalyzes the reduction of H_2O_2 to H_2O), and glutathione peroxidase (catalyzes the oxidation of glutathione by H_2O_2). O_2^- and

H_2O_2 can act as substrates for reaction with other cellular ROS (e.g., O_2^- reacts with nitric oxide to form the very reactive peroxynitrite molecule or cellular enzymes, such as MPO, which is localized in neutrophil azurophil granules and uses H_2O_2 in combination with chloride ions to generate the highly microbicidal HOCl *(21)*.

3. As of yet, no technique has been found sufficient to determine respiratory burst activity in relation to microbicidal events and tissue destruction or for determining the effects of different pharmacological agents affecting cellular responses.

4. In order for luminol to react with oxidants in an intracellular compartment, it has to cross one or more biological membranes. Currently, very little is known about the diffusion properties of luminol or the diffusion limitations in different phagocyte membranes (plasma membrane, granule membranes, and/or phagolysosomal membranes).

5. The requirement for ROS in the luminol-dependent chemiluminescence reaction is shown by the absence or very low levels of chemiluminescence obtained when stimulating cells isolated from patients with CGD (**Fig. 2**). In addition, neutrophils isolated from donors with a deficiency of the azurophil granule protein MPO produce very low levels of chemiluminescence, despite a pronounced (in fact higher than normal) production of O_2^- and H_2O_2 (as measured with other techniques described here *[22]*), suggesting that MPO is participating in the chemiluminescence-producing reaction. The fact that addition of MPO to the extracellular fluid is not sufficient to fully regenerate the chemiluminescence activity of MPO-deficient cells to that of control cells suggests that intracellular MPO (stored in the azurophil granules) also participates in the excitation of luminol by reacting with ROS generated in an intracellular compartment. This suggestion is further supported by the fact that part of the chemiluminescence response is insensitive to cell-impermeable extracellular scavengers of O_2^- and H_2O_2 *(11)*.

6. The methods described are for 1-mL assay volumes; however, they can easily be adapted to a microtiter plate format by adjusting the volumes accordingly. Based on a 200-µL assay volume for microtiter plate wells, the volumes for each step would then be reduced to one-fifth of those used in the 1-mL assay. Microtiter plate assays are quite convenient and reduce sample and reagent requirements.

7. Although the light-generating reaction is peroxidase-dependent, the amount of H_2O_2 released from the cells does not affect the activity detected *(9)*. Peroxidases (MPO and HRP) catalyze the reaction using mainly O_2^- and not H_2O_2, and samples can be analyzed in the presence and absence of SOD, in parallel, to directly determine the portion of the activity that reflects O_2^- production.

8. Future developments of new luminescent dyes that are more sensitive than isoluminol should allow measurements of respiratory burst activity in individual cells, or even measurements at the subcellular level.

9. The neutrophil respiratory burst can be activated by a number of different soluble and particulate stimuli, including chemoattractants, certain cytokines, phorbol esters, calcium ionophores, certain lectins, and various microorganisms (opsonized as well as unopsonized). All of these stimuli elicit a neutrophil chemiluminescence response in the presence of luminol *(23–30)*.

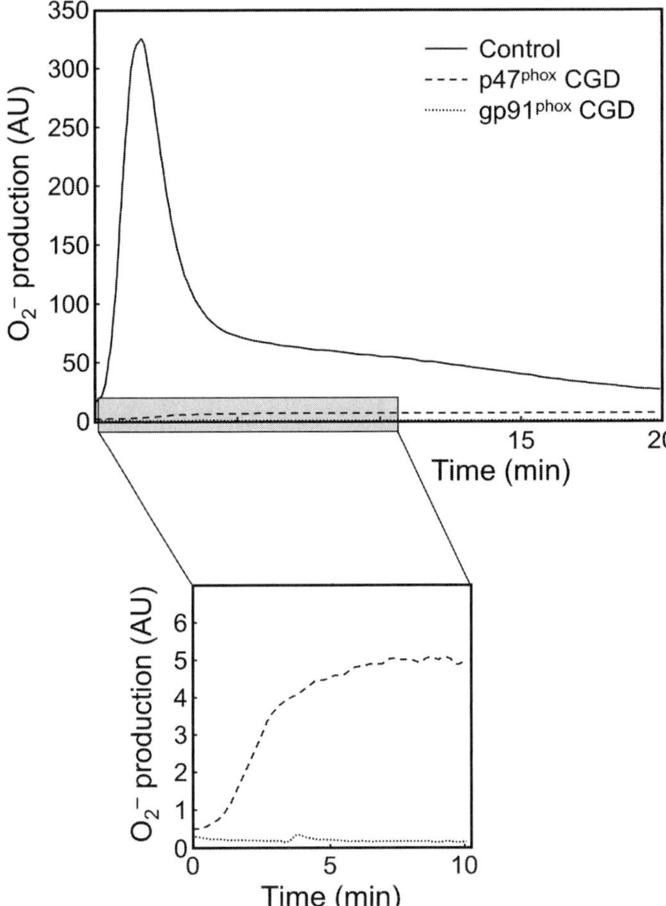

Fig, 2. Sensitivity of the luminol-enhanced chemiluminescence system. Normal neutrophils stimulated with phorbol myristate ester (PMA) give rise to a robust chemiluminescence signal (solid line). Although all forms of chronic granulomatous disease (CGD) neutrophils are clearly defective in reactive oxygen species (ROS) production when compared to control cells, neutrophils from an autosomal recessive CGD patient (p47phox-deficient) actually generate minute amounts of ROS (dashed line), whereas cells from an X-linked CGD patient (gp91phox-deficient) are completely devoid of ROS production (dotted line). The difference between these two forms of CGD can be seen using the highly sensitive luminol-enhanced chemiluminescence system. The lower panel represents a magnification of the upper panel.

10. The relation between the amount of O_2^- produced and the amount of detected light or change in absorbance is dependent on the luminometer, but can easily be determined by a direct comparison with the SOD-inhibitable reduction of cytochrome c *(31)*.

11. In order to specifically determine the amount of O_2^- produced, SOD is added to reference samples so that any part of the signal that is not SOD-inhibitable (i.e., not due to O_2^-) can be subtracted.

12. O_2^- production is often calculated and reported as the amount produced by 10^6 cells. If the number of cells in the sample is different than 10^6, then the equation will need to be adjusted by dividing by the actual number of cells in units of 10^6. For example, if 5×10^5 (0.5×10^6) or 1×10^5 (0.1×10^6) cells are in the sample, then the value obtained from **eq. 1** must be divided by 0.5 and 0.1, respectively, and so on.

13. If a microtiter plate format is used for the cytC assay, two parameters must be changed in the formula: (1) the light path length will no longer be 1 cm and should be measured as the distance from the bottom of the plate to the top of the sample liquid meniscus (usually about 0.75 cm) and (2) the assay volume is now 200 µL instead of 1 mL. As an example: O_2^- can be calculated for a 200 µL assay in a microtiter plate well with a 0.75 cm light path using **eq. 2**:

$$\Delta OD_{550} \times 12.64 = \text{nmoles } O_2^-/\text{time unit} \qquad (2)$$

This number can then be divided by the number of cells in the well to obtain O_2^- production per unit of cells (*see* **Note 12**).

14. Accumulation of H_2O_2 in the measuring system (which can occur when a large proportion of the available cytC is reduced and the H_2O_2 consuming enzymes are inhibited) can result in a re-oxidation of the ferrous form of cytC, thus giving an underestimated result.

15. In order to reflect the release of ROS using PHPA (which detects H_2O_2), SOD has to be added to the system to converte all O_2^- to H_2O_2.

16. PHPA can be replaced by other fluorogenic substrates, that when oxidized by the peroxidase-H_2O_2 complex, changes its fluorescence properties. Scopoletin (7-hydroxycoumarin) is such a substrate that in reduced form fluoresces at 460 nm when activated at 350 nm, and in oxidized form loses its fluorescence *(32)*. The scopoletin-assay system is more sensitive than the PHPA system but has the disadvantage that the fluorescence is lost in relation to H_2O_2 production, making the "measuring window" rather narrow.

17. Human neutrophils contain at least four types of granules that are mobilized (induced to fuse with the plasma membrane) hierarchically during in vivo extravasation of the cells from the blood stream to the tissue *(33)* (*see also* Chapter 4). Apparently, the membranes of the mobilizable granules store most (~95%) of the cellular content of flavocytochrome b, the membrane component of the NADPH-oxidase complex, and translocation of flavocytochrome b from the granules to the plasma membrane is associated with a reduction of the NADPH-oxidase activity in the granule fraction (as measured in a cell-free system using subcellular fractions from activated neutrophils) *(34)*.

18. The light-generating reaction is dependent on the endogenous peroxidase, MPO. It is possible for MPO to use O_2^- as well as H_2O_2 in the light-generating reaction, but the actual oxygen metabolite(s) measured, has not been identified. It is not

possible to directly quantify the amounts of O_2^- produced with this technique, and a change in luminescence could be due to either a change in O_2^-/H_2O_2 production, a change in the availability of MPO, or a change in the basic conditions (protein content; pH; ionic composition) in the intracellular compartment in which the reaction takes place. It should also be noted that luminol can in some situations inhibit the release of O_2^- *(35)*, suggesting that this dye should only be used to measure ROS generated in an intracellular compartment, and the molecular basis for any detected responses should be verified with another technique.

19. Even though NaN_3 inhibits the major H_2O_2-consuming enzymes (i.e., catalase and MPO) in neutrophils, intracellularly-generated H_2O_2 may be consumed or scavenged by other routes. Thus, the calculated amount of intracellularly-generated H_2O_2 will most likely be an underestimation of what is really produced.

Acknowledgments

This work was supported by the Swedish Medical Research Council, the Swedish Society for Medical Research and the King Gustaf V Memorial Foundation.

References

1. Quie, P. G., White, J. G., Holmes, B., and Good, R. A. (1967) In vitro bactericidal capacity of human polymorphonuclear leukocytes: diminished activity in chronic granulomatous disease of childhood. *J. Clin. Invest.* **46,** 668–679.
2. Segal, A. W. (2005) How neutrophils kill microbes. *Annu. Rev. Immunol.* **23,** 197–223.
3. Quinn, M. T. and Gauss, K. A. (2004) Structure and regulation of the neutrophil respiratory burst oxidase: comparison with non-phagocyte oxidases. *J. Leukoc. Biol.* **76,** 760–781.
4. Karlsson, A. and Dahlgren, C. (2002) Assembly and activation of the neutrophil NADPH oxidase in granule membranes. *Antioxid. Redox. Signal.* **4,** 49–60.
5. Decoursey, T. E. and Ligeti, E. (2005) Regulation and termination of NADPH oxidase activity. *Cell. Mol. Life Sci.* **62,** 2173–2193.
6. Rada, B. K., Geiszt, M., Hably, C., and Ligeti, E. (2005) Consequences of the electrogenic function of the phagocytic NADPH oxidase. *Philos. Trans.e R. Soc. Lond. B Biol. Sci.* **360,** 2293–2300.
7. Hampton, M. B., Kettle, A. J., and Winterbourn, C. C. (1998) Inside the neutrophil phagosome: oxidants, myeloperoxidase, and bacterial killing. *Blood* **92,** 3007–3017.
8. Kobayashi, T., Robinson, J. M., and Seguchi, H. (1998) Identification of intracellular sites of superoxide production in stimulated neutrophils. *J. Cell Sci.* **111,** 81–91.
9. Lundqvist, H. and Dahlgren, C. (1996) Isoluminol-enhanced chemiluminescence: a sensitive method to study the release of superoxide anion from human neutrophils. *Free Radic. Biol. Med.* **20,** 785–792.

10. Dahlgren, C. and Karlsson, A. (1999) Respiratory burst in human neutrophils. *J. Immunol. Methods* **232,** 3–14.
11. Serrander, L., Larsson, J., Lundqvist, H., et al. (1999) Particles binding β_2-integrins mediate intracellular production of oxidative metabolites in human neutrophils independently of phagocytosis. *Biochim. Biophys. Acta* **1452,** 133–144.
12. Halliwell, B. and Gutteridge, J. M. (1985) The importance of free radicals and catalytic metal ions in human diseases. *Mol. Aspects Med.* **8,** 89–193.
13. Root, R. K., Metcalf, J., Oshino, N., and Chance, B. (1975) H_2O_2 release from human granulocytes during phagocytosis. I. Documentation, quantitation, and some regulating factors. *J. Clin. Invest.* **55,** 945–955.
14. Root, R. K. and Metcalf, J. A. (1977) H_2O_2 release from human granulocytes during phagocytosis. Relationship to superoxide anion formation and cellular catabolism of H_2O_2: studies with normal and cytochalasin B-treated cells. *J. Clin. Invest.* **60,** 1266–1279.
15. Hellstrand, K., Asea, A., Dahlgren, C., and Hermodsson, S. (1994) Histaminergic regulation of NK cells. Role of monocyte-derived reactive oxygen metabolites. *J. Immunol.* **153,** 4940–4947.
16. Olofsson, P., Holmberg, J., Tordsson, J., Lu, S., Akerstrom, B., and Holmdahl, R. (2003) Positional identification of Ncf1 as a gene that regulates arthritis severity in rats. *Nat. Genet.* **33,** 25–32.
17. Hultqvist, M. and Holmdahl, R. (2005) *Ncf1* (*p47phox*) polymorphism determines oxidative burst and the severity of arthritis in rats and mice. *Cell. Immunol.* **233,** 97–101.
18. Lundqvist-Gustafsson, H. and Bengtsson, T. (1999) Activation of the granule pool of the NADPH oxidase accelerates apoptosis in human neutrophils. *J. Leukoc. Biol.* **65,** 196–204.
19. Bylund, J., Campsall, P. A., Ma, R. C., Conway, B. A., and Speert, D. P. (2005) Burkholderia cenocepacia induces neutrophil necrosis in chronic granulomatous disease. *J. Immunol.* **174,** 3562–3569.
20. Quinn, M. T., Ammons, M. C. B., and DeLeo, F. R. (2006) The expanding role of NADPH oxidases in health and disease: no longer just agents of death and destruction. *Clinical Sci.* **111,** 1–20.
21. Klebanoff, S. J. (2005) Myeloperoxidase: friend and foe. *J. Leukoc. Biol.* **77,** 598–625.
22. Dahlgren, C. and Stendahl, O. (1983) Role of myeloperoxidase in luminol-dependent chemiluminescence of polymorphonuclear leukocytes. *Infect. Immun.* **39,** 736–741.
23. Fu, H., Karlsson, J., Bylund, J., Movitz, C., Karlsson, A., and Dahlgren, C. (2006) Ligand recognition and activation of formyl peptide receptors in neutrophils. *J. Leukoc. Biol.* **79,** 247–256.
24. Karlsson, J., Fu, H., Boulay, F., Dahlgren, C., Hellstrand, K., and Movitz, C. (2005) Neutrophil NADPH-oxidase activation by an annexin AI peptide is transduced by the formyl peptide receptor (FPR), whereas an inhibitory signal is generated independently of the FPR family receptors. *J. Leukoc. Biol.* **78,** 762–771.

25. Thoren, F., Romero, A., Lindh, M., Dahlgren, C., and Hellstrand, K. (2004) A hepatitis C virus-encoded, nonstructural protein (NS3) triggers dysfunction and apoptosis in lymphocytes: role of NADPH oxidase-derived oxygen radicals. *J. Leukoc. Biol.* **76,** 1180–1186.
26. Fu, H., Belaaouaj, A. A., Dahlgren, C., and Bylund, J. (2003) Outer membrane protein A deficient Escherichia coli activates neutrophils to produce superoxide and shows increased susceptibility to antibacterial peptides. *Microbes Infect.* **5,** 781–788.
27. Almkvist, J., Dahlgren, C., Leffler, H., and Karlsson, A. (2002) Activation of the neutrophil nicotinamide adenine dinucleotide phosphate oxidase by galectin-1. *J. Immunol.* **168,** 4034–4041.
28. Lundqvist, H., Follin, P., Khalfan, L., and Dahlgren, C. (1996) Phorbol myristate acetate-induced NADPH oxidase activity in human neutrophils: only half the story has been told. *J. Leukoc. Biol.* **59,** 270–279.
29. Lock, R., Dahlgren, C., Linden, M., Stendahl, O., Svensbergh, A., and Ohman, L. (1990) Neutrophil killing of two type 1 fimbria-bearing *Escherichia coli* strains: dependence on respiratory burst activation. *Infect. Immun.* **58,** 37–42.
30. Karlsson, A., Nixon, J. B., and McPhail, L. C. (2000) Phorbol myristate acetate induces neutrophil NADPH-oxidase activity by two separate signal transduction pathways: dependent or independent of phosphatidylinositol 3-kinase. *J. Leukoc. Biol.* **67,** 396–404.
31. Foyouzi-Youssefi, R., Petersson, F., Lew, D. P., Krause, K. H., and Nusse, O. (1997) Chemoattractant-induced respiratory burst: increases in cytosolic Ca^{2+} concentrations are essential and synergize with a kinetically distinct second signal. *Biochem. J.* **322,** 709–718.
32. Boveris, A., Martino, E., and Stoppani, A. O. (1977) Evaluation of the horseradish peroxidase-scopoletin method for the measurement of hydrogen peroxide formation in biological systems. *Anal. Biochem.* **80,** 145–158.
33. Faurschou, M. and Borregaard, N. (2003) Neutrophil granules and secretory vesicles in inflammation. *Microbes Infect.* **5,** 1317–1327.
34. Dahlgren, C., Johansson, A., Lundqvist, H., Bjerrum, O. W., and Borregaard, N. (1992) Activation of the oxygen-radical-generating system in granules of intact human neutrophils by a calcium ionophore (ionomycin). *Biochim. Biophys. Acta* **1137,** 182–188.
35. Faldt, J., Ridell, M., Karlsson, A., and Dahlgren, C. (1999) The phagocyte chemiluminescence paradox: luminol can act as an inhibitor of neutrophil NADPH-oxidase activity. *Luminescence* **14,** 153–160.

24

The K-562 Cell Model for Analysis of Neutrophil NADPH Oxidase Function

Thomas L. Leto, Mark C. Lavigne, Neda Homoyounpour, Kristen Lekstrom, Gilda Linton, Harry L. Malech, and Isabelle de Mendez

Summary

Polymorphonuclear neutrophils (PMN) have a remarkable capacity for generation of large amounts of reactive oxygen species in response to a variety of infectious or inflammatory stimuli, a process known as the respiratory burst that involves activation of a multicomponent NADPH oxidase. Given their short life span, PMN are not amenable to most molecular biology methods for studying activation of this oxidant-generating system. We have explored a variety of methods for introduction of components of the phagocytic oxidase (*phox* system) into the promyelocytic erythroleukemia cell line, K-562. Here, we describe a series of cloned K-562 cell lines that were retrovirally transduced for stable production of one or more essential components of the phagocytic oxidase (*phox*) complex. We outline methods for the use of these transfectable cells for investigating structure, function, and signaling requirements for assembly and activation of the *phox* system. These versatile lines can be used to examine effects of genetic polymorphisms or mutations in *phox* components associated with chronic granulomatous disease, to serve as a system for testing gene therapy vectors designed to correct the defective oxidase, to study cross-functioning with recently described *phox* component homologs, or to explore signaling components involved in regulation of the respiratory burst.

Key Words: PMN; NADPH oxidase; *phox*; Nox2; chronic granulomatous disease; K-562 cells; superoxide; reactive oxygen species; fMLP.

1. Introduction

The unique capability of polymorphonuclear neutrophils (PMN) and other circulating phagocytes to produce large amounts of reactive oxygen species is attributed to expression of a multicomponent NADPH oxidase complex, or *phox* system *(1–3)*. The importance of this enzyme complex is evident in chronic granulomatous disease (CGD), where genetic defects or deficiencies in oxidase activity result in enhanced susceptibility to bacterial and fungal infections (*see*

From: *Methods in Molecular Biology, vol. 412: Neutrophil Methods and Protocols*
Edited by: M. T. Quinn, F. R. DeLeo, and G. M. Bokoch © Humana Press Inc., Totowa, NJ

Chapter 30 for more details). Inherited defects affecting this oxygen-dependent antimicrobial system have been described in five essential phagocyte NADPH oxidase components: gp91^{phox} (or Nox2), p22^{phox}, p47^{phox}, p67^{phox}, and Rac2. The Nox2-based enzyme is the best-characterized member of the newly recognized Nox family of NADPH oxidases *(2–4)*. The gp91^{phox} and p22^{phox} components represent subunits of a membrane-imbedded flavocytochrome b$_{558}$, while the other components are essential regulators of the enzyme. All five purified components are needed to reconstitute a superoxide-generating enzyme complex in vitro *(5)*. Early efforts aimed at exploring requirements for phagocytic oxidase function in whole cells showed that co-transfection of plasmid expression vectors encoding the three myeloid-specific *phox* components (gp91^{phox} [or Nox2], p47^{phox}, and p67^{phox}) are sufficient to reconstitute superoxide-generating activity in the promyelocytic leukemia cell line, K-562 *(6)*. The line was derived from a patient with chronic myelogenous leukemia in blast crisis and is capable of undergoing limited erythroid, megakaryocytic, or granulocytic differentiation *(7,8)*. These cells produce endogenous p22^{phox} and Rac2, which is expressed in hematopoietic cells *(6)*. Subsequently, we found that retroviral-mediated expression of the gp91^{phox} core catalytic component of the oxidase results in significantly higher levels of oxidase activity reconstitution in this model *(9)*. We and other investigators have used this system extensively to examine structural requirements for oxidase assembly and activation in intact cells *(9–15)* (*see* **Note 1**).

In this chapter, we describe the derivation and use of several retrovirally-transduced K-562 lines that were clonally selected for stable, high expression of one or more *phox* components. The fully-reconstituted cell line (K-562^{+++}) is robust and can be used to explore receptor activation and downstream signaling pathways involved in oxidase activation, while the various partially reconstituted *phox*-expressing lines allow examination of structural requirements for oxidase function in any of the transfected *phox* components (see **Note 2**). We also describe transfection methodology for K-562 cells that can be used to introduce other essential *phox* components, upstream signaling intermediates, or receptors that stimulate oxidase activation. Finally, methods are described for assessing NADPH oxidase reconstitution, including detection of the recombinant oxidase proteins, evaluation of agonist-stimulated O$_2^-$ release by whole cells, and measurement of membrane translocation of *phox* proteins following cellular activation.

2. Materials

2.1. Common Buffers/Media

 1. RPMI: RPMI 1640 medium without supplements (Gibco).

2. Complete RPMI: RPMI 1640 medium (Gibco), supplemented with 10% heat-inac-tivated (56°C for 30 min) fetal bovine serum (FBS) (Hyclone), 2 mM L-glutamine, 100 U/mL penicillin, and 100 µg/mL streptomycin (Gibco).
3. Hank's balanced-salt solution without $CaCl_2$ and $MgCl_2$ (HBSS–, Gibco).
4. Hank's balanced salt solution with $CaCl_2$ and $MgCl_2$ (HBSS+, Gibco).

2.2. Plasmid Purification

1. High Purity Plasmid Maxiprep System (Marligen Biosciences) or Qiagen® Plasmid Maxi Kit (Qiagen).
2. Polymyxin B immobilized on 4% agarose beads (Sigma). Wash with distilled water and use as a 50% slurry.
3. 70% ethanol: prepare in advance and store at –20°C.
4. High-speed refrigerated centrifuge capable of 15,000g forces with rotors capable of holding 50 mL and 500 mL Oak Ridge sealing centrifuge bottles (Nalgene) (e.g., Sorvall RC-6 Sorvall with SLA3000 and SS34 rotors).
5. 37°C bacterial (shaking) incubator.

2.3. Cell Lines

1. The wild-type K-562 cell line can be obtained from the American Type Tissue Cul-ture Collection (ATTC; #CCL 243).
2. Phox-transduced K-562 cell lines (available upon request from T. Leto): MFG-S-based, onco-retroviruses developed as gene therapy vectors for correction of p47^{phox}, p67^{phox}, and gp91^{phox} deficiencies *(16–18)* were used to transduce oxidase genes into K-562 cells and establish the cloned lines described in this chapter (*see* **Note 3**). Clones of multiply-transduced K-562 cells were selected following limit-ing dilutions (*see* **Note 4**), and those showing the highest levels of oxidase activity were further characterized by Western blotting (*see* **Fig. 1** and **Note 5**). PMA-stim-ulated O_2^- production in the most active triply-transduced K-562 cell line (clone 6B9; K-562^{+++}) was 48–75 nmole/10^7 cells/min, comparing favorably with reported rates from the same number of human neutrophils (*see* **Fig. 2A**).
3. 37°C humidified cell culture incubator, maintained with an atmosphere of 5% CO_2.
4. 80 cm^2 tissue culture flasks, with filter cap (Nunc).
5. Cell culture biosafety cabinet.

2.4. Cell Transfection (Electroporation)

1. BTX 820 T electroporator (*see* **Note 6**).
2. Gene Pulser with capacitance extender (Biorad). (For an alternative transfection protocol I, *see* **Note 7**.)
3. Nucleofector and Nucleofector Kit V (Amaxa). (For an alternative transfection pro-tocol II, *see* **Note 8**.)
4. Electroporation cell, 0.4 cm gap (BioRad).
5. 25 cm^2 tissue culture flasks.
6. Low salt/sucrose electroporation buffer: 250 mM sucrose, 10 mM sodium phos-phate, 1 mM $MgCl_2$ (pH 7.2, sterile-filtered).

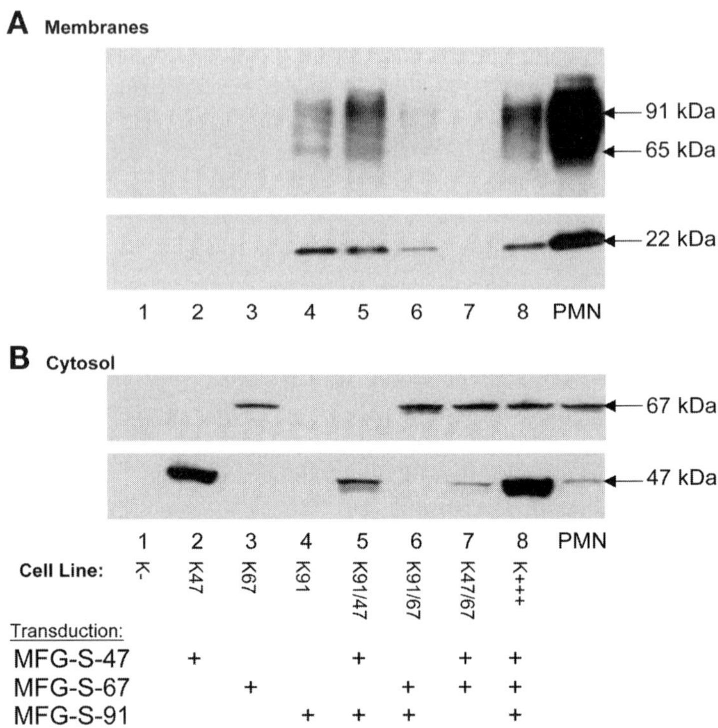

Fig. 1. Western blot detection of *phox* proteins in clonally selected K-562 cell lines stably transduced with MFG-S-*phox* expression vectors. **(A)** Blotting of membrane protein fractions (30 µg/lane) for gp91^{phox} with MoAb 48 (upper panel) detects both unglycosylated and fully glycosylated gp91^{phox} species (65-kDa and 91-kDa bands, respectively) in K-562 cells. Endogenous p22^{phox} protein detected by MoAb 449 (lower panel) closely parallel levels of transduced gp91^{phox}. **(B)** Blotting of cytosolic proteins with antibodies against p67^{phox} (#1589; upper panel) and p47^{phox} (#1588; lower panel). The *phox* components transduced in each line are described below each lane.

7. Hygromycin B (Calbiochem) for selection of some pCEP or pREP vector-transfected cells: prepare a 40X stock solution (10 mg/mL) in RPMI lacking supplements. Sterile filter and store at 4°C. (**Caution:** highly toxic. Handle with care. See material safety data sheet.)
8. G-418 (Gibco) for selection of pcDNA series vectors: prepare a 100X stock solution at 80 mg/mL in RPMI, lacking supplements. Sterile filter and store at 4°C (*see* **Note 9**).

2.5. NADPH Oxidase Assays

1. Phorbol myristate acetate (PMA): dissolve PMA (Sigma) at 1 mg/mL in DMSO, and store in small aliquots at −80°C). Dilute in HBSS− to working concentrations *immediately before use*; do not store or refreeze (*see* **Note 10**).

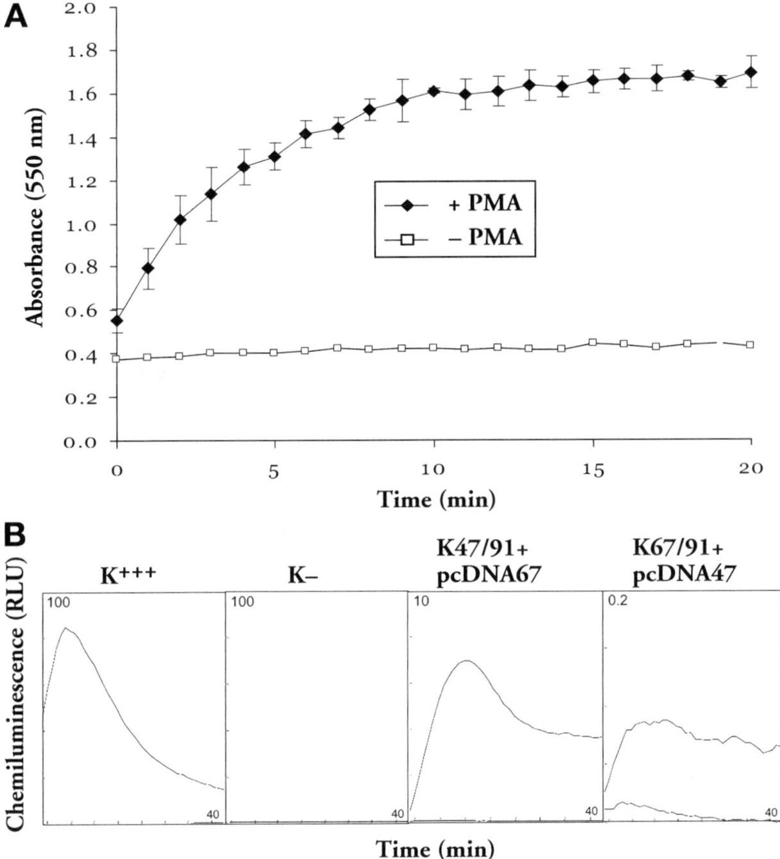

Fig. 2. Reconstitution of phagocytic NADPH oxidase activity in K-562 cells. **(A)** Kinetics of PMA-stimulated superoxide release by retrovirally transduced, *phox*-competent K-562 (K-562^{+++}) cells (5×10^5 cells), detected by ferricytochrome c reduction in response to 0.2 µg/mL phorbol myristate acetate (PMA) (closed circles) or no stimulus (open circles). Results are the average of triplicate reactions ± SD. **(B)** Reconstitution of PMA-stimulated NADPH oxidase activity in K-562 cell models of chronic granulomatous disease detected by the Diogenes chemiluminescence assay. Activities of cell lines retrovirally transduced with two *phox* components were restored by transient transfection with pcDNA 3.1 vectors encoding the third essential *phox* component. Note that the broad range of activities reconstituted in these lines can be detected by the chemiluminesence-based oxidase assay.

2. Cytochrome c: prepare a 20X stock solution by dissolving horse heart ferricytochrome c (Sigma) at 4 mM in HBSS–. Use 5 µL per 100 µL of reaction.

3. 2X cytochrome c/activator reagent: For each reaction, combine 5 µL of cytochrome c stock solution; the volume of activator required to achieve two-times the final activator concentration desired in the reaction (e.g., if 1 µg/mL PMA is desired in

the final 100 µL reactions, 5 µL of a 20 µg/mL PMA working solution are included in the 2X reagent); and HBSS+ up to a final volume of 50 µL per sample. Addition of 50 µL of 2X cytochrome c/activator reagent to the 50-µL reactions results in 1X final concentrations of cytochrome c and activator.

4. 96-well microtiter plate reader designed capable of reading at 550 nm, with thermostatic control, automatic mixing, and kinetic programming capabilities (e.g., Vesamax equipped with Softmax programming software, Molecular Devices).

5. 96-well clear, flat-bottom, tissue culture-treated microplates (Costar, Corning, or equivalent).

6. Diogenes reagent (National Diagnostics): supplied as two reagents that are dissolved separately in distilled H_2O and then combined. The reconstituted reagent can be stored at 4°C for at least 60 d.

7. Superoxide dismutase (SOD): dissolve SOD (Sigma) at 1.25 mg/mL in HBSS; use 4 µL (5 µg) per 100 µL reaction, as needed in reference reactions to determine SOD-insensitive changes.

8. Diphenylene iodonium (DPI): prepare a 1000X stock solution by dissolving DPI (Sigma) at 10 mM in dimethylsulfoxide (DMSO), and store in small aliquots at –20°C. Dilute this fresh 20-fold in HBSS, and add 1 µL of diluted DPI to 50 µL of cell suspensions in negative control reactions. NADPH oxidase-derived superoxide production is inhibited >98% following a 10-min (37°C) pretreatment of cells with 10 µM DPI (*see* **Note 11**).

9. *N*-formylmethionyl-leucyl-phenylalanine (fMLP): dissolve fMLP (Sigma) at 100 mM in DMSO. Make serial dilutions in DMSO to create a 100X stock, so that final DMSO concentration in treated cells does not exceed 1%. Prepare final dilutions of fMLP as a 2X reagent prepared in Diogenes reagent with fMLP concentrations in the 0.2–2 µM range. Add 50 µL of Diogenes/fMLP mixture to 50 µL of cell suspension. Prepare aqueous solutions of fMLP fresh and do not store.

10. For the Diogenes assays, use a microtiter plate luminometer with thermostatic (37°C) control, automatic mixer and dispensers, and kinetic mode programming capabilities (e.g., ThermoLabsystems Luminoskan Acsent or equivalent).

11. For luminescence assays of superoxide release, use opaque white flat-bottom microtiter plates or Microlite™ 1+ strips, with a 96-well holder (ThermoLabsystems) (*see* **Note 12**).

2.6. Preparation of Cell Fractions and Membrane Translocation Assays

1. Low-speed centrifuge (table-top) model, used to centrifuge and wash cells in 15- to 50-mL conical centrifuge tubes.

2. Airfuge or ultracentrifuge (Beckman Coulter), with rotor adapted to hold 250-mL tubes.

3. Microcentrifuge and Eppendorf tubes.

4. Microtip sonifier (Branson Model 150).

5. Relaxation buffer: 100 mM KCl, 3 mM NaCl, 3.5 mM MgCl$_2$, 10 mM PIPES, 1.25 mM ethylenediamine tetraacetic acid (EDTA).

6. Protease inhibitor cocktail (PIC; Sigma): dilute 1:100 in relaxation buffer. Prepare fresh and do not store.
7. 10% sucrose dissolved in Relaxation buffer.
8. Micro bicinchonic acid (BCA) protein assay kit (Pierce).

2.7. Western Blotting of NADPH Oxidase Components

1. Tris-Glycine polyacrylamide slab gels for standard sodium dodecyl sulfate (SDS)-denaturing electrophoretic separation of proteins *(19)*. We routinely use Novex pre-cast Tris-Glycine polyacrylamide gels (10 well; 1 mm thick; 8, 10, or 12% acrylamide) and commercially-prepared buffers (Invitrogen).
2. Slab gel electrophoresis chamber and electrophoretic transblotter with power supplies.
3. Nitrocellulose membranes for protein blotting.
4. Tris-buffered saline (TBS): prepare a 10X stock of 500 mM Tris-HCl, 1 M NaCl, pH 7.5. Dilute 1/10 in distilled H_2O before use as 1X TBS.
5. Tween/TBS: 0.1% Tween-20 (Sigma, v/v) diluted in TBS.
6. Blocking buffer: 5% (w/v) nonfat dried milk powder dissolved in Tween/TBS.
7. Ponceau S: prepare a 0.5% (w/v) solution in 5% acetic acid (Sigma).
8. Enhanced chemiluminescence (ECL) Plus peroxidase-based chemiluminescence immunodetection kit (Amersham/GE Healthcare). Mix 2 mL of reagent A with 50 µL of reagent B just prior to use.
9. Antibodies: Mouse anti-p47^{phox} antibody (D-10, Santa Cruz) or goat anti-p47^{phox} antiserum (#1588) *(20)*. Dilute goat antiserum at 1:1000–1:2000 in blocking solution (can be reused several times). Rabbit anti-p67^{phox} antibody (H-300, Santa Cruz) or goat anti-p67^{phox} antiserum (#1589) *(20)*. Dilute goat antiserum at 1:1000-1:2000 in blocking solution (can be reused several times). Mouse monoclonal anti-gp91^{phox} (MoAb #48) and p22^{phox} (MoAb #449) antibodies *(21)* (*See*: http://www.sanquin.nl/sanquin-eng/sqn_reagents.nsf for distributors). Dilute 1:1000 in blocking solution (can be reused several times). Horseradish peroxidase (HRP)-conjugated donkey anti-rabbit immunoglobulin (Ig)G and sheep anti-mouse IgG secondary antibodies (Amersham/ GE Healthcare) diluted 1:1000 in blocking solution. HRP-conjugated donkey anti-goat IgG reagent (Jackson ImmunoResearch) diluted 1:4000 in blocking solution.

3. Methods

3.1. Transfect of K-562 Cell Lines

Many high-copy number mammalian expression vectors are used to transfect K-562 cells. The pREP and pCEP series have been used for high expression and retention of stable episomes in cells subjected to long-term selection *(6)*; however, pcDNA3 plasmid series are smaller and are more easily engineered (*see* **Note 13**).

3.1.1. Plasmid Purification

1. Grow a 125-mL saturated (overnight) culture of bacteria in Luria-Bertani (LB) broth.
2. Purify plasmid DNA using the High Purity Plasmid Maxiprep System, strictly adhering to methods described by the manufacturer. Acceptable plasmid preparations are also obtained from the Qiagen Plasmid Maxi kits.
3. Dissolve the final DNA pellet in 200 µL of distilled H_2O at concentrations >0.5 mg/mL.
4. Add Polymyxin B-agarose bead suspension to the purified DNA (120 µL of bead suspension/200 µL of purified plasmid) to remove bacterial lipopolysaccharide, resulting in good quality, transfection-grade material.
5. Mix by rotation for 1 h at room temperature, and remove the beads by centrifugation in a microfuge (15,000g for 3 min).

3.1.2. K-562 Cell Culture

1. Grow cells in complete at 37°C in an atmosphere of 5% CO_2 (*see* **Note 14**).
2. Maintain K-562 cultures in the range of 10^5–10^6 cells/mL. This requires medium renewal or expansion of the cultures about every 3 d.
3. Freeze cells at 10^7 cells/mL: harvest healthy, exponentially growing cells, resuspend in 5% DMSO/95% FCS (v/v).
4. Freeze gradually by storing the cells at −80°C in polystyrene foam containers for the first day, and then transferring to the liquid N_2 freezer on the second day for long-term storage.

3.1.3. Transfection or Transduction of K-562 Cell Cultures

1. Maintain sterile technique throughout this protocol.
2. Collect healthy, log-phase cells (90–95% viable; usually fed 24–48 h in advance) by centrifuging for 5 min at 220g and 4°C.
3. Wash the pelleted cells by repeating the suspension and centrifugation with an equal volume of RPMI at 4°C.
4. Resuspend the final cell pellet in cold RPMI and count cells.
5. Aliquot 10^7 cells per electroporation in separate 15-mL conical centrifuge tubes.
6. Centrifuge the cells again, and resuspend each pellet (10^7 cells) in 300 µL of RPMI for each transfection. Alternatively, re-suspend cells in sucrose/low salt electroporation buffer for higher transfection efficiency.
7. Dispense 20 µg of plasmid DNA (*see* **Note 15**) into the corner of a sterile electroporation cuvette. Place the cuvet on ice.
8. Pipet 300 µL of cells into the cuvet over the top of the DNA. Perform this step in the sterile hood.
9. Incubate samples on ice for 10 min.
10. Electroporate cells with 10 pulses at 1 kV, 99 µsec for each pulse (use the high voltage instrument setting and outlet) (*see* **Notes 16** and **17**).
11. Immediately move to the sterile hood and transfer the electroporated cells into a 25-cm² tissue-culture flask by flushing the cuvette twice with 2 mL of warm com-

plete RPMI medium (37°C). Add another 10 mL of warm medium to the flask, for a total volume of 14 mL (*see* **Note 18**).

12. Incubate for 48–72 h at 37°C in a 5% CO_2 tissue culture incubator. This protocol typically results in approx 50% transfection efficiency, which may be suitable for transient functional assays after feeding and expansion of the culture for a few days. The rapid growth rate and recovery of these cells makes transient assays feasible in many experiments.

13. To favor outgrowth of transfected cells, antibiotic selection for the plasmid-encoded antibiotic resistance marker is advised.

14. Seed cultures at 10^5 cells/mL in complete RPMI medium in the presence of antibiotics [hygromycin B (250 µg/mL) or G-418 (0.8 mg/mL)] 48–72 h after electroporation.

15. Expression can be stable for weeks following an initial 7–10 d of selection in the presence of antibiotics (*see* **Note 19**).

3.2. NADPH Oxidase Assays

The fully competent K-562 line transduced with all three myeloid *phox* components produces enough O_2^- in response to PMA stimulation to be detected in a cytochrome c-based assay, such as the microtiter plate-formatted assay described below (*see* **Note 20**). In contrast, K-562 cell lines that express lower amounts of *phox* components or cells stimulated by less potent agonists do not produce enough O_2^- to be in the relatively-insensitive cytochrome c assay. To address this issue, we describe a highly-sensitive, O_2^--specific chemiluminescence assay that utilizes the luminol-based reagent, Diogenes for detection of low levels of oxidase activity in transfected K-562 cells (*see* **Notes 21**). This assay is designed for real-time measurements on reactions formatted in a 96-well microtiter plate, allowing comparisons of multiple parameters, such as several lines, transfections, dose responses, inhibitors, etc., in replicate reactions that can be monitored in parallel (*see* **Note 22**).

3.2.1. Cytochrome c-Based Assay

1. Feed cells 24–48 h in advance, as healthy, log-phage cells are needed for the assay.

2. Count cells, and collect enough for the entire experiment. This assay typically uses 1.5×10^5 K-562 cells/well, and triplicate wells are preferred. Control and reference reactions may include wells containing SOD or DPI, wells lacking cell stimulation, etc.

3. Wash cells twice with HBSS+, and resuspend at the appropriate density needed to achieve 1.5×10^5 cells/well in approx 45 µL of HBSS+.

4. Dispense cells into microtiter plate wells, and add inhibitors or other test reagents to cells, as needed. Generally, add <5 µL of test reagent and adjust the cell volumes to obtain a volume of 50 µL. For example, add 1 µL of DPI solution (10 m*M* final concentration) or 4 µL (5 µg) of SOD solution to control wells, if DPI- or SOD-insensitive readings are desired as negative controls.

5. Pre-incubate the plate in a microtiter plate reader, pre-equilibrated to 37°C.
6. Set up the microtiter plate reader using the appropriate program (i.e., SoftMax Pro) based on the number wells to be observed at 550 nm. For example, PMA-stimulated reactions are typically monitored at 1-min intervals over a total reaction time-course of 20-30 min. Preset a 5-s auto-mixing phase before the first reading and a 3-s mix between readings.
7. Prepare an adequate amount of 2X cytochrome c/activator reagent (i.e., cytochrome c/PMA reagent) for all samples, and pre-warm the reagent to 37°C (*see* **Note 23**).
8. Start the reactions by quickly adding 50 μL of the 2X cytochrome c/activator reagent to each well, and immediately initiate absorbance readings.
9. Record data as the kinetics of absorbance change at 550 nm vs time.
10. Absorbance changes can be converted to nmoles of cytochrome c reduced by O_2^-, based on the following calculation *(22)*:

$$\text{nanomoles } O_2^-/\text{well/unit time} = \Delta\text{Absorbance}_{550\text{ nm}} \times 100/6.3 \qquad (1)$$

assuming an extinction coefficient of 21,000 $M^{-1}\text{cm}^{-1}$ and a vertical light path-length of approx 0.3 cm for the height of the 100 μL solution in the microtiter plate well (*see* **Note 24**). We typically calculate the maximum rate of O_2^- production (V_{max}) from the steepest slope of the curves, using the SoftMax software.

3.2.2. Chemiluminescence-Based Assay

The more sensitive chemiluminescence-based oxidase assay is performed in a manner similar to the cytochrome c assay outlined under **Subheading 3.2.1.** above.

1. Feed cells 24–48 h in advance, as healthy, log-phage cells are needed for the assay.
2. Count the cells, wash, and aliquot into the wells of a luminometer microplate (2.5–5 × 10⁴ cells/well in 50 μL of HBSS+ (*see* **Note 25**).
3. Treated cells with various inhibitors or other reagents, as described previously.
4. Set up the microtiter plate luminometer to scan the desired number of wells. Because luminescence readings are obtained in real time, a longer dwell time may be used for higher sensitivity measurement of cells with low oxidative output. The number of wells scanned per cycle may be limited by the kinetics of oxidase activation. As an example, PMA-stimulated oxidase activity is monitored at 1-min intervals (1 s dwell time per well, if assaying <60 wells). The rapid kinetics of formyl peptide-triggered oxidase activity, however, require a faster cycling time (12-s intervals) and a dwell time of less than 1 s per well, assuming that 12 reactions or less are followed at once.
5. Add 50 μL of the prewarmed 2X Diogenes/activator reagent to start the reactions, and immediately initiate luminescence readings (*see* **Note 23**).
6. PMA-activated responses are slow enough to permit manual addition of Diogenes/activator reagent using a multi-channel pipette for a large number of reactions (*see* **Fig. 2B**).

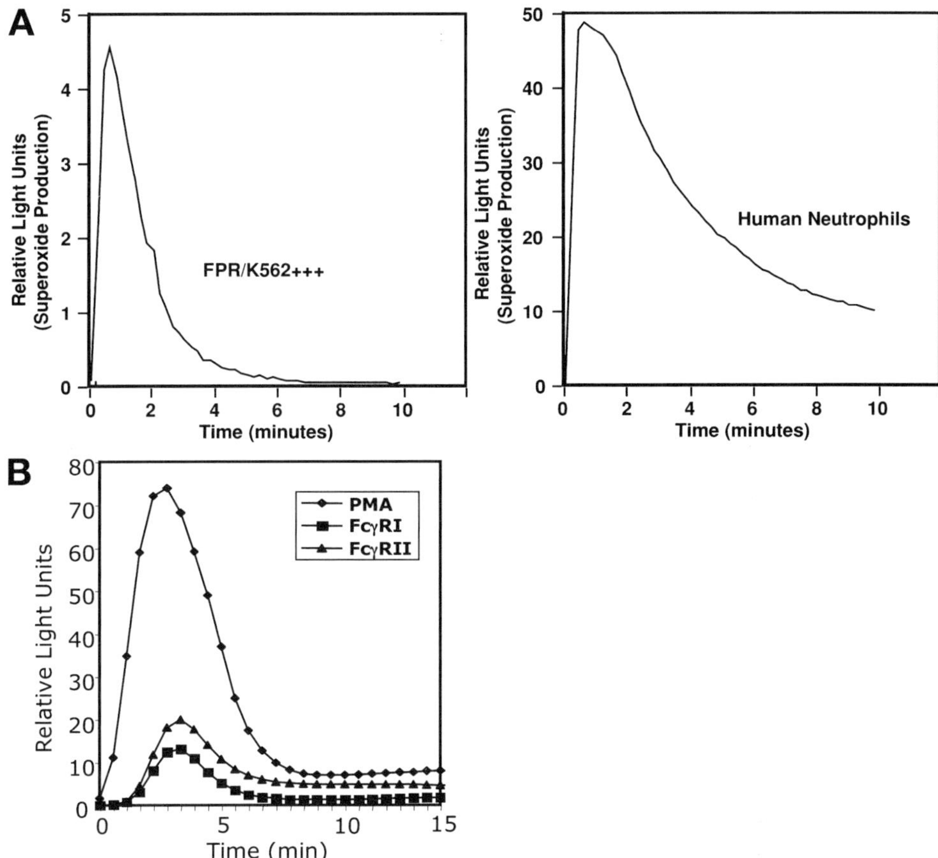

Fig. 3. Receptor-mediated NADPH oxidase activion in reconstituted K-562 cells. **(A)** Kinetics of formyl peptide receptor (FPR)-mediated oxidase activation (100 nM fMLP) in FPR-transfected *phox*-competent K-562 (FPR/K562^{+++}) cells (left) closely parallel the kinetics of oxidase activation observed in human polymorphonuclear neurtrophils (right). K-562^{+++} cells were electroporated with the human FPR1 cDNA (*29*) and selected for G-418 resistance. **(B)** Endogenous Fcγ receptor-mediated oxidase responses in *phox*-competent K-562 (K-562^{+++}) cells, compared with the response of phorbol myristate acetate-triggered cells. Bivalent Fab$_2$ fragments of monoclonal antibodies directed against the Fc-gamma receptors were bound to the cell surface at 4°C for 30 min. Ligated receptors were then cross-linked following the addition of secondary antibodies against the bound mouse Fab$_2$ fragments (*see* **Note 23**). A similar response can be elicited by serum-opsonized zymosan (not shown; *see* **Note 23**).

7. For rapidly-acting agonists (i.e., fMLP), use the instrument's automatic dispenser to add Diogenes/fMLP reagent to the cells and immediately observe luminescence on a rapidly cycling program (i.e., reading intervals of 12 s; *see* **Fig. 3A** and **Note 26**).

8. As an alternative for studies with rapidly acting agonists, the cells can be pre-chilled on ice prior to agonist addition in order to delay the initiation of oxidase activation (*see* **Fig. 3B**).
9. Calculate relative oxidase activities observed, based on the integrated relative light units observed over the course of the reaction (essentially the area under the curve of luminescence vs time). Thus, the peak in real-time luminescence values is observed at the same point as the V_{max} of cytochrome c reduction assays.
10. Any comparisons of total integrated luminescence observed between experiments should account for dwell times and the number or readings.

3.3. Preparation of Cell Fractions and Western Blotting of phox Proteins

Western blotting of NADPH oxidase components is frequently required in assessing differences in NADPH oxidase observed between various reconstituted cell lines (e.g., to confirm that variant or mutated *phox* proteins are stable or produced at comparable levels). The methods for preparation of cytosolic and membrane fractions described below were derived from protocols for fractionation of PMN *(23)*. Because K-562 cells do not contain granule constituents, this abbreviated protocol will remove nuclei and large cell fragments and provides a membrane-enriched fraction devoid of cytosolic proteins.

1. Count live cells and collect at least 1×10^7 cells in 15-mL conical tubes.
2. Centrifuge cells for 5 min at 220*g* and 4°C.
3. Wash cells three more time by resuspending the pellet in HBSS– and centrifuging (5 min, 220*g*, 4°C).
4. After the last centrifugation, resuspend cells at a minimum density of 5–10×10^7 cells/mL in Relaxation buffer containing PIC (*see* **Notes 27** and **28**), and chill the cells on ice for 10 min.
5. Sonicate the cells three times with 10-s bursts while chilling the cells on ice between each sonication. Check for adequate cell disruption by microscopic examination of the cell suspension stained with Trypan blue.
6. Spin sonicates in a microfuge for 5 min at 1400*g* and 4°C to remove unbroken cells, large cell fragments, and nuclear protein complexes.
7. A membrane-enriched pellet can be obtained by centrifugation of the supernatant from **step 6** for 45 min at 13,000–20,000*g*. However, better yield of membranes and a cleaner separation from cytosolic protein contaminants is obtained by following **steps 8** and **9** (recommended for membrane translocation assays).
8. Layer 200 µL of postnuclear supernatant fraction (**step 6**) over 50 µL of 10% sucrose-Relaxation buffer, and centrifuge for 30 min at 150,000*g* (ultracentrifuge or airfuge).
9. Wash the membrane pellets by resuspending each pellet in 100 µL of ice-cold Relaxation buffer containing PIC, and centrifuging them again for 20 min at 150,000*g*.
10. Resuspend washed pellets at a volume of 10^9 cell equivalents/mL in Relaxation buffer.
11. Perform protein assays on cytosol and membrane fractions (i.e., Micro BCA).

12. Separate 20–30 µg of cytosol and 30–40 µg of membrane on polyacrylamide gels and transfer to nitrocellulose for immunoblotting.
13. Stain the blot with Ponceau S solution (5 s), wash three times with distilled H_2O, and photograph.
14. Preblock blot by incubating in blocking buffer for at least 1 h at room temperature (or overnight at 4°C). This step also removes the Ponceau S stain.
15. Pour off blocking buffer, and wash the blot in 1X TBS for approx 5 min.
16. Incubate the blot with diluted primary antibody solution for exactly 1 h at room temperature on an orbital shaker. Higher sensitivity can be obtained with longer incubations with some antibodies, but this may increase background signals.
17. Wash the blot with TBS/Tween-20 at least two times for 3 min and three times for 10 min. Washes can be more extensive if a high background reaction is observed.
18. Incubate the blot with the appropriate species of HRP-conjugated secondary antibody solution at room temperature for exactly 1 h. A lower background may be obtained with a 45-min incubation.
19. Wash blot as before with TBS/Tween (**step 17**).
20. Using flat-tipped forceps, drain excess liquid from blot.
21. Apply 4 mL of ECL substrate to blot on flat surface, and incubate for 5 min. Drain excess ECL reagent, wrap blot in plastic wrap, and place in an X-ray film cassette.
22. Expose blot to X-ray film by applying in one motion. Expose for 2–5 s, and then develop. Adjust signal by re-exposing for longer or shorter times.

3.4. Oxidase Assembly/Membrane Translocation Assays

Fractionation procedures have been used to explore structure–function requirements for assembly of cytosolic oxidase components into activated membrane-bound oxidase complexes *(9–11)*. Membrane translocation of cytosolic *phox* components is dependent on their interactions with flavocytochrome b_{558}, as observed in stimulated PMN.

1. Count cells, and collect at least 5×10^6 cells per assay into 15-mL conical centrifuge tubes.
2. Wash cells two to three times with cold HBSS–, centrifuging between washes (5 min, $22g$, 4°C).
3. Resuspend final cell pellet at a concentration of 5×10^7 cells/mL (2.5×10^6 cells/50 µL) in HBSS+. For an unstimulated cell control, keep 5×10^6 cells on ice.
4. Prewarm cells for 5 min at 37°C.
5. Stimulate cells for 10 min at 37°C (e.g., 0.1–3 µg/mL PMA final concentration from freshly prepared PMA in HBSS+).
6. Remove tubes from the 37°C bath, and immediately place on ice and dilute with 5 mL of ice-cold HBSS–.
7. Centrifuge stimulated samples and unstimulated control cells for 5 min at $220g$ and 4°C, and resuspend them at 10^8 cells/mL in cold relaxation buffer containing PIC.
8. Sonicate cells and prepare membrane and cytosolic fractions for Western blotting, as described above (**Subheading 3.3.**). In this case, it may be necessary to wash the

"membrane" fraction up to three times (**Subheading 3.3., step 9**) to remove contaminating cytosol.

4. Notes

1. K-562 cells were used initially to explore roles of multiple SH3 domain-proline-rich target peptide interactions in membrane translocation of the cytosolic *phox* regulators *(10)*. The model was later used to confirm roles for p47^{phox} phosphorylation and Rac-p67^{phox} interactions in oxidase activation *(13,14)*. As a result of their hematopoietic origin, K-562 cells offer advantages over other easily-transfected cell models, such as Cos-7 cells *(24,25)*, since they are capable of receptor-mediated oxidase activation in the absence of other heterologously-expressed signaling intermediates. These cells also appear to express Fcγ receptors *(26)*.
2. These cell lines are versatile system and can be used to introduce other components, either transiently or in long-term transfection experiments in the presence of selection markers. These lines represent useful tools for screening gene therapy vectors designed for correction of the defective oxidase in chronic granulomatous disease *(27)*, to assess the effects of polymorphisms in oxidase components, or to examine cross-functioning of *phox* and related Nox family components *(28)*.
3. The recombinant *phox* retroviruses were harvested from culture supernatants of MFG-S transfected ψ-Crip viral packaging cell lines *(16–18)* and were used for four to six repeated rounds of transduction of K-562 cells using the "spinoculation" protocol *(18)*. Transductions with more than one *phox* component were performed sequentially, since the three vectors did not carry any selectable markers.
4. In order to isolate single cell-derived clones, the cells were serially diluted to approx 5 cells/mL and distributed into several 96-well plates (100 μL/well). The triply-transduced lines (expressing retroviral p47*phox*, p67*phox*, and gp91*phox*) were initially screened for reconstitution of PMA-stimulated oxidase activity *(6)*. Partially-reconstituted clones, transduced with only two *phox* components, were screened functionally by transfecting the third essential *phox* component encoded by an episomal expression vector *(6)*.
5. The levels of cytosolic *phox* proteins in the triply-transduced and some other lines exceeded those detected in PMN cytosol. K-562 cells produce the gp91^{phox} unglycosylated (65 kDa) species together with the fully-glycosylated species. The amount of gp91^{phox} detected in transduced K-562 cell membrane fractions on a per mg protein basis were significantly lower than those in PMN membranes (**Fig. 1A**), although on a per cell basis, it was closer to that of PMNs because K-562 cells are many times larger (not shown). Interestingly, the amount of endogenous p22^{phox} detected directly correlated with that of transduced gp91^{phox}. Thus, as in PMN, co-expression of both flavocytochrome components is required to stabilize the flavocytochrome heterodimer in K-562 cells.
6. Harvard Apparatus offers an upgraded instrument (BTX Model 830) that can be used to reproduce the protocol described using BTX Model 820.
7. *Alternative transfection protocol I:* our earlier work involved a widely available instrument (Gene Pulser, Biorad) that delivers a single exponentially decaying pulse

(6,9–12). When using this instrument, follow the same protocol given under **Subheading 3.1.3.**, but modify **step 10** as follows: electroporate cells with a single pulse at 250 V/960 µF, and then promptly transfer cells to the flask with complete medium prewarmed to 37°C, as described.

8. *Alternative transfection protocol II:* nucleofection techniques have been described for K-562 cells by Amaxa, using Nucleofector Kit V and protocol T-16. A transfection efficiency of >70% has been reported by the manufacturer, although the reagents appeared to adversely affect cell viability in our hands. The method uses only 10^6 cells, and may therefore require expansion of cultures before proceeding to functional assays.

9. G-418 (used for selection of cells transfected with the Neomycin resistance marker) is sold under the name Geneticin® (contains 726 µg G-418/mg).

10. Although earlier work suggested that 2 µg mL PMA is needed for maximum NADPH oxidase activation in K-562 cells *(6)*, we have found that 100 ng/mL PMA works comparably.

11. The effects of DPI should be interpreted cautiously, because it is not very specific for NADPH oxidase; rather, it is a general flavoenzyme inhibitor.

12. The opaque wells are critical for optimal sensitivity for luminescence assays and also prevent light transmission (cross-talk) from neighboring wells.

13. K-562 cells express the appropriate receptors for efficient retrovirus vector transduction using either gamma retrovirus or lentivirus vectors pseudotyped with any of the following envelops: VSV-G, amphotropic, GALV, or RD114 *(16–18,27)*. Some cytotoxicity is associated with VSV-G pseudotyped vectors, and among the other three, GALV is the most efficient envelope for transduction of K-562 cells.

14. Many investigators maintain K-562 cells in Iscove's modified Dulbecco's medium. See culture recommendations provided by ATCC, under cat. no. CLL 243.

15. For large plasmids such as the pREP or pCEP series, use 50 µg of DNA per transfection. When co-transfecting with more than one plasmid at once, use the same amount of total plasmid DNA (i.e., half the amount of each for two plasmids). In this instance, we have observed that a majority of transfected cells express the products of more than one co-transfected plasmid *(6)*.

16. Care should be taken to wipe the cuvet exterior dry before placing it in the cuvet holder. Be sure the metal plates of the cuvet are flush in contact with the electrode surfaces, and do not short-circuit the electrodes.

17. This square-wave pulse generator (involving the BTX 820 instrument) ensures better reproducibility and cell viability; other K-562 cell electroporation protocols have been described, using the Gene Pulsar (BioRad) or Nucleofector (Amaxa) devices (*see* **Notes 7** and **8**).

18. The immediate transfer of electroporated cells into warm complete medium is one of the most important factors for obtaining a high transfection rate and maximum cell viability. Therefore, when doing several transfections at once, the electroporation and prompt transfer to culture flasks should be done individually, not in batches.

19. Because the expression of each transgenes is heterogeneous, enrichment or isolation of cells with the highest expression may be desirable. Initiate clonal selection

of cells demonstrating highest expression by limiting dilution at least 1 wk after cells have been maintained in the presence of selection antibiotics.

20. The microtiter plate cytochrome c-based assay described here was derived from a standard microplate assay of O_2^- release by macrophages *(22)*.

21. Diogenes has been used extensively in other laboratories *(6,9–13,16–18,27,28)*. Unlike luminol alone, this reagent is specific for O_2^- (>500 more sensitive to O_2^- than H_2O_2) as a result of the presence of a redox active compound that facilitates transfer of electrons from O_2^- to luminol (*see* **Note 29**). The activity detected by this system in K-562 is abolished by exogenously added SOD (>98%) and is insensitive to catalase, indicative of its specificity for O_2^-.

22. Luminescence readings can be obtained over a very broad range of activities and can be linear with respect to cell number within limits. Typically, the relative chemiluminescence units observed over a fixed time course are integrated for quantitative comparisons between reactions, although these readings cannot be converted directly to yield of O_2^- in absolute terms.

23. For stimulation of oxidase activity through endogenous Fcγ receptors, preincubte cells on ice for 30 min with a 1:200 dilution of anti-FcRI (#32.2) or anti-FcRII (#IV.3) monoclonal F(ab')2 fragments (Medarex), wash with HBSS+, and dispense into luminescence microtiter plates. Ligand-occupied receptors are then cross-linked with a 1:200 dilution of goat anti-mouse F(ab')2 (Jackson ImmunoResearch). Alternatively, opsonized zymosan (OZ)-mediated oxidase activation can be observed using 1 μg to stimulate 1.5×10^5 cells/reaction. Prepare OZ particles by incubating 20 mg of zymosan (Sigma) per 1 mL of serum (30 min at 37°C), then wash three times in HBSS. Use 5 μL of 20 mg/ml OZ suspension per 100 μL oxidase assay.

24. This formula is based on the Beer-Lambert Law as shown in Equation 2:

$$A_{550} = \varepsilon \cdot C \cdot l \text{ or } C = A_{550} / (\varepsilon \cdot l) \tag{2}$$

where C = concentration of reduced cytochrome c $\approx O_2^-$ in M, A = the absorbance at 550 nm, ε = the molar extinction coefficient for reduced cytochrome c at 550 nm (21,000 $M^{-1}cm^{-1}$), and l = the light path length in cm. Because ε is in units of M^{-1}, the calculation must be adjusted for the sample volume of 100 μL and the desired reporting unit of nmoles. This formula is strictly dependent on the assay volume and the light path length. If these assay parameters are changed, then the formula must be adjusted accordingly.

25. The number of cells used in each assay (final volume of 100 μL) can range from $2.5–30 \times 10^4$. This would depend on several parameters, including the level of oxidase activity reconstituted and the potency of the agonist used to activate the oxidase.

26. The kinetics of NADPH activation in K-562 cells are remarkably similar to those observed in neutrophils. The 2- to 3-min lag in PMA-stimulated oxidase activation will usually permit observation of the entire burst following manual addition of the agonist (combined with the Diogenes reagent), if relatively few reactions are run at once or if a multichannel dispensing pipet is used. If many reactions are being followed concurrently or if a rapidly acting agonist is used (i.e., fMLP), the

automatic dispenser of the instrument should be used to add the Diogenes/activator component, in order to capture early phases of oxidase activation. Be sure that the dispenser is adequately primed with freshly prepared detection/ activator reagent.

27. The minimum volume required for sonication of cell suspensions in an Eppendorf tube using the microtip is 200 µL (*see* **Note 28**). Larger volumes (up to 2–3 mL) can be sonicated in 15-mL conical tubes.

28. It is best to "tune" the sonicator by establishing the minimum power output needed to attain resonance. This is done by pretesting appropriate energy levels, using the same tube and buffer volume used to sonicate the cells. Immerse the sonicator microtip approx 3–4 mm below the surface and then increase the power output until a "dancing bubble" is observed at the bottom of the tube. Avoid higher energy levels that can cause foaming.

29. Personal communication (G. Kitzler, National Diagnostics). The chemical nature of this proprietary enhancer (primary electron acceptor) component was not disclosed.

Acknowledgments

We thank Balázs Rada for assistance in preparation of figures.

References

1. Segal, B. H., Leto, T. L., Gallin, J. I., Malech, H. L., and Holland, S. M. (2000) Genetic, biochemical, and clinical features of chronic granulomatous disease. *Medicine (Baltimore)* **79,** 170–200.

2. Cross, A. R. and Segal, A. W. (2004) The NADPH oxidase of professional phagocytes—prototype of the NOX electron transport chain systems. *Biochim. Biophys. Acta* **1657,** 1–22.

3. Quinn, M. T. and Gauss, K. A. (2004) Structure and regulation of the neutrophil respiratory burst oxidase: comparison with nonphagocyte oxidases. *J. Leukoc. Biol.* **76,** 760–781.

4. Geiszt, M. and Leto, T. L. (2004) The Nox family of NAD(P)H oxidases: host defense and beyond. *J. Biol. Chem.* **279,** 51,715–51,718.

5. Rotrosen, D., Leto, T. L., Malech, H. L., and Kwong, C. H. (1992) Cytochrome b558: the flavin-binding component of the phagocyte NADPH oxidase. *Science* **256,** 1459–1462.

6. de Mendez, I. and Leto, T. L. (1995) Functional reconstitution of the phagocyte NADPH oxidase by transfection of its multiple components in a heterologous system. *Blood* **85,** 1104–1110.

7. Lozzio, C. B. and Lozzio, B. B. (1975) Human chronic myelogenous leukemia cell-line with positive Philadelphia chromosome. *Blood* **4,** 321–334.

8. Lozzio, B. B. and Lozzio, C. B. (1979) Properties and usefulness of the original K-562 human myelogenous leukemia cell line. *Leuk. Res.* **3,** 363–370.

9. de Mendez, I., Adams, A. G., Sokolic, R. A., Malech, H. L., and Leto, T. L. (1996) Multiple SH3 domain interactions regulate NADPH oxidase assembly in whole cells. *EMBO J.* **15,** 1211–1220.

10. Leto, T. L., Adams, A. G., and de Mendez, I. (1994) Assembly of the phagocyte NADPH oxidase: binding of Src homology domains to proline-rich targets. *Proc. Nat. Acad. Sci. USA* **91,** 10,650–10,654.

11. de Mendez, I., Homayounpour, N., and Leto, T. L. (1997) Specificity of 47-phox SH3 domain interactions in NADPH oxidase assembly and activation. *Mol. Cell. Biol.* **17,** 2177–2185.

12. Sathyamoorthy, M., Adams, A. G., de Mendez, I., and Leto, T. L. (1997) p40-phox down-regulates NADPH oxidase activity through interactions with its SH3 domain. *J. Biol. Chem.* **272,** 9141–9146.

13. Koga, H., Terasawa, H., Nunoi, H., Takeshige, K., Inagaki, F., and Sumimoto, H. (1999) Tetratricopeptide repeat (TPR) motifs of p67(phox) participate in interaction with the small GTPase Rac and activation of the phagocyte NADPH oxidase. *J. Biol. Chem.* **274,** 25,051–25,060.

14. Ago, T., Nunoi, H., Ito, T., and Sumimoto, H. (1999) Mechanism for phosphorylation-induced activation of the phagocyte NADPH oxidase protein p47(phox). Triple replacement of serines 303, 304, and 328 with aspartates disrupts the SH3 domain-mediated intramolecular interaction in p47(phox), thereby activating the oxidase. *J. Biol. Chem.* **274,** 33,644–33,653.

15. van Bruggen, R., Anthony, E., Fernandez-Borja, M., and Roos, D. (2003) Continuous translocation of Rac2 and the NADPH oxidase component p67phox during phagocytosis. *J. Biol. Chem.* **279,** 9097–9102.

16. Li, F., Linton, G. F., Sekhsaria, S., et al. (1994) CD34+ peripheral blood progenitors as a target for genetic correction of the two flavocytochrome b_{558} defective forms of chronic granulomatous disease. *Blood* **84,** 53–58.

17. Malech, H. L., Maples, P. B., Whiting-Theobald, N., et al. (1997) Prolonged production of NADPH oxidase-corrected granulocytes after gene therapy of chronic granulomatous disease. *Proc. Nat. Acad. Sci. USA* **94,** 12,133–12,138.

18. Weil, W. M., Linton, G. F., Whiting-Theobald, N., et al. (1997) Genetic correction of p67*phox* deficient chronic granulomatous disease using peripheral blood progenitor cells as a target for retrovirus mediated gene transfer. *Blood* **89,** 1754–1761.

19. Laemmli, U. K. (1970) Cleavage of structural proteins during the assembly of the head of bacteriophage T4. *Nature* **227,** 680–685.

20. Leto, T. L., Garrett, M. C., Fujii, H., and Nunoi, H. (1991) Characterization of neutrophil NADPH oxidase factors p47-*phox* and p67-*phox* from recombinant baculoviruses. *J. Biol. Chem.* **266,** 19,812–19,818.

21. Verhoeven, A. J., Bolscher, B. G. J. M., Meerhof, L. J., et al. (1989) Characterization of two monoclonal antibodies against cytochrome b_{558} of human neutrophils. *Blood* **73,** 1686–1694.

22. Pick, E. and Mizel, D. (1981) Rapid microassays for the measurement of superoxide and hydrogen peroxide production by macrophages in culture using an automatic immunoassay reader. *J. Immunol. Meth.* **46,** 211–226.

23. Borregaard, N., Heiple, J. M., Simons, E. R., and Clark, R. A. (1983) Subcellular localization of the b-cytochrome component of the human neutrophil microbicidal oxidase translocation during activation. *J. Cell Biol.* **97,** 52–61.

24. Price, M. O., McPhail, L. C., Lambeth, J. D., Han, C. H., Knaus, U. G., and Dinauer, M. C. (2000) Creation of a genetic system for analysis of the phagocyte respiratory burst: high-level reconstitution of of the NADPH oxidase in a nonhematopoietic system. *Blood* **99,** 2653–2661.
25. He, R., Nanamori, M., Sang, H., Yin, H., Dinauer, M. C., and Ye, R. D. (2004) Reconstitution of chemotactic peptide-induced nicotinamide adenine dinucleotide phosphate (reduced) oxidase activation in transgenic COS-phox cells. *J. Immunol.* **173,** 7462–7470.
26. Chiofalo, M. S., Teti, G., Goust, J.-M., Trifiletti, R., and La Via, M. F. (1988) Subclass specificity of the receptor for human IgG on K-562. *Cellular Immunol.* **114,** 272–281.
27. Roesler, J., Brenner, S., Bukovsky, A. A., et al. (2002) Third-generation, self-inactivating gp91phox lentivector corrects the oxidase defect in NOD/SCID mouse-repopulating peripheral blood-mobilized CD34+ cells from patients with X-linked chronic granulomatous disease. *Blood* **100,** 4381–4390.
28. Geiszt, M., Lekstrom, K., Brenner, S., et al. (2003) NAD(P)H oxidase 1, a product of differentiated colon epithelial cells, can partially replace glycoprotein 91^{phox} in the regulated production of superoxide by phagocytes. *J. I mmunol.* **171,** 299–306.
29. Hartt, J. K., Liang, T., Sahagun-Ruiz, A., Wang, J. M., Gao, J. L., and Murphy, P. M. (2000) The HIV-1 cell entry inhibitor T-20 potently chemoattracts neurophils by specifically activating the N-formylpeptide receptor. *Biophys. Biochem. Res. Commun.* **272,** 699–704.

25

Cell-Free Assays

The Reductionist Approach to the Study of NADPH Oxidase Assembly, or "All You Wanted to Know About Cell-Free Assays but Did Not Dare to Ask"

Shahar Molshanski-Mor, Ariel Mizrahi, Yelena Ugolev, Iris Dahan, Yevgeny Berdichevsky, and Edgar Pick

Summary

The superoxide ($O_2^{\bullet-}$)-generating enzyme complex of phagocytes, known as the NADPH oxidase, can be assayed in a number of in vitro cell-free (or broken cell) systems. These consist of a mixture of the individual components of the NADPH oxidase, derived from resting phagocytes or in the form of purified recombinant proteins, exposed to an activating agent (or situation), in the presence of NADPH and oxygen. $O_2^{\bullet-}$ produced by the mixture is measured by being trapped immediately after its generation with an appropriate acceptor in a kinetic assay, which permits the calculation of the linear rate of $O_2^{\bullet-}$ production over time. Cell-free assays are distinguished from whole-cell assays or assays performed on membranes derived from stimulated cells by the fact that all components in the reaction are derived from resting, nonstimulated cells and, thus, the steps of NADPH oxidase activation (precatalytic [assembly] and catalytic) occur in vitro. Cell-free assays played a paramount role in the identification of the components of the NADPH oxidase complex, the diagnosis of various forms of chronic granulomatous disease (CGD), and, more recently, the analysis of the domains present on the components of the NADPH oxidase participating in protein–protein interactions leading to the assembly of the active complex.

Key Words: Cell-free assay; broken cell-assay; reactive oxygen species; superoxide; NADPH oxidase; cytochrome b_{559}; cytosolic components; p47*phox*; p67*phox*; Rac; nucleotide exchange; isoprenylation; ferricytochrome *c* reduction; anionic amphiphile; arachidonic acid; superoxide dismutase.

1. Introduction

Phagocytic cells (neutrophils and macrophages) damage pathogenic microorganisms by a number of effector mechanisms. Among these, the production of

From: *Methods in Molecular Biology, vol. 412: Neutrophil Methods and Protocols*
Edited by: M. T. Quinn, F. R. DeLeo, and G. M. Bokoch © Humana Press Inc., Totowa, NJ

reactive oxygen species (ROS) occupies a prominent position. ROS are derived from the primordial oxygen radical, superoxide ($O_2^{\bullet-}$), which is produced in response to appropriate stimuli by a tightly regulated enzyme complex, known as the $O_2^{\bullet-}$-generating NADPH oxidase (briefly, "oxidase") (reviewed in **ref. 1**). The oxidase catalyzes the formation of $O_2^{\bullet-}$ by the NADPH-driven one-electron reduction of O_2, a process involving the transfer of electrons via several redox stations. The functionally competent oxidase complex is composed of a membrane-associated flavocytochrome b_{559} (also known as cytochrome b_{559} or flavocytochrome b_{558}; a heterodimer consisting of two subunits, gp91*phox* and p22*phox*) and four cytosolic components, p47*phox*, p67*phox*, p40*phox*, and the small GTPase Rac1/2 (reviewed in **ref. 2**). The component responsible for electron transfer from NADPH to oxygen is gp91*phox*, which contains the NADPH binding site and two redox stations, represented by flavin adenine dinucleotide (FAD) and two nonidentical hemes. In resting phagocytes, the components of the complex exist as distinct entities, and the process of oxidase activation is the consequence of the interaction of flavocytochrome b_{559} with cytosolic components. This requires translocation of the cytosolic components to the membrane environment of flavocytochrome b_{559}, a process requiring a complex set of protein–protein interactions and which is defined as oxidase assembly (reviewed in **ref. 3**).

For oxidase assembly, the decisive interaction is that of gp91*phox* with one or more of the cytosolic components, resulting in a conformational change in gp91*phox* that initiates electron flow from NADPH to O_2 and the generation of $O_2^{\bullet-}$. In one model of oxidase assembly, p67*phox* is seen as the only cytosolic component responsible for an "activating" interaction with gp91*phox* **(4)**, whereas in a different model, both p67*phox* and Rac interact directly with and have distinct roles in the regulation of electron transfer in gp91*phox* (reviewed in **ref. 5**). Because p67*phox* does not possess a membrane attachment signal of its own, it requires the assistance of p47*phox* and Rac to be brought in contact with gp91*phox* **(4,6,7)**. Because of this, p47*phox* and p67*phox* homologs associated with some of the nonphagocytic oxidases (NOXes) were named NOX organizer (NOXO)1 and NOX activator (NOXA)1, respectively (reviewed in **ref. 8**). The roles of p47*phox* and Rac in helping the association of p67*phox* with gp91*phox* are not interchangeable; under certain in vitro conditions, oxidase activation can take place in the absence of p47*phox* but not of Rac **(9,10)**. These differences in the "assistance" provided to p67*phox* are of significance in the design and interpretation of cell-free assays and will be discussed in this chapter.

The early literature on the assay of NADPH oxidase was rooted in the awareness of the multiplicity of stimulants causing ROS production by phagocytes **(11,12)** and on the finding that, independent of the nature of the stimulant,

NADPH oxidase activity could be demonstrated in the particulate (membrane) fraction of stimulated but not resting cells *(13)*. Thus, subjection of phagocytes to a stimulant, followed by cell disruption, separation of the cell homogenate into a particulate (membrane) and a cytosolic fraction, and measurement of NADPH-dependent $O_2^{\bullet-}$ production by the membrane fraction, became the standard procedure for a assessing NADPH oxidase activity *(14)*. In the course of work with the membrane fraction of stimulated phagocytes, two facts became firmly established: first, the physiological substrate of the enzyme was NADPH and not NADH *(13)*, and second, the enzyme required supplementation with FAD for optimal activity *(15)*. Working with membranes from stimulated cells was technically difficult because the activity, as expressed in reaction rates, declined rapidly over time, was inactivated at 37°C, and was sensitive to high salt concentrations and to harsh sonic disruption of the cells *(14)*. These difficulties prompted a search for an alternative procedure for the assay of NADPH oxidase. An additional stimulus for such a search was the finding that most patients with the autosomal recessive pattern of chronic granulomatous disease (CGD) possessed a normal cytochrome b_{559}, found to be missing in the X-linked form of the disease. In the early 1980s, cytochrome b_{559} was the only known component of the NADPH oxidase, and it was not known that the oxidase consisted of more than one component (reviewed in **ref. 2**). All of this led to the idea of attempting to activate the oxidase in subcellular fractions derived from resting phagocytes. This conceptual shift drew encouragement from work on the activation of adenylate cyclase in broken-cell preparations by cholera toxin, which exhibited a requirement for both membranes and a cytosolic component *(15)*. The development of cell-free oxidase activation systems, however, was not truly the result of a focused search for a better methodology but rather the consequence of a fortuitous finding (*see* **Note 1**).

A finding of paramount importance for future developments in oxidase research was that neither membrane nor cytosolic fractions of the phagocyte homogenate were, by themselves, capable of supporting fatty acid-elicited $O_2^{\bullet-}$ production, but that recombining the two fractions resulted in full reconstitution of oxidase activity *(16)*. Similar cell-free oxidase-activating systems were described within 1 yr by three other groups. These were derived from resting horse *(17)* or human neutrophils *(18,19)*, and in all cases, a requirement for cooperation between membrane and cytosol was found. An important finding made by two of these groups was that patients with the X-linked form of CGD possess a defective membrane component but a normal cytosolic component(s) *(17,19)*. This observation was followed by the reciprocal finding that most patients with CGD with the autosomal recessive mode of inheritance have a normal membrane component but a defective cytosolic component(s) *(20)*. These clinical correlates were important for "legitimizing" the cell-free system as reflecting the activity of the $O_2^{\bullet-}$-forming oxidase, known from work in intact phagocytes.

Soon after the introduction of the cell-free system, the mechanism by which fatty acids cause oxidase activation became a subject of intense investigation. One of the first breakthroughs was the discovery that fatty acids act as anionic amphiphiles, as shown by the ability of sodium or lithium dodecyl sulfates (SDS or LiDS) to replace fatty acids as activators in the cell-free system, with very similar dose–response curves *(21)*. Other anionic detergents, such as sodium cholate, sodium deoxycholate, digitonin, and saponin, containing fused aromatic rings, are inactive. Sodium dodecyl sulfonate, like SDS, consists of an aliphatic hydrophobic moiety and an anionic polar head and thus serves as an activator *(21)*. The predominant explanation for the oxidase-activating ability of some anionic amphiphiles was that they disrupted an intramolecular bond in p47*phox* between a C-terminal polybasic region and the Src homology (SH)3 tandem. In the intact phagocyte, this bond is severed by the phosphorylation of critical serines within the polybasic stretch, freeing the SH3 domains of p47*phox* for intermolecular interaction with a proline-rich region at the C-terminus of p22*phox*, a step in the assembly of the oxidase complex *(22)*. Direct experimental evidence was provided for the induction of a conformational change in p47*phox* by arachidonate and SDS, a change that is also achieved by phosphorylation of p47*phox* by protein kinase C in vitro *(23)*. Evidence was also presented for a direct effect of anionic amphiphiles on cytochrome b_{559}, leading to conformational changes that might participate in oxidase activation *(24)*. Cell-free oxidase activation can also be induced by a combination of phosphatidic acid and diacylglycerol, again requiring participation of membrane and cytosolic fractions of the phagocyte homogenate *(25)*. Finally, a cell-free system involving phosphorylation of p47*phox* by protein kinase C in vitro and mixtures of phagocyte membrane and cytosol was designed to represent as closely as possible the in vivo reality *(26)*.

A methodological but also conceptual revolution was the introduction of the semi-recombinant cell-free system. In this system, the membrane is either used in the native form or represented by a purified and relipidated cytochrome b_{559} preparation, but the cytosol is replaced by a mixture of purified recombinant components *(27)*. This system permits introduction of strict quantification of the components participating in cell-free oxidase activation, the performance of dose–response assays, and control over the ratios among cytosolic components, among these and cytochrome b_{559}, and among the activating amphiphile and the membrane and cytosolic components. The amphiphile-activated cell-free system is the most frequently used and deserves to be coined the "canonical" cell-free assay. Recently, variations of the semi-recombinant amphiphile-dependent cell-free system were introduced in which pairs of individual cytosolic components were replaced by chimeric constructs of either [p47*phox*-p67*phox*] *(28)* or [p67*phox*-Rac1] *(29)*, supplemented by the missing third component (Rac1 and p47*phox*, respectively).

The elucidation of the mechanism by which anionic amphiphiles induce oxidase activation in the cell-free system (severing the intramolecular bond in p47*phox* between the [SH3]$_2$ tandem and the polybasic C-terminus) led to the design of a new type of amphiphile-independent cell-free system. In this assay system, p47*phox* truncated at residue 286, which removes the polybasic C-terminus *(30)*, or p47*phox* engineered with mutations that cause unmasking of (SH3)$_2$ *(31)*, are used without need for an amphiphile. The need for amphiphile is circumvented because each form allows interaction between the (SH3)$_2$ of p47*phox* and the proline-rich region at the C-terminus of p22*phox*.

We developed a conceptually different amphiphile-independent cell-free system based on the use of prenylated Rac1 *(7)*, which binds to phagocyte membranes with high affinity and serves as an anchor for p67*phox*, leading to oxidase activation in the absence of amphiphile and without the need for p47*phox*. Later variations of this system are represented by a prenylated [p67*phox*-Rac1-GTP] chimera, which activates the oxidase in the absence of amphiphile and of any other component *(32)*, and by prenylated Rac-GDP, in conjunction with a guanine nucleotide exchange factor (GEF) for Rac and GTP *(33)*. Finally, yet another cell-free system was recently described, based on the finding that enrichment of phagocyte membranes with anionic phospholipids, resulting in an increase in the negative charge, enhances the affinity of Rac1 and truncated p47*phox* and p67*phox* for the phagocyte membrane. Thus, a combination of a [p67*phox*(1-210)-p47*phox* (1-286)] chimera and Rac1-GTP activates phagocyte membranes enriched in anionic phospholipids in the absence of amphiphile *(28)*.

A great variety of cell-free assays have been developed over the years (summarized in **Tables 1** and **2**), and it is quite likely that while this chapter is being published, novel methodologies are in the making. In this chapter, we describe the basic methodology for performing a variety of different cell-free oxidase assays, which are commonly used in studying the NADPH oxidase and provide details on the various uses of this type of assay.

2. Materials

2.1. Chemicals and Reagents

2.1.1. Preparation of Phagocyte Membranes

1. Paraffin oil, weight/mL = 0.83–0.86 (BDH). Used for eliciting a sterile peritoneal exudate as a source of macrophages for the preparation of membranes.

Table 1
Amphiphile-Dependent Cell-Free Systems

Activator	Membrane	$p47^{phox}$	$p67^{phox}$	Rac	[$p47^{phox}$-$p67^{phox}$] chimera	[$p67^{phox}$-Rac] chimera	[$p47^{phox}$-$p67^{phox}$-Rac] chimera	Ref.
	Native	$p47^{phox}$	$p67^{phox}$	[Rac-RhoGDI] dimer, prenylated				16
	Native	$p47^{phox}$	$p67^{phox}$	Rac, nonprenylated or prenylated				27
Anionic amphiphile	Native			Rac, nonprenylated	[$p47^{phox}$-$p67^{phox}$] chimera			28
	Native	$p47^{phox}$				[$p67^{phox}$-Rac] chimera, nonprenylated		29
	Native						[$p47^{phox}$-$p67^{phox}$-Rac] chimera, nonprenylated	a

[a]Mizrahi, A., Berdichevsky, Y., Ugolev, Y., Molshanski-Mor, S., and Pick, E., manuscript in preparation.

Table 2
Amphiphile-Independent Cell-Free Systems

Activator	Membrane	p47^{phox}	p67^{phox}	Rac	[p47^{phox}-p67^{phox}] chimera	[p67^{phox}-Rac] chimera	[p47^{phox}-p67^{phox}-Rac] chimera	Ref.
	Native		p67^{phox}	Rac, prenylated				7
	Native						[p67^{phox}-Rac] chimera, prenylated	32
	Native		p67^{phox}	Rac, prenylated + Rac GEF (Trio, Tiam1) + GTP				33
	Native	Truncated p47^{phox}	Truncated p67^{phox}	Rac, nonprenylated or prenylated				30
No activator	+ anionic phospholipid	Mutated p47^{phox}	p67^{phox}	Rac, nonprenylated				31
	+ anionic phospholipid			Rac, nonprenylated	[p47^{phox}-p67^{phox}] chimera			28,[a]
	+ anionic phospholipid	p47^{phox}	p67^{phox}	Rac, nonprenylated				[a]
	+ anionic phospholipid	p47^{phox}				[p67^{phox}-Rac] chimera, nonprenylated		[a]
	+ anionic phospholipid						[p47^{phox}-p67^{phox}-Rac] chimera, nonprenylated	[a]
	+ anionic phospholipid	p47^{phox}	p67^{phox}	[Rac-RhoGDI] dimer, prenylated				52

[a]Mizrahi, A., Berdichevsky, Y., Ugolev, Y., Molshanski-Mor, S., and Pick, E., manuscript in preparation.

2. Earle's balanced salt solution (Invitrogen): 6.8 g NaCl, 0.4 g KCl, 0.125 g NaH$_2$PO$_4$·H$_2$O, 0.2 g MgSO$_4$·7 H$_2$O, 1 g glucose, 0.2 g CaCl$_2$ (anhydrous), 1.25 g NaHCO$_3$, and H$_2$O up to 1 L.

3. Sonication buffer: 8 mM potassium, sodium phosphate buffer, pH 7.0 (made from 61 parts K$_2$HPO$_4$ and 39 parts NaH$_2$PO$_4$ stock solutions), 131 mM NaCl, 340 mM sucrose, 2 mM NaN$_3$, 5 mM MgCl$_2$, 1 mM ethylene glycol-bis(β-aminoetyl ether)-N,N,N',N'-tetraacetic acid (EGTA), 1 mM dithiothreitol (DTT), 1 mM 4-(2-amino-ethyl)-benzenesulfonyl-fluoride (AEBSF; Sigma), and 0.021 mM leupeptin hemi-sulfate (Sigma). It is best to add DTT (200 mM), AEBSF (100 mM), and leupeptin (2.1 mM) from concentrated stock solutions (concentrations are listed in paren-theses) just before using the buffer.

4. 3.5 M KCl solution in 20 mM Tris-HCl buffer, pH 7.5, for mixing with sonication buffer to reach a final concentration of 1 M KCl. It is used to wash macrophage membranes, to remove nonintegral membrane-attached proteins.

5. Octyl-β-D-glucopyranoside (octyl glucoside; Sigma).

6. Solubilization buffer: 120 mM sodium phosphate buffer, pH 7.4, 1 mM MgCl$_2$, 1 mM EGTA, 2 mM NaN$_3$, 10 μM FAD disodium salt, and 20% v/v glycerol. Add 40 mM octyl glucoside (from powder), 1 mM DTT, 1 mM AEBSF, and 0.021 mM leupep-tin from concentrated stock solutions just before using the buffer. Unused buffer can be divided in aliquots of 25–50 mL and stored frozen at −20°C. The same basic buffer, not supplemented with octyl glucoside, FAD, and AEBSF, serves for dial-ysis of solubilized membranes leading to the formation of membrane liposomes.

7. Phospholipids: 3-sn-phosphatidic acid (PA) (sodium salt, from egg yolk, 98%, Sigma) and L-α-Phosphatidyl-DL-glycerol (PG) (ammonium salt, synthetic, 99%, Sigma). Dissolve phospholipids at 5 mM in solubilization buffer containing 40 mM octyl glucoside (but lacking protease inhibitors, DTT, and FAD). Dispense into aliquots of 0.5 mL and store at −75°C.

2.1.2. Preparation and Use of Recombinant Cytosolic Components

1. Buffer for diluting recombinant cytosolic components for procedures preceding their addition to the cell-free assays: 50 mM Tris-HCl, pH 7.5, 150 mM NaCl, 4 mM MgCl$_2$, and 2 mM DTT.

2. Guanylylimidodiphosphate (GMPPNP), trisodium salt (Sigma or Roche): prepare a 10 mM stock in H$_2$O, aliquot, and store at −75°C.

3. Guanosine-5'-O-[3-(thio)triphosphate] (GTPγS), tetralithium salt (Sigma or Roche): prepare a 10 mM stock in H$_2$O, aliquot, and store at −75°C.

4. Ethylenediamine tetraacetic acid (EDTA), disodium salt (Baker): prepare a 250 mM stock in H$_2$O.

5. MgCl$_2$ solution: prepare a 1-M stock in H$_2$O.

6. Recombinant rat geranylgeranyl transferase I made in *Escherichia coli* (Calbio-chem, cat. no. 345851 or 345852). One unit transfers 1 nmol geranylgeranyl pyro-phosphate to Rho proteins per hour at 37°C, at pH 8.0.

7. Geranylgeranyl pyrophosphate (Sigma); provided as a 2-mM solution in methanol and 3 mM NH$_4$OH (Sigma).

8. Prenylation buffer: 50 mM Tris-HCl buffer, pH 7.7, 0.1 mM ZnCl$_2$, 5 mM MgCl$_2$, and 2 mM DTT.
9. Triton X-114 (Sigma): prepare a 10% (v/v) solution in H$_2$O.

2.1.3. Cell-Free Assays

1. Cytochrome c, from equine heart, 95% (Sigma).
2. β-nicotinamide adenine dinucleotide 2'-phosphate reduced, tetrasodium salt, minimum 95% (NADPH; Sigma): prepare a 5 mM stock in H$_2$O, divide in 1 mL aliquots, and store at –20°C. Avoid frequent freezing and thawing, and do not store for >2 mo.
3. LiDS, >99% (Merck): prepare a 10-mM stock in H$_2$O, and store at 4°C for unlimited periods, provided that evaporation is prevented. Unlike SDS, LiDS does not precipitate out of aqueous solutions at low temperature.
4. Superoxide dismutase (SOD), from bovine erythrocytes (Sigma): prepare a 10,000 U/mL stock in H$_2$O, aliquot in amounts of 100 μL, and store at –20°C.
5. Oxidase assay buffer: 65 mM sodium phosphate buffer, pH 7.0 (made from 61 parts K$_2$HPO$_4$ and 39 parts NaH$_2$PO$_4$ stock solutions), 1 mM EGTA, 10 μM FAD, 1 mM MgCl$_2$, 2 mM NaN$_3$, and 0.2 mM cytochrome c. The conductivity of this buffer is 7.7 mS/cm. When the concentration of LiDS to be used in a large number of assays is known, this can be dissolved in the buffer. Assay buffer with and without LiDS can be divided into batches of 100 mL and stored at –20°C for unlimited periods of time (*see* **Note 2**).

2.2. Equipment

1. Microplate spectrophotometer (SPECTRAmax 340 or 190), preferably with Path Check, fitted with SOFTmax PRO software (Molecular Devices; *see* **Note 3**).
2. Thermomixer Comfort, rotary mixer, and heater/cooler (Eppendorf).
3. Mini orbital shaker for 96-well plates (Bellco).
4. Electronic single-channel pipettors (e.g., 5–100 μL or 10–500 μL range).
5. Multipette Plus pipettor and 1-mL Combitips (Eppendorf).
6. Finnpipette digital 12-channel pipet (50–300 μL range) (Thermo).
7. Refrigerated table-top centrifuge fitted with a swing-out rotor (e.g., Sorvall RC-3B and H-6000A rotor).
8. Refrigerated high-speed centrifuge with fixed angle rotor (e.g., Sorvall RC5 and SS-34 rotor).
9. 400-W sonicator fitted with micro-tip or cup horn (e.g., Vibracell VCX, Sonics and Materials).
10. Ultracentrifuge and fixed angle rotor (e.g., Beckman L5-50 ultracentrifuge and 60-Ti rotor).

2.3. Disposable Plasticware

Ninety-six-well microplates, polystyrene, flat-bottom, and clear, with a well volume of 382 μL and a maximal height of 10.9 mm (Greiner). When the wells in these plates are filled with a 0.21 mL reaction volume, the vertical light path is 0.575 cm. The microplate must be of the low-protein-binding type. Plates intended for use in enzyme-linked immunosorbent assays (ELISAs) of medium or high hydrophilic protein-binding capacity are not recommended for use in cell-free assays.

3. Methods

Cell-free assays are used in an almost limitless variety of forms and applications (*see* **Tables 1** and **2**). In the original form of the assay, cytochrome b_{559} was represented by a total macrophage membrane preparation and the cytosolic components, by total cytosol *(16)*. In the variants developed later, a more sophisticated membrane preparation is utilized or the membrane is altogether replaced by purified relipidated cytochrome b_{559}. In all of the assays to be described, we use a modified membrane preparation, originating from guinea pig peritoneal macrophages, and the cytosol is replaced by purified and well characterized recombinant proteins (p47^{phox}, p67^{phox}, and Rac). Although a number of anionic amphiphiles can act as activators in cell-free systems, we shall limit our description to LiDS as the prototype activator. Finally, we describe a single method for the detection of ROS, based on the trapping of $O_2^{\bullet-}$ produced in the cell-free system by oxidized cytochrome c, and the spectrophotometric quantification of reduced cytochrome c by measuring the increase in absorbance at 550 nm over time *(34)*.

3.2. Membrane Component

3.2.1. Membrane Preparation

We here describe the preparation of membranes from elicited guinea pig peritoneal macrophages *(21,35)*. Considering the paucity of granules in macrophages, differential centrifugation is not used to obtain granule-free pure plasma membrane preparations in these cells. Instead, we prepare a "total" membrane fraction, defined by its sedimentation at 160,000g. The use of an uncharacterized membrane preparation is made possible by the fact that, in the cell-free assays to be described, the amounts of membrane added to the assay are based strictly on the cytochrome b_{559} content.

1. Inject guinea pigs (male or female, weighing 300 to 500 g) intraperitoneally with sterile light paraffin oil (15 mL per animal).
2. After 4 d, sacrifice the animals by CO_2 inhalation, and collect the peritoneal content via a 3-cm-long incision in the abdominal wall and repeated introduction and aspiration of 50-mL volumes of ice-cold Earle's balanced salt solution.
3. Pass the collected peritoneal lavage through a 180-μm-pore-size nylon mesh sheet (Swiss Silk Bolting Cloth Mfg. Co. Ltd.) and collect in ice-cooled Erlenmeyer flasks (*see* **Note 4**).

4. Centrifuge fluid for 20 min at 940*g* and 4°C to sediment the cells.
5. Repeat the procedure once more, and suspend the cell sediment in ice-cold distilled H_2O (20 mL to a cellular pellet derived from 400 mL lavage fluid) to lyse red cells.
6. After 3 min, add an equal volume of ice-cold 0.29 *M* NaCl solution in water (20 mL), resulting in an isotonic NaCl concentration (0.145 *M*), mix well, and re-centrifuge at 940*g*, as above. If necessary, the lysis procedure can be repeated once more.
7. Resuspend the cell pellet in Earle's solution (10 mL per guinea pig), and count cells after diluting the suspension 1:10 in 1% acetic acid. The expected cell harvest varies from 1 to 2×10^8 cells per animal.
8. Pellet the cells at 940*g*, and resuspend in sonication buffer at a concentration of 10^8 cells/mL in 16×100 mm polypropylene tubes (4 mL/tube).
9. Sonicate samples, keeping tubes immersed in ice-water, with the tip lowered into the cell suspension half-way its entire depth. Submit the cells to three cycles of sonic disruption, at an amplitude of 10%, each cycle consisting of sonication for 9 s, followed by a 1-s rest, repeated three times. Check for quality of cell disruption by phase contrast microscopy at 40× magnification (~90% cell disruption is expected).
10. Centrifuge the cell homogenate for 10 min at 3000*g* and 4°C in a swinging-bucket rotor to remove unbroken cells, aggregates, and nuclei. Collect the supernatant (postnuclear supernatant).
11. Centrifuge the supernatant for 2 h at 160,000*g* and 4°C. The supernatant from this step is collected and represents the cytosol.
12. Resuspend the membrane pellet in sonication buffer supplemented with 1 *M* KCl (*see* **Note 5**) at a volume identical to the original volume of the homogenate. Resuspend directly in the ultracentrifuge tubes (kept in ice water) by adding buffer with 1 *M* KCl in small aliquots (1 to 2 mL) and sonicating after each addition until all of the membrane is suspended (*see* **Note 6**).
13. Centrifuge the mixture at 160,000*g*, as described in **step 11**.
14. Discard the supernatant and freeze the membrane pellet by placing in a −75°C freezer. The membranes can be kept frozen indefinitely for future use. We found no reason for flash-freezing or keeping membranes at a lower temperature.

3.2.2. Preparation of Membrane Liposomes

Although membrane preparations obtained as described previously are perfectly adequate, we routinely use solubilized membrane preparations consisting of liposomes of uniform size as our source of cytochrome b_{559} in cell-free assays. Liposomes are obtained by solubilizing membranes with octyl glucoside and then removing detergent by dialysis *(35,36)*.

1. Suspend frozen membranes in ice-cold solubilization buffer at a concentration of 5×10^8 cell equivalents/mL using the original ultracentrifuge tubes as containers.
2. Stir suspension with a small magnetic bar by placing the tube in a beaker containing ice water. Continue solubilization until no intact membrane fragments remain. This might take 3–6 h, and it is important to keep the tubes ice-cooled throughout this period.

3. Centrifuge the solution for 1 h at 48,000g and 4°C in a fixed-angle rotor.
4. Transfer supernatant containing the solubilized membrane into a fresh tube (appears as a pale-yellow opalescent solution). Discard pellet.
5. Place solution in dialysis tubing with a 25,000 molecular weight cutoff (*see* **Note 7**), and dialyze against a 100-fold excess of detergent-free solubilization buffer for 18–24 h at 4°C (*see* **Note 8**). The dialyzed detergent-free solubilized membrane consists of liposomes of a size of 250–300 nm.
6. Measure the concentration of cytochrome b_{559} (*see* **Subheading 3.2.3.**).
7. Supplement the preparation with 10 µM FAD, divide into aliquots of 1–1.5 mL, and store at −75°C (*see* **Note 9**). The preparation is now ready to be used in all forms of cell-free assays (*see* **Note 10**).

3.2.3. Quantification of Cytochrome b_{559} Content

The results of cell-free assays are commonly expressed in turnover values (mol $O_2^{\bullet-}$ produced per time unit[s] per mol cytochrome b_{559} heme). Thus, it is essential that the cytochrome b_{559} content of membranes is known. Cytochrome b_{559} content is expressed by heme concentration.

1. Dilute membrane liposome preparation in solubilization buffer without octyl glucoside and FAD (1:5 or 1:10 dilution). Place diluted samples in spectrophotometer cuvet (*see* **Note 11**).
2. Place cuvet in the sample compartment of a double-beam spectrophotometer, and place a cuvet containing the buffer used for diluting the membrane preparation in the reference compartment.
3. Run an absorption spectrum scan, from 400 to 600 nm, using a band-width of 1 nm, a scanning interval of 1 nm, and a scanning speed of 100 nm/min (*see* **Note 12**). Save this as "oxidized spectrum" on the computer linked to the spectrophotometer.
4. Add a few grains of sodium dithionite (Merck), rapidly mix, and run a spectral scan again. Save this as "reduced spectrum."
5. Subtract the "oxidized spectrum" from the "reduced spectrum," and save the resulting "reduced *minus* oxidized spectrum."
6. Detect and record absorbance values of major cytochrome b_{559} peaks, located at 559/558, 529, and 427/426 nm. Detect and record absorbance at the "valley" at 411/410 nm.
7. Calculate heme content, based on the Δ extinction coefficient of the 427 nm peak/411 nm valley pair, using $\Delta\varepsilon_{427-411\ nm} = 200$ mM^{-1} cm^{-1} *(37)* (*see* **Note 13**).
8. Normally, we obtain a concentration of cytochrome b_{559} heme of 400 pmol/10^8 cell membrane equivalents. Thus, the membrane liposome suspension of 5×10^8 cell membrane equivalents/mL has a heme concentration of 2 µM (*see* **Note 14**).

3.3. Cytosolic Components

3.3.1. Preparation of Recombinant p47^{phox} and p67^{phox}

1. We routinely utilize p47^{phox} and p67^{phox} produced in baculovirus-infected cultures Sf9 cells, using a modification of the method developed by Leto et al. *(38)*. The

changes from the original procedure are related principally to the use of a different ion exchanger for the purification of p47*phox* (SP Sepharose, HiLoad; Amersham Biosciences) *(39)*.

2. As an alternative, recombinant p47*phox* and p67*phox* can be produced in *E. coli* as glutathione-*S*-transferase (GST) fusion proteins *(33,40)*.
3. The purity requirements for the performance of cell-free assays are easily achieved for baculovirus- and *E. coli*-derived preparations. Ninety percent purity or higher is adequate, and the only problem encountered with lower-purity preparations is that the actual concentrations of the components present in the assay cannot be accurately determined.
4. We measure protein concentrations by the Bradford assay *(41)* modified for use with 96-well microplates (*see* BioRad Technical Bulletin 1177 EG and **Note 15**).
5. Supplement recombinant proteins with 30% v/v glycerol, and store at −75°C. In this state, they are active for an unlimited time period.

3.3.2. Preparation of Recombinant Nonprenylated Rac1

The oxidase can be activated in the cell-free system by both Rac1 and Rac2, in the nonprenylated form. However, one should be aware that this is an artifact, because nonprenylated Rac does not exist in normal phagocytes. Translocation to the membrane of nonprenylated Rac is dependent exclusively on the net positive charge of the polybasic domain at the C-terminus. Because Rac1 contains six contiguous basic residues in this domain, whereas Rac2 contains only three that are only partially contiguous, nonprenylated Rac1 is much more active in the cell-free system than Rac2 *(6,42)*.

1. Recombinant nonprenylated Rac1 is produced as a GST fusion protein in *E. coli*, and purified as originally described by Leto et al. *(43)* and modified by us *(39)*.
2. Determine protein concentration, and store recombinant Rac1 as described above for p47*phox* and p67*phox*.
3. Native, recombinant Rac1 contains exclusively GDP *(44)*. Thus, before use in cell-free assays, Rac1 is subjected to nucleotide exchange to the nonhydrolyzable GTP analogs GMPPNP or GTPγS. For choosing between the two analogs, *see* **Note 16**.

3.3.3. Nucleotide Exchange on Rac

The procedure described here is for exchange to GMPPNP, but the same procedure is used for exchange to GTPγS. It is based on the removal of bound endogenous GDP by chelation of Mg^{2+} and replacement of the bulk of GDP by the GTP analog.

1. Decide on the size of the batch of recombinant Rac that is to be subject to guanine nucleotide exchange. For use in cell-free assays, we normally perform exchange on aliquots of 10 to 20 nmol of Rac. Place in a 1.5- or 2-mL Eppendorf polypropylene tube.

2. Add GMPPNP stock solution in a quantity representing a 10-fold molar excess over the amount of Rac. Fore example, if exchange is to be performed on 10 nmol of Rac, add 100 nmol of GMPPNP (10 µL from the 10 mM stock solution).
3. Add EDTA solution to a final concentration of 12.5 mM. Incubate for 30 min at 30°C in a rotary mixer set at 600 movements/min.
4. Stabilize the exchanged state of Rac by adding MgCl$_2$ solution to a final concentration of 25 mM.
5. Store exchanged protein frozen at −75°C (*see* **Notes 17** and **18**).

3.3.4. Prenylation of Rac In Vitro

In earlier work with prenylated (geranylgeranylated) Rac, His$_6$-tagged Rac1 was cloned into the baculovirus genome, and this was used to infect cultures of Sf9 cells *(7)*. In this procedure, prenylated Rac is expressed in the cell membrane and thus must be purified following solubilization by 21 mM sodium cholate. As described here, we recently replaced this procedure with a simpler method using in vitro enzymatic prenylation of nonprenylated recombinant Rac1 *(32)*.

1. Add 10 nmol of nonprenylated Rac (*see* **Note 19**) to a polypropylene Eppendorf tube.
2. Add 10 µL (20 nmol) of geranylgeranyl pyrophosphate stock solution, 10 µL (10 U) of geranylgeranyl transferase I stock solution, and prenylation buffer containing ZnCl$_2$ (*see* **Note 20**) to a final volume of 0.9 mL.
3. Incubate for 45 min at 37°C in a rotary mixer set at 600 movements/min (*see* **Note 21**).
4. Add 60 µL of 70 mM octyl glucoside in H$_2$O (final octyl glucoside concentration is 4.375 mM), and reincubate for 45 min under the same conditions as above.
5. Sonicate the protein in an ultrasonic processor fitted with an ice water-filled cup horn for five cycles of 10 s each at 50% amplitude.
6. Add 0.24 mL of glycerol to bring the final volume to 1.2 mL. The final concentrations of components are 8.33 µM Rac, 3.5 mM octyl glucoside, and 20% v/v glycerol (this does not take into account the glycerol carried into the reaction by nonprenylated Rac itself).
7. Prenylated Rac can be stored at −75°C. After thawing, it should be centrifuged for 15 min at 10,000g to check for the formation of aggregates. If sediment is found, use the supernatant only after measuring its protein content.

3.3.5. Checking the Degree of Rac Prenylation

Prenylation in vitro is a very reliable methodology provided that a trusted source of geranylgeranyl transferase I is available. Nevertheless, it is recommended that until enough experience is acquired, the degree of prenylation should be checked *(45,46)*.

1. Remove an aliquot from the completed prenylation mixture before adding glycerol. Because the final detection method is based on SDS-polyacrylamide gel electro-

phoresis (PAGE), one must remove sufficient protein to make detection easy. Normally, one-third of a 10 nmol Rac prenylation mixture is used for confirming prenylation (about 3 nmol Rac). The procedure is best performed in 1.5-mL conical-bottom microcentrifuge tubes (Eppendorf).

2. Add prenylation buffer up to a total volume of 0.9 mL.
3. Add 0.1 mL of 10% Triton X-114 (1% final concentration).
4. Place the tube in a mixture of ice and water for 30 min, vortexing the tube every 5 min.
5. Heat the mixture at 37°C for 10 min in a heating block or a water bath. Keep the tubes stationary—do not mix. In the stationary state, the solution will become cloudy as a result of aggregation of Triton X-114 above its cloud point. Amphiphilic proteins, such as prenylated Rac, will associate with the detergent aggregates, whereas nonprenylated Rac will remain in the aqueous phase.
6. Centrifuge the mixture at 10,000–12,000g in a microfuge at room temperature. This will result in phase separation with the upper (aqueous) phase containing non-prenylated Rac and the lower (detergent-enriched) phase containing prenylated Rac.
7. Add prenylation buffer to the lower phase to make the total volume equal to that of the upper phase, and mix well.
8. Take equal sample volumes from the two phases and subject to SDS-PAGE, followed by staining with GELCODE Blue Stain Reagent (Pierce).
9. Compare intensity of the Rac bands (21 kDa) in the two phases visually or by densitometry (*see* **Note 22**).

3.4. An Overview of Cell-Free Assay Design

For the proper design of cell-free assays, it is essential to recall a number of its theoretical aspects, as outlined here.

1. The $O_2^{\bullet-}$-producing component is the cytochrome b_{559} found in the membrane and results are to be related to the amount of cytochrome b_{559} present in the reaction.
2. All cytosolic components must be present at saturating quantities in relation to cytochrome b_{559}. These quantities are determined with dose–response experiments in which the concentration of one or all cytosolic components is varied in the presence of a constant amount of cytochrome b_{559} (membrane).
3. Normally, nonprenylated Rac1 is used in the amphiphile-dependent cell-free assay. However, identical results are obtained with prenylated Rac1, provided that p47[phox] is present in the reaction.
4. All cell-free assays comprise two stages: (a) the stage of oxidase complex assembly, in the course of which cytosolic components are translocated to the membrane, leading to the induction of conformational change in gp91[phox], and (b) the catalytic stage, initiated by the addition of NADPH, expressed in the production of $O_2^{\bullet-}$. In some forms of cell-free assay, the two stages are separated by the interruption of assembly just before the initiation of catalysis. In most assays, the assembly merges with the catalytic stage, although an effort is usually made to bring assembly to completion before the addition of NADPH.

5. Amphiphile-dependent oxidase assembly is time- and temperature-dependent (*see* **Note 23**).
6. Kinetic models of anionic amphiphile-induced oxidase assembly have been proposed both before *(47,48)* and after *(49)* the identification of the components of the oxidase. These contain useful information which, paradoxically, had relatively little influence on the design and methodological aspects of cell-free assays.
7. Oxidase activation in cell-free systems is reduced by an increase in the ionic strength of the assay buffer (*see* **Note 24**).

3.5. The Canonical Amphiphile-Dependent Cell-free Assay: "Don't Leave Home Without It"

We describe here the basic methodology for performing cell-free oxidase activation in what is called the "semi-recombinant" system. This is a modification of the original amphiphile-activated (membrane + cytosol) system *(16,21)*. Since microplate spectrophotometers became standard equipment, the performance of spectrophotometric kinetic assays in microtiter plates is now the routine procedure in most laboratories. This required adjustment of the assay to 100–300 µL volumes used with 96-well microtiter plates (96-well plates). We describe a kinetic cell-free oxidase activation assay performed in 96-well plates in which the reaction components comprise solubilized macrophage membrane liposomes, recombinant p47^{phox}, p67^{phox}, and nonprenylated Rac1.

1. Add 20 µL of solubilized membrane liposomes (50 n*M* cytochrome b_{559} heme) to the wells of a 96-well plate. This is intended to result in a final concentration of 5 n*M* cytochrome b_{559} heme in 200 µL (the total volume of the reaction before addition of NADPH) and equals 1 pmol cytochrome b_{559} heme/well.
2. Add 20 µL of a mixture of p47^{phox}, p67^{phox}, and nonprenylated Rac1-GMPPNP, each at a concentration 10-fold higher than that desired as the final concentration in 200 µL. As an example, if a final concentration of 100 n*M* is to be achieved for all three components, add 20 µL of a solution containing 1 µ*M* of each component (*see* **Note 25**). Make all dilutions of membrane and cytosolic components in oxidase assay buffer without LiDS. Dispensing of 10- or 20-µL aliquots of membrane or cytosolic components to the wells is best performed with electronic pipettors, which are accurate and fast.
3. Add 160 µL/well of assay buffer containing an optimized concentration of LiDS. We typically use a digital 12-channel pipet. For this protocol, the final concentration of LiDS causing maximal activation is 120–130 µ*M* (**Fig. 1**) (*see* **Note 26**). Because the amphiphile is diluted 1.25-fold by the volumes of membrane and cytosolic components previously added to the wells, the concentration of the amphiphile in the assay buffer must be adjusted accordingly (as an example, to achieve a final concentration of 130 µ*M*, add 160 µL of a 162.5 µ*M* solution).
4. Place the plate on an orbital shaker, and mix contents for 90 s at 500 to 600 movements/min and room temperature (*see* **Note 27**).

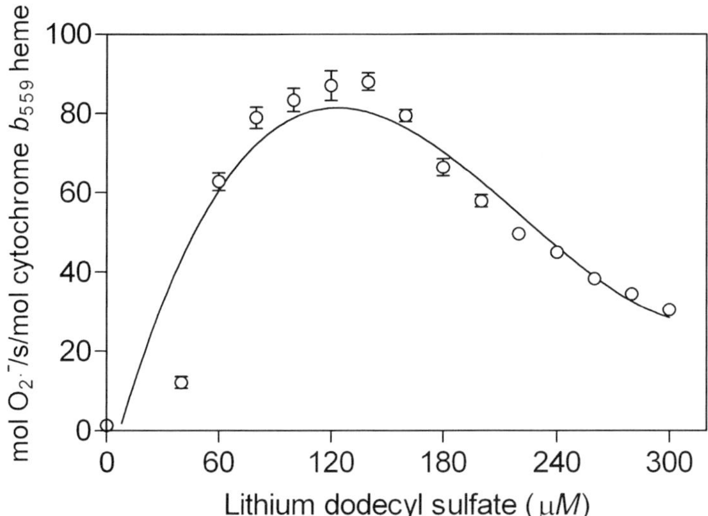

Fig. 1. Dose–response curve of lithium dodecyl sulfate (LiDS) in the amphiphile-dependent cell-free system. Assay mixtures consisting of solubilized macrophage membrane liposomes (5 nM cytochrome b_{559} heme) and recombinant p47^{phox} (100 nM), p67^{phox} (100 nM), and nonprenylated Rac1-GMPPNP (100 nM) were incubated with varied concentrations of LiDS as indicated. $O_2^{\cdot-}$ production was initiated by the addition of NADPH and measured by the kinetic cytochrome c reduction assay for 5 min. Results represent means ± SE of three experiments.

5. Dispense 10 µL of NADPH/well using an electronic pipettor. This results in a final concentration of 238 µM NADPH in a total volume of 210 µL per well, which is well above the K_m for NADPH of the oxidase in the cell-free system *(16)*.

6. Transfer the plate quickly to the microplate spectrophotometer, and mix contents for 5 s using the "automix" option of the reader. The instrument should be set to record increase in absorbance at 550 nm over a time period of 5 min with 28 readings being executed at 0.11-min intervals at room temperature (temperature regulation by the microplate spectrophotometer is set "off"). Include blank wells containing 200 µL assay buffer to which 10 µL of 5 mM NADPH were added simultaneously with its addition to the sample wells.

7. For most instruments, results in the kinetic mode are expressed in Abs$_{550\,nm}$/min units. The software of the instrument calculates these values by dividing the ΔAbs$_{550\,nm}$ or, preferably, ΔmAbs$_{550\,nm}$ over time by the number of minutes elapsed. Thus, it is essential for the increase in absorbance curve to be linear. The curves turn nonlinear whenever one or more components of the reaction is/are exhausted. Although every effort is made to prevent this from occurring by choosing the right amounts of enzyme (cytochrome b_{559} in the membrane), cytochrome c, and NADPH per well (*see* **Note 28**), it occasionally happens (*see* **Fig. 2**). In this case, the linear portion

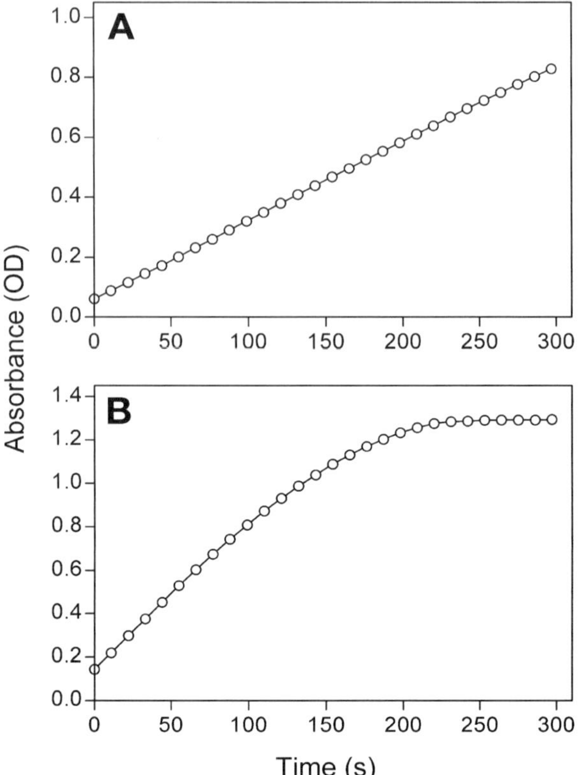

Fig. 2. Actual data displays of the results of two kinetic cell-free assays, showing the increase in absorbance at 550 nm over a 300-s interval. **(A)** Increase is linear throughout the 300 s period. **(B)** Increase is linear up to 120 s, after which time it starts leveling off. In this situation, the $Abs_{550\,nm}$/min should be recalculated for the first 120 s. Turnover values were found to be 41 mol $O_2^{\bullet-}$/s/mol cytochrome b_{559}, for panel **A**, and 106 mol $O_2^{\bullet-}$/s/mol cytochrome b_{559}, for panel **B** (recalculated value).

of the curve is chosen, and values are recalculated as $\Delta mAbs_{550\,nm}$/min for the revised time interval (shorter than 5 min). These units are transformed to nnmol cytochrome c reduced per min per well content of 210 µL, based on the extinction coefficient $\Delta E_{550} = 21$ mM^{-1} cm^{-1} for reduced minus oxidized cytochrome c (nmol $O_2^{\bullet-}$/min/well) (*see* **Note 29**).

8. Final results are expressed as "turnover": the amount of $O_2^{\bullet-}$ produced per time unit per mol membrane cytochrome b_{559} heme (mol $O_2^{\bullet-}$/s/mol cytochrome b_{559} heme; *see* y-axis of graphs in most figures in this chapter). This is easily calculated by knowing the nmol $O_2^{\bullet-}$/min/well values and the amount of cytochrome b_{559} heme per well (1 pmol, when 20 µL of membrane, containing 50 nM of cytochrome b_{559} heme, are added per well). For each experimental condition, perform the assay in

triplicate wells and make the software calculate mean values and standard deviations (*see* **Note 30**).

9. It essential to include SOD control wells in cell-free assays to assure that the reduction of cytochrome *c* is indeed due to $O_2^{\bullet-}$. This requires that parallel SOD-containing wells are included for every group of wells in which $O_2^{\bullet-}$ production is detected. Use a large excess of 100 U SOD/mL by adding 10 µL/well of a 2000 U/mL of SOD solution before the addition of NADPH. Addition of SOD is expected to prevent cytochrome *c* reduction by 95% or more (*see* **Note 31**).

10. A number of additional control reactions are requirements for the proper execution of cell-free assays, and no assay is complete without the inclusion of reaction wells in which one of the following components is omitted (i.e., anionic amphiphile, NADPH, membrane, and all or each of the individual cytosolic components).

3.6. Amphiphile-Independent Cell-Free Assays

The ability to activate the oxidase in vitro in the absence of an anionic amphiphile was first reported by Sumimoto et al. *(30)*. In this model system, the need for the amphiphile is abolished by the truncation of both p47*phox* and p67*phox* at their C-termini, allowing the establishment of novel interactions between p47*phox* and cytochrome b_{559} and between p47*phox* and the N-terminal half of p67*phox*. Amphiphile-independent systems were also described by Tamura et al. *(28)*, who used a chimeric construct consisting of truncated p67*phox* and p47*phox*, and by Kleinberg et al. *(31)*, who prevented the establishment of intramolecular bonds in p47*phox* by mutagenesis. The two groups also observed that acidification of the membrane phospholipid environment had a marked enhancing effect on oxidase activation. The mechanism of this enhancement was not established with certainty; the likely possibilities are an increase in the affinity for the membrane of either p47*phox*, via its PX domain, and/or Rac, via its polybasic tail. Rac was not prenylated in these experiments.

A conceptually distinct situation, in which oxidase activation can be achieved in the absence of amphiphile and of p47*phox*, is represented by a cell-free system consisting of membrane liposomes, p67*phox*, and prenylated Rac *(7)*. We proposed that proper targeting of p67*phox* to the membrane in conjunction with the induction of a conformational change in p67*phox* by Rac is sufficient for the initiation of electron flow in gp91*phox* *(7,32,50)*. We named this process the "propagated wave" mechanism *(51)*. Variations of this system include activation by combinations of p67*phox*, prenylated Rac, GTP, and a Rac GEF *(33)*, and the recently described amphiphile-independent oxidase activation by p67*phox* and prenylated [Rac-RhoGDI] complexes *(52)*.

We describe two methods for amphiphile-independent cell-free oxidase activation. One assay is based on the use of prenylated Rac and does not require the participation of p47*phox*; the other makes use of our ability to modify the charge of phospholipids in phagocyte membranes and works with nonprenylated Rac.

3.6.1. Amphiphile-Independent Cell-Free Oxidase Activation in Mixtures of Membrane, p67^{phox}, and Prenylated Rac1

The amphiphile-independent cell-free system is useful for investigating the role of Rac and Rac-p67^{phox} interaction in oxidase assembly. This particular aspect of assembly is more difficult to explore in the presence of p47^{phox}, which has not only an assembly-initiating function, but also a role in the stabilization of the assembled complex *(51)*. Other situations in which the amphiphile-independent cell-free system is the assay of choice are when the effects of regulators of Rac are to be explored in vitro. One example is provided by Rac GEF-dependent oxidase activation in a cell-free system consisting of membrane, p67^{phox}, prenylated Rac1-GDP, GTP, and a Rac GEF, such as Trio or Tiam1 *(33,44)*. Another example is the assessment of the ability of prenylated [Rac1-RhoGDI] complexes in conjunction with p67^{phox} to activate the oxidase when added to phagocyte membrane liposomes enriched in anionic phospholipids in the absence of an amphiphile *(52)*.

Applications of amphiphile-independent cell-free assays also comprise the study of inhibitors (proteins, peptides, phospholipids, nucleotides, detergents, drugs) on the various stages of oxidase assembly. An example is the study of the effect of the amphiphilic activator LiDS on oxidase activation by p67^{phox} and prenylated Rac1-GMPPNP, in the absence of p47^{phox}. We found LiDS to exert a marked dose-dependent inhibitory effect, in the 25–200 μM concentration range, which was relieved by the presence of p47^{phox} *(53)*. Further examples are the distinct effects of a number of compounds (GTP and GDP, a C-terminal Rac1 peptide, RhoGDI, the p21-binding domain [PBD] of p21-activated kinase [Pak]), and neomycin sulfate) on amphiphile-dependent and -independent cell-free oxidase assembly, reflecting the existence of different pathways of assembly *(53)*.

1. Subject Rac1 to nucleotide exchange with GMPPNP. It is preferable to perform this step before prenylation because, as they are hydrophobic, prenylated proteins may bind to surfaces and possibly be lost. However, when required, nucleotide exchange can also be performed after prenylation.
2. Prenylate Rac1-GMPPNP, as under **Subheading 3.3.4.**
3. Add 20 µL/well of solubilized membrane liposomes (50 nM cytochrome b_{559} heme) to the wells of a 96-well plate. This is intended to result in a final concentration of cytochrome b_{559} heme of 5 nM in 200 µL (the total volume of the reaction, before the addition of NADPH) and equals 1 pmol cytochrome b_{559} heme/well.
4. Add 20 µL of a mixture of p67^{phox} and prenylated Rac1-GMPPNP, each at a concentration 10-fold higher than that desired as the final concentration in 200 µL. If the requirements of the experiment are to add each component separately, add 10 µL of each component from a 20-fold concentrated stock solution. All dilutions of membrane and cytosolic components are made in assay buffer without LiDS (*see* **Note 32**).

5. Add 160 µL/well of assay buffer without LiDS, using a digital 12-channel pipet. Place the plate on an orbital shaker and mix for 90 s at 500 to 600 movements/min and room temperature (*see* **Note 33**).
6. Dispense 10 µL of NADPH solution/well using an electronic pipettor. This results in a final concentration of 238 µ*M* NADPH in a total volume of 210 µL per well.
7. Record activity and convert to turnover values as described for the amphiphile-dependent system (*see* **Subheading 3.5.**).

3.6.2. Amphiphile-Independent Cell-Free Oxidase Activation in Mixtures of Negatively Charged Membrane, p47*phox*, p67*phox*, and Nonprenylated Rac1

3.6.2.1. PREPARING MEMBRANE LIPOSOMES ENRICHED IN ANIONIC PHOSPHOLIPIDS

1. Dilute solubilized macrophage membrane in solubilization buffer containing 40 m*M* octyl glucoside to a concentration of cytochrome b_{559} heme of 1.2 nmol/mL.
2. Add PA or PG, both at a concentration of 5 m*M*, at a ratio of one part membrane: four parts phospholipids. This results in a final concentration of 240 pmol/ml cytochrome b_{559} heme and 4 m*M* anionic phospholipids.
3. Dialyze the membrane-phospholipid mixture (*see* **Note 7**) against a 100-fold excess of detergent-free solubilization buffer (also lacking AEBSF and FAD) for 18–24 h at 4°C (*see* **Note 8**).
4. Measure the concentration of cytochrome b_{559}, and supplement the preparation with 10 µ*M* FAD.
5. Divide into aliquots of 1–1.5 mL, and store at –75°C.

3.6.2.2. AMPHIPHILE-INDEPENDENT CELL-FREE OXIDASE ACTIVATION

This cell-free assay is a hybrid of the canonical amphiphile-dependent system (from which it borrowed the anionic charge requirement and the fact that Rac is nonprenylated) and the amphiphile-independent assay (based on the use of prenylated Rac).

1. Add 20 µL/well of membrane liposomes enriched in PA or PG (50 n*M* cytochrome b_{559} heme and about 0.8 m*M* anionic phospholipid) to the wells of a 96-well plate. This is intended to result in a final concentration of cytochrome b_{559} heme of 5 n*M* and close to 80 µ*M* anionic phospholipid in 200 µL (the total volume of the reaction, before the addition of NADPH) and equals 1 pmol cytochrome b_{559} heme/well.
2. Add 20 µL of a mixture of p47*phox*, p67*phox*, and nonprenylated Rac1-GMPPNP, each at a concentration 10-fold higher that that desired as the final concentration in 200 µL (*see* **Notes 32** and **34**). All dilutions of membrane and cytosolic components are made in assay buffer without LiDS. In most situations, concentrations of p47*phox*, p67*phox*, and nonprenylated Rac required for reaching maximal activation in this system are higher than those customary in the canonical amphiphile-dependent assay.
3. Add 160 µL/well of assay buffer without LiDS. Place the plate on an orbital shaker and mix for 90 s at 500 to 600 movements/min and room temperature.

4. Dispense 10 µL of NADPH solution/well. This results in a final concentration of 238 µ*M* NADPH in a total volume of 210 µL per well.
5. Record activity and convert to turnover values as described for the amphiphile-dependent system (*see* **Subheading 3.5.**).

3.7. Uses of Cell-Free Systems

Cell-free NADPH oxidase assays are extensively used in both basic research and clinical medicine. The continuous expansion of the field of NOXes is expected to provide further impetus for their use and for the development of variations of the assay adapted to specific NOXes. The principal uses of cell-free assays are:

1. Identification and quantification of components of the oxidase complex in whole cells and tissues by complementation with known purified components.
2. Identification, quantification, and functional assessment of oxidase components purified from cells or produced by recombinant technology.
3. Structure–function studies of recombinant oxidase components, involving mutagenesis, deletions, and chimerization.
4. Investigation of the mechanism of action of oxidase inhibitors (e.g., *see* **Subheading 3.8.**).
5. Analysis of "priming" effects (enhanced activity of the oxidase due to factors outside the actual components of the complex).
6. Diagnosis of CGD and follow-up on therapy of CGD patients.
7. Identification and quantification of NOX components in nonphagocytic cells by complementation with phagocyte oxidase components.

When properly applied, the cell-free assay is one of the most sensitive methods for the detection of components of the oxidase complex and, in addition, offers the advantages of simplicity, speed, repeatability, and the potential of being applied to high-throughput screening.

3.8. Assembly or Catalysis: Putting Inhibitors to Good Use

The cell-free systems are ideally suited for investigating the mechanism of action of compounds known or expected to exert an inhibitory effect on the NADPH oxidase.

1. By definition, inhibitors of the oxidative burst in intact phagocytes will interfere with oxidase activity in the cell-free system only if interacting with one or more of the components of the oxidase complex.
2. Lack of effect on oxidase activation under cell-free conditions usually means that the hypothetical inhibitor acts upstream of the oxidase, most likely by interfering with phagocyte receptor–ligand interaction or with a transductional step linking the receptor to the oxidase.
3. Inhibitors active in cell-free systems belong to two categories: compounds interfering with the catalytic function of gp91^{phox} (electron flow though the redox cen-

ters) or compounds interfering with the process of oxidase assembly leading to the activation of the catalytic function.

4. The cell-free system has been used to identify domains on oxidase components relevant to the process of assembly. For this purpose, extensive use was made of synthetic peptides corresponding to segments of the amino acid sequences of the various oxidase components *(42,54,55)*. Such peptides were found to inhibit amphiphile-dependent oxidase activation, and a universal finding was that most peptides inhibited only when present before the initiation of assembly and were inactive when added after the completion of assembly (*see* **Fig. 3**).

5. It is thus essential to control the sequence specificity of experimental peptides by testing the effect, under identical circumstances, of scrambled or retro-peptides, or of unrelated peptides of similar charge.

6. Careful analysis of the mechanism of inhibition of oxidase activation by Rac1 C-terminal peptides indeed revealed that the effect, far from being sequence-specific, was related exclusively to the presence of a six-residue polybasic motif. Lack of sequence specificity should not be interpreted as an automatic disqualifier of the inhibitory peptide. Charge and hydrophobicity are important parameters in protein–protein and protein–lipid interactions and much can be learned from sequence-independent inhibitory effects of peptides.

7. In order to assess the significance of peptide-inhibition results, it is recommended that dose–response studies be run in a routine manner. These should be performed within a concentration range to enable the calculation of IC_{50} values. It is also essential that the peptide does not exert a nonspecific inhibitory effect on the actual measurement of $O_2^{\bullet-}$ production. This can be easily tested by adding the peptide to a xanthine/xanthine oxidase $O_2^{\bullet-}$-generating system.

8. Peptide analogs of oxidase components are also useful in the analysis of more subtle aspects in the mechanism of oxidase activation, such as the relative importance of charge and prenylation in the translocation of Rac to the membrane (*see* **Fig. 3**).

3.9. Sense and Sensitivity in Cell-Free Assays

Here, we discuss a number of methodological issues related to the proper performance of cell-free assays. Emphasis will be placed on untested or unproven assumptions, and some "sacred cows" will be questioned.

3.9.1. LiDS, SDS, or Arachidonate?

1. Unless the purpose of performing the cell-free assay is to explore the oxidase-activating capabilities of arachidonic acid itself or of arachidonic acid oxidation products, there are few occasions justifying the use of arachidonate as an activator.

2. Arachidonic acid activates the oxidase only in its ionized salt form, and stock solutions are tedious to prepare and unstable. We recommend using LiDS or SDS as standard amphiphilic activators.

3. SDS is as good an activator as LiDS, but concentrated solutions of SDS must be kept at room temperature. In two decades of using LiDS or SDS as activators, we found that both yield more reproducible results than arachidonate. Notably, we

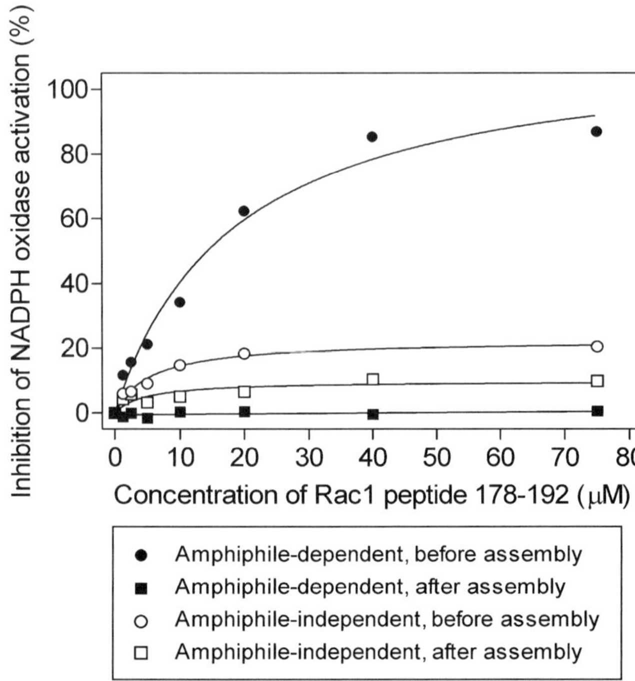

Fig. 3. Inhibition of NADPH oxidase activation in cell-free systems by a synthetic peptide, corresponding to the C-terminus of Rac1 (residues 178–192). The effect of various concentrations of the peptide (from 1.25 to 75 μ*M*) was tested in both amphiphile-dependent and amphiphile-independent systems. The amphiphile-dependent system consisting of solubilized macrophage membrane liposomes (5 n*M* cytochrome b_{559} heme), recombinant p47^{phox} and p67^{phox}, and recombinant nonprenylated Rac1 Q61L mutant (all at 100 n*M*) was incubated with 130 μ*M* lithium dodecyl sulfate, as described. The amphiphile-independent cell-free system consisting of solubilized macrophage membrane liposomes (5 n*M* cytochrome b_{559} heme), recombinant p67^{phox}, and recombinant Rac1 Q61L mutant prenylated in vitro (both at 300 n*M*) were incubated in the absence of an amphiphilic activator, as described. In all assays, $O_2^{\bullet-}$ production was initiated by the addition of NADPH and measured by the kinetic cytochrome *c* reduction assay for 5 min. The peptide was added either at time 0 (before assembly) or after 90 s of incubation (after assembly) but always preceding the addition of NADPH. The effect of the peptide is expressed as % inhibition of NADPH oxidase activation, with the activity in the absence of peptide being considered as 100%. The turnover in the amphiphile-dependent system, in the absence of the peptide was 84.42 mol $O_2^{\bullet-}$/s/ mol cytochrome b_{559}. In the amphiphile-independent system, the turnover in the absence of peptide was 65.01 mol $O_2^{\bullet-}$/s/mol cytochrome b_{559}. The results represent one characteristic experiment.

find no basis for the claim that arachidonate is to be preferred because it represents a more "physiological" form of activation.

3.9.2. To Supplement or Not to Supplement?

1. Calcium: in early experiments, we found that activation was reduced by Ca^{2+} and moderately enhanced by the Ca^{2+} chelator EGTA *(16)* (*see* **Note 35**). We re-investigated the necessity of Ca^{2+} chelation in the LiDS-activated and amphiphile-independent systems by examining the effect of EGTA, alone or in association with other supplements. As seen in **Fig. 4A,B**, EGTA had no enhancing effect on oxidase activation in the presence or absence of amphiphile.

2. FAD: early on, a flavin requirement was observed in the oxidase isolated from stimulated phagocytes *(56)*, and we also found that exogenous FAD enhanced arachidonate-activated cell-free activation *(16)*. The most likely explanation is that gp91^{phox} lost FAD during membrane preparation of membranes, leading to a need to reflavinate cytochrome b_{559}. Here, we compared cell-free oxidase activation in the presence and absence of 10 μM FAD in the assay buffer by using solubilized membrane liposomes, which are routinely supplemented with FAD. As is apparent in **Fig. 4**, FAD enhanced both amphiphile-dependent (**Fig. 4A**) and amphiphile-independent (**Fig. 4B**) oxidase activation, the effect being more pronounced on the amphiphile-dependent activation (*see* **Note 36**).

3. Magnesium: a requirement for Mg^{2+} was described very early in cell-free studies, and it was suggested that the metal interacted with a saturable oxidase component at a K_m of about 1 mM *(47)*. The identity of this component was never established, but after the discovery of the involvement of Rac in oxidase assembly, it became common belief that the requirement for millimolar concentrations of Mg^{2+} was related to its role in preventing the dissociation of GTP (or GDP) from Rac *(57)*. As shown in **Fig. 4**, supplementation of the assay buffer with 1 mM Mg^{2+} enhanced oxidase activation in both the amphiphile-dependent (**Fig. 4A**) and -independent (**Fig. 4B**) systems. Higher concentrations of Mg^{2+} (up to 5 mM) were not more effective than 1 mM (results not shown). Combining supplementation with FAD with that with Mg^{2+} did not result in an additive or synergistic effect; activities were identical to those found with FAD alone. Also, combining supplementation with FAD or Mg^{2+} with EGTA, or adding all three supplements, had no additive or synergistic effect. The almost identical ability of FAD and Mg^{2+} to improve assembly and the lack of an additive or cooperative effect suggest that they act by the same mechanism, most likely related to the stability of the gp91^{phox}–FAD bond and not to that of the Rac–GTP bond (*see* **Note 37**).

3.9.3. "Measure for Measure": The Intricacies of Dose–Response Studies With Cytosolic Oxidase Components

1. Most cell-free oxidase activation assays follow the principle of a constant amount of membrane and variable amounts of cytosolic components. This leaves open the issue of quantitative relationships among cytosolic components (*see* **Note 38**).

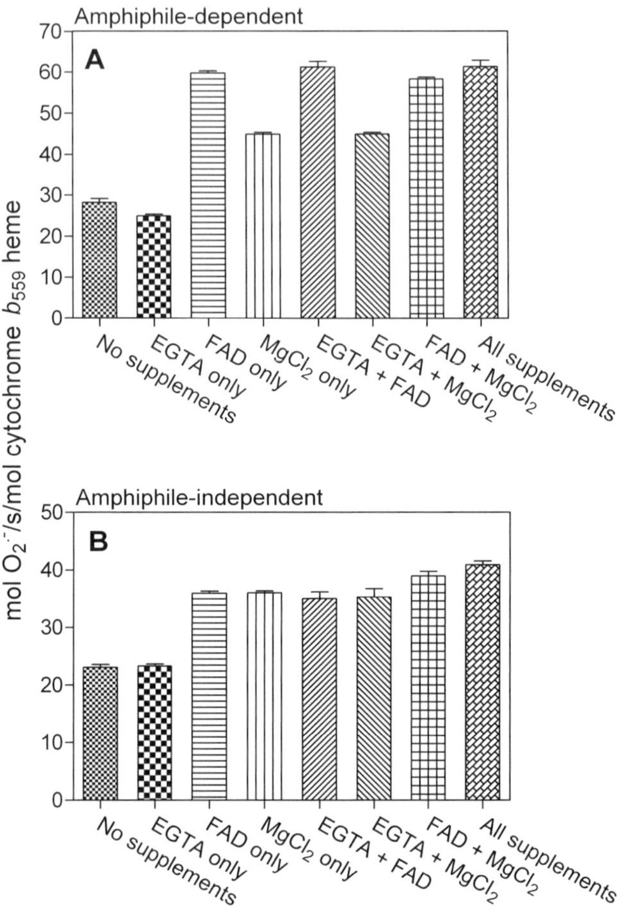

Fig. 4. Which supplements to the cell-free NADPH oxidase assay buffer are essential? Cell-free assays were performed in the canonical amphiphile-dependent system (**A**) and in the amphiphile-independent system, based on the use of prenylated Rac1 (**B**). The basic assay buffer was supplemented with our without 1 mM EGTA, 10 μM flavin adenine dinucleotide disodium salt, or 1 mM MgCl$_2$, or combinations of two or all three of these, as shown on the x-axis of panels **A** and **B**. (**A**) Amphiphile-dependent cell-free systems consisting of solubilized macrophage membrane liposomes (5 nM cytochrome b_{559} heme) and recombinant p47^{phox} (30 nM), p67^{phox} (30 nM), and nonprenylated Rac1-GMPPNP (30 nM) were incubated with 130 μM lithium dodecyl sulfate, as described. (**B**) Amphiphile-independent cell-free systems consisting of solubilized macrophage membrane liposomes (5 nM cytochrome b_{559} heme), recombinant p67^{phox} (300 nM), and recombinant Rac1-GMPPNP prenylated in vitro (300 nM) were incubated without amphiphile, as described. In both panels **A** and **B**, O$_2$·⁻ production was initiated by the addition of NADPH and measured by the kinetic cytochrome c reduction assay for 5 min. Results illustrated represent means ± SE of three experiments.

2. A problem we frequently encountered when performing cell-free assays was determining the optimal methodology for relating activity turnover values to the amounts of cytosolic proteins added to a constant amount of membrane. **Figure 5** summarizes the two main approaches used in our laboratory. In these experiments, the concentration of the membrane was constant. The concentrations of cytosolic components were either varied all in parallel or individually, in which case the other components were added at the maximal concentration in the range studied. Assays were run either in the amphiphile-dependent system (**Fig. 5A**), or in the amphiphile-independent system (**Fig. 5B**). In the amphiphile-dependent system, the concentration of LiDS was kept constant at 130 μM because the optimal activating concentration of LiDS did not vary with the concentration of cytosolic components within the 0 to 1 μM range, when using purified recombinant cytosolic components.

3. It is apparent that when all components are varied in parallel, the dose–response curve has a sigmoidal shape whereas when a single component is varied in the presence of an excess of the other component(s), the curves are hyperbolic. The highest levels of activation are seen when the concentrations of Rac1 and p47^{phox} (amphiphile-dependent system) and Rac1 (amphiphile-independent system) are varied individually, in the presence of an excess of the other component(s); the lowest activities are found when p67^{phox} is varied individually. The differences are particularly marked at lower concentrations of components.

4. When the purpose of performing cell-free assays is the detection of low concentrations of a cytosolic component, it is best to perform the assay in the presence of a clear excess of the other components, a situation generating hyperbolic curves (*see* **Note 39**).

3.9.4. To Exchange or to Add?

1. In the early period of the use of cell-free assays, it was reported repeatedly that the addition of GTP or nonhydrolyzable GTP analogs (GMPPNP, GTPγS) was an absolute requirement for the expression of oxidase activity. Many of these observations were made in cell-free systems consisting of membrane and total cytosol before the identification of Rac as the small GTPase involved in oxidase activation *(58–61)*.

2. With the advent of the semi-recombinant systems, which involved the use of recombinant Rac1 or Rac2, the "habit" of supplementing the assay buffer with GTP analogs persisted when native Rac (Rac-GDP not exchanged to GTP) was present in the reaction. The assumed explanation for this was that added GTP analogs were bound to Rac-GDP in a nucleotide exchange reaction taking place simultaneously with oxidase assembly (*see* **Note 40**).

3. Because the concentration of Mg^{2+} in the assay buffer is prohibitive for spontaneous nucleotide exchange, the ability of prenylated Rac to take up GTP from the medium points to the intervention of a GEF. In a semi-recombinant cell-free system, GEF can originate only in the membrane but its presence, identity, and quantity are unknown parameters in the vast majority of cases.

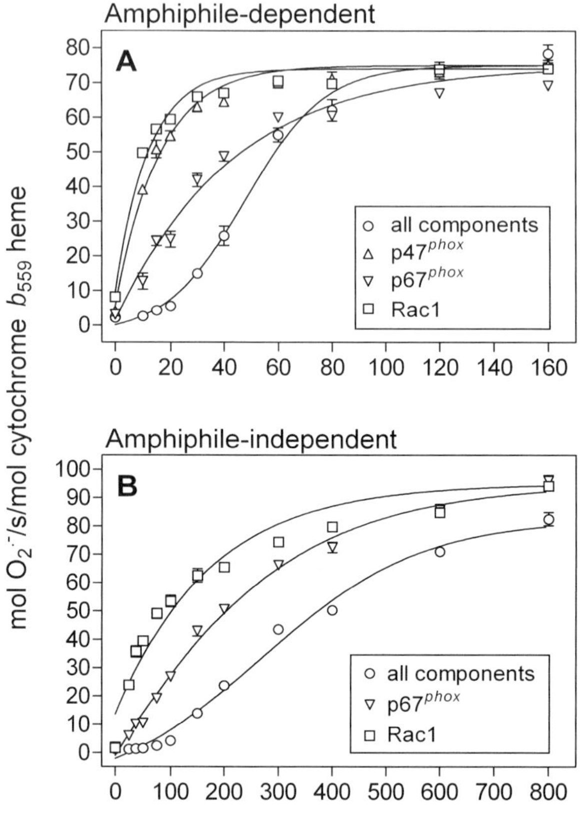

Concentration of cytosolic component(s) (n M)

Fig. 5. The effect of concentration ratios among cytosolic components on the nature of the dose–response curves in cell-free assays. (**A**) Four types of amphiphile-dependent cell-free assays, consisting of various combinations of cytosolic components, were performed. All four consisted of solubilized macrophage membrane liposomes (5 nM cytochrome b_{559} heme), recombinant p47^{phox} (varied from 0 to 160 nM), and recombinant p67^{phox} (varied from 0 to 160 nM), and recombinant nonprenylated Rac1-GMPPNP (varied from 0 to 160 nM). The four combinations of components were: 1. All three cytosolic components were present at equal concentrations (varied from 0 to 160 nM); 2. p47^{phox} was varied from 0 to 160 nM whereas p67^{phox} and Rac1 were both present at a concentration of 160 nM; 3. p67^{phox} was varied from 0 to 160 nM whereas p47^{phox} and Rac1were both present at a concentration of 160 nM, and 4. Rac1 was varied from 0 to 160 nM whereas p47^{phox} and p67^{phox} were both present at a concentration of 160 nM. In all cases, the components were incubated with 130 μM lithium dodecyl sulfate, as described. (**B**) Three types of amphiphile-independent cell-free assays, consisting of various combinations of cytosolic components, were performed. All three consisted of solubilized macrophage membrane liposomes (5 nM cytochrome

4. Another common assumption is that native Rac (Rac-GDP) is totally inactive in cell-free systems. We have shown in the past that this is true only below a certain quantitative threshold and when this is exceeded, significant activity can be achieved. Thus, in the canonical amphiphile-dependent cell-free system, the differences in V_{max} between Rac1-GDP and Rac1-GTPγS were marked at 20 nM Rac, but minimal at 200 nM Rac *(62)*.

5. **Figure 6** summarizes studies in which the influence of the following parameters on the ability of Rac1 to support oxidase activation in cell-free systems was examined: (1) GDP vs GMPPNP-bound state; (2) supplementation of the assay buffer with GTPγS; and (3) nonprenylated vs prenylated Rac, corresponding to amphiphile-dependent and -independent assay, respectively.

6. It is apparent that in the amphiphile-dependent system, when the concentration of the cytosolic components is low (30 nM) the difference in activity between native Rac1 (Rac1-GDP) and Rac1 exchanged to GMPPNP (Rac1-GMPPNP) is pronounced (**Fig. 6A**). When the concentration is raised to 100 nM, the difference in activity between Rac1-GDP and Rac1-GMPPNP is much less, which is due principally to an increase in activity of Rac1-GDP (**Fig. 6B**). What is also seen clearly in **Fig. 6A,B** is that supplementation of the assay buffer with GTPγS (10 μM) has no significant enhancing effect on the activity of Rac1-GDP and does not influence the activity of Rac1-GMPPNP (*see* **Note 41**). In the amphiphile-independent cell-free system involving the use of prenylated Rac, the difference in the ability to support oxidase activation between Rac1-GDP and Rac1-GMPPNP is marked, with practically no activity being exhibited by prenylated Rac1-GDP (**Fig. 6C**). The addition of GTPγS (10 μM) did not correct this lack of activity, in contradiction to results obtained with prenylated Rac in the presence of amphiphile *(63,64)* (*see* **Note 42**).

7. Overall, we recommend that one should never rely on "in assay" nucleotide exchange, achieved by the addition of GTP analogs to the assay buffer, and always perform quantifiable nucleotide exchange on both nonprenylated and prenylated Rac before their use in the assays. Following this advice will prevent inconsistent and poorly reproducible results due both to the lack of conversion of Rac from the GDP- to the GTP-bound form and to a possible inhibitory effect of free GTP.

Fig. 5. (*Continued*) b_{559} heme), recombinant p67^{phox} (from 0 to 800 nM), and recombinant Rac1-GMPPNP prenylated in vitro (from 0 to 800 nM). The three combinations of components were: (1) The two cytosolic components were present at equal concentrations (varied from 0 to 800 nM); (2) p67^{phox} was varied from 0 to 800 nM whereas Rac1 was present at a concentration of 800 nM, and (3) Rac1 was varied from 0 to 800 nM whereas p67^{phox} was present at a concentration of 800 nM. The components were incubated in the absence of an anionic amphiphile. In all assays, $O_2^{•-}$ production was initiated by the addition of NADPH and measured by the kinetic cytochrome c reduction assay for 5 min. Results illustrated in both panels represent means ± SE of three experiments.

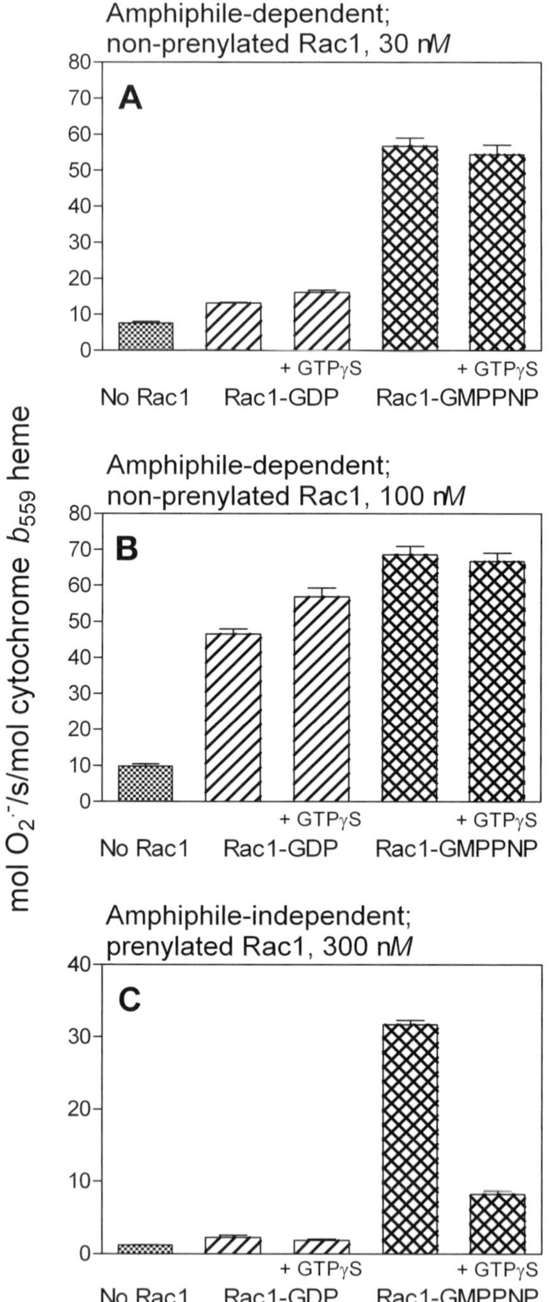

Fig. 6. Does the addition of exogenous nonhydrolyzable GTP analog (GTPγS) affect cell-free NADPH oxidase activation in the amphiphile-dependent and amphiphile-independent cell-free systems? (**A,B**) NADPH oxidase activity was measured in a cell-

3.9.5. Sibling Rivalry: Rac1 and Rac2

1. It is widely accepted today that the isoform of Rac relevant to oxidase function in neutrophils is Rac2 *(57)*, whereas Rac1 is its counterpart active in monocytes and macrophages *(65)*.
2. Comparative studies of Rac1 vs Rac2 in cell-free assays have been published (*see* **Note 43**).
3. Here, we show a comparative study of the relative abilities of Rac1 and Rac2, in nonprenylated and prenylated forms and in the absence of and following exchange to GMPPNP, to activate the oxidase in the cell-free system. Nonprenylated Rac isoforms were assayed in an amphiphile-activated system whereas prenylated Rac isoforms were used in an amphiphile-independent system. As seen in **Fig. 7A**, contrary to the dogma, both nonprenylated Rac1-GMPPNP and Rac1-GDP are capable of activation, with V_{max} values of 89 and 83 mol $O_2^{\bullet-}$/s/mol cytochrome b_{559} heme, respectively, and EC_{50} values of 5.74 nM and 9.70 nM, respectively. Rac2-GMPPNP was clearly less active than Rac1-GMPPNP, as evident in the ten-fold higher EC_{50} (53 nM). Rac2-GDP is, for practical purposes, inactive.
4. A similar study made with prenylated Rac1 and Rac2, in the absence of an amphiphilic activator and p47^{phox}, revealed that only the GMPPNP-preloaded forms of Rac1 and Rac2 are active. Again, Rac1 was more active than Rac2, as shown by the EC_{50} values of 0.13 µM and 0.73 µM, respectively (**Fig. 7B**).
5. Thus, the dogma "Rac-GTP is active; Rac-GDP is inactive" is applicable to prenylated Rac, in the absence of amphiphile and p47^{phox}, but is not strictly applicable to the canonical amphiphile- and p47^{phox}-dependent system. We believe that what determines the difference between the two situations is the presence or absence of p47^{phox} (*see* **Note 44**). At the practical level, this means that, unless the cell-free assay is specifically meant to be focused on Rac2, the use of Rac1 is recommended, based on the promise of higher oxidase activities.

Fig. 6. (*Continued*) free system (± 10 µM GTPγS, as indicated on the *x*-axis) consisting of solubilized macrophage membrane liposomes (5 nM cytochrome b_{559} heme), recombinant p47^{phox} (30 nM or 100 nM) and p67^{phox} (30 nM or 100 nM), and recombinant nonprenylated Rac1 (30 nM or 100 nM), either as native Rac1-GDP or as Rac1-GMPPNP. The components were incubated in the presence of 130 µM lithium dodecyl sulfate, as described. **(C)** NADPH oxidase activity was measured in a cell-free system (±10 µM GTPγS, as indicated on the *x*-axis) consisting of solubilized macrophage membrane liposomes (5 nM cytochrome b_{559} heme), recombinant p67^{phox} (300 nM), and recombinant Rac1 (300 nM), either as Rac1-GDP or Rac1-GMPPNP prenylated in vitro. The components were incubated in the absence of an amphiphilic activator at room temperature. In all assays, $O_2^{\bullet-}$ production was initiated by the addition of NADPH and measured by the kinetic cytochrome *c* reduction assay for 5 min. Results illustrated in all panels represent means ± SE of three experiments.

                                                                  Molshanski-Mor et al.

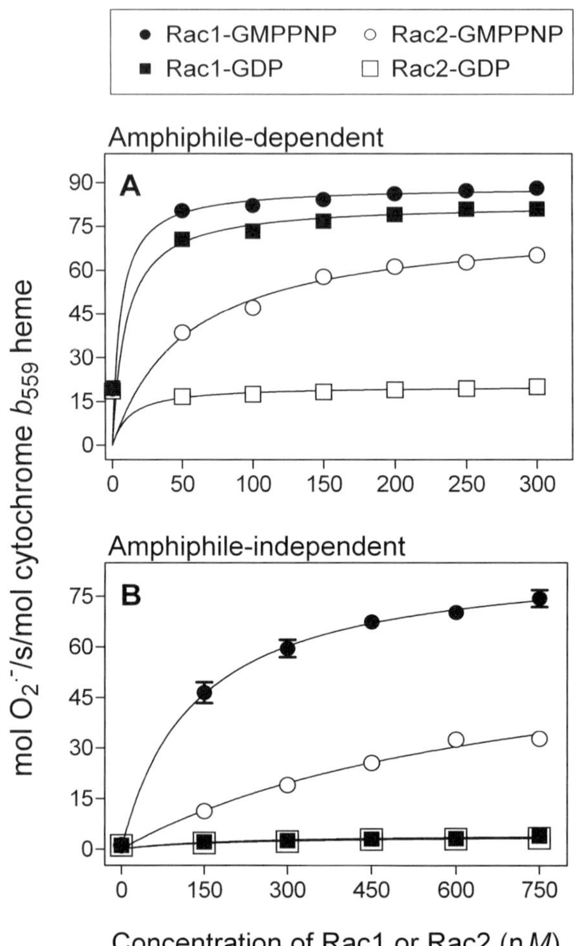

Fig. 7. Differences in the ability of Rac1 and Rac2 to support cell-free oxidase activation in relation to the nature of the bound guanine nucleotide, the absence or presence of prenylation, and the presence or absence of an amphiphilic activator. **(A)** Cell-free assays consisting of solubilized macrophage membrane liposomes (5 nM cytochrome b_{559} heme), recombinant p47^{phox} and p67^{phox} (both at a concentration of 300 nM), and recombinant nonprenylated Rac1 or Rac2, either exchanged to GMPPNP or GDP-bound (varied from 0 to 300 nM) were incubated with 130 µM LiDS, as described. **(B)** Cell-free assays consisting of solubilized macrophage membrane liposomes (5 nM cytochrome b_{559} heme), recombinant p67^{phox} (750 nM), and recombinant Rac1 or Rac2, either exchanged to GMPPNP or GDP-bound and prenylated in vitro (varied from 0 to 750 nM) were incubated in the absence of an anionic amphiphile, as described. In all assays **(A,B)**, $O_2^{\bullet-}$ production was initiated by the addition of NADPH and measured by the kinetic cytochrome c reduction assay for 5 min. Results illustrated in both panels represent means ± SE of three experiments.

3.10. "The End of the Affair"

We have been in the "cell-free business" for more than two decades. The message we would like to impart is this: running cell-free assays is not only useful, simple, and economical but, more than anything else, it is real fun. Good luck!

4. Notes

1. In studies on stimulants able to elicit $O_2^{\bullet-}$ production by macrophages, we found that procedures inhibiting endogenous phospholipase A_2 also inhibited $O_2^{\bullet-}$ production *(66)* and reasoned that fatty acids derived from the action of phospholipase A_2 on membrane phospholipids might participate in signal transduction to the NADPH oxidase. Thus, we applied long-chain unsaturated fatty acids to intact macrophages and found that arachidonic, linolenic, linoleic, and oleic acids indeed induce $O_2^{\bullet-}$ production, with arachidonic acid being the most potent activator *(66)*. Based on these results, we explored the possibility that fatty acids might activate the oxidase and found that long-chain unsaturated fatty acids elicited NADPH-dependent $O_2^{\bullet-}$ production by homogenates from resting macrophages *(16)*. It was later found that both *cis* and *trans* unsaturated fatty acids activate the oxidase in a cell-free system *(67,68)*, and shorter chain saturated acids (C12 to C16), at high concentration, are also active *(68)*.
2. This buffer was first described by us for use in the arachidonate-activated rudimentary cell-free system based on the use of membrane and total cytosol *(16)*, and was modified later *(62)*.
3. A narrow bandwidth improves the sensitivity of the cytochrome *c* reduction assay. SPECTRAmax 340 has a bandwidth of 5 nm; SPECTRAmax 190 has a bandwidth of 2 nm.
4. Analysis of the cellular connect of the lavage revealed it consisted of more than 90% macrophages at 4 to 5 d following the injection of paraffin oil.
5. The 1-M KCl concentration is achieved by mixing 2.5 volumes of sonication buffer with 1 volume of 3.5 M KCl in 20 mM Tris-HCl, pH 7.5.
6. Washing of membranes with 1 M KCl is intended to remove loosely attached cytosolic components, with special emphasis on the removal of membrane-bound Rac. Indeed, it was reported that omission of the KCl wash step resulted in an apparent decrease in the dependence of cell-free oxidase activation on added Rac, which was due to the presence of membrane-bound Rac *(63)*.
7. A dialysis membrane with a molecular weight cutoff of 25,000 is chosen because octyl glucoside is used for the solubilization of membranes at a concentration of 40 mM, which is well above its critical micellar concentration (CMC) of 25 mM. The micellar molecular weight of octyl glucoside is 8000 and a large-pore-size membrane facilitates diffusion of the detergent from the dialysis bag to the surrounding buffer.
8. The dialysis buffer used for removing octyl glucoside and generating the membrane liposomes is identical to the solubilization buffer but contains no octyl glucoside, AEBSF, leupeptin hemisulfate, and FAD. The addition of serine protease inhibi-

tors, especially AEBSF, to membrane preparations is avoided because of the reported likelihood of an interaction with cytochrome b_{559} **(69)**. FAD is omitted because it is difficult to measure the cytochrome b_{559} content accurately by sodium dithionite reduction in the presence of FAD.

9. Membrane liposomes can be kept frozen indefinitely for future use. Although keeping the material in small aliquots is recommended, freezing and thawing up to 10 times was not found to cause any damage.

10. Routine use of solubilized membranes in the form of liposomes has the built-in advantage of providing a ready-made source of a membrane preparation that elutes as a single, well defined peak by gel filtration on a Superose12 column (in the exclusion volume). This allows the easy detection, by a variety of means, of translocation of cytosolic oxidase components to the membrane in the course of cell-free activation **(7)**.

11. We use disposable $10 \times 4 \times 45$ mm cuvets (Sarstedt).

12. We use a Uvikon 943 double-beam spectrophotometer (Kontron Instruments), but any instrument with similar characteristics (scanning ability, narrow band-width, and data analysis capability) should be suitable.

13. Quite a number of alternative Δ extinction coefficients for calculating the cytochrome b_{559} heme concentration have been published. Most of these are centered on the 559/558 nm peak. Although this peak is more specific for cytochrome b_{559}, the fact that it is much lower than the 427/426 nm peak increases the chances of error, and we prefer to base our calculations on the 427/426 nm peak. We found no evidence for the presence of significant amounts of b-type cytochromes, other than b_{559}, in macrophage membranes **(70)**.

14. The high concentration of cytochrome b_{559} heme (2 μM) might raise the question of whether freeing the solubilized membrane of octyl glucoside by dialysis is a necessary step in preparing membranes intended for the performance of cell-free assays. In these, the concentration of cytochrome b_{559} heme is 5 to 10 nM, and one might assume that just diluting the membrane 200- to 400-fold is sufficient to reduce the concentration of octyl glucoside well below the CMC without dialysis. It is our experience, however, that this is not the case. Solubilized membrane preparations, dialyzed free of detergent, are clearly superior in cell-free assays to preparations not subjected to dialysis.

15. It is desirable to use the same protein standard throughout in order to compare protein concentrations over long time periods. No external protein standard reflects the true protein concentration of the recombinant proteins, but it is more important to maintain the same level of "error."

16. In the Rho GTPase folklore, it is commonly believed that GMPPNP is more resistant to the intrinsic GTPase activity than GTPγS but that the affinity for GTPγS is higher than that for GMPPNP. For the last 5 yr, we have employed exclusively GMPPNP for generating the GTP-bound form of Rac1 or Rac2.

17. Rac exchanged to GMPPNP was found, in our hands, to be stable for several weeks, and we found no reason to perform the exchange a short time before using the protein in cell-free assays. However, when larger amounts of Rac are subjected to

exchange, it is wise to divide it in smaller aliquots, in order to avoid repeated freezing and thawing.

18. It is obvious that, when using this methodology, unbound GMPPNP is not removed and is present in the Rac preparations at a concentration roughly 10-fold higher than that of the protein. This means that free GMPPNP is transferred to the cell-free assays, which must be taken into consideration. When removal of unbound nucleotide is desired to eliminate possible unwanted effects of free GMPPNP, the nucleotide-exchanged Rac is subjected to buffer exchange by centrifugal ultrafiltration, using Ultra-4 centrifugal filter units fitted with 10-kDa-pore-size cutoff membranes (Millipore). Three volumes of 4 mL each are filtered, using the buffer in which Rac is found, supplemented with 30% v/v glycerol, and the sample is returned to its original volume. Protein concentration is measured again, to check for some unavoidable loss in the course of ultrafiltration.

19. Prenylation can be performed after and before nucleotide exchange to a no-hydrolyzable GTP analog, but we prefer to prenylate after nucleotide exchange.

20. The presence of $ZnCl_2$ is an essential cofactor of geranylgeranyl transferase I.

21. This method can be applied to larger amounts of Rac, provided that nonprenylated Rac is sufficiently concentrated. However, the heated rotary mixer usually accommodates tubes with maximal volumes of 1.5 or 2 mL, which are also convenient for storage.

22. Two additional bands of 48 and 43 kDa are visible in the aqueous phase. These represent the α and β subunits of the enzyme geranylgeranyl transferase I.

23. This was first shown by Ligeti et al. *(58)*. Thus, at 25°C, full assembly was achieved after 5 min of exposure to the amphiphile whereas, at close to 0°C, the process took 30 min. Note, however, that at both temperatures, the same level of assembly is achieved.

24. This observation was first made by Pilloud et al. *(48)* in a system composed of membrane and whole cytosol, and was thought to be related to the effect of salt on the micellar state of the anionic amphiphile. More recent work in a semi-recombinant system demonstrated that increasing ionic strength prevents binding of nonprenylated Rac1 (via its polybasic tail) to negatively charged phospholipids in the membrane *(42)*. No effect of ionic strength was seen on Rac2, which possesses a lesser positive charge at its C-terminus. These facts are of practical importance for the design of an optimal assay buffer and are also of relevance for the mechanism of amphiphile-independent cell-free activation of membranes enriched in anionic phospholipids.

25. If the requirements of the experiment are to add each component separately or in groupings of two, add 10 µL of each component from a 20-fold concentrated stock solution or 20 µL of a 10-fold concentrated mixture of two components and 10 µL of a 20-fold concentrated stock of the third component. This will bring the total volume of the added components to 30 µL and requires the reduction of the volume of added membrane preparation to 10 µL and the proportional increase in the concentration of the membrane stock preparation to 20-fold.

26. The optimization of amphiphile concentration was first discussed in 1989 *(48)* in a paper well worth reading even today. Because anionic amphiphiles act on both

cytosolic *(23)* and membrane *(24)* components, optimization might be a complex issue when setting up radically new conditions and/or when nonpurified components are used, some of which might bind the anionic amphiphile. It is a much simpler procedure when using purified recombinant cytosolic components.

27. Optimization of the time required for the amphiphile-dependent assembly of the oxidase complex is discussed in *(58)*. We found 90 s to be sufficient in the overwhelming majority of cases. However, it is important to point out that prolonging the time of assembly for up to 5 min might be advantageous and is, most definitely, not damaging. Thus, when in the preliminary stages of a project involving cell-free activation, exploring longer assembly times is recommended. Once the minimal time assuring full activation is found, this can be used routinely, but adding to it a "safety time supplement" is a wise move.

28. Turnovers in the amphiphile-activated cell-free system rarely exceed 100 mol $O_2^{\bullet-}$/s/mol cytochrome b_{559} heme. This corresponds to 100 pmol $O_2^{\bullet-}$/s/pmol cytochrome b_{559} heme. Thus, in each well containing 1 pmol cytochrome b_{559} heme, 100 pmol $O_2^{\bullet-}$ are produced per s, which means 6 nmol per min, and 30 nmol per 5 min. The total amount of cytochrome c present in the well is 40 nmol, which is sufficient for binding all the $O_2^{\bullet-}$ produced in 5 min. A total of 48 nmol NADPH are added to the well, which based on a stoichiometry of 1 mol NADPH supporting the production of 2 mol of $O_2^{\bullet-}$, is more than sufficient for the production of 30 nmol of $O_2^{\bullet-}$. In the presence of sufficient cytochrome c, no need was found for the addition of catalase to prevent reoxidation of cytochrome c by H_2O_2 originating in the spontaneous dismutation of $O_2^{\bullet-}$ that escaped scavenging by cytochrome c.

29. The extinction coefficient for the absorbance at 550 nm of reduced minus oxidized cytochrome c, as applied to a 1-cm path length, must be modified for the vertical path length of the microplate wells. This varies with the dimension and shape of the wells and the volume of the reaction mixture present in the well. Newer microplate spectrophotometers (e.g., SPECTRAmax 190, Molecular Devices) have a "PathCheck" sensor, allowing the normalization of absorbance values to a 1-cm path length. This allows the use of the canonical extinction coefficient to calculate the concentration of reduced cytochrome c in the well, without the need for any correction. This has only to be translated into the total amount of reduced cytochrome c in 210 µL, which permits the calculation of the turnover. With instruments not having the "PathCheck" option, the length of the vertical path length in the well must be determined by other means. Once it is known, the following equation will allow the direct calculation of nmol $O_2^{\bullet-}$ per min per well:

$$\text{nmol } O_2^{\bullet-} / \text{min} / \text{well} = \Delta m Abs_{550\,nm} / \text{min} \times 0.047619 \times \text{reaction volume (in mL)} / \text{path length (in cm)}$$

As an example, when then the reaction volume is 0.21 mL and the path length is found to be 0.575 cm, nmol $O_2^{\bullet-}$/min / well = $\Delta m Abs_{550\,nm}$ / min $\times$ 0.047619 $\times$ 0.21/0.575 = $\Delta m Abs_{550\,nm}$ / min $\times$ 0.01739.

30. The proper way to express results of cell-free oxidase activation assays is as turnover values. Unless there is a compelling reason for not doing so, oxidase activities should be related to the heme content of cytochrome b_{559} present in the membrane and not to cell number equivalents, total membrane protein, or the protein concentration of one or the other of the cytosolic components.

31. From the advent of semi-recombinant cell-free assays and the increasing rarity of the use of whole cytosol or partially purified cytosolic fractions, the need for the SOD control has been drastically reduced. In our laboratory, in which cell-free assays are performed routinely, we have not encountered a single occasion of nonspecific cytochrome *c* reduction for the last decade or so. This is to be expected because nonspecific cytochrome *c* reduction is seen mostly in end-point assays and rarely in kinetic assays, in which the presence of a reducing agent would cause cytochrome *c* reduction already at time zero. Of course, SOD controls should be used when cell-free assays are utilized as a diagnostic means on unpurified biologic material and in novel experimental situations.

32. Dispensing 10–20 µL aliquots to the wells is best performed with electronic pipettors. In most situations, the concentrations of p67*phox* and prenylated Rac required for reaching maximal activation, in the absence of amphiphile and p47*phox*, are higher than those necessary for amphiphile-dependent activation in the presence of identical amounts of membrane.

33. In the amphiphile-independent cell-free system, we found on some occasions that prolonging the incubation to up to 5 min resulted in increased oxidase activities. This time interval should, however, not be exceeded.

34. The presence of p47*phox* is not an absolute requirement for oxidase activity to be detected in this assay, but turnover values are higher in its presence.

35. This effect was thought to be due to an elevation of the Krafft point of the fatty acids in the presence of Ca^{2+} *(71)*.

36. These results support the recent proposal that the stability of the FAD-gp91*phox* bond is enhanced by anionic amphiphile-induced changes in gp91*phox* and by the process of assembly with the cytosolic components *(72)*.

37. In the experiments described in **Fig. 4**, both nonprenylated and prenylated Rac1 were exchanged to GMPPNP and the exchange stabilized by 25 m*M* MgCl$_2$. Thus, it is unlikely that significant dissociation of GMPPNP from Rac can occur upon dilution of exchanged Rac preparations in assay buffer during the 90-s time interval of oxidase assembly. However, assay buffer is also used for intermediary dilution steps of recombinant cytosolic components, and it is possible that the diluted proteins are kept in assay buffer for longer periods of time. In the particular case of Rac, it is recommended to make such dilutions in the buffer listed (*see* **Subheading 2.1.2., step 1**), which contains 4 m*M* MgCl$_2$.

38. There is good evidence for equimolar and simultaneous translocation of the cytosolic components in neutrophils stimulated by two elicitors of a respiratory burst, and it was also found that this translocation corresponded temporally with the generation of $O_2^{\bullet -}$ *(73)*.

39. It is of interest that sigmoidal dose–response curves were described in the early cell-free assays when amounts of total cytosol were related to activity in the presence of a constant amount of amphiphile *(47,48)*.
40. This assumption was examined by Heyworth et al. *(64)* and Fuchs et al. *(63)*, who showed that exchange with GTP added to the assay buffer only takes place with prenylated Rac and that preloading with GTP by Mg^{2+} chelation is required for nonprenylated Rac to work in the amphiphile-dependent cell-free assay.
41. This confirms the contention that no significant nucleotide exchange on nonprenylated Rac takes place upon addition of GTP analogs to the assay buffer *(63,64)*.
42. Another distinguishing feature of this system is the paradoxical inhibitory effect of added (free) GTPγS (10 μ*M*) on oxidase activation by prenylated Rac exchanged to GMPPNP. The inhibitory effect of free GTP and GTPγS on amphiphile-independent oxidase activation was described by us in the past, but was never satisfactorily explained *(53)*.
43. In one study, it was concluded that nonprenylated Rac1 and Rac2 were poorly active in an amphiphile-dependent cell-free assay in the absence of nucleotide exchange to GTPγS *(61)*. In another study, it was shown that the nonprenylated form of Rac1 exchanged to GTPγS was much superior to Rac2 exchanged to GTPγS *(6)*. This difference was clearly attributed to the lesser positive charge at the C-terminus of Rac2, a suggestion also supported by the inhibitory effect of increased ionic strength on the activity of Rac1, but not Rac2, and by the enhancing effect of an increase in the negative charge of the membrane on the effect of Rac1, but not Rac2 *(6)*. These authors also found that prenylated Rac1 and Rac2, when preloaded with GTPγS, support oxidase activation in an amphiphile-dependent system. In those studies, Rac1 was more efficient at activating the oxidase.
44. We proposed recently that because both p47[phox] and Rac function as NOX organizers, thereby contributing to the establishment of an optimal (more stable?) interaction between p67[phox] and gp91[phox], the presence of one organizer lessens the dependence on the other. This is expressed in the ability to activate the oxidase in the absence of p47[phox] *(7,9,10,32)* and in a more "liberal" interpretation of the requirement for Rac to be in the GTP-bound form when p47[phox] is present *(51)*.

Acknowledgments

The research described in this report was supported by the Julius Friedrich Cohnheim-Minerva Center for Phagocyte Research, the Ela Kodesz Institute of Host Defense against Infectious Diseases, Israel Science Foundation Grants 428/01 and 19/05, the Roberts-Guthman Chair in Immunopharmacology, and the Walter J. Levy Benevolent Trust. Edgar Pick would like to thank his fellow scientists, too many to name, who provided materials and invaluable advice, for making this work possible. There is no greater satisfaction than the realization that, on so many occasions, what started as collaboration evolved into friendship.

References

1. Klebanoff, S. J. (1999) Oxygen metabolites from phagocytes, in *Inflammation: Basic Principles and Clinical Correlates* (Gallin, J. I. and Snyderman, R., eds.), Lippincott, Williams & Wilkins, Philadelphia, pp. 721–768.
2. Cross, A. R. and Segal, A. W. (2004) The NADPH oxidase of phagocytes –prototype of the NOX electron transport chain systems. *Biochim. Biophys. Acta* **1657,** 1–22.
3. Quinn, M. T. and Gauss, K. A. (2004) Structure and regulation of the neutrophil respiratory burst oxidase: comparison with nonphagocyte oxidases. *J. Leukoc. Biol.* **76,** 760–781.
4. Han, C.-H., Freeman, J. L. R., Lee, T., Motalebi, S., and Lambeth, J. D. (1998) Regulation of the neutrophil respiratory burst oxidase. Identification of an activation domain. *J. Biol.* Chem. **273,** 16,663–16,668.
5. Bokoch, G. M. and Diebold, B. A. (2002) Current models for NADPH oxidase regulation by Rac GTPase. *Blood* **100,** 2692–2696.
6. Kreck, M. L., Freeman, J. L., and Lambeth, J. D. (1996) Membrane association of Rac is required for high activity of the respiratory burst oxidase. *Biochemistry* **35,** 15,683–15,692.
7. Gorzalczany, Y., Sigal, N., Itan, M., Lotan, O., and Pick, E. (2000) Targeting of Rac1 to the phagocyte membrane is sufficient for the induction of NADPH oxidase assembly. *J. Biol. Chem.* **275,** 40,073–40,081.
8. Lambeth, J. D. (2004) NOX enzymes and the biology of reactive oxygen. *Nat. Rev. Immunol.* **4,** 181–189.
9. Freeman, J. L. and Lambeth, J. D. (1996) NADPH oxidase activity is independent of p47^{phox} in vitro. *J. Biol. Chem.* **271,** 22,578–22,582.
10. Koshkin, V., Lotan, O., and Pick, E. (1996) The cytosolic component p47^{phox} is not a *sine qua non* participant in the activation of NADPH oxidase but is required for optimal superoxide production. *J. Biol. Chem.* **271,** 30,326–30,329.
11. Cheson, B. D., Curnutte, J. T., and Babior, B. M. (1977) The oxidative killing mechanism of the neutrophil. *Prog. Clin. Immunol.* **3,** 1–65.
12. Pick, E. and Keisari, Y. (1981) Superoxide anion production and hydrogen peroxide production by chemically elicited macrophages—induction by multiple non-phagocytic stimuli. *Cell. Immunol.* **59,** 301–318.
13. Babior, B. M., Curnutte, J. T., and McMurrich, B. J. (1976) The particulate super-oxide-forming system from human neutrophils: properties of the system and further evidence supporting its participation in the respiratory burst. *J. Clin. Invest.* **58,** 989–996.
14. Markert, M., Andrews, P. C., and Babior, B. M. (1984) Measurement of $O_2^{\bullet-}$ production by human neutrophils. The preparation and assay of NADPH oxidase-containing particles from human neutrophils. *Methods Enzymol.* **105,** 358–365.
15. Flores, J., Witkum, P., and Sharp, G. W. G. (1976) Activation of adenylate cyclase by cholera toxin. *J. Clin. Invest.* **57,** 450–458.
16. Bromberg, Y. and Pick, E. (1984) Unsaturated fatty acids stimulate NADPH- dependent superoxide generation by cell-free system in macrophages. *Cell. Immunol.* **88,** 213–221.

17. Heynemann, R. A. and Vercauteren, R. E. (1984) Activation of a NADPH oxidase from horse poymorphonuclear leukocytes in a cell-free system. *J. Leukoc. Biol.* **36,** 751–759.

18. McPhail, L. C., Shirley, P. S., Clayton, C. C., and Snyderman, R. (1985) Activation of the respiratory burst enzyme from human neutrophils in a cell-free system. *J. Clin. Invest.* **75,** 1735–1739.

19. Curnutte, J. T. (1985) Activation of human neutrophil nicotinamide adenine dinucleotide phosphate reduced (triphosphopyridine nucleotide, reduced) oxidase by arachidonic acid in a cell-free system. *J. Clin. Invest.* **75,** 1740–1743.

20. Curnutte, J. T. and Babior, B. M. (1987) Chronic granulomatous disease. *Adv. in Human Genetics* **16,** 229–297.

21. Bromberg, Y. and Pick, E. (1985) Activation of NADPH-dependent superoxide production in a cell-free system by sodium dodecyl sulfate. *J. Biol. Chem.* **260,** 13,539–13,545.

22. Groemping, Y., Lapouge, K., Smerdon, S. J., and Rittinger, K. (2003) Molecular basis of phosphorylation-induced activation of the NADPH oxidase. *Cell* **113,** 343–355.

23. Swain, S. D., Helgerson, S. L., Davis, A. R., Nelson, L. K., and Quinn, M. T. (1997) Analysis of activation-induced conformational changes in p47^{phox} using tryptophan fluorescence spectroscopy. *J. Biol. Chem.* **272,** 29,502–29,510.

24. Foubert, T. R., Burritt, J. B., Taylor, R. M., and Jesaitis, A. J. (2002) Structural changes are induced in human neutrophil cytochrome *b* by NADPH oxidase activators, LDS, SDS and arachidonate: Intermolecular resonance energy transfer between trisulfopyrenyl-wheat germ agglutinin and cytochrome b_{558}. *Biochim. Biophys. Acta* **1567,** 221–231.

25. Qualliotine-Mann, D., Agwu, D. E., Ellenburg, M. D., McCall, C. E., and McPhail, L. C. (1993) Phosphatidic acid and diacylglycerol synergize in a cell-free system for activation of NADPH oxidase from human neutrophils. *J. Biol. Chem.* **268,** 23,843–23,849.

26. Park, J.-W., Hoyal, C. R., El Benna , J., and Babior, B. M. (1997) Kinase-dependent activation of the leukocyte NADPH oxidase in a cell-free system—Phosphorylation of membranes and p47^{phox} during oxidase activation. *J. Biol. Chem.* **272,** 11,035–11,043.

27. Abo, A., Boyhan, A., West, I., Thrasher, A. J., and Segal, A. W. (1992) Reconstitution of neutrophil NADPH oxidase activity in the cell-free system by four components: p67^{phox}, p47^{phox}, p21rac1, and cytochrome b$_{-245}$. *J. Biol. Chem.* **267,** 16,767–16,770.

28. Ebisu, K., Nagasawa, T., Watanabe, K., Kakinuma, K., Miyano, K., and Tamura, M. (2001) Fused p47^{phox} and p67^{phox} truncations efficiently reconstitute NADPH oxidase with higher activity than the individual components. *J. Biol. Chem.* **276,** 24,498–24,505.

29. Alloul, N., Gorzalczany, Y., Itan, M., Sigal, N., and Pick, E. (2001). Activation of the superoxide-generating NADPH oxidase by chimeric proteins consisting of seg-

ments of the cytosolic component p67^{phox} and the small GTPase Rac1. *Biochemistry* **40,** 14,557–14,566.

30. Hata, K., Ito, T., Takeshige, K., and Sumimoto, H. (1998) Anionic amphiphile-independent activation of the phagocyte NADPH oxidase in a cell-free system by p47^{phox} and p67^{phox}, both in C terminally truncated forms. Implications for regulatory Src himology 3 domain-mediated interactions. *J. Biol. Chem.* **273,** 4232–4236.

31. Peng, G., Huang, J., Boyd, M., and Kleinberg, M. E. (2003) Properties of phagocyte NADPH oxidase p47^{phox} mutants with unmasked SH3 (Src homology 3) domains: Full reconstitution of oxidase activity in a semi-recombinant cell-free system lacking arachidonic acid. *Biochem. J.* **373,** 221–229.

32. Gorzalczany, Y., Alloul, N., Sigal, N., Weinbaum, C., and Pick, E. (2002) A prenylated p67^{phox}-Rac1 chimera elicits NADPH-dependent superoxide production by phagocyte membranes in the absence of an activator and of p47^{phox}. Conversion of a pagan NADPH oxidase to monotheism. *J. Biol. Chem.* **277,** 18,605–18,610.

33. Mizrahi, A., Molshanski-Mor, S., Weinbaum, C., Zheng, Y., Hirshberg, M., and Pick, E. (2005) Activation of the phagocyte NADPH oxidase by Rac guanine nucleotide exchange factors in conjunction with ATP and nucleoside diphosphate kinase. *J. Biol. Chem.* **280,** 3802–3811.

34. Babior, B. M., Kipnes, R. S., and Curnutte, J. T. (1973) Biological defense mechanisms. The production by leukocytes of superoxide, a potential bactericidal agent. *J. Clin. Invest.* **52,** 741–744.

35. Pick, E., Bromberg, Y., Shpungin, S., and Gadba, R. (1987) Activation of the superoxide forming NADPH oxidase in a cell-free system by sodium dodecyl sulfate. Characterization of the membrane-associated component. *J. Biol. Chem.* **262,** 16,476–16,483.

36. Shpungin, S., Dotan, I., Abo, A., and Pick, E. (1989) Activation of the superoxide forming NADPH oxidase in a cell-free system by sodium dodecyl sulfate. Absolute lipid dependence of the solubilized enzyme. *J. Biol. Chem.* **264,** 9195–9203.

37. Light, D., Walsh, C., O'Callaghan, A. M., Goetzl, E. J., and Tauber, A. I. (1981) Characteristics of the cofactor requirements for the superoxide-generating NADPH oxidase of human polymorphonuclear leukocytes. *Biochemistry* **20,** 1468–1476.

38. Leto, T. L., Garrett, M. C., Fujii, H., and Nunoi, H. (1991) Characterization of neutrophil NADPH oxidase factors p47^{phox} and p67^{phox} from recombinant baculoviruses. *J. Biol. Chem.* **266,** 19,812–19,818.

39. Koshkin, V., Lotan, O., and Pick, E. (1997) Electron transfer in the superoxide-generating NADPH oxidase complex reconstituted in vitro. *Biochim. Biophys. Acta* **1319,** 139–146.

40. Wientjes, F. B., Panayotou, G., Reeves, E., and Segal, A. W. (1996) Interactions between cytosolic components of the NADPH oxidase: p40^{phox} interacts with both p67^{phox} and p47^{phox}. *Biochem. J.* **317,** 919–924.

41. Bradford, M. M. (1976) A rapid and sensitive method for the quantitation of microgram quantities of protein utilizing the principle of protein-dye binding. *Anal. Biochem.* **72,** 248–254.
42. Kreck, M. L., Uhlinger, D. J., Tyagi, S. R., Inge, K. L., and Lambeth, J. D. (1994) Participation of the small molecular weight GTP-binding protein Rac1 in cell-free activation and assembly of the respiratory burst oxidase. Inhibition by a carboxyl-terminal Rac peptide. *J. Biol. Chem.* **269,** 4161–4168.
43. Kwong, C. H., Malech, H. L., Rotrosen, D., and Leto, T. L. (1993) Regulation of the human neutrophil NADPH oxidase by rho-related G-proteins. *Biochemistry* **32,** 5711–5717.
44. Sigal, N., Gorzalczany, Y., Sarfstein, R., Weinbaum, C., Zheng, Y., and Pick, E. (2003) The guanine nucleotide exchange factor Trio activates the phagocyte NADPH oxidase in the absence of GDP to GTP exchange on Rac. "The emperor's new clothes". *J. Biol. Chem.* **278,** 4854–4861.
45. Bordier, C. (1981) Phase separation of integral membrane proteins in Triton X-114 solution. *J. Biol. Chem.* **256,** 1604–1607.
46. Gutierrez, L., Magee, A. I., Marshall, C. J., and Hancock, J. F. (1989) Post-translational processing of p21^{ras} is two-step and involves carboxyl-methylation and carboxy-terminal proteolysis. *EMBO J.* **8,** 1093–1098.
47. Babior, B. M., Kuver, R., and Curnutte, J. T. (1988) Kinetics of activation of the respiratory burst oxidase in a fully soluble system from human neutrophils. *J. Biol. Chem.* **23,** 1713–1718.
48. Pilloud, M.-C., Doussiere, J., and Vignais, P. V. (1989) Parameters of activation of the membrane-bound O_2^--generating oxidase from neutrophils in a cell-free system. *Biochem. Biophys. Res. Comm.* **159,** 783–790.
49. Cross, A. R., Erickson, R. W., and Curnutte, J. T. (1999) Simultaneous presence pf p47^{phox} and flavocytochrome b_{-245} are required for activation of NADPH oxidase by anionic amphiphiles. Evidence for an intermediate state of oxidase activation. *J. Biol. Chem.* **274,** 15,519–15,525.
50. Sarfstein, R., Gorzalczany, Y., Mizrahi, A., et al. (2004) Dual role of Rac in the assembly of NADPH oxidase, tethering to the membrane and activation of p67^{phox}. A study based on mutagenesis of p67^{phox}-Rac1 chimeras. *J. Biol. Chem.* **279,** 16,007–16,016.
51. Mizrahi, A., Berdichevsky, Y., Ugolev, Y., et al. (2006) Assembly of the phagocyte NADPH oxidase complex: chimeric constructs derived from the cytosolic components as tools for exploring structure–function relationships. *J. Leukoc. Biol.* **79,** 881–895.
52. Ugolev, Y., Molshanski-Mor, S., Weinbaum, C., and Pick, E. (2006) Liposomes comprising anionic but not neutral phospholipids cause dissociation of [Rac(1 or 2)-RhoGDI] complexes and support amphiphile-independent NADPH oxidase activation by such complexes. *J. Biol. Chem.* **281,** 19,204–19,279.
53. Sigal, N., Gorzalczany, Y., and Pick, E. (2003) Two pathways of activation of he superoxide-generating NADPH oxidase of phagocytes in vitro—Distinctive effects of inhibitors. *Inflammation* **27,** 147–159.

54. Rotrosen, D., Kleinberg, M. E., Nunoi, H., Leto, T., Gallin, J. I., and Malech, H. L. (1990) Evidence for a functional cytoplasmic domain of phagocyte oxidase cytochrome b_{558}. *J. Biol. Chem.* **265,** 8745–8750.
55. Nauseef, W. M., McCormick, S., Jan, R., Leidal, K. G., and Clark, R. A. (1993) Functional domains in an arginine-rich carboxyl-terminal region of p47*phox*. *J. Biol. Chem.* **268,** 23,646–23,651.
56. Babior, B. M. and Kipnes, R. S. (1977) Superoxide-forming enzyme from human neutrophils: evidence for a flavin requirement. *Blood* **50,** 517–524.
57. Knaus, U. G., Heyworth, P. G., Kinsella, B. T., Curnutte, J. T., and Bokoch, G. M. (1992) Purification and characterization of Rac2. A cytosolic GTP-binding protein that regulates human neutrophil NADPH oxidase. *J. Biol. Chem.* **267,** 23,575–23,582.
58. Ligeti, E., Doussiere, J., and Vignais, P. V. (1988) Activation of the $O_2^{\bullet-}$-generating oxidase in plasma membrane from bovine polymorphonuclear neutrophils by arachidonic acid, a cytosolic factor of protein nature, and nonhydrolyzable analogues of GTP. *Biochemistry* **27,** 193–200.
59. Seifert, R., Rosenthal, W., and Schultz, G. (1986) Guanine nucldeotides stimulate NADPH oxidase in membranes of human neutrophils. *FEBS Lett.* **105,** 161–165.
60. Gabig, T. G., English, D., Akard, L. P., and Schell, M. J. (1987) Regulation of neutrophil NADPH oxidase activation in a cell-free system by guanine nucleotides and fluoride. Evidence for participation of a pertussis and cholera toxin-insensitive G protein. *J. Biol. Chem.* **262,** 1685–1690.
61. Aharoni, I. and Pick, E. (1990) Activation of the superoxide-generating NADPH oxidase of macrophages by sodium dodecyl sulfate in a soluble cell-free system. Evidence for involvement of a G protein. *J. Leukoc. Biol.* **48,** 107–115.
62. Toporik, A., Gorzalczany, Y., Hirshberg, M., Pick, E., and Lotan, O. (1998) Mutational analysis of novel effector domains in Rac1 involved in the activation of nicotinamide adenine dinucleotide phosphate (reduced) oxidase. *Biochemistry* **37,** 7147–7156.
63. Fuchs, A., Dagher, M.-C., Jouan, A., and Vignais, P. V. (1994) Activation of the O_2^--generating NADPH oxidase in a semi-recombinant cell-free system. Assessment of the function of Rac in the activation process. *Eur. J. Biochem.* **226,** 587–595.
64. Heyworth, P. G., Knaus, U. G., Xu, X., et al. (1993) Requirement for posttranslational processing of Rac GTP-binding proteins for activation of human neutrophil NADPH oxidase. *Mol. Biol. Cell* **4,** 261–269.
65. Zhao, X., Carnevale, K. A., and Cathcart, M. K. (2003) Human monocytes use Rac1, not Rac2, in the NADPH oxidase complex. *J. Biol. Chem.* **278,** 40,788–40,792.
66. Bromberg, Y. and Pick, E. (1983) Unsaturated fatty acids as second messengers of superoxide generation by macrophages. *Cell. Immunol.* **79,** 243–252.
67. Seifert, R. and Schultz, G. (1987) Fatty acid-induced activation of NADPH oxidase in plasma membranes of human neutrophils depends on neutrophil cytosol and is potentiated by stable guanine nucleotides. *Eur. J. Biochem.* **162,** 563–569.

68. Tanaka, T., Makino, R., Iizuka, T., Ishimura, Y., and Kanegasaki, S. (1988) Activation by saturated and monounsaturated fatty acids of the O_2^--generating system in a cell-free preparation from neutrophils. *J. Biol. Chem.* **263,** 13,670–13,676.
69. Diatchuk, V., Lotan, O., Koshkin, V., Wikstroem, P., and Pick, E. (1997) Inhibition of NADPH oxidase activation by 4-(2-aminoethyl)-benzenesulfonyl fluoride and related compounds. *J. Biol. Chem.* **272,** 13,292–13,301.
70. Pick, E. and Gadba, R. (1988) Certain lymphoid cells contain the membrane-associated component of the phagocyte-specific NADPH oxidase. *J. Immunol.* **140,** 1611–1617.
71. Yamaguchi, T., Kaneda, M., and Kakinuma, K. (1986) Effect of saturated and unsaturated fatty acids on the oxidative metabolism of human neutrophils. The role of calcium ion in the extracellular medium. *Biochim. Biophys. Acta* **861,** 440–446.
72. Hashida, S., Yuzawa, S., Suzuki, N. N., et al. (2004) Binding of FAD to cytochrome b_{558} is facilitated during activation of the phagocyte NADPH oxidase, leading to superoxide production. *J. Biol. Chem.* **279,** 26,378–26,386.
73. Quinn, M. T., Evans, T., Loetterle, L. R., Jesaitis, A. J., and Bokoch, G. M. (1993) Translocation of Rac correlates with NADPH oxidase activation. Evidence for equimolar translocation of oxidase components. *J. Biol. Chem.* **268,** 20,983–20,987.

26

Immunoaffinity Purification of Human Phagocyte Flavocytochrome b and Analysis of Conformational Dynamics

Ross M. Taylor and Algirdas J. Jesaitis

Summary

The heterodimeric integral membrane protein flavocytochrome b (Cyt b) is the catalytic core of the phagocyte NADPH oxidase, an enzyme complex that initiates a cascade of reactive oxygen species critical for the elimination of infectious agents. Many fundamental questions remain concerning the structure and catalytic mechanism of Cyt b, largely because of the inability to isolate this protein in quantities required for both biochemical analysis and meaningful attempts at high-resolution structure determination. In order to facilitate the direct analysis of Cyt b, the following method describes a rapid and efficient procedure for the immunoaffinity purification of Cyt b (under nondenaturing conditions) from neutrophil membrane fractions. The protocol presented here contains a number of steps that have been optimized and improved since the original description of this Cyt b isolation method. In order to address questions concerning the mechanism of superoxide generation by the NADPH oxidase complex, methods are additionally presented for analysis of conformational dynamics of immunoaffinity-purified Cyt b by resonance energy transfer.

Key Words: Phagocyte; reactive oxygen species; flavocytochrome b; immunoaffinity purification; conformational change; resonance energy transfer.

1. Introduction

In response to inflammatory stimuli, phagocytic leukocytes (neutrophils, monocytes, and macrophages) migrate to sites of infection and tissue injury, where they are activated to release massive quantities of superoxide through the action of the multiprotein NADPH oxidase complex *(1)*. The catalytic core of the phagocyte NADPH oxidase (flavocytochrome b [Cyt b]) is a heterodimeric, integral membrane protein composed of a heavily glycosylated, 570-residue large subunit (gp91^{phox}; Nox2) and a 195-residue small subunit (p22^{phox})

From: *Methods in Molecular Biology, vol. 412: Neutrophil Methods and Protocols*
Edited by: M. T. Quinn, F. R. DeLeo, and G. M. Bokoch © Humana Press Inc., Totowa, NJ

(2,3). Because Cyt b is currently isolated from whole human blood in limited quantities, our group developed a rapid, single-step purification of Cyt b from neutrophil membranes using a monoclonal antibody (44.1) that binds p22*phox* *(4,5)*. Following detergent solubilization of neutrophil membranes and specific binding of Cyt b to the monoclonal antibody 44.1 affinity matrix, high-yield elution of Cyt b is achieved by the addition of an epitope-mimicking peptide (Ac-PQVRPI-CONH$_2$) that was derived by phage-display analysis *(5)*. This immunoaffinity purification procedure is conducted under nondenaturing conditions, reproducibly results in final Cyt b yields of approx 30–45% *(4)*, and generates a Cyt b sample that can be concentrated to relatively high levels for direct analysis of the detergent-solubilized protein.

Assuming that an appropriate fluorescence donor/acceptor pair can be generated, resonance energy transfer (RET) methods provide a highly sensitive assay for measuring both distance and conformational changes in biological molecules *(6,7)*. Because few methods have been reported for the direct analysis of activation-induced conformational states of Cyt b, our group has biochemically labeled Cyt b-specific monoclonal antibodies with fluorescent probes (donor molecules) that are efficiently quenched when brought into close proximity to the Cyt b heme prosthetic groups (acceptor molecules). The resonance energy transfer system described here provides a sensitive method that has been used to monitor antibody binding and characterize lipid-induced conformational changes in immunoaffinity-purified Cyt b that may be critical for oxidase activation *(8)*.

2. Materials

2.1. Construction of the Monoclonal Antibody 44.1 Affinity Matrix

1. Antibody binding matrix: protein A-Sepharose (Amersham Biosciences).
2. Hybridoma culture supernatant containing the p22*phox*-specific monoclonal antibody 44.1 is generated in-house by standard hybridoma culture technology and stored at 4°C.
3. 250-mL polypropylene centrifuge tubes (Corning).
4. Disposable 5-inch polystyrene chromatography columns (Evergreen Scientific).
5. Phosphate-buffered saline (PBS): 10 m*M* sodium phosphate (pH 7.4), 100 m*M* NaCl.
6. Relax Buffer: 10 m*M* HEPES (pH 7.4), 100 m*M* KCl, 10 m*M* NaCl, 1 m*M* ethylenediamine tetraacetic acid (EDTA).
7. Nonionic detergent dodecylmaltoside (DDM): prepare in deionized water (dH$_2$O) as a 20% stock (w/v), and store for no more than 1 mo at room temperature.

2.2. Immunoaffinity Purification of Cyt b

1. Human neutrophil membrane fractions: prepare as previously described *(2)*, and store at −70°C.

2. 5 *M* NaCl stock: prepare in dH$_2$O and store at room temperature.
3. Hand-operated, 40 mL glass homogenizer (Kimble/Kontes).
4. Relax buffer and the DDM stock are prepared as under **Subheading 2.1.**
5. Protease inhibitors: chymostatin (Sigma-Aldrich) prepared at 20 mg/mL in dimethylsulfoxide (DMSO) and stored at −20°C; phenylmethylsulfonylfluoride (PMSF) (CalBiochem) prepared as a 200 m*M* stock in ethanol and stored at 4°C; mammalian protease inhibitor cocktail (Sigma-Aldrich) purchased as a 1000-fold stock in DMSO and stored at −20°C.
6. Dithiothreitol (DTT; CalBiochem): prepare fresh as a 100 µ*M* stock in dH$_2$O.
7. 50 Sonic dismembrator probe sonicator (Fisher Scientific).
8. Elution peptide Ac-PQVRPI-CONH$_2$ (Macromolecular Resources): store lyophilized at −20°C.
9. Absorption spectrophotometer with quartz microcuvets (~60 µL minimum sample volume).

2.3. Concentration and Desalting of Cyt b

1. Centricon YM-100 (100,000 molecular weight cutoff) concentration devices (Millipore).
2. Econo-Pac 10 DG desalting columns (30 × 10 mL; BioRad) equilibrated in Relax buffer containing the desired amount of DDM (typically 0.03–0.1%) at 4°C. Use for removal of elution peptide from purified Cyt b samples.
3. Relax buffer and the DDM stock prepared as under **Subheading 2.1.**

2.4 Cascade Blue Labeling of Cyt b-Specific Monoclonal Antibodies

1. Cascade Blue (CCB) acetyl azide (Molecular Probes): prepare as a 10 mg/mL stock in methanol, and store in desiccant at −20°C.
2. Antibody purification matrix: GammaBind Plus-Sepharose (Amersham Biosciences).
3. Labeling reaction quenching buffer: 1 *M* Tris-HCl (pH 7.4, room temperature).
4. Econo-Pac 10 DG desalting columns (30 × 10 mL; BioRad): equilibrate in PBS at room temperature. Use for removal of free dye from monoclonal antibody labeling reactions.
5. BCA Protein Assay Kit (Pierce).

2.5. Resonance Energy Transfer Analysis

1. Phospholipids: purchase as chloroform stocks (Avanti Polar Lipids), and store at −20°C in ReactiVial Small Reaction Vials containing Teflon coated lids (Pierce). Briefly overlay lipid stocks with argon prior to storage to minimize oxidation.
2. Cylindrical, ultraviolet (UV)-transparent microcuvets (Sienco) that are continuously stirred with a microstirbar (Sigma-Aldrich). The cuvet holders were manufactured in-house.
3. Polyethersulfone syringe filters (0.2 µ*M*; Fisher Scientific).
4. TLA-100 ultracentrifuge with thickwall polycarbonate tubes in a TLA-100.2 rotor (Beckman Instruments).
5. Fluorescence spectrophotometer capable of real-time emission scans.

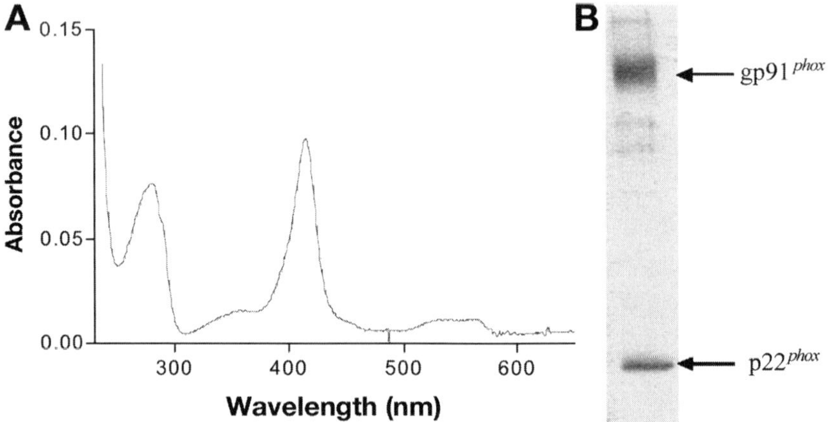

Fig. 1. Immunoaffinity purification of flavocytochrome b from human neutrophil membrane fractions. (**A**) Absorption spectrum of purified flavocytochrome b showing both the protein absorption and Soret band for oxidized heme. (**B**) Silver-stained sodium dodecyl sulfate-polyacylamide gel showing the gp91^{phox} and p22^{phox} subunits of immunoaffinity-purified Cyt b. (Adapted from **ref. 4**.)

3. Methods

The following methods outline the preparation of purified, detergent-solubilized Cyt b (*see* **Fig. 1**) for analysis of conformational changes by resonance energy transfer (*see* **Fig. 2**). Several modifications to the original protocol for immunoaffinity purification of Cyt b (*4*) that result in improved preparations are described, including: (1) washing the affinity matrix in a column (rather than batch) format; (2) elution of Cyt b at detergent concentrations near the critical micelle concentration; (3) concentration of eluted Cyt b samples using a 100-kDa cutoff membrane to minimize the simultaneous concentration of detergent; and (4) the rapid removal of elution peptide (and buffer exchange if desired) on a desalting column. The entire immunoaffinity purification procedure can be readily conducted in a single day and generates ample material for conducting RET measurements.

3.1 Construction of the Monoclonal Antibody 44.1 Affinity Matrix

1. One day prior to the Cyt b purification, equilibrate approx 1.5 mL of protein A-Sepharose beads in PBS and then rotate the beads overnight at 4°C with 150 mL of culture supernatant derived from the monoclonal antibody 44.1 hybridoma cell line (a 250-mL centrifuge tube is used for this incubation).
2. Centrifuge the above mixture for 5 min at approx 200g and 4°C in a clinical centrifuge, and carefully aspirate off the hybridoma culture supernatant without disturbing the packed beads.

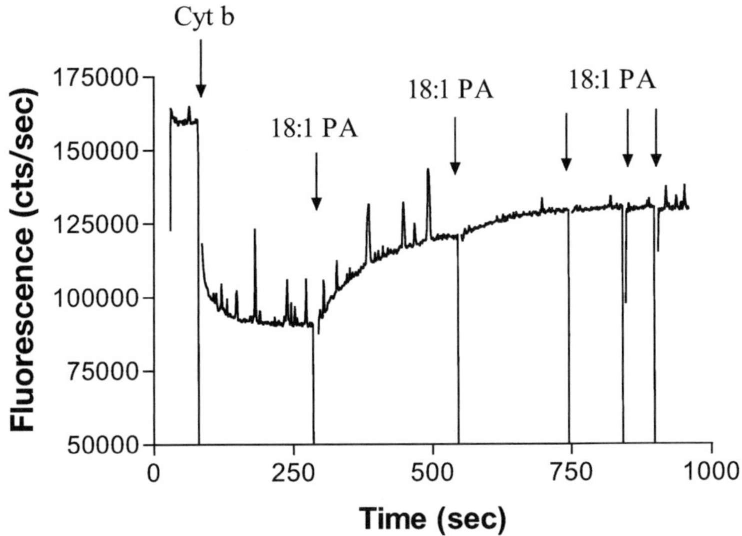

Fig. 2. Analysis of conformational dynamics in purified flavocytochrome b (Cyt b) by resonance energy transfer. In this system, the binding of detergent-solubilized Cyt b to the Cascade Blue (CCB)-labeled monoclonal antibody 44.1 is measured by the quenching of the initial fluorescence emission and conformational changes in Cyt b are measured as a change in fluorescence quenching that occurs upon addition of anionic lipid. The saturable relaxation of fluorescence quenching observed when the CCB-44.1:Cyt b complex is titrated with 18:1 phosphatidic acid (PA). Arrows indicate the addition of both Cyt b (30 nM final) and DOPA (25 μM additions) to the fluorescence cuvet containing CCB-44.1 (10 nM). (Reproduced from **ref. 8**, with permission from Elsevier Science.)

3. Transfer the beads to a disposable 5-inch polystyrene chromatography column, and wash the beads at room temperature with Relax buffer until the absorption of the flow-through fraction is at baseline.
4. Equilibrate the beads with 10 mL of Relax buffer containing 0.1% DDM at room temperature.

3.2. Immunoaffinity Purification of Cyt b

1. Thaw 20 mL of neutrophil membranes (~1 × 10^{10} cell equivalents) and bring the solution to 1 M NaCl (by addition from a 5 M NaCl stock).
2. Centrifuge the mixture at 114,000g for 30 min at 4°C.
3. Resuspend the resulting membrane pellet by homogenization in 19 mL of Relax buffer, 0.1 mM DTT, 10 µg/mL chymostatin, 2 mM PMSF, and 1 µl/mL P8340 (*see* **Note 1**).
4. Add the nonionic detergent DDM to 0.75% final from a 20% stock.
5. Homogenize the extract and then briefly sonicate (3 × 5 s, setting 3).
6. Gently tumble the extract at room temperature for 50 min.

7. Centrifuge the membrane extract at 114,000g for 30 min at 22°C.
8. Remove the soluble fraction, and analyze by absorption spectroscopy (200–700 nm). The Cyt b content is determined using $\varepsilon_{414} = 131,000\ M^{-1}cm^{-1}$ for the heme Soret band with the assumption of two heme prosthetic groups per Cyt b.
9. Add the DDM-membrane extract to the equilibrated affinity matrix (in a 50-mL Falcon tube) and rotate for 1.5 h at room temperature.
10. Pour the mixture into a disposable column, and analyze the flow fraction by absorption spectroscopy (414 nm) to assess binding of Cyt b to the affinity matrix.
11. Wash the packed beads with Relax buffer containing 0.25% DDM (~10 mL), and then wash with Relax buffer containing 0.05% DDM until absorbance of the flow-through fraction is at baseline.
12. For elution of Cyt b, rotate the affinity matrix with 3.5 mL of Relax buffer, 0.05% DDM, and 1.5 mM elution peptide AC-PQVRPI-CONH$_2$ for 5 min at 37°C. Warm the elution buffer to 37°C prior to addition to the affinity matrix (*see* **Note 2**).
13. Collect the elution fractions, place on ice, and analyze the Cyt b content by absorption spectroscopy.
14. Repeat the above elution two more times with 2 mL of elution buffer.
15. Pool desired elution fractions (*see* **Notes 3** and **4**).

3.3. Concentration and Desalting of Cyt b

1. Equilibrate a Centricon YM-100 concentrator (*see* **Note 5**) by passing approx 1 mL of Relax buffer (the concentrator membrane is relatively fragile and cannot be centrifuged over 1000g).
2. Concentrate the pooled Cyt b elution fractions to approx 0.5 mL, with occasional gentle pipetting of the retained sample to minimize Cyt b adsorption to the concentrator membrane.
3. Measure the absorption spectrum of the flow-through fraction to ensure that Cyt b was retained on the concentrator.
4. For removal of elution peptide, pass the concentrated Cyt b sample over an Econo-Pac 10 DG desalting column at 4°C. Collect 0.5-mL fractions, and analyze fractions for Cyt b content by absorption spectroscopy.
5. Following desalting and pooling of samples, concentrate Cyt b to the desired level, and store at 4°C for up to 2 wk prior to resonance energy transfer measurements (*see* **Fig. 1** and **Notes 6** and **7**).

3.4. Cascade Blue Labeling of Cyt b-Specific Monoclonal Antibodies

1. Purify monoclonal antibody 44.1 on GammaBind-Sepharose as previously described (*8*), and store at 4°C in PBS, 0.02% NaN$_3$.
2. Add a 40-fold molar excess of CCB acetyl azide to the antibody (0.5 mg of antibody 44.1 in 400 μL PBS), and rotate this mixture for approx 90 min (*see* **Note 8**) at room temperature. Cover the microfuge tube in aluminum foil to protect the reaction from light.
3. Terminate the labeling reaction by adding Tris-HCl buffer to 75 mM final from a 1 M stock (pH 7.4), and incubate the mixture at room temperature for 10 min.

4. To remove free dye, desalt the mixture on an Econo-Pac 10 DG desalting column equilibrated in PBS at room temperature. The eluted antibody is detected by absorption spectroscopy.
5. Pool the desalted antibody and dialyze against 2000 volumes of PBS at 4°C. Following dialysis, analytical gel filtration chromatography is used to confirm the absence of free label in the antibody conjugate.
6. Add NaN_3 to 0.02% final, and store labeled antibody for up to a year in the dark at 4°C.
7. For determination of antibody labeling stoichiometry, determine protein content by the BCA method, and quantify CCB by absorption spectroscopy using $\varepsilon_{400} = 28{,}000\ M^{-1}cm^{-1}$.

3.5. Resonance Energy Transfer Analysis

1. Perform fluorescence measurements in continuously stirred, UV-transparent microcuvets at room temperature with the excitation monochromator set at 376 nm and the emission monochromator set at 418 nm. Set the slit width for both monochromators at 2 nm.
2. Dilute the CCB-labeled antibody (*see* **Note 9**) to 10 n*M* in PBS containing 0.03% DDM, and filter prior to use. Keep antibody working solution on ice.
3. Add CCB-labeled antibody to the stirred cuvet, and monitor fluorescence until a stable steady-state emission is achieved (*see* **Note 10**).
4. Add DDM-solubilized Cyt b (typically 30 n*M* final) directly to the stirred microcuvet, and monitor fluorescence emission until a steady state is achieved (*see* **Note 11**).
5. Add lipids and amphiphiles, as desired, directly to the microcuvet, and continue monitoring fluorescence emission until a steady state is achieved.
6. In our studies *(8)*, a relaxation of fluorescence quenching (not attributable to a direct effect of lipids on antibody fluorescence emission, dissociation of the CCB-44.1:Cyt b complex or loss of Cyt b heme prosthetic groups) has been taken as a measure of conformational changes in Cyt b (*see* **Fig. 2** and **Note 12**).

4. Notes

1. Following the 1 *M* NaCl wash, membrane fractions can vary in the ease with which they are dispersed and this homogenization step should be carried out until the membranes appear optimally dispersed to increase the extraction efficiency.
2. Although difficulties have yet to be observed by elution at 37°C, recent studies have also indicated efficient elution of Cyt b from the affinity matrix at room temperature.
3. Following peptide elution from the affinity matrix, monoclonal antibody 44.1 is occasionally observed as a minor contaminant in Cyt b preparations (as confirmed by Western blot analysis and matrix-assisted laser desorption/ionization peptide mass mapping). Recent experiments have shown the potential utility CNBr-Sepharose immobilized antibodies as a potential alternative if antibody contamination appears to be a persistent problem.

4. Following the Cyt b purification, the protein A-Sepharose beads can be regenerated for use in multiple preparations by washing with PBS, elution with 0.5 *M* acetic acid (pH 3.0), and storage in PBS, 0.02% NaN$_3$ at 4°C.
5. Concentration of purified Cyt b samples in the 100-kDa cutoff membrane proves critical for control of final detergent levels as free DDM micelles have been shown to pass through the membrane and do not concentrate with purified protein–detergent complex.
6. The low critical micelle concentration of DDM creates difficulties for functional reconstitution of purified Cyt b into lipid vesicles by dialysis and dilution methods, and we reported a low-yield method of reconstitution using SM-2 BioBeads in our original publication *(4)*. The generation of Cyt b samples in 10 m*M* HEPES (pH 7.4), and 0.1% DDM at the desalting step allows for binding of Cyt b to heparin-Agarose and exchange into octylglucoside for subsequent reconstitution by previously described dialysis and dilution methods *(9)*.
7. If Cyt b is stored at 4°C for long periods of time, samples should be examined by absorption spectroscopy and clarified by high-speed centrifugation if aggregation (increased light scattering in the absorption spectrum) is observed. Although freshly prepared Cyt b samples have been primarily examined to this point, it serves to note that the absorption spectrum of detergent-solubilized Cyt b looks nearly identical to the original material following a single freeze/thaw cycle.
8. A labeling stoichiometry of approx 5.7:1 CCB:antibody (molar ratio) has been used in our recent studies and the precise labeling time should be empirically determined with each monoclonal antibody and batch of CCB to achieve the desired degree of labeling.
9. RET studies have also been successfully conducted using Cyt b-specific monoclonal antibodies other than 44.1. Preliminary studies have also shown the utility of monoclonal antibodies that have been labeled with Alexa Fluor 350 succinimidyl ester in these assays.
10. With some samples, the fluorescence is observed to gradually decrease (not achieve a constant emission) for relatively long periods of time, possibly because the labeled antibody has adhered to the cuvet. If this occurs, a new cuvet and CCB-antibody sample are simply used in the assay.
11. RET analysis has recently been conducted on relipidated Cyt b in the absence of detergent, and these results will be reported in a subsequent study.
12. In addition to the analysis of conformational changes, gross distance approximation (monoclonal antibody epitope to Cyt b hemes) should be possible using fluorescence lifetime measurements in conjunction with monoclonal antibodies labeled at defined sites.

References

1. Cross, A. R. and Segal, A. W. (2004) The NADPH oxidase of professional phagocytes-prototype of the NOX electron transport chain systems. *Biochim. Biophys. Acta* **1657,** 1–22.

2. Parkos, C. A., Allen, R. A., Cochrane, C. G., and Jesaitis, A. J. (1987) Purified cytochrome b from human granulocyte plasma membrane is comprised of two polypeptides with relative molecular weights of 91,000 and 22,000. *J. Clin. Invest.* **80,** 732–742.

3. Quinn, M. T. and Gauss, K. A. (2004) Structure and regulation of the neutrophil respiratory burst oxidase: comparison with non-phagocyte oxidases. *J. Leukoc. Biol.* **76,** 760–781.

4. Taylor, R. M., Burritt, J. B., Foubert, T. R., et al. (2003) Single-step immuno-affinity purification and characterization of dodecylmaltoside-solubilized human neutrophil flavocytochrome b. *Biochim. Biophys. Acta* **1612,** 65–75.

5. Burritt, J. B., Busse, S. C., Gizachew, D., et al. (1998) Antibody imprint of a membrane protein surface: phagocyte flavocytochrome b. *J. Biol. Chem.* **273,** 24,847–24,852.

6. dos Remedios, C. G. and Moens, P. D. (1995) Fluorescence resonance energy transfer spectroscopy is a reliable "ruler" for measuring structural changes in proteins. Dispelling the problem of the unknown orientation factor. *J. Struct. Biol.* **115,** 175–185.

7. Heyduk, T. (2002) Measuring protein conformational changes by FRET/LRET. *Curr. Opin. Biotechnol.* **13,** 292–296.

8. Taylor, R. M., Foubert, T. R., Burritt, J. B., Baniulis, D., McPhail, L. C., and Jesaitis, A. J. (2004) Anionic amphiphile and phospholipid-induced conformational changes in human neutrophil flavocytochrome b observed by fluorescence resonance energy transfer. *Biochim. Biophys. Acta* **1663,** 201–213.

9. Knoller, S., Shpungin, S., and Pick, E. (1991) The membrane-associated component of the amphiphile-activated, cytosol-dependent superoxide-forming NADPH oxidase of macrophages is identical to cytochrome b559. *J. Biol. Chem.* **266,** 2795–2804.

VII

ANALYSIS OF NEUTROPHIL GENE EXPRESSION AND TRANSCRIPTION FACTORS

27

Genome-Scale Transcript Analyses in Human Neutrophils

Scott D. Kobayashi, Dan E. Sturdevant, and Frank R. DeLeo

Summary

Transcriptome analyses of single- and multicellular organisms have changed fundamental understanding of biological and pathological processes across multiple scientific disciplines. Over the past 5 yr, studies of polymorphonuclear leukocyte (or neutrophil) (PMN) gene expression on a global scale have provided new insight into the molecular processes that promote resolution of infections in humans. Herein, we present methods to analyze gene expression in human neutrophils using Affymetrix oligonucleotide microarrays, which include isolation of high-quality RNA, generation and labeling of cRNA, and GeneChip hybridization and scanning. Notably, the procedures utilize commercially available reagents and materials and thus represent a standardized approach for evaluating PMN transcript levels.

Key Words: Neutrophil; microarray; gene expression; Affymetrix; transcript; phagocyte; polymorphonuclear leukocytes.

1. Introduction

Mature polymorphonuclear leukocytes (granulocytes or neutrophils [PMN]) are fully equipped to destroy invading microorganisms, and inhibitors of transcription and translation fail to have an immediate effect on phagocytosis and microbicidal capacity *(1–4)*. As such, it was widely presumed that new gene transcription played no significant role in neutrophil function. However, extended incubation with actinomycin D ($\geq$1 h), an inhibitor of RNA synthesis, significantly reduces granulocyte phagocytosis, superoxide production, and bactericidal activity *(1–4)*, suggesting that new gene transcription is important for maintaining PMN function.

From: *Methods in Molecular Biology, vol. 412: Neutrophil Methods and Protocols*
Edited by: M. T. Quinn, F. R. DeLeo, and G. M. Bokoch © Humana Press Inc., Totowa, NJ

Methods for measuring gene expression in human granulocytes have existed for at least 40 yr. In 1966, Cline first reported that phagocytosis increases RNA synthesis in human granulocytes *(4,5)*, and subsequent studies by Jack and Fearon showed that peripheral blood neutrophils constitutively synthesize a limited repertoire of mRNA transcripts and proteins *(6)*. These early studies used radiolabel-based techniques, such as incorporation of H^3-uridine or C^{14}-orotic acid into RNA, to estimate overall increases in transcript levels *(5,6)*. Northern blotting has also been a widely used method for evaluating neutrophil mRNA levels *(7,8)*. Inasmuch as these methods are suited to evaluate a limited number of transcripts or provide an indication of general RNA synthesis, they are not appropriate for genome-wide analyses of transcript levels. More recently, studies by Itoh et al. and Subrahmanyam et al. investigated patterns of gene expression in neutrophils by sequencing of 3'-directed cDNA libraries or cDNA display, respectively *(9,10)*. Such studies were the first to use global approaches to investigate neutrophil gene expression. The work by Itoh et al. and Subrahmanyam et al. provided a segue into current PMN transcriptome analyses, which almost exclusively use microarray-based approaches *(11–22)*. Importantly, the microarray-based work has significantly enhanced our understanding of neutrophil biology and function. For example, studies of the PMN transcriptome during phagocytosis provided new insight into the role of PMN in the resolution of inflammation and infection *(23,24)*.

In this chapter, we provide a method for analyzing human PMN gene expression with Affymetrix oligonucleotide microarrays, arguably to most universal platform for microarray-based research in human tissues. With the exception of human neutrophils, all of the materials are commercially available. Therefore, the method described in the following pages produces results that can be compared readily to those generated by other laboratories using the same methodology.

2. Materials

2.1. Isolation of Neutrophils and Analysis of Purity (see Chapters 2 and 3 for a Comprehensive List of Materials)

1. Ficoll-Paque^{PLUS}.
2. Injection- or irrigation-grade H_2O (Abbott Laboratories).
3. Dulbecco's phosphate-buffered saline (DPBS), pH 7.2 (Gibco/Invitrogen).
4. Injection- or irrigation-grade 0.9% NaCl solution (w/v) (Fisher Scientific).
5. 3% Dextran solution (w/v): dissolve 30 g of Dextran 500 (Amersham Biosciences) into 1 L (final volume) of injection- or irrigation-grade 0.9% NaCl solution. Filter-sterilize using a 0.2 μm bottle-top filter and test for endotoxin.
6. 1.7% NaCl solution (w/v): dissolve 17 g of RNase-free NaCl into 1 L (final volume) of injection- or irrigation-grade water. Filter-sterilize using a 0.2-μm bottle-top filter and test for endotoxin.

2.2. Purification of RNA From Neutrophils

1. Buffered RPMI 1640 medium: Add 5 mL of sterile, endotoxin-tested 1 *M* HEPES (Sigma) to a new 500-mL bottle of RPMI 1640 medium (Gibco/Invitrogen).
2. 12-well, flat-bottom, nonpyrogenic polystyrene cell culture plates (Corning).
3. Aerosol-resistant micropipet tips, RNase-free and nonpyrogenic (Molecular Bio-Products).
4. 1.7-mL microcentrifuge tubes, RNase-free, nonstick (Ambion).
5. RNeasy mini kit and QiaShredder columns (Qiagen).
6. 14.5 *M* 2-mercaptoethanol (2-ME) (Sigma).
7. Diethylpyrocarbonate (DEPC)-treated or nuclease-free H_2O (Ambion).

2.3. Determination of RNA Quantity

1. Quant-iT RiboGreen RNA Assay (Molecular Probes).
2. 96-well black microtiter plates (Corning).

2.4. Analysis of RNA Quality

1. RNA 6000 Nano LabChip (Agilent Technologies).
2. RNA 6000 Ladder (Ambion).

2.5. Generation of Labeled cRNA

1. Affymetrix GeneChip one-cycle target labeling and control reagents kit (Affymetrix) (*see* **Note 1**).
2. 1 *M* Tris-acetate solution: dissolve 121.14 g Trizma-base in nuclease-free H_2O and adjust pH to 8.1 with glacial acetic acid.
3. 5X cRNA fragmentation buffer: mix 4.0 mL 1 *M* Tris-acetate solution, 0.64 g MgOAc (Sigma), and 0.98 g KOAc (Sigma). Make up to a final volume of 20 mL with nuclease-free H_2O. Filter-sterilize and store at room temperature.

2.6. Hybridization of cRNA to Affymetrix GeneChips

1. Control oligo B2, 3 n*M* (Affymetrix).
2. Amber microcentrifuge tubes (Eppendorf).
3. Tough-spots label dots (Diversified Biotech).
4. 5 *M* NaCl solution, RNase- and DNase-free (Ambion).
5. 20X SSPE solution: 3 *M* NaCl, 0.2 *M* NaH_2PO_4, 0.02 *M* ethylenediamine tetraacetic acid (EDTA) (Cambrex).
6. ImmunoPure Streptavidin (Pierce).
7. Herring sperm DNA (Promega, Madison, WI).
8. EDTA disodium salt, 0.5 *M* solution (100 mL) (Sigma).
9. Goat immunoglobulin (Ig)G, reagent grade (Sigma).
10. Biotinylated goat anti-streptavidin antibody (Vector Laboratories).
11. 12X MES stock solution: dissolve 64.61 g MES hydrate and 193.3 g of MES sodium salt in 800 mL of nuclease-free water. Adjust volume to 1 L; the solution pH should fall between 6.5 and 6.7. Filter-sterilize and store shielded from light at 2–8°C.

12. Wash buffer A (nonstringent): combine 6X SSPE (from 20X stock) and 0.01% Tween-20 (v/v) (from 10% stock; Surfact-Amps 20, Pierce Biotechnology). Filter-sterilize and store at room temperature.
13. Wash buffer B (stringent): Combine 100 mM MES (from 12X stock), 0.1 M [Na+], 0.01% Tween-20. This buffer is prepared from 12X MES, 5 M NaCl, and 10% Tween-20 stocks, passed through a 0.2-µm filter, and stored at room temperature.
14. 2X hybridization buffer (200 mM MES, 2 M [Na+], 40 mM EDTA, 0.02% Tween-20). Use 12X MES, 5 M NaCl, 0.5 M EDTA and 10% Tween-20 stocks to prepare this solution. The solution is stored at 2–8°C and shielded from light.
15. 2X stain buffer (200 mM MES, 2 M [Na+], 0.1% Tween-20). Solution is prepared from 12X MES, 5 M NaCl, and 10% Tween-20 stocks, passed through a 0.2-µm filter, and stored shielded from light at 2–8°C.
16. Steptavidin-phycoerythrin solution (SAPE): prepare fresh by mixing 1X MES (from 2X stock), 2 mg/mL bovine serum albumin (BSA; from 50 mg/mL stock; Invitrogen), and 10 µg/mL streptavidin-phycoerythrin (from 1 mg/mL stock prepared in DPBS; Molecular Probes).
17. Antibody solution: prepare fresh by mixing 1X MES (from 2X stock), 2 mg/mL BSA (from 50 mg/mL stock), 0.1 mg/mL goat IgG (from 10 mg/mL stock prepared in 150 mM NaCl), and 5 µg/mL biotinylated anti-streptavidin antibody (0.5 mg/mL stock).
18. Hybridization cocktail for single probe array (200 µL/array). Each hybridization reaction will contain 1X hybridization buffer (2X stock), 50 pM B2 Control Oligo (3 nM stock), 0.1 mg/mL Herring Sperm DNA (10 mg/mL stock), 0.5 mg/mL BSA (50 mg/mL stock), 7.8% dimethylsulfoxide (DMSO) (100% stock), 15 µg fragmented and labeled cRNA, 20X spike-in.

2.7. GeneChip Processing, Scanning, and Conversion of Image Files

1. Affymetrix GeneChip Scanner 3000, enabled for high-resolution scanning.
2. Affymetrix GeneChip Operating Software (GCOS) including the GeneChip Scanner 3000 high-resolution scanning patch.

3. Methods

3.1. Isolation of Neutrophils

1. Human PMN can be isolated using Hypaque-Ficoll gradient separation described by Nauseef in Chapter 2 or Siemsen et al. under **Subheading 3.8.** of Chapter 3. The factors most critical for transcriptome analyses are high-purity of PMN preparations, i.e., ≤1% contaminating mononuclear cells (typically lymphocytes and monocytes), and isolation of neutrophils that are not primed or activated. Techniques to assess cell purity and viability are described in Chapters 2 and 3.

3.2. Purification of RNA From Neutrophils

1. PMN assays (e.g., phagocytosis) are routinely performed in 12-well tissue culture plates containing 10^7 PMN/well. The PMN are lysed directly in the wells by add-

ing 600 μL RLT (Qiagen) + 2-ME (143 mM final) per 10^7 PMN (*see* **Note 2**). Lysate should be pipetted to ensure complete lysis of PMN and to provide maximum protection to RNA. When performing time course experiments, lysate can be stored at –20 or –80°C to facilitate sample processing.

2. Homogenize lysate by passage over QIAshredder column and centrifuge ($30,000g$) in a microcentrifuge for 2 min.
3. Add 600 μL of 70% ethanol and mix by pipetting. Load one-half of the lysate onto Qiagen RNeasy column and centrifuge 15 s ($30,000g$). Decant flow-through and repeat with the second half of the lysate.
4. Place the column in a new collection tube and add 500 μL of buffer RPE (Qiagen). Spin for 15 s in a microcentrifuge ($30,000g$), decant flow-through, and repeat.
5. Place the column in a new collection tube and centrifuge for 2 min ($30,000g$) to remove any residual wash. Place the column in a new RNase-free elution tube.
6. Elute RNA by addition of 50 μL of DEPC-treated H_2O (70°C), allow to sit for 1 min, and then centrifuge ($30,000g$) for 1 min. Repeat a second time for maximum elution of RNA, and pool the two eluates (100 μL total).
7. Second purification of RNA is critical for complete cDNA synthesis (*see* **Note 3**). To the 100 μL of RNA eluate, add 350 μL of RLT + 2-ME (143 mM final). Add 250 μL of 100% ethanol and mix by pipetting. Apply mixture to a new RNeasy column and spin ($30,000g$) for 15 s.
8. Decant flow-through and add 500 μL of RPE. Spin for 15 s ($30,000g$), decant flow-through, and repeat.
9. Place the column in a new tube and spin for 1 min at the maximum rpm to remove residual wash. Place the column in an elution tube, add 50 μL of DEPC-treated H_2O (70°C), allow to sit for 1 min, and centrifuge ($30,000g$) for 1 min. Repeat a second time for maximum elution of RNA. Store RNA at –80°C. The expected yield of total RNA prepared by this method is approx 2.0–5.0 μg.

3.3. Determination of RNA Quantity

1. RNA quantity can be measured either by absorbance at 260 nm or by fluorescence. The following protocol assumes the use of a microplate spectrofluorometer (e.g., Molecular Devices Gemini XPS, Sunnyvale, CA) and the Molecular Probes Quant-iT RiboGreen RNA Assay. The excitation maximum for RiboGreen bound to RNA is 500 nm and the emission maximum is 525 nm (*see* **Note 4**).
2. A 2X working stock of RiboGreen is prepared from a 1:200 dilution of the manufacturer's supplied stock and is used at a concentration of 1X per assay.
3. Each microtiter plate should contain two blank wells (1X RiboGreen) and a serial dilution of control rRNA (supplied). Each assay should contain 1 μL of PMN total RNA, 1X RiboGreen, and 1X TE buffer. The expected final concentration of total RNA as prepared above (from 10^7 PMN) is approx 20 μg/mL (0.02 μg/mL in the assay). Thus, the final concentration of the rRNA for the standard curve in the RiboGreen assay should cover a range from 0 to 1 μg/mL.
4. The assay is incubated for 2 to 5 min at room temperature and protected from light.

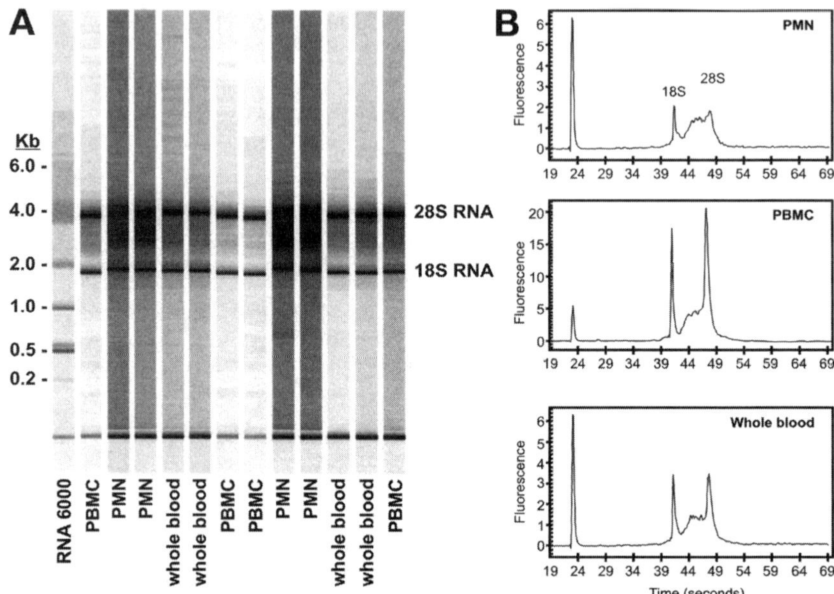

Fig. 1. Analysis of RNA quality using a microfluidics-based automated electrophoresis system. (**A**) 1 µL of total RNA (typically 20–30 ng) purified from whole blood, peripheral blood mononuclear cells (PBMC), or polymorphonuclear neutrophils (PMN) was evaluated with an Agilent 2100 Bioanalyzer. RNA 6000 indicates an RNA 6000 Ladder (Ambion) comprised of 6 transcripts of varied size as shown. (**B**) Histograms of representative RNA samples.

5. Following incubation, the plate is scanned in the microplate reader and the concentration of PMN total RNA is calculated by extrapolating from the standard rRNA curve (*see* **Note 5**).

3.4. Analysis of RNA Quality

1. The isolation of high quality RNA is essential for microarray analysis. RNA integrity of low-yield samples can be assessed on microfluidics-based automated electrophoresis systems (**Fig. 1**). The following protocol assumes the use of an Agilent 2100 Bioanalyzer (Agilent Technologies) or similar device and the RNA 6000 Nano LabChip (Agilent).

2. The gel–dye mix is prepared by centrifuging 400 µL RNA gel matrix through the supplied spin filter at 1500*g* for 10 min. The filtered gel matrix is stored at 4°C and must be used within 4 wk. To make the working stock of gel matrix–dye reagent, 2 µL of RNA dye concentrate is added to 130 µL of the filtered gel mix and thoroughly vortexed. Remaining gel–dye mix is stored protected from light at 4°C and should be used within 1 wk.

3. Place a new chip on the priming station and load 9 µL prepared gel-dye mix into the well labeled "G" (third row, black circle). Close the priming station and depress

the plunger until it engages with the clip. After 30 s, release the clip and check the back of the chip for air bubbles.

4. Add 9 µL prepared gel–dye mix to the remaining two "G" wells (rows 1 and 2, no circle). Load 5 µL of the supplied marker buffer to the ladder well and all 12 sample wells. Add 1 µL heat-denatured (95°C, 5 min) RNA 6000 ladder to the ladder well, and load 1 µL of purified PMN RNA to each sample well.
5. Place the loaded chip into the adapter in the supplied vortexer, and mix for 1 min. The chip must be run within 5 min.
6. The chip is read using the integrated eukaryotic total RNA algorithm, and RNA integrity is assessed by the appearance of two well defined peaks denoting the 18S and 28S rRNA subunits (approx 2:1 ratio) (**Fig. 1**).

3.5. Generation of Labeled cRNA

1. The volume of PMN total RNA is reduced to 8 µL in a centrifugal vacuum concentrator, or alternatively desiccated and suspended in 8 µL of RNase-free water. cDNA synthesis, cleanup, and cRNA labeling are performed with the GeneChip one-cycle target labeling and control reagents kit (*see* **Note 6**).
2. Preparation of poly-A RNA spike-in controls. The RNA controls consist of four *Bacillus subtilis* genes (*lys*, *phe*, *thr*, and *dap*) that are absent from eukaryotic samples and are used to monitor the labeling efficiency of target RNA. The final spike-in control concentration is based on the amount of total PMN RNA (~1–2 µg). Three serial dilutions are prepared from the Affymetrix poly-A control stock. For the first dilution, add 2 µL poly-A control stock to 38 µL of poly-A control dilution buffer (1:20). The second and third dilutions are prepared at 1:50 into the dilution buffer. Two microliters of the final dilution are used in each cDNA synthesis reaction.
3. First-strand cDNA synthesis, primer hybridization. Add 2 µL T7-Oligo (dT) primer (50 μM), 8 µL total PMN RNA, and 2 µL diluted poly-A controls to a 0.5-mL RNase-free tube. Incubate the reaction for 10 min at 70°C and cool to 4°C.
4. First-strand cDNA synthesis. Add 2 µL dithiothreitol (DTT) (0.1 M), 4 µL 5X first-strand cDNA synthesis buffer, and 1 µL dNTP (10 mM) and incubate for 2 min at 42°C. Add 1 µL SuperScript III reverse-transcriptase and incubate at 42°C for 60 min.
5. Second-strand cDNA synthesis. To each tube from the first strand cDNA reaction, add 130 µL master mix (91 µL DEPC-H_2O, 30 µL second-strand buffer, 3 µL dNTPs [10 mM], 1 µL DNA ligase [10 U/µL], 4 µL DNA Polymerase I [10 U/µL], and 1 µL RNase H [2 U/µL]). Incubate the reactions at 16°C for 2 h. Add 2 µL T4 DNA polymerase to each reaction and incubate 16°C for 15 min. Stop the reaction by addition of 10 µL 0.5 M EDTA, pH 8.0.
6. Purification of cDNA, sample cleanup module. Add 600 µL of cDNA binding buffer to the double-stranded cDNA reaction mix and vortex for 3 s. Apply 500 µL of the sample to the cDNA cleanup spin column (save the remaining sample), and centrifuge for 1 min at 8000g. Discard the flow-through. Reload the spin column with the remaining sample and centrifuge for 1 min at 8000g. Discard the flow-through and collection tube. Transfer the spin column to a new 2-mL collection tube. Add

750 µL of the cDNA wash buffer to the spin column. Centrifuge for 1 min at 8000*g*. Discard the flow-through. Cut the cap off of the spin column (label the side of the column) and centrifuge for 5 min at >25,000*g*. Discard the flow-through and the collection tube. Transfer the spin column to a new collection tube and add 14 µL of cDNA elution buffer directly onto the spin column membrane. Incubate 1 min at room temperature and centrifuge 1 min at >25,000*g*. The expected volume of cDNA following cleanup is 12 µL.

7. Synthesis of labeled cRNA, GeneChip in vitro transcription (IVT) labeling. The entire 12 µL of cDNA (from 10^7 PMN) is used in the IVT labeling reaction. Prepare a master mix containing 4 µL 10X IVT labeling buffer, 12 µL IVT labeling NTP mix, 8 µL RNAse-free water, and 4 µL IVT labeling enzyme mix per reaction. Add 28 µL of the IVT master mix to 12 µL cDNA in 0.5 mL PCR tubes. Incubate the reaction at 37°C for 16 h. Store labeled cRNA at −70°C or proceed to cleanup.

8. Cleanup of labeled cRNA. The sample cleanup module is used for this step. Add 60 µL of RNase-free water to the IVT reaction and mix by vortexing for 3 s. Add 350 µL IVT cRNA binding buffer to the sample and vortex for 3 min. Add 250 µL 100% ethanol and mix well by pipetting. Apply the sample to the IVT cRNA cleanup spin column and centrifuge for 15 s at 8000*g*. Discard the flow through and collection tube. Transfer the spin column to a new collection tube and pipe 500 µL IVT cRNA wash buffer onto the column. Centrifuge for 15 s at 8000*g* and discard flow-through. Pipet 500 µL 80% (v/v) ethanol onto the spin column and centrifuge for 15 s at 8000*g* and discard flow-though. Cut the cap off of the spin column (label the side of the column) and centrifuge for 5 min at >25,000*g*. Transfer the spin column to a new collection tube and pipe 11 µL of RNase-free water directly onto the spin column membrane. Centrifuge 1 min at >25,000*g* to elute. Pipet 10 µL of RNase-free water directly onto the spin column membrane and centrifuge 1 min at >25,000*g*. Store cRNA at −70°C.

9. Determination of cRNA quantity. The quantity of cRNA is determined similar to that of total RNA as described under **Subheading 3.3.** by use of RiboGreen. However, the expected yield of cRNA generated from the IVT reaction is 10-fold the starting amount of total RNA. Therefore, the cRNA sample should be diluted 10-fold in order to fall within the linear range of the rRNA standard curve. *Optional*: labeled cRNA samples can be analyzed with Agilent 2100 Bioanalyzer as described under **Subheading 3.4.** A typical signature for PMN cRNA is a broad smear of rather than distinct bands (**Fig. 2**).

10. Fragmentation of labeled cRNA. Fragmentation of target cRNA prior to hybridization is necessary for optimal assay sensitivity. cRNA is desiccated in a centrifugal vacuum concentrator and suspended to 1 µg/µL in RNase-free water. Add 8 µL of 5X fragmentation buffer and 17 µL of RNase-free water to 15 µL of cRNA. The cRNA is fragmented by incubation at 94°C for 35 min followed by cooling on ice.

3.6. Hybridization of cRNA to Affymetrix GeneChips

1. The following protocol is designed for the analysis of human PMN RNA transcripts on Affymetrix 49 format (standard) arrays such as the U133 Plus 2.0.

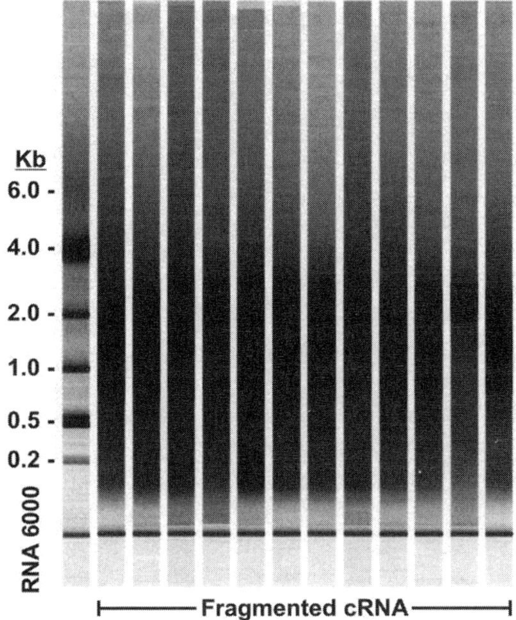

Fig. 2. Analysis of labeled polymorphonuclear neutrophil cRNA using an Agilent 2100 Bioanalyzer.

2. Remove chips from 4°C storage and acclimate to room temperature (≥60 min) prior to hybridization. Remove reagents from cold storage and allow to thaw at room temperature. Set heat blocks to 65 and 99°C. Set the hybridization oven to 45°C.

3. Place the 20X Eukaryotic hybridization controls tube into the 65°C heat block for 5 min.

4. Mix the hybridization cocktail for each chip by using 15 μg of cRNA mix with 5 μL of control oligonucleotide B2 at 3 n*M*, 15 μL of 20X Eukaryotic hybridization controls that have been heated to 65°C for 5 min, 3 μL of herring sperm (10 mg/mL), 3 μL of BSA (50 mg/mL), 150 μL of 2X hybridization buffer, 30 μL of DMSO, and 54 μL RNase-free water for a final volume of 300 μL.

5. Incubate the hybridization cocktail at 99°C for 5 min.

6. Place the chips on the bench so that the back is facing up, insert a 200-μL pipet tip into one of the septa, and fill the chip with 200 μL of 1X hybridization buffer through the remaining septum.

7. Incubate the filled chips in the 45°C hybridization oven for 10 min at 60 rpm.

8. The hybridization cocktail is incubated at 45°C for 5 min, and centrifuged for 5 min at the maximum rpm.

9. Remove the chips from the hybridization oven and place a 200-μL pipet tip in one of the septa. Then using the other septum, remove the 1X buffer and add 200 μL of the appropriate hybridization cocktail.

10. Place the chips back into the 45°C hybridization oven for 16 h at 60 rpm.

3.7. GeneChip Processing, Scanning, and Conversion of Image Files

1. The following protocol requires the use of an Affymetrix GeneChip Scanner 3000, enabled for high-resolution scanning, and the GCOS including the GeneChip Scanner 3000 high-resolution scanning patch.
2. It is important to process the Affymetrix GeneChips directly following hybridization. The preparation of staining and washing reagents and the priming of the Affymetrix workstation must occur prior to completion of the hybridization step.
3. Turn on the fluidics station and verify that tubing is appropriately connected to wash bottles A and B—water and waste.
4. Turn on the workstation and open GCOS.
5. Click on the fluidics icon and prime the workstation(s) intended for use.
6. For each chip, using an amber tube, mix 600 µL of 2X stain buffer, 48 µL BSA (50 mg/mL), 12 µL streptavidin-phycoerythrin (1 mg/mL) 540 µL of distilled water. This is the SAPE solution mix.
7. Remove 600 µL of the SAPE solution and place into another amber tube.
8. For each chip, using a clear tube, mix 300 µL of 2X stain buffer, 24 µL BSA (50 mg/mL), 6 µL Goat IgG stock (10 mg/mL), 3.6 µL biotinylated antibody (0.5 mg/mL), and 266.4 µL of distilled water. This is the antibody solution mix.
9. Remove the chips from the hybridization oven and place a 200-µL pipet tip into one of the septa. Remove the hybridization cocktail and place into the appropriate sample tube. These samples can be run on other chips at a later date. Store at −70°C.
10. Fill the chip with 250 µL of wash buffer A without air bubbles. If processing more chips than allowed in the fluidics station, the remaining filled chips can be stored up to 4 h at 4°C.
11. In GCOS, click on the "experiments" icon and create an experiment (*.EXP) for each chip.
12. In GCOS, click on "Tools, filters" and set the filters so that your experiments can be viewed.
13. In the GCOS fluidics window, select the appropriate protocol for each module and run.
14. Place the appropriate chip into the fluidics modules and place the SAPE solution mix tubes into sample holder positions 1 and 3 for each module.
15. Place the antibody solution mix tube into sample holder position 2 for each module.
16. The protocol should take approx 90 min to finish. Once the fluidics station displays "remove cartridge," take the chip out and inspect for air bubbles. If any are present, place the clip back into the fluidics module and let it refill. If there are still bubbles, then the chip must be filled manually with wash buffer A.
17. Place a tough spot over each septum of the chips to be scanned.
18. Gently wipe the glass of the chip with a kimwipe in one direction. Do not change direction, or the paraffin at the edges may coat the glass, making scanning difficult.
19. Scanning. The scanning protocol is written using the autoloader. If an autoloader is not present, the protocol must be altered to accommodate single-chip scanning.
20. Press the "start" button on the scanner.

21. Open GCOS and verify that the connection between the scanner and workstation is complete.
22. Allow the scanner to warm up for 10 min.
23. Place chips into the autoloader starting with position one.
24. In GCOS, press the start icon and while the scanner is performing the automated tasks, click on "view" and "scan in progress" to visualize the results once scanning begins.
25. A successful scan will create the *.DAT and *.CEL files. The *.CEL files are now ready for analysis.

4. Notes

1. The one-cycle target labeling kit contains reagents for cDNA synthesis, IVT labeling, sample cleanup, and control reagents. Reagents are provided for 30 reactions.
2. Do not exceed 10^7 PMN per Qiagen RNA mini-column. Numbers in excess do not increase RNA quantity and effectively reduce the purity of final RNA. Larger quantities of total RNA can be obtained by pooling multiple preparations or, alternatively, using larger RNA purification columns.
3. The second purification of PMN total RNA on RNeasy columns is essential for obtaining high-quality RNA. Omission of this step may lead to incomplete cDNA synthesis and subsequent poor (5':3') ratios of labeled target cRNA.
4. The optimal excitation and emission spectra of the RiboGreen fluorophore should be empirically determined on the fluorometer to ensure that the sample readings remain in the detection range.
5. Residual DNA from the PMN total RNA preparations will contribute to the overall signal obtained in the RiboGreen assay. We have found that the levels of contaminating DNA present in the RNA samples do not affect the overall yield of labeled target cRNA. The concentration of cRNA will also be obtained in the determination of RNA quality (**Subheading 3.4.**) and can be used to verify the RiboGreen results.
6. Unless stated otherwise, the incubation steps for all cDNA and target labeling reactions are performed in a thermocycler.

Acknowledgments

This work was supported by the National Institutes of Health (NIH)-National Center for Research Resources grant P20RR015587 (S. D. K.), and the Intramural Program of the National Institutes of Allergy and Infectious Diseases, NIH (F. R. D.).

References

1. Kasprisin, D. O. and Harris, M. B. (1977) The role of RNA metabolism in polymorphonuclear leukocyte phagocytosis. *J. Lab. Clin. Med.* **90,** 118–124.
2. Chang, F. Y. and Shaio, M. F. (1990) In vitro effect of actinomycin D on human neutrophil function. *Microbiol. Immunol.* **34,** 311–321.

3. Kasprisin, D. O. and Harris, M. B. (1978) The role of protein synthesis in polymorphonuclear leukocyte phagocytosis II. *Exp. Hematol.* **6,** 585–589.

4. Cline, M. J. (1966) Phagocytosis and synthesis of ribonucleic acid in human granulocytes. *Nature* **212,** 1431–1433.

5. Cline, M. J. (2002) Ribonucleic acid biosynthesis in human leukocytes effects of phagocytosis on RNA metabolism. *Blood* **28,** 188–200.

6. Jack, R. M. and Fearon, D. T. (1988) Selective synthesis of mRNA and proteins by human peripheral blood neutrophils. *J. Immunol.* **140,** 4286–4293.

7. Newburger, P. E., Dai, Q., and Whitney, C. (1991) *In vitro* regulation of human phagocyte cytochrome *b* heavy and light chain gene expression by bacterial lipopolysaccharide and recombinant human cytokines. *J. Biol. Chem.* **266,** 16,171–16,177.

8. Newburger, P. E., Ezekowitz, C., Whitney, J., et al. (1988) Induction of phagocyte cytochrome b heavy chain gene expression by interferon gamma. *Proc. Natl. Acad. Sci. USA* **85,** 5215–5219.

9. Itoh, K., Okubo, K., Utiyama, H., et al. (1998) Expression profile of active genes in granulocytes. *Blood* **92,** 1432–1441.

10. Subrahmanyam, Y. V. B. K., Yamaga, S., Prashar, Y., et al. (2001) RNA expression patterns change dramatically in human neutrophils exposed to bacteria. *Blood* **97,** 2457–2468.

11. Borjesson, D. L., Kobayashi, S. D., Whitney, A. R., et al. (2005) Insights into pathogen immune evasion mechanisms: *Anaplasma phagocytophilum* fails to induce an apoptosis differentiation program in human neutrophils. *J. Immunol.* **174,** 6364–6372.

12. Kobayashi, S. D., Voyich, J. M., Buhl, C. L., et al. (2002) Global changes in gene expression by human polymorphonuclear leukocytes during receptor-mediated phagocytosis: Cell fate is regulated at the level of gene expression. *Proc. Natl. Acad. Sci. USA* **99,** 6901–6906.

13. Kobayashi, S. D., Voyich, J. M., Somerville, G. A., et al. (2003) An apoptosis-differentiation program in human polymorphonuclear leukocytes facilitates resolution of inflammation. *J. Leukoc. Biol.* **73,** 315–322.

14. Kobayashi, S. D., Voyich, J. M., Braughton, K. R., and DeLeo, F. R. (2003) Down-regulation of proinflammatory capacity during apoptosis in human polymorphonuclear leukocytes. *J. Immunol.* **170,** 3357–3368.

15. Kobayashi, S. D., Braughton, K. R., Whitney, A. R., et al. (2003) Bacterial pathogens modulate an apoptosis differentiation program in human neutrophils. *Proc. Natl. Acad. Sci. USA* **100,** 10,948–10,953.

16. Kobayashi, S. D., Voyich, J. M., Braughton, K. R., et al. (2004) Gene expression profiling provides insight into the pathophysiology of chronic granulomatous disease. *J. Immunol.* **172,** 636–643.

17. Kobayashi, S. D., Voyich, J. M., Whitney, A. R., and DeLeo, F. R. (2005) Spontaneous neutrophil apoptosis and regulation of cell survival by granulocyte macrophage-colony stimulating factor. *J. Leukoc. Biol.* **78,** 1408–1418.

18. Fessler, M. B., Malcolm, K. C., Duncan, M. W., and Worthen, G. S. (2002) Lipopolysaccharide stimulation of the human neutrophil—an analysis of changes

in gene transcription and protein expression by oligonucleotide microarrays and proteomics. *Chest* **121,** 75S–76S.

19. Theilgaard-Mönch, K., Knudsen, S., Follin, P., and Borregaard, N. (2004) The transcriptional activation program of human neutrophils in skin lesions supports their important role in wound healing. *J. Immunol.* **172,** 7684–7693.
20. Tsukahara, Y., Lian, Z., Zhang, X. Q., et al. (2003) Gene expression in human neutrophils during activation and priming by bacterial lipopolysaccharide. *J. Cell. Biochem.* **89,** 848–861.
21. Kluger, Y., Tuck, D. P., Chang, J. T., et al. (2004) Lineage specificity of gene expression patterns. *Proc. Natl. Acad. Sci. USA* **101,** 6508–6513.
22. Zhang, X. Q., Kluger, Y., Nakayama, Y., et al. (2004) Gene expression in mature neutrophils: early responses to inflammatory stimuli. *J. Leukoc. Biol.* **75,** 358–372.
23. Kobayashi, S. D. and DeLeo, F. R. (2004) An apoptosis differentiation programme in human polymorphonuclear leucocytes. *Biochem. Soc. Trans.* **32,** 474–476.
24. DeLeo, F. R. (2004) Modulation of phagocyte apoptosis by bacterial pathogens. *Apoptosis* **9,** 399–413.

28

Fast and Accurate Quantitative Analysis of Cytokine Gene Expression in Human Neutrophils by Reverse Transcription Real-Time PCR

Nicola Tamassia, Marco A. Cassatella, and Flavia Bazzoni

Summary

In recent years, several studies have brought forward new and exciting discoveries in polymorphonuclear neutrophil research by establishing that the release of inflammatory cytokines constitutes a novel and important aspect of the biology of this cell. At present, neutrophils should no longer be regarded as cells that only release preformed mediators, but instead as fundamental contributors for many events that control and regulate inflammatory, immune, and other relevant responses. In this context, a correct methodological analysis of this novel neutrophil function represents a critical step toward a better understanding of how the release of cytokines by neutrophils may influence pathophysiological processes in vivo. We now describe and discuss methods we have recently developed to more rapidly characterize the pattern of cytokine expression in in vitro-activated human neutrophils. The validation of the real-time PCR assay as a suitable strategy for an accurate, sensitive, reliable, and *bona fide* analysis of cytokine gene expression in human neutrophils overcomes several problems strictly specific to neutrophils and offers an important tool for high-throughput analysis of gene expression in neutrophils.

Key Words: Neutrophils; real-time PCR; cytokines; SYBR¨ Green; gene expression.

1. Introduction

Neutrophils constitute a first line of defense against pathogens by virtue of their ability to release a set of preformed cytotoxic enzymes and generate reactive oxygen-derived species. Neutrophils can also produce, upon appropriate stimulation in vitro and in vivo, a variety of proteins, including cytokines, chemotactic molecules, and other mediators that are involved in various effector functions *(1)*. This latter function is currently the subject of a new wave of enthusiastic research, in that it shows neutrophils may also act as key regulators of the in vivo cross-talk between immune, endothelial, stromal, and parenchymal cells.

From: *Methods in Molecular Biology, vol. 412: Neutrophil Methods and Protocols*
Edited by: M. T. Quinn, F. R. DeLeo, and G. M. Bokoch © Humana Press Inc., Totowa, NJ

For a fairly exhaustive description of experimental conditions leading to the production of individual cytokines by neutrophils, as well as the molecular regulation and potential biological relevance of this process, the reader is referred to recent reviews *(1,2)*.

There is an increasing interest in the identification and molecular characterization of protein/cytokine expression and production by neutrophils. In most studies aimed at characterizing neutrophil-derived proteins, granulocytes are isolated from fresh blood or buffy coats, suspended in culture medium containing up to 10% serum, and then cultured at 37°C in a 5% CO_2 humidified atmosphere. Other investigators suspend neutrophils in serum-free medium or in buffered solutions containing various concentrations of serum or bovine serum albumin. Although this is appropriate for short-term incubations, it is not recommended for experiments requiring several hours in culture, such as the determination of neutrophil gene expression or protein release. In our studies, highly purified neutrophil populations are routinely suspended at no more than 10^7 cells/mL in medium containing 10% low-endotoxin fetal calf serum. Under these conditions, granulocytes can be cultured in tissue culture plasticware for up to 44 h if appropriate neutrophil survival factor(s) are added to the cultures. Cytokine synthesis and/or release by polymorphonuclear neutrophils (PMN) can be measured in cell-free supernatants (or in cell pellets) by using different methods, such as enzyme-linked immunosorbent assay (ELISA), radioimmunoassay (RIA), immunoprecipitation after metabolic labeling, bioassays, and also by intracellular staining followed by flow cytometry and immunohistochemistry. Because the induction of cytokine production and release by neutrophils is usually preceded by an increased accumulation of the related mRNA transcripts, a number of molecular biology techniques, such as Northern blotting, ribonuclease protection assays (RPA), reverse-transcription (RT)-PCR, *in situ* hybridization, and microarray analysis can be also performed (e.g., *see* Chapter 13). Northern blots and RPA have the advantage of making differences in cytokine mRNA expression quantitatively measurable. A limiting factor in high-throughput analysis of gene expression modulation in neutrophils is represented by the very low RNA content of these cells (average 1 μg/10^7 cells). The development of real-time PCR has revolutionized the measurement of gene expression, because it requires less RNA template than other methods of gene expression analysis and allows a reliable quantification of RT-PCR products. Real-time PCR assays are 10,000–100,000-fold more sensitive than RNase protection assays *(3)*, 100-fold more sensitive than dot blot *(4)*, and can even detect a single copy of a specific transcript *(5)*. On the other hand, because of its extremely high sensitivity, real-time PCR can easily amplify cytokine mRNA from a few (<0.5%) contaminating monocytes, lymphocytes, or eosinophils. This represents a major concern, because peripheral blood mononuclear cells (PBMCs) possess 10–20 times more RNA per cell than neutrophils, so that a contamination of only 1% PBMCs can translate into an up to 20% of contamination at the RNA level. Therefore, it is critical to work with highly purified PMN populations (>99.5%) and to generate substantial evidence that cytokine mRNA expression can be directly attributed to neutrophils, as opposed to contaminating PBMCs or even eosinophils. Based on reports from our group *(6)* as well as from other laboratories *(7,8)*, the absence of detectable interleukin (IL)-6 mRNA in lipopolysaccharide (LPS)-stimulated neutrophils represents a good control of neutrophil purity. In addition, correct methodological analysis and detailed investigation of the pattern of cytokine gene expression by neutrophils in vitro represent necessary and important steps. Finally, data obtained by real-time PCR analysis should be validated by comparable results obtained by Northern blot or RPA. In this chapter, we describe validated methods currently used in our laboratory for the optimal application of the real-time PCR strategy in the analysis of human neutrophil cytokine gene expression. There are several types of detection chemistry available for the real-time PCR assay (DNA binding dyes, Taqman probes, hybridization probes, hydrolysis probes, molecular beacons, scorpions, etc.). Among the two most commonly used (SYBR Green and Taqman probes), we describe here the SYBR Green-based assay. SYBR Green is a fluorescent DNA intercalator dye that binds to double-stranded PCR products that accumulate during cycling. With well designed primers, SYBR Green produces results with the sensitivity and range of quantification comparable to those obtained with Taqman, making this strategy very flexible and less expensive as compared to other chemistries. Finally, and most importantly, the data obtained by this real-time PCR strategy have been validated by comparable results obtained with other methods for gene expression analysis (e.g., *see* Chapter 13).

2. Materials

2.1. Miscellaneous

1. Cell culture medium: RPMI-1640 medium supplemented with 10% low-endotoxin fetal bovine serum (FBS; LPS content must be lower than 5–10 pg/mL, certified by the company or verified by a *Limulus Amebocytes Lysate* [LAL] assay).
2. Agonists: Ultra Pure LPS, extracted by successive enzymatic hydrolysis steps and purified by the phenol-TEA-DOC extraction, as described by Hirschfeld et al. *(9)* (commercially available from different sources) and recombinant human interferon (IFN)-γ.
3. All plasticware utilized to work with RNA (pipet tips, polypropylene tubes, syringes, 20-G needles) must be RNase-free, autoclaved, and kept in a separate dedicated area. All disposable plastics utilized for RT-PCR and real-time PCR, including pipet tips or tubes, must be certified for the absence of DNA contamination, RNase, DNase, or other PCR inhibitors, and kept in a dedicated area.

4. Whenever indicated, use sterile, pyrogen- and RNase-free H_2O approved for clinical use. One-milliliter aliquots are autoclaved and used for RNA, cDNA, PCR, and real-time PCR.
5. 70% (v/v) ethanol: dilute ethanol in RNase-free H_2O, and use it for RNA manipulation only.

2.2. Total RNA Purification

1. RNeasy mini kit (Qiagen): contains RNeasy mini spin columns, RLT Buffer (lysis buffer) (supplement with 143 mM β-mercaptoethanol immediately before use), RW1 Buffer (first washing buffer), and RPE Buffer (second washing buffer) (dilute with 4 vol of 96–100% ethanol immediately before use) (*see* **Note 1**).
2. RNase-Free DNase Set (Qiagen): contains RNase-free water; DNase I, which must be dissolved in 550 μL of RNase-free water, aliquoted, and stored at −20°C (*see* **Note 2**); RDD Buffer for on-column DNA digestion (stored at 4°C); and DNase I incubation mix, which must be prepared by adding 10 μL of a DNase I stock solution to 70 μL of RDD buffer (*see* **Note 1**).

2.3. RNA Quantitation

1. RiboGreen RNA Quantitation Kit (Molecular Probes): contains RiboGreen RNA quantitation reagent (store at −20°C protected from the light); 20X TE (200 mM Tris-HCl, 20 mM ethylenediamine tetraacetic acid [EDTA], pH 7.5), which must be stored at 4°C and diluted to 1X with RNase-free water before use; ribosomal RNA Standard (2 μg/mL), which must be stored at 4°C or −20°C for short- or long-term periods, respectively (*see* **Note 1**).
2. Black, 96-well microplates optimized for fluorescence readings (Packard Instruments).
3. Fluorescence microplate reader equipped with approx 480-nm excitation and approx 520-nm emission filters (e.g., Packard FluoroCount).

2.4. RNA Reverse Transcription

1. Reverse Transcriptase: Superscript™ II Reverse Transcriptase (200 U/μL), supplied with 5X First-Strand Buffer, and 0.1 M dithiotreithol (DTT) (Invitrogen). Store all reagents at −20°C.
2. Recombinant Ribonuclease Inhibitor: RNaseOUT™ Recombinant enzyme (acidic protein) from Invitrogen. Store at −20°C.
3. Keep Superscript II and RNaseOUT in a refrigerated bench cooler (−20°C) when performing reactions.
4. 10 mM dNTP mix: aliquot and store at −20°C.
5. Random primers (hexameric oligonucleotides): dilute at 100 ng/μL with RNase-free H_2O. Aliquot and store at −20°C.

2.5. PCR for Glyceraldehyde-3-Phosphate Dehydrogenase

1. Taq DNA polymerase (from any commercial source): aliquot (20 μL) and store at −20°C.

2. 10X PCR buffer (Sigma): aliquot and store at $-20°C$.
3. 25 mM MgCl$_2$ solution: aliquot and store at $-20°C$.
4. Forward and reverse primers for glyceraldehyde-3-phosphate dehydrogenase (GAPDH) or other housekeeping genes suspended in autoclaved H$_2$O at the final concentration of 100 μM. Dilute primers at a working concentration of 20 μM with autoclaved H$_2$O. Aliquot and store at $-20°C$.
5. 10 mM dNTP mix: aliquot and store at $-20°C$.

2.6. Real-Time PCR

1. Thermoycler for real-time PCR: The protocol described here is optimized for the DNA Engine Opticon® 2 system, including the Opticon monitor software (MJ Research) (*see* **Note 3**).
2. SYBR Green real-time master mix (2X concentrated solution; e.g., SYBR Premix Ex Taq™ [Takara Bio Inc.]). Store at $-20°C$ protected from light. Once thawed, store master mix at $4°C$ for up to 3 mo. This solution contains all necessary reagents to perform a real-time PCR (i.e., appropriate buffer, dNTPs, SYBR Green, and Taq DNA polymerase), except primers (*see* **Note 4**).
3. Forward and reverse primers: dissolve in autoclaved H$_2$O at a final concentration of 100 μM. Dilute primers at the working concentration of 20 μM. Aliquot and store at $-20°C$.
4. White, 96-well microplates or tube strips equipped with ultraclear cap strips suitable for a thermocycler. All plastics must be stored in a clean place to avoid any possible contamination by external DNA.

3. Methods

3.1. Total RNA Purification

The extraction of total RNA from neutrophils can be performed by different methods. However, depending on the method chosen, the quantity and/or the quality of the RNA extracted from neutrophils may greatly differ. It is essential that RNA for use in RT real-time PCR not be degraded, contain as little contaminating salts as possible, and be free of genomic DNA. For RT real-time PCR, small quantities of total RNA are required (from 1 ng to 1 μg), unlike other methodologies, including the Northern Blotting or RPA, which need at least 5–10 μg of RNA for each condition. Thus, the RNA yield does not represent a limiting factor for this technique.

RNA purification methods that rely on selective RNA binding properties of silica-gel-based membranes are, in our opinion, the best because they are fast, are reliable, allow rapid and simultaneous processing of many samples, and do not need a phenol/chloroform extraction step. If one desires to obtain 1 μg of total RNA per sample (which is the optimum RNA amount that can be used for RT-PCR studies), 10^7 neutrophils/condition are necessary, given that 10^6 neutrophils contain approx 0.1 μg of total RNA. 10^7 cells correspond to the maximum number of cells per sample that can be processed by an RNeasy mini kit (*see* **Note 5**). We process and purify our samples exactly as described in the "spin protocol for animal cells" reported in the RNeasy mini kit instruction manual. The protocol is detailed, comprehensive, and easy to follow, so that no detailed explanation needs to be given herein. The volumes used to resuspend cells in the initial steps of the procedures are specified as follows.

1. Resuspend purified neutrophils at 5×10^6 cells/mL in RPMI supplemented with 10% low-endotoxin FBS (*see* **Note 6**).
2. Plate 10^7 cells in a six-well tissue culture plate (2 mL/well), stimulate as appropriate, and incubate at 37°C in a 5% CO_2 atmosphere for the desired time.
3. Collect neutrophils into 2-mL tubes, and pellet at $300g$ for 5 min.
4. Carefully remove culture supernatants, and loosen cell pellets thoroughly by flicking the tube several times.
5. Resuspend pellets in 600 µL of RLT buffer, and vortex (*see* **Note 7**).
6. Homogenize lysates by passing them through a 20-G needle fitted to a syringe at least five times or until the solution becomes less dense. At this stage, lysates can be further processed or stored at −70°C for several months without any change in RNA integrity (*see* **Note 8**).
7. We do not perform the optional step 6 of the RNeasy mini kit protocol, because we perform the on-column DNase digestion (*see* **Note 9**). Instead, the optional step 8a of the RNeasy mini kit protocol is performed. Otherwise, there are no further changes to the procedures described in the manual supplied with the kit.
8. Elute RNA from the column with 30 µL of RNase-free water. The expected RNA yield from 10^7 neutrophils is approx 1 µg.
9. Concentrate RNA with a Speedvac concentrator, and resuspend dry RNA in 12 µL of RNase-free water. From this step onward, RNA must be kept on ice or stored at −20° to −70°C.

3.2. RNA Quantitation

RNA quantitation is a crucial step for optimal real-time PCR results, and we recommend performing the reverse transcription reaction with equivalent amounts of RNA from each sample. This limits differences in RT efficiency among the samples and favours subsequent normalization of real-time PCR results. The most used technique for measuring nucleic acid concentration is the determination of the RNA absorbance at 260 nm (A_{260}). However, a major disadvantage of the absorbance-based method is relative insensitivity of the assay (an A_{260} of 0.1 corresponds to a 4 µg/mL RNA solution). Because very low amounts of total RNA are usually extracted from 10^7 human neutrophils (≤1 µg), we do not recommend absorbance-based quantitation. The use of more sensitive methods, such as staining of nucleic acid with fluorescent dyes, overcomes this problem. We routinely use the RiboGreen RNA quantitation reagent that consists of an ultrasensitive fluorescent nucleic acid stain. This reagent detects as little as 1 ng/mL of RNA in solution. The following method is adapted from the instructions included in the RiboGreen RNA quantitation kit (*see* **Note 1**).

1. Prepare a stadard curve by diluting the standard RNA solution (provided with the kit and usually corresponding to 2 µg/mL) with TE.
2. Prepare 1000, 500, 50, and 20 ng/mL standards and blanks in duplicate. Pipet 100 µL of the prepared dilutions into 96-well black microtiter plates.
3. Dilute RNA samples directly into the microtiter plates by adding 1 µL of RNA sample to 99 µL of TE. Make one or more replicates.
4. Dilute (200-fold) the RiboGreen reagent with TE, and add 100 µL of the diluted RiboGreen reagent to each well.
5. Gently shake the plate, and incubate it in the dark for 5 min.
6. Measure the fluorescence in each well with a fluorescence microtiter plate reader (excitation ~480 nm, emission ~520 nm) (*see* **Note 10**). To ensure that the sample readings remain in the detection range of the fluorometer, set the instrument's gain so that the sample containing the highest RNA concentration yields a fluorescence intensity near the fluorometer's maximum.
7. Subtract the blank signal from the fluorescence value of each sample, and generate a standard curve of fluorescence vs RNA concentration.
8. Determine the RNA concentration of the various samples by linear regression using the standard curve as reference.

3.3. RNA Reverse Transcription

RT is carried out with SuperScript II Reverse Transcriptase. The following procedure is based on the protocol included in the Invitrogen manual for this enzyme (*see* **Notes 1** and **11**).

1. For each sample, prepare the following RNA/primer mix directly in a 0.2-mL tube (keep samples on ice) (*see* **Note 12**): 1 µg of total RNA, 1 µL of random primers from 100 ng/µL stock, and 1 µL of dNTP mix from 10 m*M* stock. Make up to 13 µL final volume with RNase-free H_2O.
2. Prepare identical samples for controls that will not receive reverse transcriptase (–RT) (*see* **Note 13**).
3. Centrifuge the tubes briefly (quick-spin).
4. For optimal RNA/random primer annealing, incubate samples at 65°C for 5 min and then at 25°C for 10 min in a preset thermocycler.
5. Prepare RT reaction master mix. For each reaction add: 4 µL of 5X First-Strand Buffer, 2 µL of DTT from 0.1 *M* stock, 0.5 µL RNAaseOUT from 40 U/µL stock, and 0.5 µL of SuperScript II from 200 U/µL stock). Prepare a volume of master mix of greater than what is necessary. For the –RT controls, prepare the RT reaction master mix in the same manner, but omit SuperScript II.
6. At the end of the annealing step, pause the thermocycler and add 7 µL of RT reaction master mix to each tube. Mix by pipetting and, if bubbles form, perform a quick-spin with the tubes.

7. Incubate tubes at 42°C for 1 h, then heat-inactivate the reactions at 95°C for 5 min.
8. At this stage, samples are single-strand cDNA and can be stored at –20°C or immediately used as template for amplification in PCR.

3.4. PCR for GAPDH

We usually check the RT reaction by performing a standard PCR before setting up the real-time assay. This gives us information on the quality and quantity of the cDNA synthesized in the RT reaction for each sample and allows us to check for the eventual presence of genomic DNA. We usually amplify GAPDH as a housekeeping gene (*see* **Note 14**).

1. Prepare PCR master mix. For each reaction, add: 2.5 μL of 10X PCR buffer, 1.5 μL of $MgCl_2$ from 25 mM stock, 0.5 μL of dNTP mix from 10 mM stock, 0.5 μL of forward primer from 20 μM stock, 0.5 μL of reverse primer from 20 μM stock, 0.25 μL Taq polymerase from 4 U/μL stock, and 18.25 μL of autoclaved H_2O (24 μL final volume).
2. Prepare a volume of master mix large enough to include at least four more samples (one positive control, one negative control, the –RT control, and an extra one to ensure enough master mix for all the samples). Keep all reagents and samples on ice during the entire procedure.
3. Aliquot 24 μL of PCR master mix to each 0.2-mL tube.
4. Add 1 μL of cDNA. For the positive control, use 1 μL of a cDNA sample previously tested for GAPDH. For the negative control (to check PCR contamination), add 1 μL of autoclaved H_2O instead of cDNA.
5. Centrifuge tubes briefly (quick-spin), place them in a preset thermocyler, and start the program. A standard PCR thermal cycling program is 95°C for 4 min (denaturation), 95°C for 30 s (denaturation), 60°C for 30 s (annealing), 72°C for 30 s (extension), repeat for 25 cycles, and run at 72°C for 4 min (final extension).
6. At the end of the PCR reaction, remove the tubes from the thermocycler, and run 10 μL from each reaction sample on a 2% agarose gel to check for the presence of a single PCR product of correct size/length.

3.5. Real-Time PCR

The following real-time PCR procedure is adjusted for use of SYBR Premix Ex Taq as real-time master mix in combination with a DNA Engine Opticon 2 thermocycler (*see* **Notes 3** and **4**). In keeping with the high-throughput capacity of real-time PCR, thermocycling conditions are constant for all assays.

1. Edit the following program in the real-time thermocycler:
   a. Step 1: 95°C for 10 s (initial denaturation) (*see* **Note 4**).
   b. Step 2: 95°C for 10 s (melting); 60°C for 45 s (annealing); fluorescence read. Repeat for 45 cycles. An extension step is not required, because all of the PCR products are 50–250 bp.
   c. Step 3: Melting Curve, set the final sample heating from 60°C to 90°C and have the fluorescence reading recorded every 0.5°C (*see* **Note 15**).

2. Prepare real-time PCR master mixes based upon the number of genes to be analyzed. Analysis of the expression of a housekeeping gene must be always included in the same plate for normalization (*see* **Note 16**). Each sample must be tested in triplicate, and negative controls must be included.

3. For a 20 µL-reaction, use the following components for each well or tube (*see* **Note 14**): 10 µL of 2X SYBR Premix Ex Taq, 0.2 µL of forward primer from 20 µ*M* stock, 0.2 µL of reverse primer from 20 µ*M* stock, 7.6 µL of autoclaved H_2O (18 µL final volume).

4. Dilute all cDNA samples to be analyzed to a concentration of 10 ng/µL (*see* **Notes 17** and **18**).

5. Aliquot sequentially 18 µL of real-time master mix and 2 µL of sample or H_2O (for the "no template" control) or –RT (for the negative control) into each well or tube. This step must be carried out very rapidly, as SYBR Green is light-sensitive.

6. Carefully close the wells or the tubes with optical cap strips, paying great attention not to touch the inner part of the cap and not to dirty or scratch the outer part of each.

7. Briefly centrifuge the plate or the tubes, and then start the reaction in the real-time thermocycler.

8. Remove the plate or the tubes from the thermocycler after the real-time PCR is finished.

9. Separate 5 µL of each sample by electrophoresis on a 2% agarose gel to check for quality of PCR products. If the reaction proceeded correctly, a single amplified product of the expected length should be visible in the gel.

10. At the end of the reaction, check the melting curve analysis to verify whether abnormal amplification plots or any bimodal dissociation curves are present (*see* **Note 15** and **Fig. 1A**).

11. At the end of the run, the DNA Engine Opticon 2 system software shows raw data (fluorescence value of each sample for each real-time PCR cycle) that must be analyzed. The expression level of the gene of interest will be reported in arbitrary units relative to an endogenous housekeeping reference RNA.

3.6 Analysis of Real-Time PCR Data

In this section, we specify the software used for analysis of real-time PCR data. A detailed description is referred in each software manual.

1. Set the baseline. The initial cycles of PCR where there is little fluorescence signal are termed "baseline" and should be eliminated (**Fig. 1B**). The baseline should be wide enough to eliminate background, but should not overlap with the area in which the amplification signal begins to rise above background (*see* **Note 19**).

2. Set the threshold. The threshold is an arbitrary fluorescence value that must intercept the PCR amplification curve in the exponential phase of the reaction (*see* **Note 20** and **Fig. 1B**).

3. Calculate the efficiency of the real-time PCR reaction for each gene assayed. There are several methods of calculating amplification efficiency on raw data collected

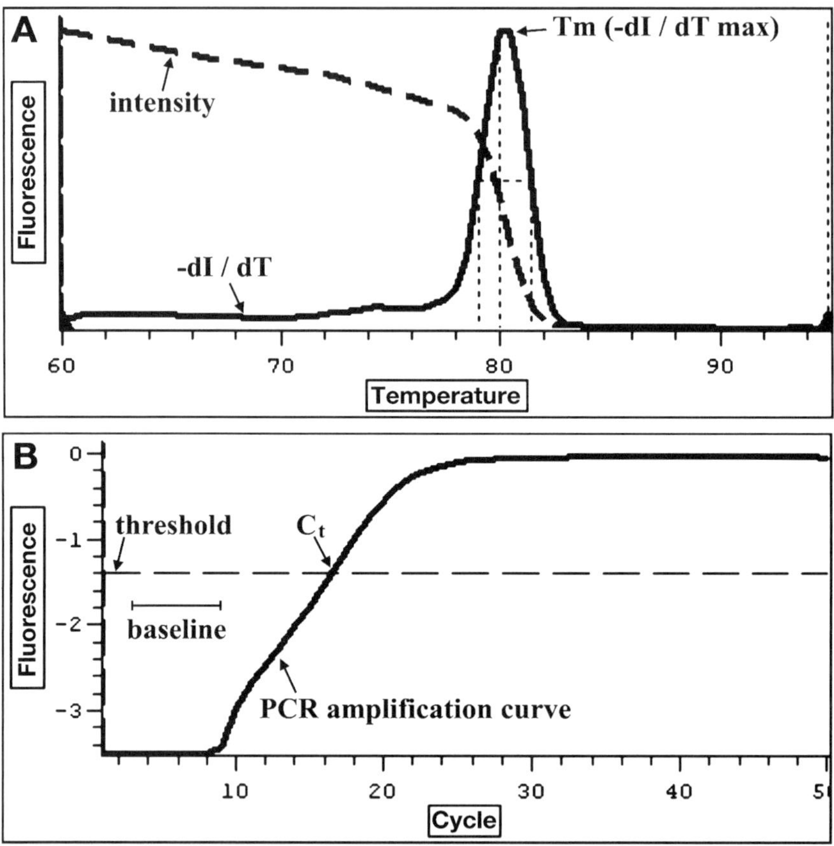

Fig. 1. Melting and PCR amplification curves as shown by the Opticon Monitor software. (**A**) The melting curve is plotted as fluorescence intensity (dot line) or as first derivative (−dI/dT) (solid line) of the PCR product as function of temperature. The melting temperature (Tm) of the PCR product is defined as the temperature at which −dI/dT reach the maximum value. (**B**) PCR amplification curve. The fluorescence intensity is plotted as a function of cycle number. The baseline, threshold, and threshold cycle are indicated.

during PCR. We routinely calculate the amplification efficiency with LinRegPCR software *(10)*. The program uses linear regression analysis to find the best-fit straight line through the PCR data set. From this line, PCR efficiency of each individual sample is calculated. The amplification efficiency value that we take for each gene is the average of the single efficiencies calculated with LinRegPCR (*see* **Note 21**).

4. Export threshold cycles (C_t) values to the Q-Gene Excel datasheet. The cycle at which the fluorescence signal from the reaction crosses the threshold line is defined as the C_t. Once the threshold line is set, the real-time software automatically gen-

erates a C_t value for each sample. The C_t is a reliable indicator of the initial copy number of each amplified cDNA.

5. Perform the final quantitation with Q-Gene software (*see* **Note 22**). Insert the following values in the Q-Gene Excel datasheet: (1) the C_ts generated for each sample, both for the target gene and for the housekeeping gene by the real-time thermocycler software; (2) the calculated amplification efficiency for the gene of interest and for the housekeeping gene; and (3) the calculation procedure 2, corresponding to equation 3 described in Muller *(11)*.
6. The Q-Gene software calculates the Mean Normalized Expression (MNE) with standard errors for each sample, corrected for amplification efficiencies. Examples of the results that can be obtained are shown in (**Figs. 2** and **3**).

4. Notes

1. See accompanying instruction manual for further details.
2. Once thawed, reconstituted DNase I is stable for up to 6 wk at 4°C. Do not refreeze the aliquots after thawing. Never vortex DNase I, because this enzyme is ultra-sensitive to physical denaturation.
3. If real-time PCR thermocyclers from other suppliers are going to be used, appropriate modifications to our protocol should be preliminarily determined (for instance, the reaction volumes, the PCR reaction settings, usage of reference dyes in the master mix, and so forth).
4. Many real-time master mix solutions are commercially available. If using real-time master mix purchased from a different company, we suggest carefully reviewing the accompanying product insert. Changes in the reaction preparation and/or in the PCR reaction (for instance, the initial denaturation step or the annealing/elongation step) may vary, as the length of the denaturation step depends on the polymerase included in the master mix.
5. Among the commercially available kits, we have experimentally verified that the "RNeasy mini kit" works well, because it facilitates purification of high-quality RNA (not degraded, salts- and genomic DNA-free), whose integrity and purity is easily assessed by denaturing agarose gel electrophoresis and ethidium bromide staining (*see also* **Chapter 13**). In addition, this kit permits use of an on-column DNase digestion step, thereby avoiding potential DNase carry-over.
6. It is very important to work with highly pure neutrophil preparations. A rapid and easy way to check the cell purity is May-Grünwald-Giemsa staining. The percentage of contaminating eosinophils may vary from donor to donor (usually from 1 to 8%). This should be noted if the gene(s) of interest is/are expressed in this latter cell type. More critical is the potential contamination of mononuclear leukocytes (monocytes and lymphocytes). The presence of >1% contaminating monocytes may greatly influence the results attributed to neutrophils *(7)*. As shown in **Fig. 2**, a parallel analysis of cytokine gene expression in neutrophils and monocytes purified from the same donor allows potential identification of false-positive results due to the presence of contaminating monocytes in the neutrophils culture.

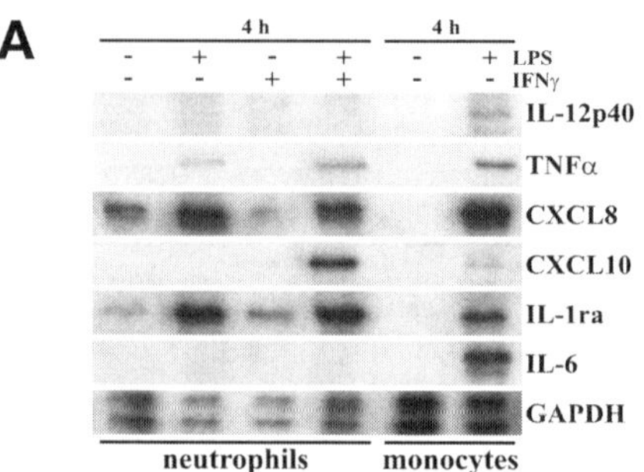

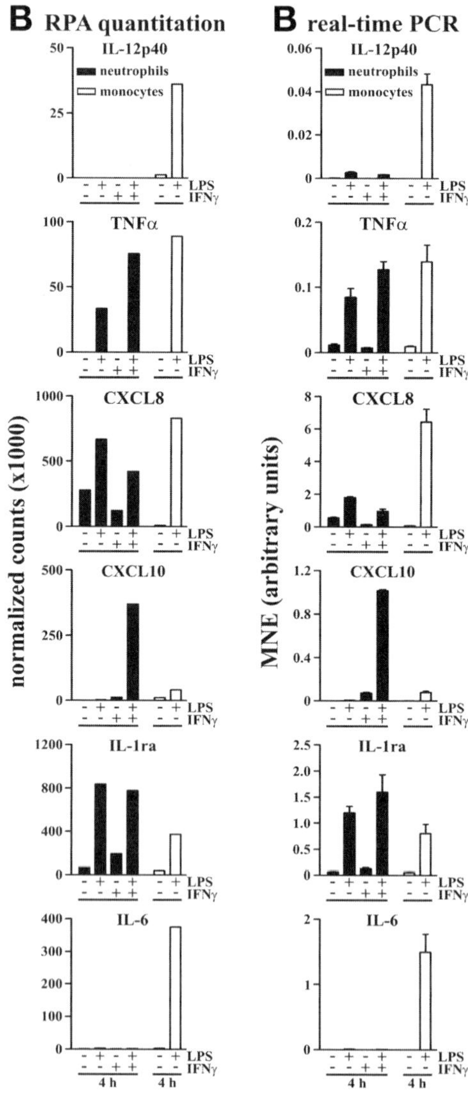

7. Some neutrophil agonists, for instance LPS or tumor necrosis factor (TNF)-α, cause strong cell adherence to the culture vessel, especially within the first 2 h of incubation. In this case, we suggest removing culture medium and adding RLT buffer directly into the well.

8. The quality of the cDNA template depends on the integrity of the RNA. All possible precautions must be taken in order to maintain an RNase-free environment when handling/manipulating RNA. Hands and dust particles may carry common sources of RNase contamination, such as bacteria and molds. Always wear gloves and change them frequently. Because RNases are less active at cold temperature, we suggest keeping samples chilled on ice, especially following RNA extraction.

9. DNA digestion is a necessary step in all cases in which contaminating genomic DNA can interfere with subsequent applications. The RNeasy kit offers the optional step of "on column" DNase treatment, which we always perform. If a different kit for RNA purification is used, genomic DNA should be removed by a DNase digestion step following RNA isolation. To avoid the risk of a possible DNase carry-over, which could be detrimental for the subsequent cDNA synthesis step, we recommend performing phenol-chloroform treatment of all RNA samples after DNase treatment.

10. The excitation maximum for RiboGreen reagent bound to RNA is approx 500 nm and the emission maximum is approx 525 nm.

11. Because even a small amount of contaminants can greatly impact results of the assay, use of molecular biology-grade water, RNase/DNase/nucleic acid-free tubes, aerosol-barrier pipet tips, and dedicated pipettors of all types (i.e., pipettors used only for RNA or PCR applications, which are kept out of areas used for plasmid or genomic DNA work) is strongly recommended for all steps. The use of a small PCR hood may be very helpful to avoid PCR contamination. If a hood is not available, perform PCR in a dedicated space, possibly at distance from the thermocycler or spaces in which the electrophoresis equipment are located.

Fig. 2. (*Opposite page*) Quantitative analysis of cytokine gene expression in activated neutrophils and monocytes performed in ribonuclease protection assay (RPA) and reverse-transcription (RT) real-time PCR. Neutrophils and autologous monocytes were cultured with 100 ng/mL Ultra Pure LPS, without or (in the case of neutrophils) with 100 U/mL interferon-γ. After 4 h, total RNA was extracted and analysed for interleukin (IL-12p40, tumor necrosis factor-α, CXCL8, CXCL10, IL-1ra, and IL-6 mRNA accumulation either by RPA, using a custom kit (Pharmingen), or by the RT real-time PCR procedure (described in this chapter). (**A**) autoradiography of the RPA assay; 5 μg of total RNA/sample were processed according to the manufacturer's protocol. (**B**) quantitative analysis of the expression of each gene shown in panel (**A**). The extent of hybridization signal was quantified with InstantImager (Packard Instruments) and plotted after glyceraldehyde-3-phosphate dehydrogenase normalization. (**C**) results obtained by RT real-time PCR, in which 1 μg RNA/condition was processed. Expression levels of cytokine mRNAs is reported as mean normalized expression (MNE), calculated after normalization (*see* **Note 16**). Data demonstrate that these two techniques produce similar quantitative results.

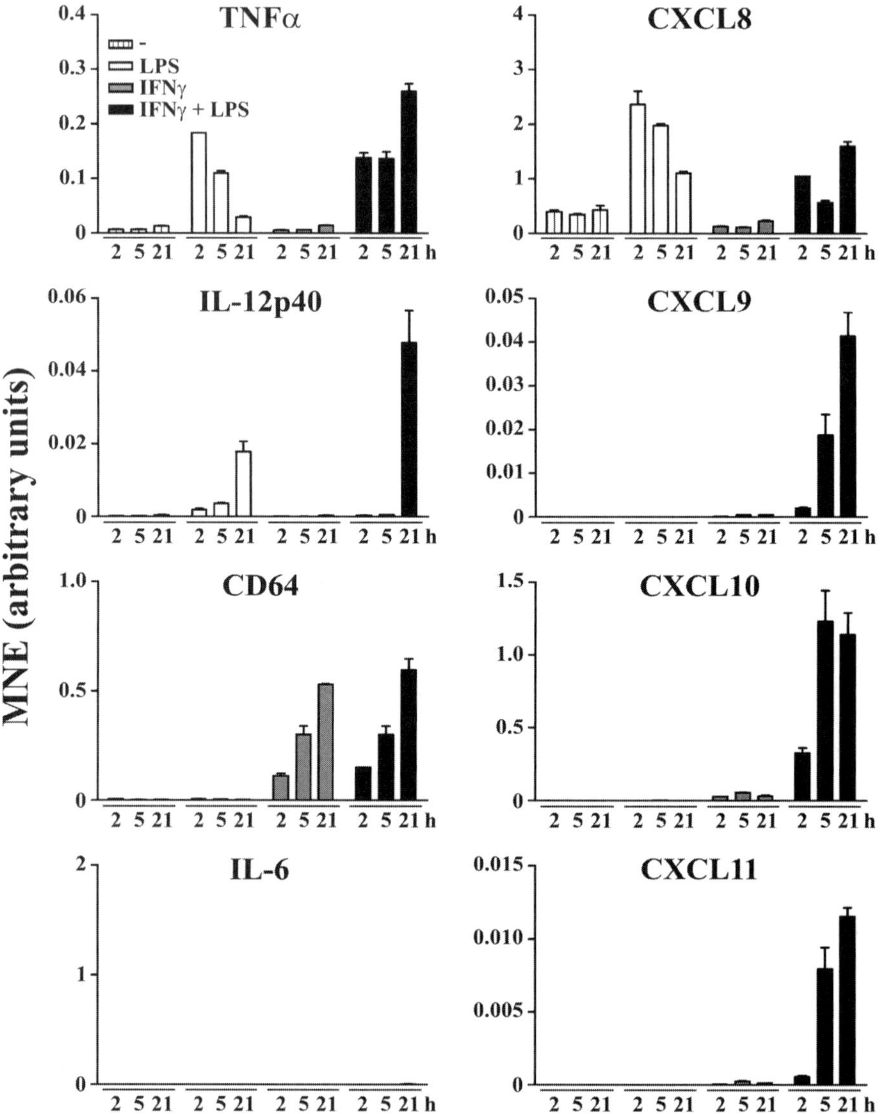

Fig. 3. Time-dependent expression of tumor necrosis factor (TNF)-α, interleukin (IL)-12p40, CD64, IL-6, CXCL8, CXCL9, CXCL10, and CXCL11 mRNAs in activated neutrophils. Neutrophils were cultured for the times indicated with 100 ng/mL Ultra Pure LPS and/or 100 U/mL interferon-γ. Total RNA was extracted, and levels of TNF-α, IL-12p40, CD64, IL-6, CXCL8, CXCL9, CXCL10, and CXCL11 transcript were analyzed by reverse-transcription real-time PCR, as described. mRNA accumulation of the various genes is depicted as Mean Normalized Expression units, calculated after normalization (*see* **Note 16**). The figure depicts results that are typically obtained from such type of experiment.

12. We suggest processing all samples from each experiment at the same time, starting from the same amount of RNA for each sample. This reduces differences in the quality and quantity of cDNAs obtained and increases the reliability of the final real-time PCR results. If RNA recovery is low following the RNA purification step, RT can be performed with <1 µg of RNA without major problems. However, we suggest that the RT reaction be performed with ≥0.1 µg of RNA.

13. Because most of the primers we use in real-time PCR span an exon junction, the −RT control is not needed for every sample. We usually perform −RT controls only for new primer pairs, and occasionally on random samples.

14. Primer designing is a crucial step for performance of real-time PCR. Primers should be designed according to standard PCR guidelines. A good primer (1) is 19–24 bases in length; (2) preferably has a base composition of 50–60% in G+C; (3) preferably has a primer ending at the 3'-end that is a G or C, or CG or GC—this prevents "breathing" of ends and increases efficiency of priming; (4) has no runs of three or more Cs or Gs at the 3'-end of primers, which may promote mispriming at G or C-rich sequences (because of stability of annealing) and should be avoided; (5) has 3'-ends that are not complementary (i.e., base pair), as otherwise primer dimers will be synthesized preferentially to any other product; and (6) has no primer self-complementarity (ability to form 2° structures such as hairpins). To optimize efficiency, the product of real-time PCR must be approx 80–250 bp. Melting temperature (Tm) of the primers should be 60–64°C; it is important that all the primers function at the same annealing temperature, thereby allowing analysis of different targets in the same plate. When designing primers for RT-PCR, we suggest identifying at least one primer that overlaps an exon–exon boundary, in order to prevent the amplification of genomic DNA. Alternatively, the two primers should anneal to two different exons on each side of an intron. A public database for the selection of primer sequences suitable for RT real-time PCR is available at the following website: http://medgen.ugent.be/rtprimerdb/index.php. Our primers are designed with the PerlPrimer software *(12)*, which is free at http://perlprimer.sourceforge.net.

15. Specificity of the real-time PCR reaction can be confirmed by melting curve analysis, where the presence of different PCR products is reflected in the number of first-derivative melting peaks. The melting curve analysis is set as final step of the real-time PCR program by gradually heating the real-time PCR product (from 60°C to 95°C) and recording the fluorescence value every 0.5°C. As the temperature increases from 60°C to 95°C, the final double-strand PCR product is denatured, intercalating SYBR Green is released, and the fluorescence intensity decreases (*see* **Fig. 1A**). The fluorescence intensity vs temperature plotted as first derivative (−dI/dT) generates the melting curve. The presence of more than one peak indicates that more than one PCR product has been formed and that the PCR reaction is not specific. The negative control, containing PCR primers but no template, could give rise to a peak in the melting curve that usually has a low Tm and is formed in the last PCR cycles. This peak is usually representative of primer dimer formation.

16. In order to more accurately quantitate the expression levels of a given mRNA, it is very important to select a stable housekeeping gene that must be preliminarily identified depending on the cell type. For neutrophils, we and other groups *(13–15)* suggest using the β2-microglobulin mRNA. However, GAPDH or Ubiquitin C (UBC) mRNA can be used as well.
17. It is possible to use 1 pg to 100 ng of cDNA in a real-time PCR reaction, and the amount available will likely depend on the quantity of total RNA isolated from neutrophils. We experimentally determined that 20 ng of cDNA is an optimal amount and permits quantification of low- and high-expressed transcripts.
18. Note that there is a risk for pipetting errors associated with manipulating very small vol (<2 μL) that could cause differential input of template.
19. For most of the cytokine genes, baseline can be set at cycle range 3–9. For low-expressed genes this value should be extended to 3–20.
20. In our conditions, the threshold is usually set at a fluorescence value of 0.04.
21. Amplification efficiency of the reaction is an important consideration when performing relative quantitation. An efficiency value of 2 means that the PCR product doubles during every cycle within the exponential phase of the reaction. However, many PCR reactions do not have ideal amplification efficiencies, and calculation without an appropriate correction factor may overestimate starting concentration. We use as amplification efficiency for each gene analyzed (including the housekeeping gene) the average of amplification efficiencies calculated by LinRegPCR for each single sample. Data input and output are through an Excel spreadsheet. The LinRegPCR software is freely available on request at bioinfo@ amc.uva.nl.
22. Q-Gene is Excel-based software available free for download at: www.biotechniques. com *(11)*.

Acknowledgments

This work was supported by grants from M.I.U.R. (PRIN 2005 and 60% funds), AIRC, and "Progetto Sanità, Fondazione CARI-VR-VI-BL-AN."

References

1. Cassatella, M. A. (1999) Neutrophil-derived proteins: selling cytokines by the pound. *Adv. Immunol.* **73,** 369–509.
2. Cassatella, M. A. (ed.) (2003) *Chem Immunol Allergy*, vol. 83. Karger, Basel, Switzerland.
3. Wang, T. and Brown, M. J. (1999) mRNA quantification by real time TaqMan polymerase chain reaction: validation and comparison with RNase protection. *Anal Biochem.* **269,** 198–201.
4. Malinen, E., Kassinen, A., Rinttila, T., and Palva, A. (2003) Comparison of real-time PCR with SYBR Green I or 5'-nuclease assays and dot-blot hybridization with rDNA-targeted oligonucleotide probes in quantification of selected faecal bacteria. *Microbiology* **149,** 269–277.

5. Palmer, S., Wiegand, A. P., Maldarelli, F., et al. (2003) New real-time reverse transcriptase-initiated PCR assay with single-copy sensitivity for human immunodeficiency virus type 1 RNA in plasma. *J. Clin. Microbiol.* **41,** 4531–4536.
6. Bazzoni, F., Cassatella, M. A., Laudanna, C., and Rossi, F. (1991) Phagocytosis of opsonized yeast induces tumor necrosis factor-alpha mRNA accumulation and protein release by human polymorphonuclear leukocytes. *J. Leukoc. Biol.* **50,** 223–228.
7. Wang, P., Wu, P., Anthes, J. C., et al. (1994) Interleukin-10 inhibits interleukin-8 production in human neutrophils. *Blood* **83,** 2678–2683.
8. Takeichi, O., Saito, I., Tsurumachi, T., Saito, T., and Moro, I. (1994) Human polymorphonuclear leukocytes derived from chronically inflamed tissue express inflammatory cytokines in vivo. *Cell. Immunol.* **156,** 296–309.
9. Hirschfeld, M., Ma, Y., Weis, J. H., Vogel, S. N., and Weis, J. J. (2000) Cutting edge: repurification of lipopolysaccharide eliminates signaling through both human and murine toll-like receptor 2. *J. Immunol.* **165,** 618–622.
10. Ramakers, C., Ruijter, J. M., Deprez, R. H., and Moorman, A. F. (2003) Assumption-free analysis of quantitative real-time polymerase chain reaction (PCR) data. *Neurosci. Lett.* **339,** 62–66.
11. Muller, P. Y., Janovjak, H., Miserez, A. R., and Dobbie, Z. (2002) Processing of gene expression data generated by quantitative real-time RT-PCR. *Biotechniques* **32,** 1372–1374, 1376, 1378–1379.
12. Marshall, O. J. (2004) PerlPrimer: cross-platform, graphical primer design for standard, bisulphite and real-time PCR. *Bioinformatics* **20,** 2471–2472.
13. Hayashi, F., Means, T. K., and Luster, A. D. (2003) Toll-like receptors stimulate human neutrophil function. *Blood* **102,** 2660–2669.
14. Zhang, X., Ding, L., and Sandford, A. J. (2005) Selection of reference genes for gene expression studies in human neutrophils by real-time PCR. *BMC. Mol. Biol.* **6,** 4.
15. Vandesompele, J., De Preter, K., Pattyn, F., et al. (2002) Accurate normalization of real-time quantitative RT-PCR data by geometric averaging of multiple internal control genes. *Genome Biol.* **3,** research0034.1-0034.11.

29

Detection of Intact Transcription Factors in Human Neutrophils

Patrick P. McDonald and Richard D. Ye

Summary

The crucial contribution of neutrophils to innate immunity extends well beyond their traditional role as professional phagocytes. Indeed, it is now well established that neutrophils generate a plethora of inflammatory cytokines and chemokines that are profoundly involved in the onset and evolution of the inflammatory reaction. Several recent studies have also shown that neutrophils can represent an important source of inflammatory cytokines in a number of pathophysiological settings. The inflammatory cytokines produced by neutrophils are generally encoded by immediate-early response genes, which in turn depend on the activation of transcription factors such as those belonging to the nuclear factor (NF)-κB and signal transducers and activators of transcription (STAT) families. We have shown in the past that such factors are expressed and inducible in neutrophils stimulated by physiological agonists. However, the detection of *intact* (i.e., undegraded) transcription factors in neutrophils requires special precautions and an alternative protocol, as a result of the huge amounts of endogenous proteases present in these cells. This protocol is the focus of this chapter.

Key Words: Transcription factors; NF-κB; STAT; nuclear extracts; electrophoretic mobility shift assay; granulocytes.

1. Introduction

Neutrophils are the most abundant circulating leukocytes and play a crucial role in innate immunity. They are usually the first cells to infiltrate inflamed tissues, where they exert a series of antimicrobial responses culminating in the elimination or inactivation of the infectious agent. In addition to their traditional role as professional phagocytes, neutrophils also express a plethora of genes in response to pro-inflammatory cytokines and bacterial products such as lipopolysaccharide (LPS) and *N*-formyl peptides *(1–5)*. Examples include cytokines such as tumor necrosis factor (TNF)-α, interleukin (IL)-1, IL-12; chemokines

From: *Methods in Molecular Biology, vol. 412: Neutrophil Methods and Protocols*
Edited by: M. T. Quinn, F. R. DeLeo, and G. M. Bokoch © Humana Press Inc., Totowa, NJ

such as IL-8/CXCL8, Mip-1α/CCL3, Mip-1β/CCL4, MIG/CXCL9, IP-10/
CXCL10, and I-TAC/CXCL11; and receptors such as MHCII and the high-
affinity receptor for immunoglobulin (Ig)G, FcγRI/CD64. A common characteristic of these molecules (aside from their pivotal implication in the inflammatory reaction) is that they are encoded by immediate-early response genes,
which are under the control of the nuclear factor (NF)-κB and/or signal transducers and activators of transcription (STAT) transcription factors. We have
previously shown that the same stimulatory conditions allowing for the expression of the aforementioned mediators and receptors also lead to the activation
of the NF-κB or STAT transcription factors in neutrophils *(6–11)*.

Both NF-κB and STAT factors are normally sequestered in the cytoplasm and
are mobilized to the nucleus upon cell activation, where they can bind to their
cognate sequences on the promoter region of target genes, and thereby initiate
or enhance gene transcription (for recent reviews, *see* **refs.** *12–15*). The activation of NF-κB is a highly regulated process, which usually involves the engagement of cell surface receptors, thereby generating various intracellular signals
that eventually converge on the κB kinase (IKK) complex. This results in the
phosphorylation and activation of IKK subunits, which can in turn phosphorylate the inhibitor protein IκB-α, thereby targeting it for ubiquitin-dependent
degradation. The proteolysis of IκB-α, which is coupled to NF-κB dimers in
the cytoplasm of resting cells, unmasks nuclear localization sequences on NF-
κB subunits that allow the NF-κB complexes to translocate to the nucleus. By
comparison, the STAT proteins are latent cytoplasmic proteins that are rapidly
recruited to Src homology (SH)2 domains within the cytoplasmic tail of cytokine receptors upon cell stimulation, where they become tyrosine-phosphorylated by Janus kinases (JAKs). Tyrosine-phosphorylated STAT proteins can then
dimerize into various DNA-binding configurations, forming homo- or heterodimers, and translocate to the nucleus.

The detection of nuclear transcription factors involves cell disruption followed by the preparation of nuclear extracts, which are then incubated with an
excess of radiolabeled oligonucleotide probe that contains the cognate DNA
binding sequence for a given factor. The resulting mixtures are then migrated
on native gels to separate transcription factor complexes bound to labeled probe
from unbound probe. Conventional protocols typically achieve cell lysis by
solubilizing the cytoplasmic membrane with nonionic detergents *(16)*, or sometimes by repeated freeze–thaw cycles or homogenization. In neutrophils, however, these approaches are problematic because of the huge quantities of proteolytic enzymes that are present in neutrophil cytoplasmic granules. Indeed,
repeated freeze–thaw cycles break open the protease-rich neutrophil granules
(17,18). We have shown that detergent lysis similarly results in the solubilization of granule-bound proteases, resulting in partially degraded NF-κB or STAT

complexes and individual constituent proteins, even in the presence of an elaborate cocktail of protease inhibitors *(8)*. Accordingly, transcription factor complexes that are detected following neutrophil disruption by conventional procedures almost invariably migrate faster than authentic complexes and no longer react with certain antibodies (extensively reviewed in **ref. *19***). In contrast, nitrogen cavitation represents a gentle yet efficient way to disrupt human neutrophils while preserving the integrity of intracellular organelles such as granules and nuclei *(17)*. We have adapted this technique to the preparation of neutrophil nuclear extracts, and this approach consistently yields *intact* (i.e., undegraded) NF-κB and STAT complexes *(6,8,10)*. Our cavitation-based protocol is the focus of this chapter.

2. Materials

2.1. Neutrophils

Peripheral blood neutrophils are obtained from healthy volunteers, whose informed consent is ensured in accordance with the relevant institutional review board. Neutrophils must be prepared under endotoxin-free conditions to avoid inadvertent activation during isolation. The isolation procedure we use *(20)* is one of the many variations of the original Boyum protocol *(21)*. Details on neutrophil isolation can be found in Chapter 2. Regardless of the exact variation, isolated neutrophils should contain less than 1% contaminating mononuclear cells, and display at least 98% viability.

2.2. Buffers

1. Relaxation buffer: 10 m*M* PIPES, pH 7.30, 30 m*M* NaCl, 3.5 m*M* MgCl$_2$, 0.5 m*M* EGTA, 0.5 m*M* ethylenediamine tetraacetic acid (EDTA), and 1 m*M* dithiothreitol (DTT). Relaxation buffer is supplemented with an anti-protease cocktail (1 m*M* diisopropylfluorophosphate [DFP], 1 m*M* 4-[2-aminoethyl]-benzenesulfonyl-fluoride [AEBSF], 1 m*M* phenylmethylsulfonylfluoride [PMSF], and 10 µg/mL each of aprotinin, leupeptin and pepstatin A) and phosphatase inhibitors (10 m*M* NaF, 1 m*M* Na$_3$VO$_4$, 10 m*M* Na$_4$P$_2$O$_7$). This buffer is designed for the generation of nuclear extracts from cavitated neutrophils (*see* **Note 1**).
2. Nuclear extraction buffer: relaxation buffer containing 10% (v/v) glycerol and the anti-protease cocktail. Phosphatase inhibitors may also be used, with the exception of Na$_3$VO$_4$, which interferes with protein quantitation by the method of Bradford *(22)*.
3. Binding buffer: 20 m*M* Tris-Base, pH 7.50, 50 m*M* KCl, 1 m*M* EDTA, 1 m*M* DTT, 300 µg/mL µg acetylated bovine serum albumin (BSA), 50 µg/mL poly (dI-dC), 0.1% NP-40 (v/v), and 5% glycerol (v/v). This is the buffer in which transcription factors present in nuclear extracts are allowed to interact with labeled DNA probes (*see* **Note 2**).

2.3 Nitrogen Cavitation Vessel

A cell disruption bomb of any model can be used. We favor Model 4635 (Parr Instrument Company, Moline, IL), custom fitted with four discharge valves, as this makes it possible to process many samples simultaneously.

2.4. Electrophoretic Mobility Shift Assay

1. Double-stranded oligonucleotide probes:
   a. For NF-κB binding, use any probe containing a consensus NF-κB sequence. In this chapter, we use an oligonucleotide containing tandemly-repeated NF-κB sites (capitalized) identical to those of the HIV promoter (5'-gatca GGGACTTT CC gctg GGGACTTTCC-3').
   b. For the binding of STAT proteins, use various commercially available probes, depending on the nature of the STAT complex. In this chapter, we use the a oligo- nucleotide probe corresponding to the gamma response region (GRR) of the CD64 promoter (5'-CTT TTC TGG GAA ATA CAT CTC AAA TCC TTG AAA CAT GCT-3'), which can bind STAT1-, STAT3- and STAT5-containing complexes.
2. [γ-^{32}P]ATP with specific activity of at least 3000 Ci/mmol for probe labeling.
3. Sephadex G-50 spin columns for the separation of labeled oligonucleotide probe from unincorporated [γ-^{32}P]ATP, used according to the manufacturer's instructions.

2.5 Native Polyacrylamide Gel Electrophoresis

1. TBE buffer: 89 m*M* Tris, 89 m*M* boric acid, 0.2 m*M* EDTA, pH 8.3. Use the same batch of TBE buffer for gel preparation and subsequent electrophoresis. A 5X TBE stock can be made and stored at room temperature (*see also* **Note 3**).
2. Nondenaturing polyacrylamide gel preparation: using a 25:1 mixture of acryl- amide:bis acrylamide ensures that large protein complexes migrate through the gel without undue hindrance. Although 5% gels are generally used, varying the total acrylamide content of the gels can yield better resolution. Use a freshly pre- pared ammonium persulfate solution to ensure optimal polymerization. Always let the gels completely polymerize (at least 1.5 h) to ensure the best possible resolu- tion. Finally, it is advisable to use a gel size of 16 × 14 cm (w × h), as it allows for better sample separation than mini-gels. A gel thickness of 1–1.5 mm works well.

3. Methods

Neutrophil stimulation with several physiological agonists results in a greatly enhanced detection of nuclear NF-κB-binding complexes in electrophoretic mobility shift assay (EMSA), and in the onset of nuclear STAT-containing DNA-binding activities (reviewed in **ref. *19***). However, the detection of intact (i.e., unproteolyzed) transcription factors in neutrophils requires an alternative proto- col that we developed, which is based on nitrogen cavitation of the cells. This protocol consistently yields NF-κB and STAT complexes that are indistinguish-

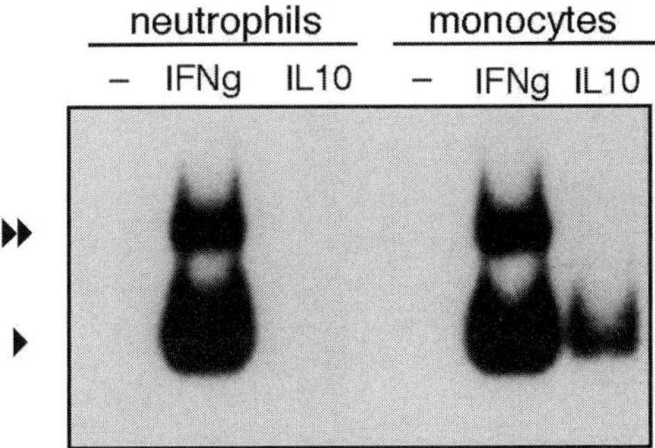

Fig. 1. Induction of signal transducers and activators of transcription (STAT)1 multimers in neutrophils and monocytes. Human neutrophils and autologous monocytes were stimulated with 100 U/mL interferon-γ, 1000 U/mL interleukin-10, or diluent control, for 15 min at 37°C. Cells were cavitated, and nuclear extracts were prepared and analyzed in electrophoretic mobility shift assay using a gamma response region probe. The single arrowhead represents STAT1 dimers, and the double arrowheads represent STAT1 tetramers, as previously reported *(10)*.

able from those isolated from other cell types, such as human monocytes or Jurkat cells *(8)* (*see also* **Fig. 1** for STAT multimers and **Figs. 3** and **4** for NF-κB dimers). In the case of STAT complexes, conventional detergent-based methods additionally produce artifacts such as the detection of a constitutive DNA binding in nuclear extracts from resting neutrophils *(8,23)*, which are not observed when the neutrophils are cavitated instead *(8)* (*see also* **Fig. 1**). The above considerations make it clear that nitrogen cavitation offers unmatched advantages for the preparation of neutrophil nuclear extracts. Because few granules are ever broken, protease inhibitors can effectively neutralize the small quantities of proteases, which may be released during preparation (as opposed to the huge amounts released using conventional approaches).

Over the years, several refinements were made to the procedure that we originally described *(6,8)*, which we have incorporated in the following protocol. Because crucial aspects of the protocol are the disruption of neutrophils and preparation of nuclear extracts, we will mainly focus on these aspects. The subsequent EMSA analysis is essentially the same for neutrophil nuclear extracts as for extracts originating from other cell types.

3.1. Preparation of Neutrophil Nuclei

1. Stimulate neutrophils either in suspension or plated in tissue culture-treated plasticware and then stimulated (both approaches will yield a comparable outcome). Use at least 3×10^7 cells per experimental condition to offset the loss of material due to the significant void volume of the cavitation device (*see* **Note 4**).

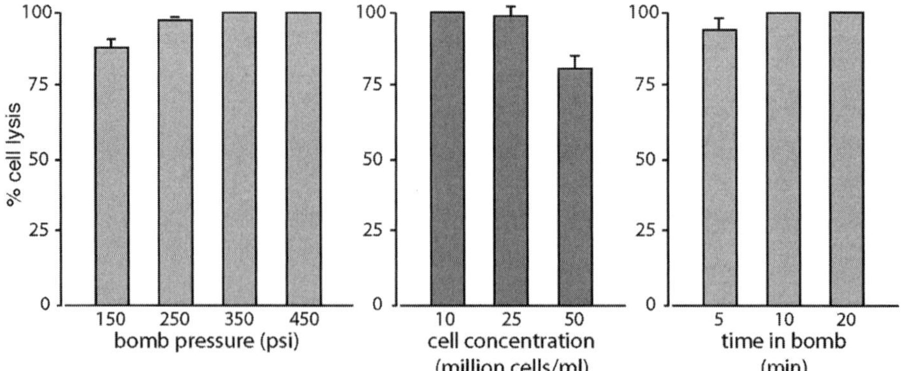

Fig. 2. Nitrogen cavitation set-ups for human neutrophils. Left panel: neutrophils (10^7/mL) were cavitated at the indicated pressures for 10 min. Middle panel: neutrophils were cavitated (350 psi, 10 min) at the indicated cell concentrations. Right panel: neutrophils (10^7/mL) were cavitated at 350 psi for the indicated times. Mean ± SEM of at least three independent experiments.

2. For suspension cells, a maximum concentration of 10^7 cells/mL is recommended. Keep the cells in suspension by gentle swirling at 2- to 3-min intervals (or with constant swirling in an orbital water bath) during the stimulation period.
3. For plated cells, neutrophils can be cultured in large tissue culture-treated Petri dishes in an incubator. Cell density is less important, but care must be taken so that neutrophils do not stack on each other when they sediment to the bottom of the dish.
4. For suspension or plated cells, add an equal volume of ice-cold buffer to stop the incubation (the same buffer used for cell incubations, such as Hank's balanced salt solution [HBSS], Krebs-Ringer phosphate buffer containing dextrose [KRPD], or culture medium) containing 2 mM DFP and phosphatase inhibitors (10 mM NaF, 1 mM Na_3VO_4, 10 mM $Na_4P_2O_7$).
5. For plated neutrophils, collect cells by gentle resuspension using a pipet fitted with a 1-mL tip that has been cut at the end to minimize turbulence.
6. Centrifuge collected cells (300g, 5 min, 4°C) and resuspend in ice-cold Relaxation buffer supplemented with protease and phosphatase inhibitors in 100×16 mm round-bottom polypropylene centrifuge tubes (Sarstedt or Falcon). The final cell concentration should be 2×10^7/mL or less (**Fig. 2**, middle panel) to ensure total cell lysis.
7. Place samples into a nitrogen bomb containing crushed ice. This keeps the samples cold, and also helps position the four tubes into the four-outlet bomb (*see* **Note 4**).
8. Place a small magnetic stir bar in each tube, and place the nitrogen bomb itself on a magnetic stirrer to prevent cells from aggregating during the cavitation process (*see* **Note 5**).
9. Pressurize samples with N2 (350 psi) for 10 min at 4°C (*see* **Note 6** and **Fig. 2**).
10. Collect each sample by inserting the bomb's outlet tube halfway into a 15-mL conical polypropylene centrifuge tube and carefully opening the discharge valve.

11. Collect cavitates drop-wise into the tube, as opposed to releasing the sample all at once. This step requires some practice, because it must be performed correctly and swiftly to avoid leaving the last sample for too long in the bomb (*see* **Note 7**).
12. Place tubes on ice immediately after each cavitate is collected.
13. Centrifuge the cavitates (1500*g*, 4°C, 10 min) to pellet the nuclei.
14. Collect the resulting supernatants and recentrifuge under the same conditions to pellet the remaining nuclei.
15. Combine the nuclear pellets, wash once with 500 μL Relaxation buffer, and use immediately for preparation of nuclear extracts (*see* **Note 8**).

3.2. Preparation of Nuclear Extracts

1. *Gently* resuspend nuclear pellets in Nuclear Extraction buffer using a 100-μL pipet tip that has been cut at the end to avoid breaking the fragile nuclei during pipetting. We found that using a volume (in microliters) that is 1.5 times the number of million cell-equivalents in the nuclear pellet works best.
2. Add NaCl from a concentrated NaCl solution (made in Relaxation buffer) to yield a final concentration of 400 m*M* (*see* **Notes 9–11**).
3. Immediately mix samples by flicking the Eppendorf tubes a few times.
4. Perform nuclear extraction on wet ice for 20 min, with occasional flicking of the tubes (*see* **Note 12**).
5. Pellet samples in a microfuge (top speed, 10 min, 4°C).
6. Set aside a small volume of the resulting supernatants (the nuclear extracts) for protein content determination, and immediately snap-freeze the extracts in liquid nitrogen.
7. Store frozen extracts at –80°C (*see* **Note 13**). Nuclear extracts prepared in this manner should contain 1–1.5 μg/μL of protein.

3.4. Radiolabeling of Oligonucleotide Probe

1. End label annealed oligonucleotide probes with either T4 polynucleotide kinase or the Klenow fragment according to the supplier's instructions.
2. Proper labeling generates probes with at least 2×10^6 cpm/pmol probe (corrected for the reference date of the radiolabel).
3. A labeled probe has a useful life of about 4 wk, although using freshly labeled probe obviously reduces the autoradiography exposure times.

3.5. Binding Reaction

The basic protocol for nuclear extract binding to a labeled probe is described first. This is followed by variations in which competition binding or antibody binding is also involved.

1. For each sample, mix 2–5 μg of the nuclear extract proteins with 10 μL of Binding buffer by gentle flicking of the Eppendorf tubes (*see* **Note 14**).
2. Adjust the final volume to 15 μL.

3. Incubate the samples at room temperature for 10 min.
4. Add approx 20 fmol of ^{32}P-labeled oligonucleotide probe (30,000–50,000 cpm) in a volume of 2 μL, and incubate the reaction mixture for an additional 10 min at room temperature.
5. Immediately load the samples on native polyacrylamide gels (**Subheading 3.8.**).

3.6. Confirming Specificity

To confirm the specificity of the interaction between the DNA-binding complexes and the DNA sequence of interest, add increasing amounts of unlabeled oligonucleotide probe to the binding mix (typically, 10X, 25X, and 50X excess cold probe) to compete with the radiolabeled probe for binding to the transcription factor. Any specific binding should be displaced using 10X or 25X cold probe (as shown for neutrophil NF-κB DNA-binding activities in **Fig. 3**).

1. Allow the nuclear extracts to interact with unlabeled competitor probes for 10 min at room temperature in Binding buffer.
2. Add labeled probe, and incubate for another 10 min at room temperature. Load the samples on native polyacrylamide gels.
3. As a negative control, perform the same competition experiments using an oligonucleotide with a mutated binding site for the transcription factor of interest, which should only displace nonspecific complexes (as shown in **Fig. 3**).

3.7. Identification of DNA-Binding Complex Constituents

To directly identify the constituents of a specific DNA-binding complex, antibodies raised against potential candidates can be included in the binding mix. Binding of antibody to its target protein will increase the size of the DNA–protein complex, resulting in further retardation of its gel mobility or "supershift" as shown in **Figs. 4** and **5** for neutrophil NF-κB and STAT DNA-binding activities, respectively (*see* **Note 15**).

1. Allow the nuclear extracts to interact with one or more antibodies (about 2 μg each) for 20 min at room temperature in Binding buffer. As a negative control, use isotype-matched control antibodies at the same concentration (*see* **Note 15**).
2. Add labeled probe, and incubate for another 10 min at room temperature. Load the samples on native polyacrylamide gels.

3.8. Sample Analysis by Polyacrylamide Gel Electrophoresis and Autoradiography

1. Pre-run the native gels made in TBE or in an alternative buffer (*see* **Note 3**) for 90 min at 10 V/cm. This ensures a completely isocratic buffer system for optimal migration.
2. Load the samples immediately after the binding reaction. Also load 10 μL xylene cyanol (0.1% w/v in Binding buffer) in the first and last lanes of the gel as a tracking dye.

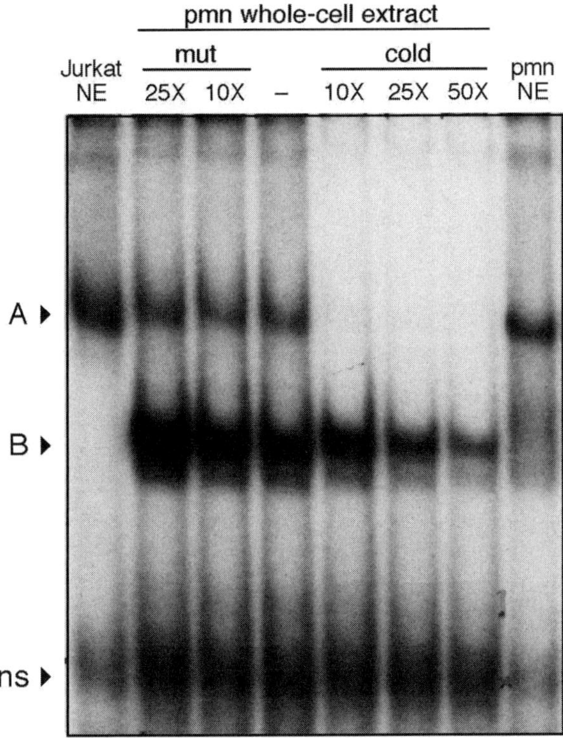

Fig. 3. Human neutrophils were stimulated for 10 min with 100 U/mL tumor necrosis factor (TNF)-α, and whole-cell extracts were prepared by supplementing raw cavitates with glycerol and NaCl (10% v/v and 400 m*M* final concentrations, respectively) and incubating on ice for 20 min, prior to centrifugation. Extracts were then incubated in Binding buffer in the absence ("−") or presence of increasing concentrations of either unlabeled nuclear factor (NF)-κB probe ("cold") or of a mutated NF-κB probe ("mut") in which the consensus sequence was changed to 5'-<u>AAT</u>ACTTTCC (the mutated nucleotides are underlined), prior to addition of labeled NF-κB probe and subsequent electrophoretic mobility shift assay analysis. For comparison, a nuclear extract from the same neutrophil preparation was loaded in the last lane ("pmn NE"), and a nuclear extract from TNF-activated Jurkat cells ("Jurkat NE") was loaded in the first lane as a positive control. "A" denotes the specific, inducible NF-κB complex; "B" denotes a constitutive complex showing some specificity which is present in whole-cell extracts (but mostly absent from the corresponding nuclear extracts); "ns" denotes a constitutive nonspecific complex.

3. Run the gels at 12–14 V/cm for an additional 2–3 h. The total run time depends on how far the tracking dye has migrated. Empirically determine what distance corresponds to the optimal migration of the transcription factor complexes under study).

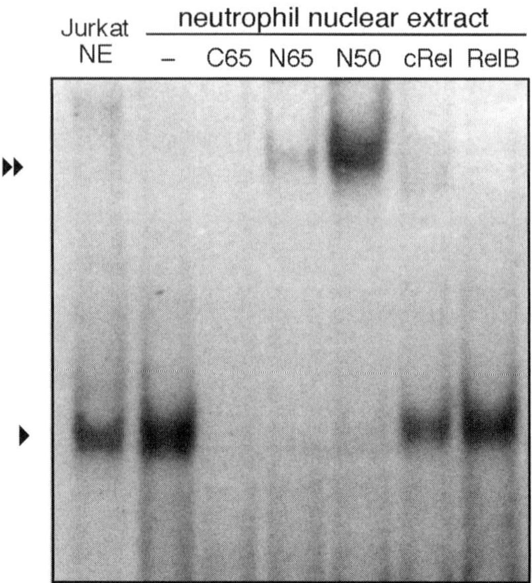

Fig. 4. Human neutrophils were stimulated for 10 min with 100 ng/mL lipopolysaccharide, and nuclear extracts were incubated in Binding buffer in the absence ("–") or presence of antibodies directed at a C-terminal sequence within RelA that abuts the Rel homology domain ("C65"), the N-terminus of RelA ("N65"), the N-terminus of p50 ("N50"), the C-terminus of c-Rel ("cRel"), or the N-terminus of RelB ("RelB"), prior to addition of labeled nuclear factor (NF)-κB probe and subsequent electrophoretic mobility shift assay analysis. A nuclear extract from TNF-activated Jurkat cells ("Jurkat NE") was loaded in the first lane as a positive control. The single arrowhead denotes the specific, inducible NF-κB complex; double arrowheads indicate the supershifted complexes.

4. Transfer the gels to Whatman DE-81 paper, and heat dry under vacuum in a gel dryer for 60 min, or until the gels are completely dry.
5. Expose the dried gels to autoradiographic film at −80°C with intensifying screens. Alternatively, gels can be exposed to PhosphorScreens (GE Healthcare) at RT prior to being read on a STORM 860 PhosphorImager (Molecular Devices).

4. Notes

1. Our Relaxation buffer formulation represents a modification of the original Relaxation buffer described by Borregaard *(17)*, which among other things contained too much salt, resulting in the extraction of nuclear proteins during cell disruption. Our formulation overcomes this limitation. We no longer include the protease inhibitors, phosphoramidon and bestatin, because they do not alter the degradation of NF-κB or STAT complexes by purified neutrophil granule proteases.
2. This formulation gives very consistent results with either NF-κB or STAT transcription factors. For other factors, various parameters may need to be changed,

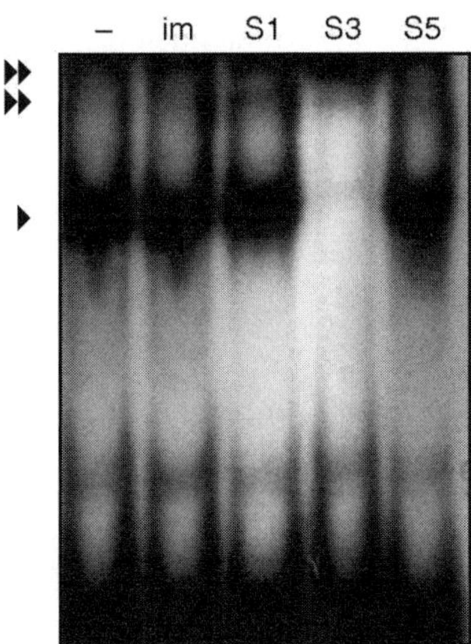

Fig. 5. Human neutrophils were stimulated for 10 min with 1000 U/mL granulocyte colony-stimulating factor, and nuclear extracts were incubated in Binding buffer in the absence ("–") or presence of antibodies directed at STAT1 ("S1"), STAT3 ("S3"), STAT5 ("S5"), or an isotype-matched control Ab ("im"), prior to the addition of labeled gamma response region (GRR) probe and subsequent electrophoretic mobility shift assay analysis. The single arrowhead denotes the specific, inducible GRR complex; double arrowheads indicate the supershifted complexes.

such as the amounts of DTT, acetylated BSA, and poly(dI-dC); the presence of $MgCl_2$; and so forth.

3. TGE buffer (50 mM Tris-HCl, pH 7.5, 380 mM glycine, 2 mM EDTA, pH 8.0) may be used instead of TBE. Again, use the same buffer for gel preparation and electrophoresis. Because of the high ionic strength of this buffer, the concentration of carrier DNA may be reduced.
4. The device we use for nitrogen cavitation is a Model 4635 Cell Disruption Bomb fitted with four outlets (*see* **Subheading 2.3.**). The length of each dip tube used for sample collection is about 12 cm, with a diameter of 1 mm. This translates into a void volume of about 0.5 mL, which makes it difficult to work with small volumes without incurring a significant loss of cavitated sample. A smaller cell disruption device (Model 4639) is also available from the same company, featuring nearly no void volume, but it does not allow custom fitting of additional discharge valves for the simultaneous fractionation of multiple samples.
5. Small magnetic stir bars ("fleas") can be bought from Sigma-Aldrich (cat #Z11, 884-2).
6. As shown in **Fig. 2**, neutrophils are efficiently lysed over a wide pressure range. However, it must be pointed out that although 450 psi achieves total cell lysis, the resulting nuclear extracts usually contain fewer DNA-binding transcription factor complexes. Thus, cavitating between 250 and 350 psi is recommended. **Figure 2**

also shows that a cavitation time of 10 min is sufficient to achieve total cell lysis; 5 min also yields good results, albeit with significant variability.

7. One can practice using distilled water instead of a cell suspension. The entire step should not last more than 10 min for the collection of four samples. It is also sometimes necessary to inject more nitrogen into the bomb, if the pressure goes down to 250 psi after collecting two to three samples.

8. No further purification of the nuclei is required, as the current protocol does not yield any unbroken cells that would otherwise co-sediment with the nuclei. Nuclei thus prepared are free from cytoplasmic contamination *(6,24)*.

9. The final NaCl concentration must not exceed 500–600 m*M*, as nuclear lysis is likely to occur.

10. We found that adding NaCl after the nuclear pellets have been resuspended (instead of including it in the Nuclear Extraction buffer, as with conventional protocols) improves the yield of extracted transcription factors significantly. By comparison, resuspension of nuclear pellets in the presence of high NaCl in the Nuclear Extraction buffer results in some degradation of the NF-κB and STAT proteins. This possibly reflects the impossibility of adding enough protease inhibitors to neutralize the large quantities of proteases released in a very small volume as the nuclear-bound granules break, presumably during resuspension of the nuclei.

11. We observed that our modified Relaxation buffer, unlike the original formulation, results in the association of about 10–15% of the granules with the nuclei (as determined by presence of myeloperoxidase and lactoferrin). These granules are still intact, as evidenced by the consistent lack of transcription factor degradation achieved using our protocol.

12. Nuclei should not be allowed to extract for longer times, as significant nuclear lysis already occurs at 30 min.

13. Simply placing the samples at −80°C results in a noticeable loss of activity, and this loss increases after each subsequent freeze–thaw cycle. By contrast, snap-freezing almost completely prevents this loss.

14. The volume of nuclear extract in the binding mix must not exceed 5 μL or the salt concentration will be too high, resulting in unpalatable smearing on the gels. If the samples are too dilute, they can be re-concentrated and desalted using spin columns with a molecular weigh cutoff of about 100,000, such as the Microcon-100 (Millipore).

15. Sometimes, the antibody used binds at, or near to, a transcription factor's DNA-binding domain. In this particular case, the labeled DNA probe can no longer interact with the resulting DNA-binding complex, and there will consequently be no observable supershift, but rather a disappearance of the band (as shown in **Fig. 4**, lane 3).

16. The signal can sometimes be a little stronger with the control antibody. This has been observed in many studies and is considered normal.

Acknowledgments

This work was supported by grants to P. P. McD. from the Canadian Institutes for Health Research and the Arthritis Foundation of Canada, and to R. D. Y.

from the National Institutes of Health. P. P. McD. is a Scholar from the Fonds de la recherche en santé du Québec (FRSQ).

References

1. Cassatella, M. A. (1999) Neutrophil-derived proteins: selling cytokines by the pound. *Adv. Immunol.* **73,** 369–509.
2. Scapini, P., Lapinet-Vera, J. A., Gasperini, S., Calzetti, F., Bazzoni, F., and Cassatella, M. A. (2000) The neutrophil as a cellular source of chemokines. *Immunol. Rev.* **177,** 195–203.
3. Ellis, T. N. and Beaman, B. L. (2004) Interferon-gamma activation of polymorphonuclear neutrophil function. *Immunology* **112,** 2–12.
4. Galligan, C. and Yoshimura, T. (2003) Phenotypic and functional changes of cytokine-activated neutrophils. *Chem. Immunol. Allergy* **83,** 24–44.
5. Cheng, S. S. and Kunkel, S. L. (2003) The evolving role of the neutrophil in chemokine networks. *Chem. Immunol. Allergy* **83,** 81–94.
6. McDonald, P. P., Bald, A., and Cassatella, M. A. (1997) Activation of the NF-κB pathway by inflammatory stimuli in human neutrophils. *Blood* **89,** 3421–3433.
7. Bovolenta, C., Gasperini, S., and Cassatella, M. A. (1996) Granulocyte colony-stimulating factor induces the binding of STAT1 and STAT3 to the IFNγ response region within the promoter of the FcγRI/CD64 gene in human neutrophils. *FEBS Lett.* **386,** 239–242.
8. McDonald, P. P., Bovolenta, C., and Cassatella, M. A. (1998) Activation of distinct transcription factors in neutrophils by bacterial LPS, interferon-γ, and GM-CSF and the necessity to overcome the action of endogenous proteases. *Biochemistry* **37,** 13,165–13,173.
9. McDonald, P. P. and Cassatella, M. A. (1997) Activation of transcription factor NF-κB by phagocytic stimuli in human neutrophils. *FEBS Lett.* **412,** 583–586.
10. Bovolenta, C., Gasperini, S., McDonald, P. P., and Cassatella, M. A. (1998) High affinity receptor for IgG (FcγRI/CD64) gene and STAT protein binding to the IFNγ response region (GRR) are regulated differentially in human neutrophils and monocytes by IL-10. *J. Immunol.* **160,** 911–919.
11. McDonald, P. P., Russo, M. P., Ferrini, S., and Cassatella, M. A. (1998) Interleukin-15 induces NF-κB activation and IL-8 production in human neutrophils. *Blood* **92,** 4828–4835.
12. Hayden, M. S. and Ghosh, S. (2004) Signaling to NF-κB. *Genes Dev* **18,** 2195–2224.
13. Schmitz, M. L., Mattioli, I., Buss, H., and Kracht, M. (2004) NF-κB: a multifaceted transcription factor regulated at several levels. *Chembiochem* **5,** 1348–1358.
14. Levy, D. E. and Darnell, J. E. Jr. (2002) STATs: transcriptional control and biological impact. *Nat. Rev. Mol. Cell. Biol.* **3,** 651–662.
15. Kisseleva, T., Bhattacharya, S., Braunstein, J., and Schindler, C. W. (2002) Signaling through the JAK/STAT pathway, recent advances and future challenges. *Gene* **285,** 1–24.

16. Dignam, J. D., Lebovitz, R. M., and Roeder, R. G. (1983) Accurate transcription initiation by RNA polymerase II in a soluble extract from isolated mammalian nuclei. *Nucleic Acids Res.* **11,** 1475–1489.

17. Borregaard, N., Heiple, J. M., Simons, E. R., and Clark, R. A. (1983) Subcellular localization of the b-cytochrome component of the human neutrophil microbicidal oxidase: translocation during activation. *J. Cell Biol.* **97,** 52–61.

18. Nachman, R., Hirsch, J. G., and Baggiolini, M. (1972) Studies on isolated membranes of azurophil and specific granules from rabbit polymorphonuclear leukocytes. *J. Cell Biol.* **54,** 133–140.

19. McDonald, P. P. (2004) Transcriptional regulation in neutrophils: Teaching old cells new tricks. *Adv. Immunol.* **82,** 1–48.

20. Cloutier, A., Ear, T., Borissevitch, O., Larivee, P., and McDonald, P. P. (2003) Inflammatory cytokine expression is independent of the c-Jun N-terminal kinase/AP-1 signaling cascade in human heutrophils. *J. Immunol.* **171,** 3751–3761.

21. Boyum, A. (1968) Isolation of mononuclear cells and granulocytes from human blood. *Scand J. Clin. Lab. Invest.* **Suppl. 97,** 77–89.

22. Bradford, M. M. (1976) A rapid and sensitive method for the quantitation of microgram quantities of protein utilizing the principle of protein-dye binding. *Anal. Biochem.* **72,** 248–254.

23. Epling-Burnette, P. K., Zhong, B., Bai, F., et al. (2001) Cooperative regulation of Mcl-1 by Janus kinase/STAT and phosphatidylinositol 3-kinase contribute to granulocyte-macrophage colony-stimulating factor-delayed apoptosis in human neutrophils. *J. Immunol.* **166,** 7486–7495.

24. Ear, T., Cloutier, A., and McDonald, P. P. (2005) Constitutive nuclear expression of the IκB kinase complex and its activation in human neutrophils. *J. Immunol.* **175,** 1834–1842.

VIII

Neutrophil Defects and Diagnosis

30

Disorders of Neutrophil Function

An Overview

Mary C. Dinauer

Summary

Primary disorders of neutrophil function result from impairment in neutrophil responses that are critical for host defense. This chapter summarizes inherited disorders of neutrophils that cause defects in neutrophil adhesion, migration, and oxidative killing. These include leukocyte adhesion deficiencies, actin defects, and other disorders of chemotaxis; hyperimmunoglobulin E syndrome; Chédiak-Higashi syndrome; neutrophil specific granule deficiency; chronic granulomatous disease; and myeloperoxidase deficiency. Diagnostic tests and treatment approaches are also summarized for each neutrophil disorder.

Key Words: *Aspergillus* species; Chédiak-Higashi syndrome; Chemotaxis; Chronic Granulomatous Disease; Hyperimmunoglobulin E; Leukocyte adhesion deficiency; Mac-1; Myeloperoxidase; NADPH oxidase; Neutrophil granule; *Staphylococcus aureus*.

1. Introduction

This chapter provides a brief overview of disorders of neutrophil function. Neutrophils play an essential role in the initial response to invading bacteria and fungi; thus, patients with defects in neutrophil function typically present in infancy or childhood with recurrent and/or difficult to treat bacterial infections *(1,2)*. Infections typically involve the skin, mucosa, gums, lung, or draining lymph nodes, or cause deep tissue abscesses. The microorganisms causing these infections are often unusual or opportunistic pathogens. Many of the disorders have characteristic clinical and microbiological features that are related to the specific nature of the defect in neutrophil function. Note that congenital disorders affecting neutrophil function represent at most 20% of reported primary immune deficiencies. Thus, patients with suspected disorders of host defense should also be screened for defects in antibody production, T-cell function,

From: *Methods in Molecular Biology, vol. 412: Neutrophil Methods and Protocols*
Edited by: M. T. Quinn, F. R. DeLeo, and G. M. Bokoch © Humana Press Inc., Totowa, NJ

and the complement system. **Figure 1** summarizes key steps in the response of neutrophils to invading microbes, which include initial adhesion to the endothelium, subsequent migration toward a site of infection, ingestion of the pathogen, and pathogen killing via oxidative metabolites, proteases, and other toxic peptides in neutrophil granules. Disorders of neutrophil function can affect one or more of these pathways.

In the diagnostic work-up of a patient with bacterial and fungal infections, a clinician must decide whether a complete evaluation of neutrophil function is warranted. Four key aspects of the patient history with infections should be considered in reaching this decision; namely, frequency, severity, and location of infections and the identity of the causal infectious agent. The presence of unusual features should trigger consideration of a neutrophil disorder. The patient's age and other medical conditions should also be taken into account. For example, one might suspect neutrophil dysfunction if an infectious agent that commonly affects children is observed in an older patient. Distinctive clinical findings can provide helpful guidelines in determining which patients merit further testing, and which tests are appropriate. A family history can also provide useful clues.

2. Disorders of Adhesion and Chemotaxis

The ability of neutrophils to adhere to the endothelium, tissue matrix, and to invading microbes is essential for their migration from the bloodstream to sites of infection, where they eliminate pathogens. Interactions between cell surface glycoproteins expressed on neutrophils and endothelial cells are fundamental to this process. The initial steps of adhesion require expression of E-selectins on the surface of endothelial cells, which bind to fucosylated proteins on leukocytes. The next step involves induction and activation of cell surface integrins on leukocytes, which mediate tight adhesion between neutrophils and endothelial cells. Subsequent migration from capillaries into tissues uses additional cell-surface receptors. Defects in these interactions and/or chemotaxis result in impaired recruitment of neutrophils into sites of infection or inflammation, often with relative neutrophilia in peripheral blood yet poor formation of pus.

2.1. Leukocyte Adhesion Deficiency Type I

Leukocyte adhesion deficiency (LAD) I is an autosomal recessive disorder characterized by deficiency of three leukocyte glycoproteins that are members of the integrin superfamily of cell surface adhesion molecules (**Table 1**) (*1,3, 4*). This disorder results from genetic defects in CD18, the common chain of the β_2 integrin family, which is required for stable expression of three distinct β_2 integrins; CD11a/CD18 (lymphocyte function antigen [LFA]-1), CD11b/CD18 (Mac-1), and CD11c/CD18 (p150,95). In LAD I, mutations in the common

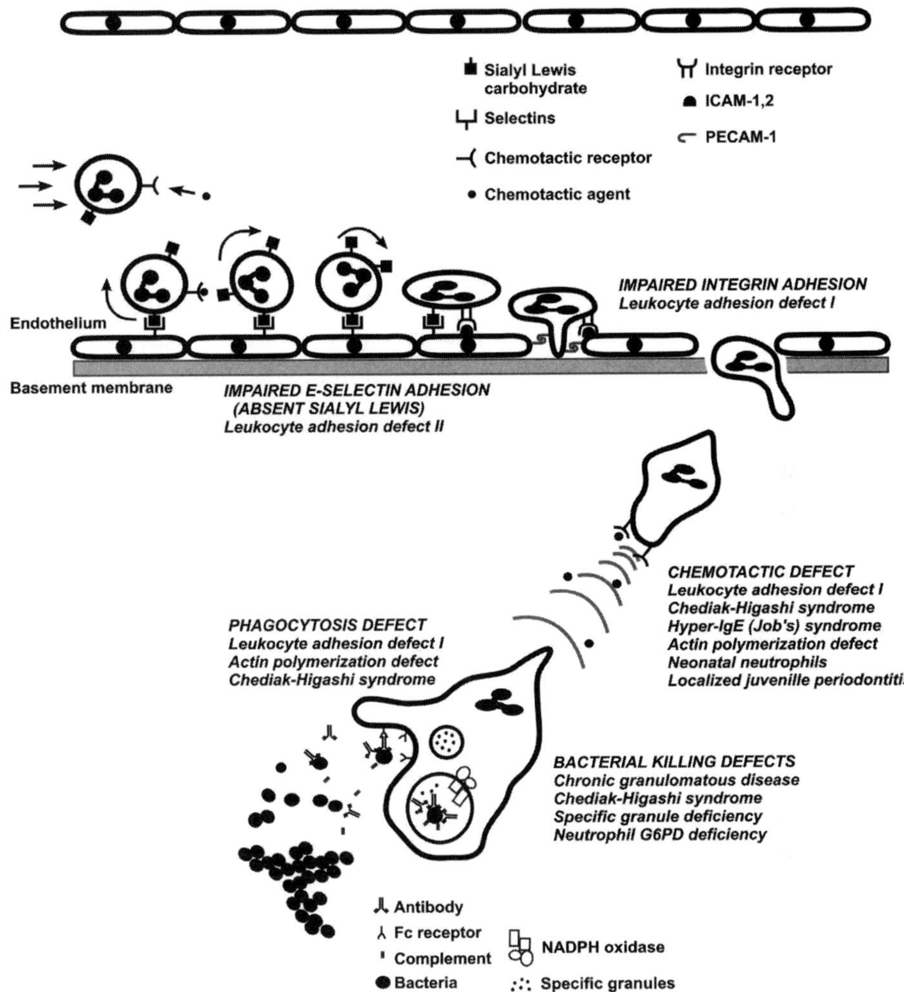

Fig. 1. Steps in the response of circulating neutrophils to infection or inflammation. The adhesion molecule E-selection is upregulated on endothelial cells in response to inflammatory mediators (interleukin [IL]-1, endotoxin, and tumor necrosis factor [TNF]-α). Neutrophils interact with E-selectins on endothelial cells through sialyl Lewis carbohydrates, resulting in rolling attachment and margination. Chemoattractants, such as IL-8 released by endothelial cells, upregulate neutrophil β2 integrins, which in turn mediate tight adhesion to intercellular adhesion molecule-1 and platelet-endothelial cell adhesion molecule-1 on endothelial cells. Activated neutrophils detect small changes in the chemoattractant gradient, which causes them to move toward the site of infection. Neutrophils phagocytose bacteria opsonized by antibody and complement. Both oxidative and nonoxidative antimicrobial mechanisms mediate bacterial killing. Disorders of phagocyte function associated with each of these steps are listed. (Reprinted from **ref. *1***, with permission from Elsevier.)

CD18 chain typically eliminate expression of these three leukocyte glycoprotein complexes. β_2 integrins interact with intercellular adhesion molecule (ICAM)-1 proteins expressed on endothelial cells, which are upregulated in response to inflammatory cytokines. Thus, neutrophil adhesion to the endothelium is severely defective in LAD I. In addition, Mac-1 is the major receptor for the opsonic complement fragment C3bi, an important trigger for phagocytosis of complement-opsonized microbes. Mac-1 also mediates binding to fibrinogen. Finally, binding to Mac-1 provides an important co-stimulatory signal for other pathways important for adhesion, degranulation, and activation of reactive oxidant production *(5,6)*. Because of the multiple defects in adhesion-related functions, LAD I patients develop recurrent bacterial and fungal infections, typically with *Staphylococcus aureus* or Gram-negative enteric microbes. Characteristic clinical features include frequent skin and periodontal infections, delayed separation of the umbilical cord and omphalitis, and deep tissue abscesses. Neutrophilia with paucity of neutrophils at inflamed or infected sites is characteristic.

Mutations in the CD18 gene in patients with LAD I are heterogeneous. Typically, cell surface expression of the CD11/CD18 heterodimers is absent, although variants with low-level expression have been described; the latter phenotype is associated with reduced severity of disease.

Diagnosis of LAD I is straightforward, and can be made by flow cytometry of peripheral blood leukocytes, most often with a monoclonal antibody directed to the CD11b/CD18 heterodimer, Mac-1. Patients with LAD I also have diminished neutrophil migration in vivo. This can be studied with the Rebuck skin window; however, this test is no longer used for diagnostic purposes, because of the availability of flow cytometric methods.

Because of the severity of infectious complications, bone marrow transplantation is generally recommended for patients with severe LAD I. Patients with partial expression of LAD I have a longer life expectancy without bone marrow transplantation, but require good supportive care including prophylactic antibiotics and scrupulous dental hygiene. Nevertheless, morbidity from periodontal disease, bacterial infection, and delayed healing are still problematic.

2.2. Leukocyte Adhesion Deficiency Types II and III

Several unusual variants of leukocyte adhesion deficiency associated with distinct clinical and genetic defects have been described in a small number of patients (**Table 1**) *(3,4)*. LAD II, caused by mutations in the membrane transporter for fucose, is associated with loss of expression of fucosylated glycans on the cell surface. This disorder is also known as congenital disorder of glycosylation (CDG)-IIc. Fucosylated proteins, such as sialyl-Lewis X (CD15s), are ligands for endothelial selectins and are important for the early phases of adhesion to endothelial cells. Patients with LAD II also have leukocytosis and form

Table 1
Leukocyte Adhesion Deficiency

	Genetic defect	Clinical presentation	Diagnosis
LAD I	CD18 subunit of β_2 integrins, resulting in impaired adhesion, chemotaxis	Skin infection, soft tissue abscesses, delayed separation of umbilical cord and omphalitis, periodontal disease	Flow cytometry for CD11b/CD18 (Mac1)
LAD II	GDP-fucose transporter 1, resulting in impaired expression of fucosylated proteins, including SLeX ligand for selectins	Similar infections to LAD I but not as severe; developmental delay, short stature	Flow cytometry for leukocyte CD15s (SLeX) Bombay (hh) phenotype in red blood cell typing
LAD III	Integrin activation defect, resulting in impaired leukocyte and platelet adhesion	Similar to LAD I; also bleeding tendency	Functional assays for neutrophil and platelet adhesion

LAD, leukocyte adhesion deficiency.

pus poorly, although infections tend to be less severe in LAD II patients compared to LAD I patients. In addition, LAD II patients have severe mental retardation and short stature, and the absence of fucosylated proteins on the surface of red blood cells results in the Bombay (hh) red cell phenotype. LAD II is readily diagnosed by flow cytometry for expression of leukocyte CD15s. Prolonged therapy with oral fucose has been beneficial in some but not all LAD II patients.

LAD III has been described in four patients *(3)*. LAD III is characterized by defects in cell signaling that interfere with activation of multiple classes of integrins downstream of G protein-coupled receptors, and appears to be an autosomal recessive disorder. The phenotype of this disorder is similar to LAD I, but it can also be associated with defects in integrin-mediated platelet aggregation and excessive bleeding *(7)*. One LAD III patient is reported to have a defect in activation of Rap1 GTPase *(8)*. The diagnosis of LAD III should be considered in patients with the clinical features of LAD I, but normal expression of β_2 integrin. Comprehensive evaluation of such patients requires specialized functional assays that are usually carried out in research laboratories.

2.3. Defect in Adhesion and Chemotaxis Due to Mutation in Rac2 GTPase

A dominant-negative mutation in the Rac2 GTPase was described in an infant with impaired neutrophil adhesion and motility, along with decreased neutrophil glucose-6-phosphate dehydrogenase (NADPH) oxidase activation and degranulation in response to chemoattractants *(9,10)*. Symptoms similar to LAD I developed in the first months of life. Leukocyte β_2 integrin expression was normal, and the phenotype presumably results from interference by the mutant Rac2 in Rac-regulated signaling pathways. This patient successfully underwent bone marrow transplantation. A similar patient with severe, recurrent skin and tissue abscesses, normal leukocyte β_2 integrin expression, and similar neutrophil function defects in response to chemoattractants died in infancy without determination of the genetic defect *(11)*.

2.4. Other Genetic Disorders of Chemotaxis

Several rare human diseases characterized primarily by prominent defects in neutrophil chemotaxis have been described. In general, the neutrophils in these patients exhibit normal adhesive properties. Patients typically present with recurrent, severe skin and soft tissue infections in infancy. One patient exhibited markedly abnormal polymerization of neutrophil actin and higher than normal expression of an actin binding protein *(12,13)*. A young female patient is reported to have a heterozygous point mutation in β-actin and impaired binding to actin-regulatory proteins *(14)*. The phenotype of this patient includes recurrent infections, mental retardation, and photosensitivity.

Localized Juvenile Periodontitis (LJP) is a rare disorder of unknown etiology that presents in children and adolescents with chronic and recurrent periodontal infections and severe alveolar bone loss *(15,16)*. This disorder is heterogeneous and is associated with defective neutrophil chemotaxis in vitro. Some cases are sporadic, but others appear to be familial and thus linked to genetic defects. Because periodontal disease is a frequent symptom of neutropenia or of neutrophil function disorders, these diagnostic categories should also be considered in teenagers who present with periodontitis. LJP is diagnosed by its characteristic clinical features, the absence of nonperiodontal infections, normal expression of β_2 integrins, normal oxidative burst, and normal neutrophil morphology.

Hyperimmunoglobulin (hyper-Ig)E syndrome is characterized by elevated serum levels of IgE (i.e., $\geq$200 IU-mL) and recurrent *Staphylococcal* infections of the skin and lung, pneumatoceles, chronic dermatitis, and skeletal and dental abnormalities *(17,18)*. Localized *Staphylococcal* infections are indolent and lack the usual characteristics of inflammation ("cold" abscesses). Chronic candidiasis of the mucosal surface and nail beds, hyperextensible joints, and scoliosis are also frequently observed, and delay or failure to shed the primary teeth is not uncommon. Many patients have coarse facial features by early adolescence. The disease is rare and the inheritance is variable, with both sporadic and familiar forms reported, and autosomal dominant transmission has been documented in some families. The molecular basis of hyper-IgE syndrome is unknown. High serum IgE levels may reflect a T-leukocyte imbalance, which causes abnormally high production of IgE and abnormally low production of interferon (IFN)-γ and tumor necrosis factor. Patients with this syndrome have variable defects in neutrophil chemotaxis, which are independent of fluctuations in serum IgE. The relationship between the immunological abnormalities, chemotactic defects, and the other features of this disorder is not understood. The hallmark laboratory finding in hyper-IgE syndrome is the marked elevation of serum IgE, especially in children or young adults. Antibiotic prophylaxis for Staphylococcal infection is beneficial in patients with hyper-IgE syndrome. Atopic dermatitis is the major differential diagnosis, as it can also be associated with eosinophilia and elevated serum IgE.

2.5. Noninherited Disorders of Neutrophil Chemotaxis

A chemotactic abnormality, detected by in vitro assay, has been reported in neonatal neutrophils *(19,20)*. This abnormality may be due in part to defects in cellular adhesion and impaired mobilization and activation of intracellular adhesion proteins at the cell surface. Reduced neutrophil chemotaxis has also been described in burn victims and in patients with bacterial sepsis or diabetes. However, the chemotactic defects in these settings are variable, and their contribution to associated infectious complications is uncertain.

3. Disorders of Ingestion and Degranulation

Following phagocytosis, neutrophil granules fuse with phagosomal membranes and release proteases, enzymes, and antibacterial proteins into the phagosomal lumen. This process greatly facilitates microbial killing. In LAD I patients, C3bi-obsonized microbes fail to be ingested, because CD11b/CD18 (Mac-1), the receptor for C3bi, is not expressed. This receptor also provides an important co-stimulatory signal for Fcγ receptor-mediated phagocytosis. Deficiency of opsonization with complement is seen in serum complement deficiencies, and primary B-cell deficiencies, such as Bruton's agammaglobulinemia, result in defective opsonization with specific immunoglobulins. Thus, these patients can present with recurrent infections with pyogenic bacteria, such as *S. aureus*, *Pneumococci*, and *Haemophilus influenzae*. In contrast, primary neutrophil defects causing disorders of ingestion and/or degranulation are very rare. Two such genetic disorders affecting neutrophil granules have been described (*see* **Subheadings 3.1.** and **3.2.**), both of which are associated with increased frequency of bacterial infections.

3.1. Chédiak-Higashi Syndrome

Chédiak-Higashi syndrome is a rare autosomal recessive disorder associated with widespread defects in granule morphogenesis and multiorgan pathology, including ineffective granulopoiesis, moderate neutropenia, and delayed and incomplete degranulation *(21,22)*. Neutrophils from patients with Chédiak-Higashi syndrome have giant granules that appear to be a coalescence of azurophilic and specific granules. The giant granules are often more prominent in bone marrow neutrophils than in peripheral blood neutrophils. Giant granules are also seen in lymphocytes and natural killer cells from patients with Chédiak-Higashi syndrome. Patients with Chédiak-Higashi syndrome experience frequent *S. aureus* infections in lung and skin, and develop gingivitis and periodontitis. Patients also have partial oculocutaneous albinism, a mild tendency to bleed, and neuropathies. These manifestations are all related to abnormal granule morphogenesis. Chédiak-Higashi syndrome is caused by mutations in *CHS1*, which encodes a large protein thought to regulate lysosomal and granule trafficking. Supportive care with prophylactic antibiotics is a mainstay of treatment. A frequent complication of the disease is the so-called "accelerated phase," during which patients experience fever, lymphadenopathy, progressive pancytopenia, and lymphohistiocytic infiltration. This phenomenon may be related to granule defects in natural killer cells and lymphocytes and precipitated by infection with Epstein-Barr Virus.

3.2. Neutrophil-Specific Granule Deficiency

Neutrophil-specific granule deficiency (SGD) is a very rare disorder characterized by the absence of specific or secondary granules in developing neutrophils *(23,24)*. Neutrophils from SGD patients also typically have morphologically abnormal bilobed nuclei. SGD neutrophils are markedly deficient in many important microbicidal granule proteins, including lactoferrin and the defensins. SGD neutrophils also demonstrate relatively severe chemotactic defects, which are thought to result from a decrease in the pool of intracellular leukocyte adhesion molecules normally mobilized to the cell surface in response to inflammatory stimuli. SGD patients present with recurrent and difficult to treat bacterial and fungal infections, primarily involving the skin and lungs. *S. aureus*, enteric Gram-negative bacteria, *Pseudomonas aeruginosa*, and *Candida albicans* are the major pathogens. This disorder appears to be inherited in an autosomal recessive manor. The molecular defect responsible for some cases of SGD involves the myeloid transcription factor C/EBPε *(25,26)*. C/EBPε plays an important role as a gene-specific transcriptional regulator during late promyelocyte and early myelocyte development. The diagnosis of SGD is made if microscopic examination of neutrophils reveals the absence of specific granules. The diagnosis can be confirmed by assessing expression of granule-specific proteins (i.e., lactoferrin or gelatinase) by staining or other assays. Treatment of SGD is supportive, with prophylactic antibiotics, and prompt and prolonged treatment of infections.

4. Disorders of Oxidative Metabolism

Reactive oxygen species (ROS) play an important role in killing microbial pathogens. The major source of ROS is the phagocyte respiratory burst pathway, which is activated in response to phagocytosis or soluble inflammatory stimuli. The initial reaction in this pathway is catalyzed by an NADPH oxidase, which is found in plasma and phagosome membranes of neutrophils, monocytes/macrophages, and eosinophils. NADPH oxidase produces the superoxide free radical by catalyzing the transfer of electrons from NADPH to molecular oxygen. Superoxide is converted to hydrogen peroxide, and, in the presence of neutrophil myeloperoxidase, HOCl, in addition to numerous other microbicidal oxidants that synergize with granule proteins to kill microbes in the neutrophil phagosome. Thus, patients lacking key steps in the respiratory burst pathway are deficient in microbial killing and develop recurrent and severe infections. These infections typically involve microorganisms for which oxidant-mediated killing is particularly critical for effective host defense.

4.1. Chronic Granulomatous Disease

Chronic granulomatous disease (CGD) is a group of inherited disorders caused by defects in the phagocyte NADPH oxidase complex. Genetic defects in four critical subunits of NADPH oxidase cause CGD, and these mutations lead to complete or, less frequently, partial loss of NADPH oxidase activity *(1,27)*.

The estimated incidence of CGD is approx 1 in 200,000 live births, and it is the most common clinically significant inherited disorder of neutrophil function. Because respiratory burst oxidants are an important component of microbial killing mechanisms, patients with CGD suffer recurrent, often life-threatening bacterial and fungal infections. An additional and distinctive hallmark of CGD is the formation of inflammatory granulomas.

Four polypeptide subunits are essential for NADPH oxidase function, and mutations in the corresponding genes are responsible for four genetic subgroups of CGD (**Table 2**). These NADPH subunits are referred to by their apparent molecular mass (kDa) and the designation *phox* (phagocyte oxidase). Approximately 70% of CGD cases result from defects in the X-linked gene encoding the gp91^{phox} subunit of flavocytochrome b_{558}, a membrane heterodimer that is the redox center of NADPH oxidase. This subunit contains both flavoprotein and heme-binding domains. An uncommon autosomal recessive form of CGD is associated with mutations in the gene encoding p22^{phox}, which mediates translocation of two regulatory subunits of NADPH oxidase, p47^{phox} and p67^{phox}. Mutations in genes encoding p47^{phox} and p67^{phox} are affected in two other autosomal recessive subgroups of CGD. p47^{phox} binds p22^{phox} and acts as an adapter protein for recruitment and positioning of the p67^{phox} subunit, which contains an activation domain that stimulates electron transport through flavocytochrome *b*. Mutations in the genes encoding the flavocytochrome *b* subunits and p67^{phox} are heterogenous and cause deficient expression of the affected subunit. In patients with p47^{phox} mutations, a deletion of a GT dinucleotide at the start of Exon 2 in the p47^{phox} gene is frequently seen, which results from recombination with a closely-linked pseudogene. This deletion causes a frameshift and premature stop codon that terminates translation of the p47^{phox} mRNA transcript *(28)*.

Although not associated with cases of CGD, NADPH oxidase activity also requires activated Rac, which binds to and activates p67^{phox}. p40^{phox} is a third *phox* subunit and interacts with p67^{phox}, which may play a role in regulating oxidant production during phagocytosis. As described previously, a dominant-negative mutation in the Rac2 GTPase has been reported, which inhibits NADPH oxidase activity in response to certain agonists.

Patients with CGD typically present in infancy or early childhood with recurrent infections involving the skin, lung and draining lymph nodes; liver abscesses and osteomyelitis also develop as a result of hematogenous spread of microorganisms *(1)*. *S. aureus* is the most common microbe involved, but patients with CGD are also prone to infection by opportunistic pathogens such as *Burkholderia cepacia*, *Serratia marcescens*, *Aspergillus* species, and *Nocardia*. Many of these organisms express catalase, which removes microbe-generated hydrogen peroxide and thus inhibits ROS-mediated killing in the phagosomes of CGD cells. The lymph nodes, liver, and spleen of patients with CGD can become

Table 2
Genetic Defects in Neutrophil Glucose-6-Phosphate Dehydrogenase Oxidase Subunits in Chronic Granulomatous Disease (CGD)

Subunit	Inheritance	Gene locus	Function	Mutations	Frequency in CGD
gp91^{phox}	X	Xp21.1	Integral membrane glycoprotein; contains flavoprotein domain and heme groups for electron transport	Heterogeneous, most with absent flavocytochrome b_{558}	approx 70%
gp22^{phox}	AR	16p24	Integral membrane protein, flavocytochrome subunit; contains docking site for p47^{phox}	Heterogeneous, most with absent flavocytochrome b_{558}	approx 5%
p47^{phox}	AR	7q11.23	Cytosolic protein, activated by phosphorylation, mediates translocation of p67^{phox} to flavocytochrome b_{558}	Majority with GT deletion in Exon 2; absent expression of p47^{phox}	approx 25%
p67^{phox}	AR	1q25	Cytosolic protein, activates electron transport after translocation	Heterogeneous, most with absent expression of p67^{phox}	approx 5%

AR, autosomal recessive.

enlarged. Patients with CGD can also develop chronic inflammatory granulomas, which are a hallmark of this subgroup of neutrophil disorders. Chronic inflammatory granulomas can obstruct the gastric outlet or urinary tract or cause granulomatous colitis and the symptoms of inflammatory bowel disease. These chronic inflammatory complications are thought to reflect an important role for superoxide in downregulating the inflammatory response. Patients with X-linked CGD or autosomal recessive CGD involving the p22^{phox} or p67^{phox} subunits have a more severe clinical course than individuals with p47^{phox}-deficient CGD. In the latter subgroup, very low levels of superoxide activity are often still detectable *(29)*, which may explain the typically milder clinical course. Polymorphisms in oxygen-independent antimicrobial defense mechanisms or other components of the innate immune response are also thought to modify the severity of CGD in some patients *(30)*.

The diagnosis of CGD is made by measuring respiratory burst activity in neutrophils. The nitroblue tetrazolium (NBT) test using peripheral blood neutrophils is the classic method, in which the soluble yellow dye NBT is reduced by activated neutrophils, forming insoluble dark purple formazan deposits. However, a flow cytometry assay, based on oxidation of dihydrorhodamine 123 (DHR) by activated neutrophils, is increasingly replacing the NBT test for diagnosis in clinical laboratories *(29,31)*. Both the NBT and DHR tests identify two populations of neutrophils in female carriers of X-linked CGD. Chemiluminescence-based assays for respiratory burst activity are also occasionally used for clinical diagnosis. Family history and carrier testing will frequently discriminate between X-linked and autosomal recessive forms of CGD. However, approx 10% of patients with X-linked CGD carry new mutations. Molecular tests for defects in the gene encoding gp91^{phox} and for p47^{phox} are commercially available. For other forms of CGD, genetic diagnosis is generally based on analysis of protein expression, which is carried out in research laboratories.

All forms of CGD are treated similarly. Prophylactic trimethoprim-sulfamethoxazole (or dicloxacillin in sulfa-allergic patients) as antibiotic prophylaxis is a mainstay of treatment. Recent studies show that daily itraconazole reduces the incidence of *Aspergillus* infections, a serious complication of CGD *(32)*. Thus, prophylactic itraconazole is now recommended for routine use. Prophylactic IFN-γ, which is thought to enhance nonoxidative phagocyte functions, is also recommended for routine use, based on reports that it substantially decreases infectious complications in patients with CGD *(33,34)*. With optimal medical management, including aggressive treatment of acute infections, the mortality for CGD is now estimated to be approx 2% per patient per year *(34)*. Bone marrow transplantation has also been used with success to treat CGD *(35,36)*. However, use of this procedure is limited, because a human leukocyte antigen

(HLA)-matched donor must be found, and it is associated with the risks of immune suppression, graft vs host disease, and other transplant-associated complications. Gene replacement therapy targeted at hematopoietic stem cells is currently under development.

4.2. Myeloperoxidase Deficiency

Myeloperoxidase (MPO) deficiency is the most common inherited disorder of phagocytes, with complete deficiency occurring in approx 1 in 4000 individuals *(37)*. However, MPO deficiency is rarely associated with clinical symptoms. MPO is inherited in an autosomal recessive manner, and is caused by mutations in the MPO gene on chromosome 17. Mutations in the MPO gene are frequently point mutations that cause defective posttranslational processing of the MPO precursor protein. Deficiency in MPO inhibits formation of hypochlorous acid from chloride and hydrogen peroxide. There is a remarkable lack of clinical symptoms in most individuals with MPO deficiency, despite in vitro defects in the ability to kill *C. albicans* and *Aspergillus fumigatus* hyphae. Bacterial killing in vitro is also slower than normal. Nevertheless, patients with MPO deficiency rarely develop symptoms unless they also suffer from diabetes mellitus, which leads to disseminated candidiases and other fungal infections. Diagnosis of MPO deficiency is straightforward using histochemical analysis of peroxidase activity. Because patients with MPO deficiency are typically asymptomatic, no prophylactic antibiotics are prescribed. Individuals with MPO deficiency and diabetes are usually treated aggressively to prevent fungal infections.

4.3. Neutrophil Glucose-6-Phosphate Dehydrogenase Deficiency

NADPH, the primary substrate for the superoxide-generating NADPH oxidase, is generated by the first two reactions of the hexose monophosphate shunt pathway, catalyzed by glucose-6-phosphate dehydrogenase (G6PD) and 6-phosphogluconate dehydrogenase (6PGD). The leukocyte and erythrocyte G6PD are encoded by the same gene. However, the only symptom of vast majority of individuals with inherited G6PD deficiency is red cell hemolysis triggered by oxidative stress *(38)*. Most G6PD mutations cause a gradual decay in G6PD, which has little or no impact in short-lived neutrophils; in contrast, longer-lived erythrocytes are severely affected by loss of G6PD activity. Even most of the more unstable G6PD variants that cause chronic nonspherocytic hemolytic anemia even in the absence of oxidative stress do not result in critically low levels of NADPH in neutrophils. A few rare G6PD mutations lead to extremely low levels of G6PD in both neutrophils and erythrocytes, resulting in chronic, severe hemolytic anemia and CGD-like symptoms; these combined features clearly distinguish this very rare form of NADPH oxidase deficiency from CGD.

References

1. Dinauer, M. C. and Coates, T. D. (2005) Disorders of phagocyte function and number, in *Hematology: Basic Principles and Practices*, (Hoffman, R., Benz, J., Shattil, S., et al., eds.), Elsevier Churchill Livingstone, Philadelphia, pp. 787–830.
2. Lekstrom-Himes, J. A. and Gallin, J. I. (2000) Immunodeficiency diseases caused by defects in phagocytes. *N. Engl. J. Med.* **343,** 1703–1714.
3. Etzioni, A. and Alon, R. (2004) Leukocyte adhesion deficiency III: a group of integrin activation defects in hematopoietic lineage cells. *Curr. Opin. Allergy Clin. Immunol.* **4,** 485–490.
4. Bunting, M., Harris, E. S., McIntyre, T. M., Prescott, S. M., and Zimmerman, G. A. (2002) Leukocyte adhesion deficiency syndromes: adhesion and tethering defects involving beta 2 integrins and selectin ligands. *Curr. Opin. Hematol.* **9,** 30–35.
5. Brown, E. (1997) Neutrophil adhesion and the therapy of inflammation. *Semin. Hematol.* **34,** 319–326.
6. Lowell, C. A. and Berton, G. (1999) Integrin signal transduction in myeloid leukocytes. *J. Leukoc. Biol.* **65,** 313–320.
7. McDowall, A., Inwald, D., Leitinger, B., et al. (2003) A novel form of integrin dysfunction involving beta1, beta2, and beta3 integrins. *J. Clin. Invest.* **111,** 51–60.
8. Kinashi, T., Aker, M., Sokolovsky-Eisenberg, M., et al. (2004) LAD-III, a leukocyte adhesion deficiency syndrome associated with defective Rap1 activation and impaired stabilization of integrin bonds. *Blood* **103,** 1033–1036.
9. Ambruso, D. R., Knall, C., Abell, A. N., et al. (2000) Human neutrophil immunodeficiency syndrome is associated with an inhibitory Rac2 mutation. *Proc. Natl. Acad. Sci. USA* **97,** 4654–4659.
10. Williams, D. A., Tao, W., Yang, F., et al. (2000) A dominant negative mutation of the hematopoietic-specific RhoGTPase, Rac 2, is associated with a human phagocyte immunodeficiency. *Blood* **96,** 1646–1654.
11. Roos, D., Kuijpers, T. W., Mascart-Lemone, F., et al. (1993) A novel syndrome of severe neutrophil dysfunction: unresponsiveness confined to chemotaxim-induced functions. *Blood* **81,** 2735–2743.
12. Coates, T. D., Torkildson, J. C., Torres, M., Church, J. A., and Howard, T. H. (1991) An inherited defect of neutrophil motility and microfilamentous cytoskeleton associated with abnormalities in 47-Kd and 89-Kd proteins. *Blood* **78,** 1338–1346.
13. Howard, T., Li, Y., Torres, M., Guerrero, A., and Coates, T. (1994) The 47-kD protein increased in neutrophil actin dysfunction with 47- and 89-kD protein abnormalities is lymphocyte-specific protein. *Blood* **83,** 231–241.
14. Nunoi, H., Yamazaki, T., Tsuchiya, H., et al. (1999) A heterozygous mutation of beta-actin associated with neutrophil dysfunction and recurrent infection. *Proc. Natl. Acad. Sci. USA* **96,** 8693–8698.
15. Van Dyke, T. E. and Vaikuntam, J. (1994) Neutrophil function and dysfunction in periodontal disease. *Curr. Opin. Periodontol.* 19–27.
16. De Nardin, E. (1996) The molecular basis for neutrophil dysfunction in early-onset periodontitis. *J. Periodontol.* **67,** 345–354.

17. Grimbacher, B., Holland, S. M., Gallin, J. I., et al. (1999) Hyper-IgE syndrome with recurrent infections—an autosomal dominant multisystem disorder [see comments]. *N. Engl. J. Med.* **340,** 692–702.
18. Grimbacher, B., Holland, S. M., and Puck, J. M. (2005) Hyper-IgE syndromes. *Immunol. Rev.* **203,** 244–250.
19. Koenig, J. M. and Yoder, M. C. (2004) Neonatal neutrophils: the good, the bad, and the ugly. *Clin. Perinatol.* **31,** 39–51.
20. Hill, H. R. (1987) Biochemical, structural, and functional abnormalities of polymorphonuclear leukocytes in the neonate. *Pediatr. Res.* **22,** 375–382.
21. Huizing, M., Anikster, Y., and Gahl, W. A. (2001) Hermansky-Pudlak syndrome and Chediak-Higashi syndrome: disorders of vesicle formation and trafficking. *Thromb. Haemost.* **86,** 233–245.
22. Ward, D. M., Shiflett, S. L., and Kaplan, J. (2002) Chediak-Higashi syndrome: a clinical and molecular view of a rare lysosomal storage disorder. *Curr. Mol. Med.* **2,** 469–477.
23. Boxer, L. A. and Smolen, J. E. (1988) Neutrophil granule constituents and their release in health and disease. *Hematol. Oncol. Clin. North Am.* **2,** 101–135.
24. Gallin, J. I. (1985) Neutrophil specific granule deficiency. *Annu. Rev. Med.* **36,** 263–274.
25. Lekstrom-Himes, J. A., Dorman, S. E., Kopar, P., Holland, S. M., and Gallin, J. I. (1999) Neutrophil-specific granule deficiency results from a novel mutation with loss of function of the transcription factor CCAAT/enhancer binding protein epsilon. *J. Exp. Med.* **189,** 1847–1852.
26. Gombart, A. F. and Koeffler, H. P. (2002) Neutrophil specific granule deficiency and mutations in the gene encoding transcription factor C/EBP(epsilon). *Curr. Opin. Hematol.* **9,** 36–42.
27. Heyworth, P. G., Cross, A. R., and Curnutte, J. T. (2003) Chronic granulomatous disease. *Curr. Opin. Immunol.* **15,** 578–584.
28. Roesler, J., Curnutte, J. T., Rae, J., et al. (2000) Recombination events between the p47^{phox} gene and its highly homologous pseudogenes are the main cause of autosomal recessive chronic granulomatous disease. *Blood* **95,** 2150–2156.
29. Vowells, S. J., Fleisher, T. A., Sekhsaria, S., Alling, D. W., Maguire, T. E., and Malech, H. L. (1996) Genotype-dependent variability in flow cytometric evaluation of reduced nicotinamide adenine dinucleotide phosphate oxidase function in patients with chronic granulomatous disease. *J. Pediatr.* **128,** 104–107.
30. Foster, C. B., Lehrnbecher, T., Mol, F., et al. (1998) Host defense molecule polymorphisms influence the risk for immune-mediated complications in chronic granulomatous disease. *J. Clin. Invest.* **102,** 2146–2155.
31. Vowells, S. J., Sekhsaria, S., Malech, H. L., Shalit, M., and Fleisher, T. A. (1995) Flow cytometric analysis of the granulocyte respiratory burst: a comparison study of fluorescent probes. *J. Immunol. Meth.* **178,** 89–97.
32. Gallin, J. I., Alling, D. W., Malech, H. L., et al. (2003) Itraconazole to prevent fun-gal infections in chronic granulomatous disease. *N. Engl. J. Med.* **348,** 2416–2422.

33. Marciano, B. E., Wesley, R., De Carlo, E. S., et al. (2004) Long-term interferon-gamma therapy for patients with chronic granulomatous disease. *Clin. Infect. Dis.* **39,** 692–699.

34. The International Chronic Granulomatous Disease Cooperative Study Group. (1991) A controlled trial of interferon gamma to prevent infection in chronic granulomatous disease. *N. Engl. J. Med.* **324,** 509–516.

35. Horwitz, M. E., Barrett, A. J., Brown, M. R., et al. (2001) Treatment of chronic granulomatous disease with nonmyeloablative conditioning and a T-cell-depleted hematopoietic allograft. *N. Engl. J. Med.* **344,** 881–888.

36. Seger, R. A., Gungor, T., Belohradsky, B. H., et al. (2002) Treatment of chronic granulomatous disease with myeloablative conditioning and an unmodified hemopoietic allograft: a survey of the European experience, 1985-2000. *Blood* **100,** 4344–4350.

37. Nauseef, W. M. (1998) Insights into myeloperoxidase biosynthesis from its inherited deficiency. *J. Mol. Med.* **76,** 661–668.

38. Beutler, E. (1994) G6PD deficiency. *Blood* **84,** 3613–3636.

31

Diagnostic Assays for Chronic Granulomatous Disease and Other Neutrophil Disorders

Houda Zghal Elloumi and Steven M. Holland

Summary

Inasmuch as neutrophils are the primary cellular defense against bacterial and fungal infections, disorders that affect these white cells typically predispose individuals to severe and recurrent infections. Therefore, diagnosis of such disorders is an important first step in directing long-term treatment/care for the patient. Herein, we describe methods to identify chronic granulomatous disease (CGD), leukocyte adhesion deficiency (LAD), and neutropenia. The assays are relatively simple to perform, cost-effective, and can be performed with equipment available in most laboratories.

Key Words: Chronic granulomatous disease (CGD); chemiluminescence; leukocyte adhesion deficiency (LAD); neutropenia; nitroblue tetrazolium; reactive oxygen species; superoxide.

1. Introduction

Neutrophils play a key role in protecting the body, utilizing a wide array of oxygen-dependent and oxygen-independent microbicidal weapons to destroy and remove infectious agents *(1)*. Oxygen-dependent mechanisms involve the production of reactive oxygen species (ROS), which can be microbicidal *(2)*, while oxygen-independent mechanisms include most other neutrophil functions, such as chemotaxis, phagocytosis, degranulation, and release of lytic enzymes and bactericidal peptides *(3)*. Thus, it is important to consider and evaluate various neutrophil functions when patients present with recurrent- and/or hard-to-treat bacterial and fungal infections *(4)*.

The importance of neutrophil ROS production is exemplified by chronic granulomatous disease (CGD), a disorder first described in 1957 *(5)*. CGD is characterized by recurrent, life-threatening bacterial and fungal infections and tissue granuloma formation. CGD is caused by mutation in any of the four components of the NADPH oxidase (gp91^{phox}, p22^{phox}, p47^{phox}, p67^{phox}) and is

From: *Methods in Molecular Biology, vol. 412: Neutrophil Methods and Protocols*
Edited by: M. T. Quinn, F. R. DeLeo, and G. M. Bokoch © Humana Press Inc., Totowa, NJ

diagnosed by analysis of neutrophil function using one of several assays that measure the capability of neutrophils to produce superoxide anion, hydrogen peroxide, or bleach. As detailed in this chapter, neutrophil ROS production can be measured spectrophotometrically by ferricytochrome c reduction, microscopically using nitroblue tetrazolium reduction, with a flow cytometer, and by other methods based on chemiluminescence or electron-spin resonance. Kinetics of O_2^- production can also be followed by ferricytochrome-c reduction or chemiluminescence assays. Immunoblotting is typically used to determine which gene of the NADPH oxidase complex (i.e., gp91^{phox}, p47^{phox}, p22^{phox}, or p67^{phox}) is defective *(6)*. Genomic or complementary DNA sequencing is then used to specify mutations. Because protein-positive mutations exist in each gene, immunoblotting alone is a good, but not infallible, guide to identifying the affected gene.

Leukocyte adhesion deficiency (LAD) is caused by defects in cellular adhesion and function (*see* Chapter 30). LAD-1 is an autosomal recessive disorder resulting from mutations in the gene encoding CD18 (*ITGB2*), the common β subunit for all the β_2 integrins *(7)*. CD18 associates with CD11a to form lymphocyte function antigen (LFA)-1, with CD11b to form Mac-1, and with CD11c to form p150,95. β_2 integrins are expressed as inactive adhesion receptors that undergo conversion to their active form with stimulation to mediate leukocyte motility and adhesion. We describe diagnostic assays to measure β_2 integrin expression, cell adhesion, and chemotaxis.

Neutropenia is characterized by reduction of the absolute neutrophil count (ANC) below 1500 neutrophils/mm^3 *(8)*. Cyclic and severe congenital neutropenias are characterized by early maturation arrest of myelopoiesis *(9)*, and mutations in the gene encoding neutrophil elastase have been associated with both forms *(10)*. Neutropenia in peripheral blood can be associated with retention of mature neutrophils in bone marrow, as in the case of the autosomal dominant warts, hypogammaglobulinemia, immunodeficiency, myelokathexis (WHIM) syndrome, which is due to heterozygous mutations of the chemokine receptor CXCR4 *(11,12)*. Neutropenia can also be autoimmune, as a result of the formation of autoantibodies against a variety of human neutrophil antigens (HNAs). These autoantibodies can be detected in patient serum by the granulocyte immunofluorescence test (GIFT) *(13)*.

2. Materials

2.1. Diagnostic Assays for Chronic Granulomatous Disease

1. Phorbol 12-myristate 13-acetate solution: make 1 mg/mL phorbol myristate acetate (PMA) (Sigma) stock solution in dimethylsulfoxide (DMSO). Aliquot and store at –20°C.

2. Dulbecco's modified Eagle's medium (DMEM; Gibco/Invitrogen).
3. Phosphate-buffered saline, pH 7.2 (PBS, Gibco/Invitrogen).
4. HBSS–: Hank's balanced salt solution (HBSS) without Ca^{2+}, Mg^{2+}, and phenol red (Gibco/Invitrogen).
5. HBSS+: HBSS with Ca^{2+} and Mg^{2+} (Gibco/Invitrogen).
6. RPMI 1640 media (Gibco/Invitrogen).
7. 1% safranin O solution or Giemsa stain (VWR).
8. Nitroblue tetrazolium (NBT) solution (100 mg/mL) in 70% dimethylformamide (v/v) (Roche).
9. Potassium hydroxide (KOH) solution: dissolve 56 g of KOH (Sigma) in sterile H_2O (500 mL final volume) to form a 2 M solution.
10. Superoxide dimutase (SOD; Boehringer Mannheim): make stock solution of 400 μg/mL in phosphate-buffered saline (PBS) or HBSS–. Aliquot and store at –20°C.
11. 2-(4-iodophenyl)-3-(2,4-disulfophenyl)-5-2H-tetrazolium, monosodium salt (WST-1; Dojindo Laboratories): make a 10 mM stock solution in PBS or HBSS–. Protect from light.
12. Horse heart ferricytochrome c: make 1 mM stock solution in PBS or HBSS–. Aliquot and store at –20°C.
13. Catalase (Sigma): make 1 mg/mL stock solution in PBS or HBSS–. Aliquot and store at –20°C.
14. Lucigenin (Molecular Probes): make 250 μM stock solution in PBS.
15. Incubation buffer: RPMI 1640 supplemented with 10 mM HEPES, pH 7.4, and 10% fetal calf serum.
16. Lysis buffer: dissolve 829 mg ammonium chloride, 100 mg potassium bicarbonate, and 3.7 mg ethylenediamine tetraacetic acid (EDTA) in 100 mL of distilled H_2O.
17. Flow buffer: 500 mL of HBSS–, 0.5 g of albumin (human fraction V), and 1 mL of 0.5 M EDTA.
18. Dihydrorhodamine-123 (DHR) solution: dissolve DHR (Sigma) in DMSO at a concentration of 1 M. For experiments, make a working stock by diluting 10 μL of the stock solution in 990 μL PBS for a final concentration of 1 mM.
19. 2'7'-dichlorofluorescein 3',6'-diacetate (DCFH-DA): dissolve DCFH-DA (Sigma) in dimethyl sulfoxide (DMSO) at a concentration of 1 M. For experiments, make a working stock by diluting 10 μL of the stock solution in 490 μL PBS for a final concentration of 2 mM.
20. 12- and 24-well tissue culture plates (Corning).
21. 5-(diethoxyphosphoryl)-5-methyl-1-pyrroline N-oxide (DEPMPO) (Sigma).

2.2. Diagnostic Assays for Leukocyte Adhesion Deficiency

1. PBS, pH 7.2 (Gibco/Invitrogen).
2. Bovine serum albumin (BSA), fraction V (Sigma).
3. Fluorescence-activated cell sorting (FACS) buffer: PBS containing 3% BSA (w/v) and 0.1% w/v sodium azide.

4. 2,7-bis-(2-carboxyethyl)-5-(and-6)-carboxyfluorescein, acetoxymethyl ester fluorescent (BCECF) dye (Molecular probes).
5. Polycarbonate permeable transwell filters with 5-μm pore size and a surface area of 0.33 cm^2 (Costar).
6. Polystyrene round-bottom tubes (e.g., Falcon).
7. Anti-CD18 monoclonal antibody: can be selected from a variety of sources (BD Bioscience, Abcam, Beckman Coulter, etc.) and can be unlabeled or conjugated directly with fluorochrome.
8. Mouse IgG or IgM (Sigma).
9. Fluorophore-conjugated secondary antibody: can be selected from a variety of sources (BioRad, Sigma, Jackson Immunoresearch) (*see* **Note 1**).
10. Adhesion buffer: RPMI 1640 supplemented with 10 m*M* HEPES, pH 7.4 and 10% fetal calf serum (v/v).
11. MgCl$_2$/EDTA solution: 100 m*M* MgCl$_2$ and 20 m*M* EGTA in PBS.
12. MnCl$_2$ solution: 10 m*M* MnCl$_2$ in PBS.

2.3. Diagnostic Assays for Neutropenia

1. Vacutainer tubes containing EDTA (Becton-Dikinson).
2. PBS, pH 7.2 (Gibco/Invitrogen).
3. PBS-EDTA buffer: PBS, 0.1 m*M* EDTA, and 0.02% sodium azide.
4. Paraformaldehyde (PFA) solution: 1% paraformaldehyde (v/v) in PBS.
5. Fluorescein isothiocyanate (FITC)-conjugated anti-human IgG antibody F(ab')2 specific for IgG or IgM (e.g., Abcam).
6. Anti-FcγRIII (CD16) monoclonal antibody (e.g., Abcam).
7. Lysis buffer: PBS, 1% Triton X-100, 3% BSA, and protease inhibitors (2 m*M* phenylmethylsulfonyl fluoride, 5 m*M* EDTA, 0.2 μg/mL soybean trypsin inhibitor).
8. Goat anti-mouse IgG.
9. Horseradish peroxidase (HRP)-conjugated goat anti-human IgG.
10. HRP substrate.
11. Microtiter plate reader (e.g., Molecular Devices SpectraMax).

3. Methods

3.1. Diagnostic Assays for CGD

3.1.1. Nitroblue Tetrazolium Assay

The NBT assay relies on the accumulation of blue-black formazan precipitate. The membrane permeable, water-soluble, yellow NBT is reduced to blue-black formazan by O_2^-, which is generated upon activation of phagocytes, typically by PMA (**Fig. 1**) *(14)*. PMA activates NADPH oxidase by enhancing protein kinase C (PKC) and thus stimulates production of ROS. This assay is easily read by light microscopy, allowing the recognition of individual cells that have reduced NBT.

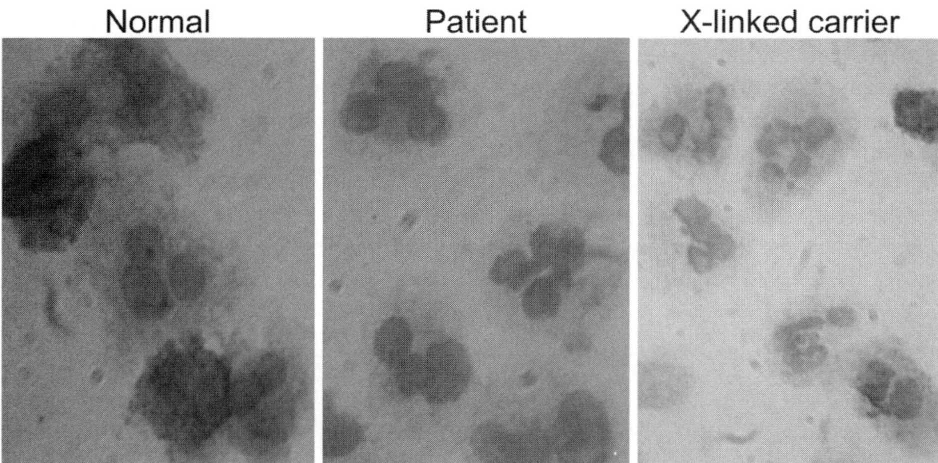

Fig. 1. Nitroblue tetrazolium (NBT) assay. Following phorbol myristate acetate stimulation, reduction of yellow NBT to blue black formazan occurred in the healthy control cells (left), but is completely absent in cells from a patient with X-linked chronic granulomatous disease (center). The mother of the patient is a carrier of the mutation and has both NBT-positive and -negative cell populations (right). (color insert will appear after page 300)

More than 95% of normal neutrophils will reduce NBT, showing orange nuclei with clumps of blue-black precipitate in the cytoplasm. CGD cells have no or greatly diminished NBT reduction. Mothers of patients with X-linked CGD typically show mosaicism of peripheral blood neutrophil NBT reduction, because X-inactivation is thought to occur randomly in hematopoietic precursors. This mosaic pattern can range from 0.001 to 97% of neutrophils, although the range of NBT reduction for cells from most carriers is 20–80%. At the extremes of lyonization, X-linked carriers may be indistinguishable from patients with CGD or normal individuals, respectively. The X-linked form of CGD is the most common (about 65% of cases). Thus, analysis of the mother's blood at the time of her son's NBT test can identify disease, and if the mother is a carrier, strongly suggest X-linked inheritance (*see* **Note 2**).

1. Isolate neutrophils from peripheral blood of a patient and a healthy control individual as described in Chapter 2.
2. Place glass coverslips in a 12-well tissue culture plate.
3. Aliquot cells into the wells at 5×10^5 neutrophils per well in 400 µL of DMEM.
4. Allow cells to adhere to the coverslips for 1 h at 37°C.
5. Add 100 µL of NBT solution to each well.
6. Add 1 µL of PMA stock (200 ng/mL final) or buffer (unstimulated control) to the wells, and incubate plate for 20 min at 37°C.
7. Wash the cells twice with warm (37°C) PBS.

8. Allow the coverslips to air-dry.
9. Fix the cells with methanol, and counterstain with 1% safranin O solution. Alternatively, stain with Giemsa stain for 15 min. Wash and air-dry coverslips.
10. View coverslips under microscope to determine the percentage of cells containing blue-black formazan particles.
11. Express results as the number of neutrophils with reduced NBT/total number of neutrophils counted.
12. This assay can also be performed on regular microscope slides.

3.1.2. Modified Colorimetric NBT Assay

The NBT test is semi-quantitative and does not precisely reflect the amount of O_2^- produced. A new colorimetric assay is more sensitive, quantitative, and can detect even small amounts of O_2^- production. This assay dissolves cells after stimulation allowing for plate-based absorbance measurements to reflect the amount of NBT reduced by neutrophils *(15)*.

1. Isolate neutrophils from peripheral blood as described in Chapter 2.
2. Aliquot cells into the wells of a 24-well plate at 1×10^5 neutrophils per well in 400 μL of DMEM.
3. Add 100 μL of NBT solution to each well.
4. Add 1 μL of PMA stock (200 ng/mL final) or buffer (unstimulated control) to the wells, and incubate plate for 30 min at 37°C.
5. Wash the cells twice with warm (37°C) PBS. It is critical to remove extracellular NBT.
6. Fix cells with methanol, and air-dry plates.
7. Add 120 μL of KOH solution per well to solubilize the cell membranes.
8. Dissolve the blue formazan deposited inside the cells by adding 140 μL of DMSO per well and gently shaking for 10 min at room temperature.
9. Transfer the dissolved NBT solution to a 96-well plate, and measure the absorbance at 620 nm with a microtiter plate reader.
10. Compare absorbance at 620 nm between the NBT + PMA and NBT − PMA conditions to quantify O_2^- production in each cell sample.

3.1.3. WST-1 Assay

In recent years, newer tetrazolium salt derivatives have been developed to measure O_2^- production. For example, the sulfonated tetrazolium salt, (WST-1, can be used for diagnostic measurement of O_2^- *(16–18)*. WST-1 is efficiently reduced by O_2^- to a stable, water-soluble formazan with high molar absorptivity. In addition, WST-1 generates a twofold greater increase in absorbance than ferricytochrome c *(16)*. SOD is added to control samples to demonstrate specificity for O_2^- *(17)*. The methods described are based on previously-reported microtiter plate-based WST-1 assay systems *(16,18)*.

1. Isolate neutrophils from peripheral blood as described in Chapter 2.
2. Resuspend neutrophils at 1×10^6 cells/mL in HBSS+ containing glucose (2 mg/mL).
3. Aliquot 100 µL of cells into wells of a 96-well microtiter plate (*see* **Note 3**).
4. Add 4 µL of WST-1 stock solution (500 µ*M* final) and 4 µL of catalase stock solution (20 µg/mL final), with or without 10 µL SOD stock (20 µg/mL final).
5. Add 81 µL of HBSS+ for a final volume of 199 µL.
6. Equilibrate samples for 3 min at 37°C in the microplate reader and monitor baseline absorbance at 450 nm.
7. Add 1 µL of PMA stock (200 ng/mL final) or buffer (unstimulated control) to the wells, and monitor absorbance at 450 nm for 20 min at 37°C (*see* **Note 4**).
8. Calculate O_2^- equivalents using the molar extinction coefficient for WST-1 (37.0×10^3 M^{-1} cm^{-1} at 450 nm).
9. Calculate specific O_2^- generation by subtracting signal obtained with SOD from that obtained without SOD for each pair of samples.

3.1.4. Ferricytochrome c Reduction

Specific detection of O_2^- production can be monitored by measuring the SOD-inhibitable reduction of ferricytochrome c to ferrocytochrome c, as shown in **eq. 1**:

$$\text{Fe}^{3+}\text{-cyt c} + O_2^- \rightarrow 3\ \text{Fe}^{2+}\text{-cyt c} + O_2 \tag{1}$$

This spectrophotometric reaction is monitored at 550 nm, because ferricytochrome c and ferrocytochrome c have different extinction coefficients (0.89×10^4 *M*/cm and 2.99×10^4 *M*/cm, respectively). Reduction of cytochrome c is not absolutely specific for O_2^-, and can also be caused by other cellular reductants, such as ascorbate and glutathione, thus SOD must be added to control samples to demonstrate specific generation of O_2^- *(17)*. The ferricytochrome c assay has been readily adapted to a microtiter plate format *(19)* (*see* **Note 5**).

1. Isolate neutrophils from peripheral blood as described in Chapter 2.
2. Resuspend neutrophils at 1×10^6 cells/mL in HBSS+ containing glucose (2 mg/mL).
3. Aliquot 100 µL of cells into wells of a 96-well microtiter plate (*see* **Note 3**).
4. Add 16 µL of ferricytochrome c stock (80 µ*M* final), 4 µL of catalase stock solution (20 µg/mL final), with or without 10 µL SOD stock (20 µg/mL final).
5. Add 69 µL of HBSS+ for a final volume of 199 µL.
6. Equilibrate samples for 3 min at 37°C in the microplate reader and monitor baseline absorbance at 550 nm.
7. Add 1 µL of PMA stock (200 ng/mL final) or buffer (unstimulated control) to the wells, and monitor absorbance at 450 nm for 20 min at 37°C (*see* **Note 4**).
8. Calculate O_2^- equivalents using the molar extinction coefficient for ferrocytochrome c (21.1×10^3 M^{-1} cm^{-1} at 550 nm).
9. Calculate specific O_2^- generation by subtracting signal obtained with SOD from that obtained without SOD for each pair of samples.

3.1.5. Chemiluminescence Assays

Reaction with O_2^- can cause a variety of reagents to luminesce, and these reagents can provide higher sensitivity and analysis of both intracellular and extracellular O_2^- *(20–22)* (*see* Chapter 14). Among the most commonly used chemiluminescent reagents for detection of O_2^- are lucigenin (bis-*N*-methylacridinium nitrite), luminol, isoluminol, and L-012 *(20–22)*. One of the problems with chemiluminescence detection systems has been the presence of artifacts due to redox cycling of the detection reagents (e.g., lucigenin) and lack of specificity for distinct radical species (O_2^-, H_2O_2, OH^-, etc.) *(21,23)*. More recently, improved chemiluminescence-based reagents have emerged, such as Diogenes (National Diagnostics) which has been shown to be a O_2^--specific chemiluminescence reagent, with 10^6 greater response to O_2^- vs H_2O_2 *(24)*.

3.1.5.1. LUCIGENIN

1. Isolate neutrophils from peripheral blood as described in Chapter 2.
2. Resuspend neutrophils at 5×10^5 cells in 480 µL of incubation buffer in plastic tubes.
3. Add 10 µL of lucigenin stock solution (50 µ*M* final) with or without 10 µL SOD stock (20 µg/mL final).
4. Add 1 µL of PMA stock (200 ng/mL final) or buffer (unstimulated control) to the tubes, and monitor chemiluminescence with a luminometer at 30 s intervals for 5 min (*see* **Note 6**).
5. Subtract signal obtained with SOD from that obtained without SOD for each pair of samples.

3.1.5.2. LUMINOL

1. Isolate neutrophils from peripheral blood as described in Chapter 2.
2. Resuspend neutrophils at 1×10^6 cells/mL in HBSS+ containing glucose 2 mg/mL.
3. Aliquot 100 µL of cells into wells of a 96-well black microtiter plate (*see* **Note 3**).
4. Add 10 µL of luminol stock (500 µ*M* final) with or without 10 µL SOD stock (20 µg/mL final).
5. Add 90 µL of HBSS+ for a final volume of 200 µL.
6. Equilibrate samples for 3 min at 37°C in the luminescence microplate reader to monitor baseline luminescence.
7. Add 1 µL of PMA stock (200 ng/mL final) or buffer (unstimulated control) to the wells, and monitor luminescence for 15 min at 37°C (*see* **Note 4**).
8. Subtract signal obtained with SOD from that obtained without SOD for each pair of samples.

3.1.5.3. L-012 CHEMILUMINESCENCE ASSAY FOR WHOLE BLOOD

1. Collect blood from patients and healthy donors in vacutainer tubes.
2. Dilute blood 1:10 in HBSS, and aliquot 180 µL of this diluted blood into wells of a 96-well black microtiter plate.
3. Add 10 µL L-012 stock (200 µ*M* final) with or without 10 µL SOD stock (20 µg/mL final).

4. Equilibrate samples for 3 min at 37°C in the luminescence microplate reader to monitor baseline luminescence.
5. Add 1 μL of PMA stock (200 ng/mL final) or buffer (unstimulated control) to the wells, and monitor luminescence for 15 min at 37°C (*see* **Note 4**).
6. Subtract signal obtained with SOD from that obtained without SOD for each pair of samples.

3.1.6. Flow Cytometry

Flow cytofluorometric assays using dyes sensitive to the generation of ROS are more sensitive and rapid and are replacing NBT and chemiluminescence methods *(25)*. The first flow cytometry-based assay reported by Bass et al. *(26)* required granulocyte isolation prior to adding the dye 2'7'-dichlorofluorescein 3',6'-diacetate (DCFH-DA), which is highly sensitive for measuring H_2O_2 *(27)*. DCFH-DA is a small nonpolar, nonfluorescent molecule that is able to diffuse into cells, where it is stable for a few hours. After entering the cell, DCFH-DA is hydrolyzed by nonspecific intracellular esterases to liberate the polar molecule 2'7'-dichlorofluorescin (DCFH). Upon reaction with ROS, DCFH is converted to the highly fluorescent 2',7'-dichlorofluorescein (DCF), which is measured after excitation by blue light (488 nm). The resulting fluorescence is linearly related to the activity of the respiratory burst. However, other molecules, such as nitric oxide (NO) have been shown to excite DCFH. DCFH can also be auto-oxidized, resulting in a spontaneous increase in fluorescence following light exposure *(28)*. For these reasons, newer fluorescence probes specific for the generation of different ROS species have been developed. For example, DHR freely enters the cell and, after oxidation by ROS to rhodamine-123 (RHO), emits a bright fluorescent signal exclusively localized inside the cell *(29)*. DHR can be used with as little as 0.2 mL of whole blood and shows the greatest fluorescent intensity and separation of fluorescent signals between CGD and normal granulocytes (**Fig. 2**) *(25)*.

DCFH and DHR can react with various ROS, including O_2^-, H_2O_2, and NO. In order to minimize the artifactual activation of cells prior to testing, Alvarez-Larran and colleagues *(30)* developed a method using whole blood, DHR, and the fluorescent dye DRAQ-5 (BioStatus). The latter stains nucleated cells, avoiding the erythrocyte lysis and wash steps, thus reducing neutrophil depletion. However, the lysis method may lead to significantly higher DHR values, with a higher PMA activation index.

1. Obtain freshly drawn, heparin-anticoagulated whole blood from patients and control.
2. Incubate 150 μL of blood in a propylene tube with 2 mL of prewarmed (37°C) lysis buffer.
3. Mix tubes by inversion, and let stand 5 min at room temperature.
4. Centrifuge tubes at 800*g* for 5 min at room temperature to pellet the leukocytes. Discard the supernatant and blot tubes dry.

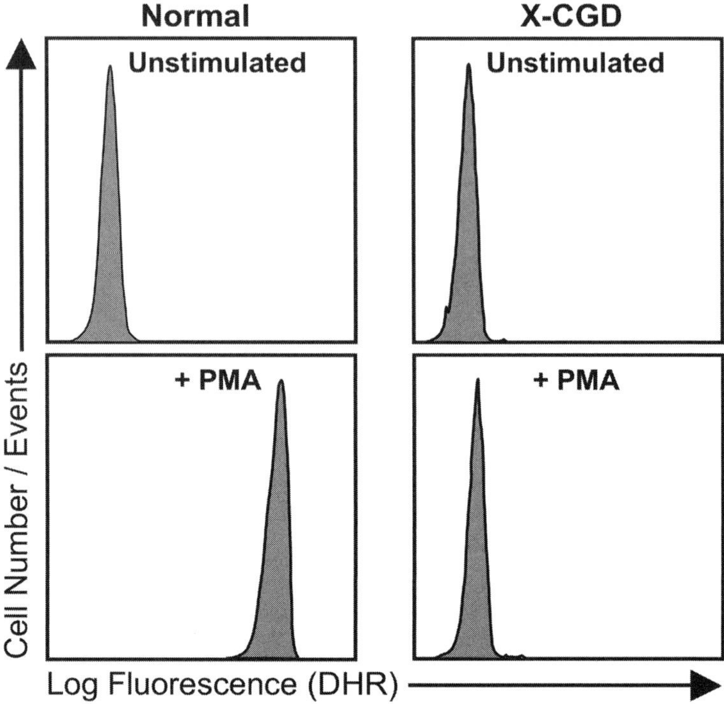

Fig. 2. Dihydrorhodamine-123 (DHR) assay. In the top panels are DHR histograms from unstimulated cells from both a healthy control and a patient with X-linked chronic granulomatous disease (X-CGD). The lower panels show DHR oxidation after phorbol myristate acetate stimulation. Whereas the normal shows shift of the peak to the right, the X-CGD patient shows no shift with stimulation, corresponding to a lack of super-oxide production.

5. Wash the cell pellet with flow buffer, and centrifuge again ($800g$ for 5 min).
6. Add 400 µL flow buffer and mix gently to resuspend the pellet.
7. Add 5 µL of catalase solution to each tube.
8. Add 2 µL of DHR stock (5 µ*M* final) or DCFH-DA stock (10 µ*M* final), and incubate tubes at 37°C in a shaking water bath for 5 min. Keep tubes covered with foil.
9. Add 1 µL of PMA stock (200 ng/mL final) or suspension buffer (unstimulated control) to the tubes, and incubate for 15 min at 37°C in a shaking water bath. Keep tubes covered with foil.
10. Immediately analyze cells by flow cytometry. In the granulocyte gate, collect non-fluorescent measurements (forward- and side-scatter) as well as 585 nm fluorescence for 10,000 events.
11. Compare healthy subjects and patients on the basis of the stimulation index (SI), which represents the fold increase in signal (mean channel fluorescence of stimulated neutrophils/mean channel fluorescence of unstimulated neutrophils) (*see* **Note 7**).

3.1.7. Electron Spin Resonance

O_2^- formation can be measured by electron paramagnetic resonance (EPR) spectroscopy, one of the most sensitive and quantitative methods *(31,32)*. The most widely used spin trap for the detection of O_2^- is 5,5-dimethyl-1-pyrroline-*N*-oxide (DMPO) *(31)*. However, the O_2^- adduct DMPO-OOH is unstable, with a half-life of less than 1 min and further decomposes to form DMPO-OH. A new spin trap, DEPMPO, is reported to have improved properties for O_2^- detection, with a good efficiency for trapping O_2^- *(32)* (*see* **Note 8**).

1. Isolate neutrophils from peripheral blood as described in Chapter 2.
2. Resuspend neutrophils at 4×10^6 cells/mL in PBS containing glucose (2 mg/mL), and aliquot into plastic or polypropylene test tubes (400 µL/tube).
3. Add 2 µL of PMA stock (200 ng/mL final) or 40 µL of opsonized zymosan (1 mg/mL final). As an unstimulated control add buffer to the tubes (*see* **Note 4**).
4. Add DEPMPO at a final concentration of 10 m*M*.
5. Incubate the reaction mixture at 37°C in a shaking water bath for 15 min.
6. Transfer to a capillary or quartz flat cell and measure with an EPR spectrometer (e.g., Magnettech Miniscope or Bruker ER 300).

3.1.8. Prenatal Diagnosis

Prenatal diagnosis of CGD can be performed by measuring NADPH oxidase activity in cells from fetal blood *(33)*. NBT staining can be performed on blood samples obtained from fetal placental-vessel puncture *(34)* or from cultured amniotic fibroblasts *(35)*, and luminol-enhanced chemiluminescence has been adapted for prenatal diagnosis of CGD *(35)* (*see* **Note 9**).

3.2. Diagnostic Assays for LAD-1

The diagnosis of LAD-1 is typically made by flow cytometric analysis of neutrophils for the expression of CD18-containing molecules. The clinical phenotype reflects the level of β_2 integrins expressed. Those individuals with 1–10% of normal CD18 expression have a milder course than those with <1% expression. Neutrophils from heterozygotes have 40–60% expression of β2 integrins and are clinically normal *(36)*. Many antibodies are currently available for flow cytometry. For example, CD18 monoclonal antibodies L130 (BD Biosciences) and MEM-48 (Abcam) work well for measuring cell-surface β_2 integrin expression by flow cytometry (*see* **Note 10**).

3.2.1. Flow Cytometry

1. Isolate neutrophils from peripheral blood as described in Chapter 2.
2. Resuspend neutrophils at 1×10^6 cells/mL in FACS buffer, and aliquot into polystyrene round-bottom tubes (50 µL/tube).

3. Add desired primary antibody (0.1–10 µg/mL), and incubate for 30 min at room temperature. Set up a parallel control labeling reaction using an isotype-matched immunoglobulin (e.g., mouse IgG).
4. Centrifuge cells (400*g* for 5 min), and wash three times with FACS buffer.
5. Resuspend cell pellet in ice-cold FACS buffer.
6. Dilute the fluorochrome-labeled secondary antibody in FACS buffer at the optimal dilution (according to the manufacturer's instructions), and then resuspend the cells in this solution (*see* **Note 1**).
7. Incubate for at least 30 min on ice. Cover tubes with foil. This incubation must be done in the dark.
8. Wash the cells three times as in **step 4**. Resuspend final pellet in 200 µL FACS buffer.
9. The cells are now ready for analysis. Keep the cells on ice until analysis.
10. Analyze by flow cytometry. Perform a parallel control labeling using control isotype- and species-matched antibody.

3.2.2. Cell Adhesion Assay

The $\beta 2$ cytoplasmic tail interacts with a number of actin-binding proteins through which the adhesion properties of LFA-1 and Mac-1 are regulated. LFA-1-mediated adhesion can therefore be assayed on the immobilized ligand, intercellular adhesion molecule (ICAM)-1). The integrins bind divalent cations such as Mg^{2+}or Mn^{2+} in order to function. Specificity of adhesion can be tested by running the assay in the presence of a function-blocking antibody *(37)*.

3.2.2.1. LOADING NEUTROPHILS WITH BCECF

1. Isolate neutrophils from peripheral blood as described in Chapter 2.
2. Resuspend neutrophils at 1×10^5 cells/mL in adhesion buffer (*see* **Note 3**).
3. Add 3.0 m*M* BCECF, and incubate for 20 min at 37°C.
4. Wash labeled cells once, and resuspend at 1×10^5 cells/mL in adhesion buffer.

3.2.2.2. ANALYSIS OF LFA-1 ADHESION TO ICAM-1

1. Coat wells of a microtiter plate with 100 µL/well of goat anti-human IgG at 5 µg/mL in 50 m*M* sodium bicarbonate buffer for 16 to 20 h at 4°C.
2. Block nonspecific binding sites with PBS containing 0.5% (w/v) BSA for 30 min at 37°C.
3. Coat each well with 50 µL of the soluble recombinant form of human ICAM1/Fc chimera (R&D systems), used at a final concentration of 1 µg/mL in PBS containing 0.1% BSA (w/v).
4. Incubate plate for 2 h at room temperature.
5. Load 200 µL of neutrophils (2×10^4 cells) into each well of the ligand-coated plates, and incubate for 30 min at 37°C in 5% CO_2 incubator.
6. Add 10 µL of $MgCl_2$/EGTA solution to activate LFA-1-mediated adhesion, and incubate 5 min at 37°C.

7. Wash the plate three times and then measure bound cell fluorescence using a fluorescence plate reader.

3.2.2.3. ANALYSIS OF MAC-1 MEDIATED ADHESION

1. Coat wells of a microtiter plate with 100 µL/well of 50 µg/mL BSA in 50 mM sodium bicarbonate buffer for 24 h at 4°C.
2. Block nonspecific binding with 150 µL/well of 0.2% (w/v) polyvinyl-pyrrolidone in PBS for 30 min at 37°C.
3. Wash coated plates twice with adhesion buffer.
4. Load 200 µL of neutrophils (2×10^4 cells) into each well of the ligand-coated plates, and incubate for 30 min at 37°C and 5% CO_2.
5. Add 10 µL of $MnCl_2$ solution to activate Mac-1-mediated adhesion, and incubate for 5 min at 37°C.
6. Wash the plate three times with wash buffer, and measure bound cell fluorescence using a fluorescence microtiter plate reader.

3.2.3. Neutrophil Transmigration Assay

1. Isolate neutrophils from peripheral blood as described in Chapter 2 (*see* **Note 11**).
2. Resuspend neutrophils at 1×10^7 cells/mL in HBSS+.
3. Set up transwell plates with 0.5 mL of HBSS+ in the lower chamber.
4. Aliquot 150 µL of cell suspension into the upper chamber of the transwell.
5. Initiate transmigration by adding chemoattractants (e.g., 0.1 µM N-formylmethionyl-leucyl-phenylalanine [fMLP], 0.4 µg/mL interleukin [IL]-8, or 0.1 µM LTB4) into the lower chamber.
6. Cover and incubate the plate for 2 h at 37°C and 5% CO_2.
7. Quantify neutrophil migration into the lower chamber by counting cells or solubilizing cells with 1% Triton X-100 in order to do a colorimetric assay for myeloperoxidase (MPO) (*see* Chapter 32) (*see* **Note 12**).
8. Time-course assays can be performed by moving the transwell to a fresh well every 30 min. The same concentration of chemoattractant is used in the lower well (*see* **Note 3**).

3.3. Diagnostic Assays for Neutropenia

3.3.1. Granulocyte Immunofluorescence Test Using Flow Cytometry

GIFT using flow cytometery, originally described by Verheugt in 1977 (*38*), detects antibodies against neutrophil surface antigens. Flow cytometric analysis using a panel of donor neutrophils to consistently quantify neutrophil-associated antibodies has allowed testing of patient serum samples against a panel of normal donor neutrophils already typed for the antigens HNA-1a, -1b, -1c, -2a, and -5a (*39*).

1. Collect blood from normal donors in vacutainer tubes.
2. Isolate neutrophils from peripheral blood as described in Chapter 2.

3. Wash cells twice in 2 mL of PBS-EDTA, centrifuging the cells between each wash at 300*g* for 5 min.
4. Resuspend the pellets in 200 µL of PFA solution, and incubate for 5 min at 22°C to fix the cells.
5. Wash fixed cells with PBS-EDTA. Centrifugation at 300*g* for 5 min between washes.
6. Store serum or EDTA-anticoagulated plasma from neutropenic patients. A negative control serum is also included in the study.
7. Incubate donor neutrophils with patient serum or control serum in polystyrene test tubes for 30 min at 20–22°C.
8. Wash the cells three times with PBS-EDTA, as in **step 5**.
9. Add 100 µL of FITC-labeled anti-human IgG or anti-human IgM antiserum, and incubate for 30 min at 4°C.
10. Centrifuge the tubes at 300*g* for 5 min, discard the supernatant, and wash once with 3 mL of PBS-EDTA.
11. Resuspend cells in 0.5 mL of PBS-EDTA.
12. Measure antibody binding by flow cytometry, and grade the results from – to +++ on the basis of the fluorescence intensity.
13. The ratio of the mean fluorescence channel of each sample to that of control serum is expressed as relative fluorescence intensity (RFI) *(39,40)*.
14. Report increased binding of IgG or IgM over the normal control sera.

3.3.2. Monoclonal Antibody Immobilization of Granulocyte Antigens Assay

Neutrophil antigens can be targets for autoimmune attack. This assay detects neutrophil targets of autoantibodies by allowing binding of antibodies to antigen and then detecting these after incubation with an anti-human antibody bound to an enzyme-linked immunosorbent assay (ELISA) plate.

1. Collect blood from normal donors in vacutainer tubes.
2. Isolate neutrophils from peripheral blood as described in Chapter 2. Resuspend at 5×10^6 cells/mL in PBS/EDTA buffer.
3. Incubate 20 µL of cell suspension (10^6 cells) with 60 µL of serum for 1 h at room temperature.
4. Wash fixed cells 3 times with PBS-EDTA. Centrifugation at 300*g* for 5 min between washes.
5. Resuspend cells in PBS/BSA containing anti-FcγRIII (CD16) monoclonal antibody to block Fc receptors, and incubate for 1 h at room temperature.
6. Wash the cells two times with PBS-EDTA, as in **step 4**.
7. Add 200 µL of lysis buffer, and incubate for 30 min at 4°C to lyse the cells.
8. Add cell lysates to wells of a microtiter plate coated with goat anti-mouse IgG to capture neutrophil antigen–serum antibody complexes.
9. Incubate overnight at 4°C, and wash the wells with PBS/EDTA.
10. Add 100 µL of horseradish peroxidase-labeled goat anti-human IgG, and incubate for 2 h at 4°C.

11. Wash plate with PBS/EDTA, and add HRP substrate.
12. Read the plate at 405 nm with a microtiter plate reader to quantify the HRP signal.
13. Use sera with known HNA-1a and HNA-1b specificity and sera from AB-positive controls as positive and negative assay controls, respectively.
14. Results for a patient's serum are considered positive if the mean optical density (OD) value obtained is greater than or equal to to twice the mean OD obtained with the normal serum control *(38)*.

4. Notes

1. Primary antibodies can often be purchased directly conjugated to the desired fluorophore. If a directly-conjugated antibody is used, then the secondary antibody is not required.
2. The NBT assay has numerous limitations. For example, neutrophils from patients with CGD, such as some patients who are protein-positive and gp91^{phox}-mutant and many of those with p47^{phox} deficiency, may exhibit some residual O_2^- production and reduce small amounts of NBT. This makes differentiation from normal individuals difficult. In addition, because the NBT test is based on visual inspection of a limited number of cells, it is not ideal for X-linked carrier screening. Finally, drugs such as d-penicillamine can act as O_2^--generating sources when oxidized and reduce NBT.
3. The number of cells per well can be adjusted to achieve the desired signal strength.
4. PMA concentration should be adjusted empirically to achieve maximal responses.
5. Any SOD-independent ferricytochrome c reduction can be minimized by reducing the concentration of ferricytochrome c without losing signal. O_2^- does not effectively cross biological membranes, thus the ferricytochrome c assay is not useful for measurement of ROS generated inside cells. If measurement of intracellular O_2^- is desired, other detection reagents are utilized (i.e., chemiluminescent and fluorescent dyes) (*see* Chapter 14).
6. This assay is easily adapted to a microtiter plate format if a luminescence microtiter plate reader is available.
7. PMA-stimulated DHR oxidation yields between 50- and 200-fold greater mean channel fluorescence than in unstimulated cells. In contrast, CGD neutrophils usually show less than 10-fold augmentation of DHR oxidation by PMA. X-linked carriers show two distinct populations of neutrophils, those that oxidize normally in the presence of PMA, and those that do not *(41)*.
8. In the presence of O_2^-, DEPMPO is transformed into DEPMPO-OOH with no significant decomposition to DEPMPO-OH *(32)*. When neutrophils are stimulated by PMA or zymosan in the presence of DEPMPO, the major signal observed is the O_2^- adduct DEPMPO–OOH, with a small amount of DEPMPO–OH (less than 8% at 20 min). In the absence of stimulant, no signal is observed. The ESR signal is completely inhibited by SOD, whereas catalase has no effect on either DEPMPO–OOH or DEPMPO–OH.
9. Prenatal diagnosis can also be accomplished by restriction fragment length polymorphism (RFLP) *(42)*. Where available, identification of specific mutations is preferred

(43). Genomic DNA from chorionic villi can be analyzed by denaturing high-performance liquid chromatography (DHPLC), followed by sequence analysis *(44)*.

10. Other neutrophil disorders are also characterized by impaired chemotaxis, such as Chédiak-Higashi syndrome (CHS) and specific granule deficiency (SGD). CHS is a rare autosomal recessive disorder characterized by severe immune deficiency with neutropenia *(45)*. CHS neutrophils contain giant granules that are easily appreciated under the light microscope. In SGD, another rare congenital disorder, neutrophils show abnormalities in nuclear morphology with an atypical bi-lobed nuclei and lack expression of secondary and tertiary granule proteins. The diagnosis of SGD can be made by electron microscopy, as well as by measuring lactoferrin and/or defensins *(46)*. Commercial ELISA kits are available.

11. It is critical to use freshly isolated polymorphonuclear neutrophils (PMN), and avoid any treatment inducing cell priming. Activation of PMN dramatically affects their chemotactic transmigration *(47)*.

12. An alternative method for quantification of transmigrated cells is to use a colorimetric lactate dehydrogenase (LDH) assay (e.g., CytoTox 96 Non-Radioactive Cytotoxicity Assay; Promega). The LDH measurements can then be converted to absolute cell numbers by comparison of the values to standard curves created by aliquotting a range of control cell standards into wells containing no transwell inserts and processing identically to the sample wells with inserts *(48)*.

References

1. Witko-Sarsat, F., Rieu, P., Descamps-Latscha, B., Lesavre, P., and Halbwachs-Mecarelli, L. (2000) Neutrophils: molecules, functions and pathophysiological aspects. *Lab. Invest.* **80,** 617–653.

2. Roos, D., Van Bruggen, R., and Meischl, C. (2003) Oxidative killing of microbes by neutrophils. *Microbes. Infect.* **5,** 1307–1315.

3. Ganz, T. (2004) Antimicrobial polypeptides. *J. Leukoc. Biol.* **75,** 34–38.

4. Rosenzweig, S. D. and Holland, S. M. (2004) Phagocyte immunodeficiencies and their infections. *J. Allergy Clin. Immunol.* **113,** 620–626.

5. Berendes, H., Bridges, R. A., and Good, R. A. (1957) A fetal granulomatous disease of childhood: the clinical study of a new syndrome. *Minn. Med.* **40,** 309–312.

6. Levy, R., Rotrosen, D., Nagauker, O., Leto, T., and Malech, H. (1990) Induction of the respiratory burst in HL-60 cells, correlation of function and protein expression. *J. Immunol.* **145,** 2595–2601.

7. Etzioni, A. (1996) Adhesion molecules—their role in health and disease. *Pediatr. Res.* **39,** 191–198.

8. Badolato, R., Fontana, S., Notarangelo, L. D., and Savoldi, G. (2004) Congenital neutropenia: advances in diagnosis and treatment. *Curr. Opin. Allerg. Immunol.* **4,** 513–521.

9. Kostman, R. (1975) Infantile genetic agranulocytosis. A review with presentation of ten new cases. *Acta. Paediatr. Scand.* **64,** 362–368.

10. Horwitz, M., Benson, K. F., Person, R. E., Aprikyan, A. G., and Dale, D. C. (1999) Mutations in ELA2 encoding neutrophil elastase, define a 21-day biological clock in cyclic haematopoiesis. *Nat. Genet.* **23**, 433–436.

11. Zuelzer, W. W. (1964) 'Myelokathexis': a new form of chronic granulocytopenia. *N. Engl. J. Med.* **270**, 699–704.

12. Hernandez, P. A., Gorlin, R. J., Lukens, J. N., et al. (2003) Mutations in the chemokine receptor gene CXCR4 are associated with WHIM syndrome, a combined immunodeficiency disease. *Nat. Genet.* **34**, 70–74.

13. Stroncek, D. F., Skubitz, K. M., and McCullough, J. (1990) Biochemical nature of the neutrophil-specific antigen NB1. *Blood* **75**, 744–755.

14. Baehner, R. L., Boxer, L. A., and Davis, J. (1976) The biochemical basis of nitroblue tetrazolium reduction in normal human and chronic granulomatous disease polymorphonuclear leukocytes. *Blood* **48**, 309–313.

15. Choi, H. S., Kim, J. W., Cha, Y. N., and Kim, C. (2006) A quantitative nitroblue tetrazolium assay for determining intracellular superoxide anion production in phagocytic cells. *J. Immunoass. Immunochem.* **27**, 31–44.

16. Tan, A. S. and Berridge, M. V. (2000) Superoxide produced by activated neutrophils efficiently reduces the tetrazolium salt, WST-1 to produce a soluble formazan: a simple colorimetric assay for measuring respiratory burst activation and for screening anti-inflammatory agents. *J. Immunol. Meth.* **238**, 59–68.

17. Tarpey, M. M., Wink, D. A., and Grisham, M. B. (2004) Methods for detection of reactive metabolites of oxygen and nitrogen: in vitro and in vivo considerations. *Am. J. Physiol. Regul. Integr. Comp. Physiol.* **286**, R431–444.

18. Peskin, A. V. and Winterbourn, C. C. (2000) A microtiter plate assay for superoxide dismutase using a water-soluble tetrazolium salt (WST-1). *Clin. Chim. Acta* **293**, 157–166.

19. Björquist, P., Palmer, M., and Ek, B. (1994) Measurement of superoxide anion production using maximal rate of cytochrome (III) C reduction in phorbol ester stimulated neutrophils, immobilised to microtiter plates. *Biochem. Pharmacol.* **48**, 1967–1972.

20. Liu, L., Dahlgren, C., Elwing, H., and Lundqvist, H. (1996) A simple chemiluminescence assay for the determination of reactive oxygen species produced by human neutrophils. *J. Immunol. Meth.* **192**, 173–178.

21. Hasegawa, H., Suzuki, K., Nakaji, S., and Sugawara, K. (1997) Analysis and assessment of the capacity of neutrophils to produce reactive oxygen species in a 96-well microplate format using lucigenin- and luminol-dependent chemiluminescence. *J. Immunol. Meth.* **210**, 1–10.

22. Daiber, A., August, M., Baldus, S., et al. (2004) Measurement of NAD(P)H oxidase-derived superoxide with the luminol analogue L-012. *Free Radic. Biol. Med.* **36**, 101–111.

23. Skatchkov, M. P., Sperling, D., Hink, U., et al. (1999) Validation of lucigenin as a chemiluminescent probe to monitor vascular superoxide as well as basal vascular nitric oxide production. *Biochem. Biophys. Res. Commun.* **254**, 319–324.

24. Stielow, C., Catar, R. A., Muller, G., et al. (2006) Novel Nox inhibitor of oxLDL-induced reactive oxygen species formation in human endothelial cells. *Biochem. Biophys. Res. Commun.* **344,** 200–205.

25. Vowells, S. J., Sekhsaria, S., Malech, H. L., Shalit, M., and Fleisher, T. A. (1995) Flow cytometric analysis of the granulocyte respiratory burst: a comparison study of fluorescent probes. *J. Immunol. Meth.* **178,** 89–97.

26. Bass, D. A., Parce, W., Dechatelet, L. R., Szejda, P., Seeds, M. C., and Thomas, M. (1983) Flow cytometry studies of oxidative product formation by neutrophils: a graded response to membrane stimulation. *J. Immunol.* **130,** 1910–1917.

27. Keston, A. S. and Brandt, R. (1965) The fluorometric analysis of ultramicro quantities of hydrogen peroxide. *Anal. Biochem.* **11,** 1–5.

28. Hempel, S. L., Buettner, G. R., O'Malley, Y. Q., Wessels, D. A., and Flaherty, D. M. (1999) Dihydrofluorescein diacetate is superior for detecting intracellular oxidants: comparison with 2',7'-dichlorodihydrofluorescein diacetate, 5(and 6)-carboxy-2',7'-dichlorodihydrofluorescein diacetate, and dihydrorhodamine 123. *Free Radical. Biol. Med.* **27,** 146–159.

29. Emmendörffer, A., Nakamura, M., Rothe, G., Spiekermann, K., Lohmann-Matthes, M. L., and Roesler, J. (1994) Evaluation of flow cytometric methods for the diagnosis of chronic granulomatous disease variants under routine laboratory conditions. *Cytometry* **18,** 147–155.

30. Alvarez-Larran, A., Toll, T., Rives, S., and Estella, J. (2005) Assessment of neutrophil activation in whole blood by flow cytometry. *Clin. Lab. Haem.* **27,** 41–46.

31. Pou, S., Rosen, G. M., Bntigan, B. E., and Cohen, M. S. (1989) Intracellular spintrapping of oxygen centered radicals generated by human neutrophils. *Biochim. Biophys. Acta* **991,** 459–464.

32. Roubaud, V., Sankarapandi, S., Kuppusamy, P., Tordo, P., and Zweier, J. L. (1997) Quantitative measurement of superoxide generation using the spin trap 5-(diethoxyphosphoryl)-5-methyl-1-pyrroline *N*-oxide. *Anal. Biochem.* **247,** 404–411.

33. Huu, T. P., Dumez, Y., Marquetty, C., Durandy, A., Boue, J., and Hakim, J. (1987) Prenatal diagnosis of chronic granulomatous disease (CGD) in four high risk male fetuses. *Prenat. Diagn.* **7,** 253–260.

34. Newburger, P. E., Cohen, H. J., Rothchild, S. B., Hobbins, J. C., Malawista, S. E., and Mahoney, M. J. (1979) Prenatal diagnosis of chronic granulomatous disease. *N. Engl. J. Med.* **300,** 178–181.

35. Matthay, K. K., Golbus, M. S., Wara, D. W., and Mentzer, W. C. (1984) Prenatal diagnosis of chronic granulomatous disease. *Am. J. Med. Genet.* **17,** 731–739.

36. Anderson, D. C., Schmalsteig, F. C., Finegold, M. J., et al. (1985) The severe and moderate phenotypes of heritable Mac-1, LFA-1 deficiency: their quantitative definition and relation to leukocyte dysfunction and clinical features. *J. Infect. Dis.* **152,** 669–689.

37. Tan, S. M., Hyland, R. H., Al-shamkhani, A., Douglass, W. A., Shaw, J. M., and Law, S. K. (2000) Effect of integrin beta 2 subunit truncations on LFA-1 (CD11a/CD18) and Mac-1 (CD11b/CD18) assembly, surface expression, and function. *J. Immunol.* **165,** 2574–2581.

38. Verheugt, F. W., Von dem Borne, A. E., Decary, F., and Engelfriet, C. P. (1977) The detection of granulocyte alloantibodies with an indirect immunofluorescence test. *Br. J. Haematol.* **36,** 533–544.
39. Curtis, B. R., Reno, C., and Aster, R. H. (2005) Neonatal alloimmune neutropenia attributed to maternal immunoglobulin G antibodies against the neutrophil alloantigen HNA-1c (SH): a report of five cases. *Transfusion* **45,** 1308–1313.
40. Wikman, A., Olsson, I., Shanwellt, A., and Lundahl, J. (2001) Detection by flow cytometry of antibodies against surface and intracellular granulocyte antigens. *Scand. J. Clin. Lab. Invest.* **61,** 307–316.
41. Holland, S. M. (2006) Neutropenia and neutrophil defects, in *Manual of Molecular and Clinical Laboratory Immunology.* 7th Edition. (Detrick, B., Hamilton, R. G., and Folds, J. D., eds.), ASM, Washington, DC, pp. 924–932.
42. Lindlöf, M., Kere, J., Ristola, M., et al. (1987) Prenatal diagnosis of X-linked granulomatous disease using restriction fragment length polymorphism analysis. *Genomics* **1,** 87–92.
43. Heyworth, P. G. and Curnutte, J. T. (2006) Molecular Diagnosis of Chronic Granulomatous Disease, in *Manual of Molecular and Clinical Laboratory Immunology.* 7th Edition. (Detrick, B., Hamilton, R. G., and Folds, J. D., eds.), ASM, Washington, DC, pp. 262–271.
44. Chien, S. C., Lee, C. N., Hung, C. C., Tsao, P. N., Su, Y. N., and Hsieh, F. J. (2003) Rapid prenatal diagnosis of X-linked chronic granulomatous disease using a denaturing high performance liquid chromatography (DHPLC) system. *Prenat. Diagn.* **23,** 1092–1096.
45. Introne, W., Boissy, R. E., and Gahl, W. A. (1999) Clinical, molecular, and cell biological aspects of Chediak-Higashi syndrome. *Mol. Genet. Metab.* **68,** 283–303.
46. Tamura, A., Agematsu, K., Mori, T., et al. (1994) A marked decrease in defensin mRNA in the only case of congenital neutrophil-specific granule deficiency reported in Japan. *Int. J. Hematol.* **59,** 137–142.
47. Zen, K., Reaves, T. A., Soto, I., and Liu, Y. (2006) Response to genistein: assaying the activation status and chemotaxis efficacy of isolated neutrophils. *J. Immunol. Meth.* **309,** 86–98.
48. Hanson, A. J. and Quinn, M. T. (2002) Effect of fibrin sealant composition on human neutrophil chemotaxis. *J. Biomed. Mater. Res.* **61,** 474–481.

32

Diagnostic Assays for Myeloperoxidase Deficiency

William M. Nauseef

Summary

Polymorphonuclear neutrophils (PMN) represent the dominant cell in the acute response to microbial infection. Effective antimicrobial activity reflects the combined action of soluble agents in plasma with PMN-derived reactive oxygen species and granule proteins, including the azurophilic granule protein myeloperoxidase (MPO). The inhibition or absence of the MPO-H_2O_2-halide system results in marked reduction in PMN killing of a variety of microbes, implicating its relative prominence in the hierarchy of PMN antimicrobial systems. Although the most profound clinical defects are manifested in patients lacking the capacity to generate reactive oxygen species, as seen in chronic granulomatous disease, an inherited deficiency of MPO can also increase the frequency or severity of clinical infections.

As a first step in evaluating the MPO-dependent system, determination of peroxidase activity of isolated leukocytes enriched for PMN coupled with histochemical staining of leukocytes allows assessment of functional MPO within PMN. Immunoblotting of the isolated leukocytes for MPO provides additional information, supplying insight into the molecular basis of the observed absence of functional enzyme.

Key Words: Peroxidase; myeloperoxidase; eosinophil peroxidase; azurophilic granules.

1. Introduction

As noted in Chapter 1, optimal microbicidal action of human polymorphonuclear neutrophils (PMN) relies on the integrated activity of reactive oxygen species generated by the NADPH oxidase and of an array of granule proteins delivered to the phagosome during phagosome–granule fusion. Granule contents include proteins that exert direct antimicrobial activity, such as bactericidal permeability-increasing protein, as well as proteases such as elastase and cathepsins, which indirectly contribute to the killing and degradation of ingested microbes. Restricted to PMN and monocytes is the azurophilic granule protein myeloperoxidase (MPO), a heme protein present in high concentrations in PMN

From: *Methods in Molecular Biology, vol. 412: Neutrophil Methods and Protocols*
Edited by: M. T. Quinn, F. R. DeLeo, and G. M. Bokoch © Humana Press Inc., Totowa, NJ

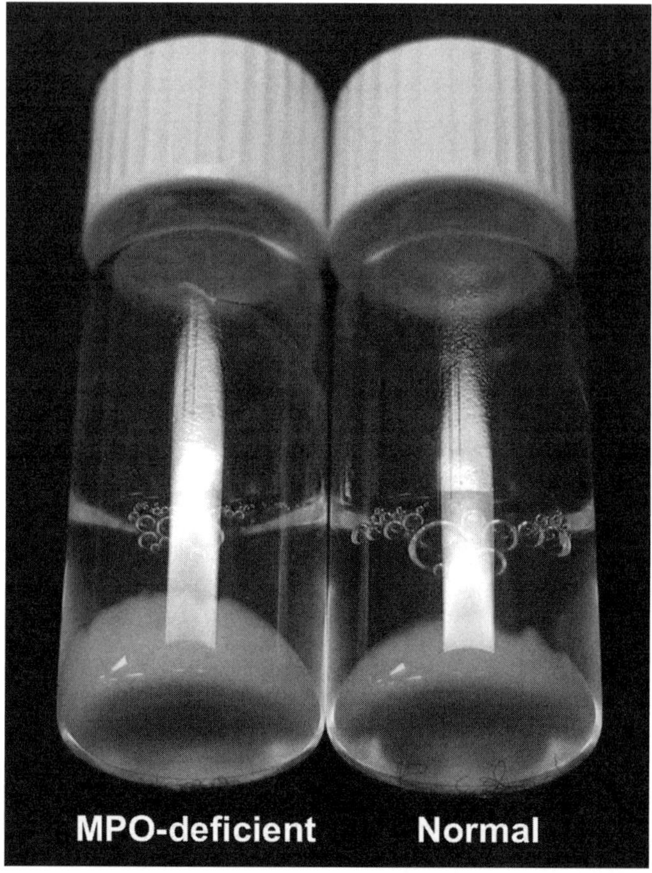

Fig. 1. Comparison of polymorphonuclear neutrophils (PMN) from PMN from healthy subjects (Normal) and those with myeloperoxidase (MPO) deficiency (MPO-deficient). Primary human PMN were isolated from peripheral blood using the protocol provided in Chapter 2. (color insert will appear after page 300)

and responsible for the yellow-green color of pus (**Fig. 1**). Although a member of the family of animal peroxidases, which includes eosinophil peroxidase, lactoperoxidase, and thyroid peroxidase, MPO has the unique capacity to oxidize Cl^- to Cl^+ at physiologic pH, thereby generating HOCl, a potent microbicidal agent.

Thus the MPO-H_2O_2-chloride system represents a major contributor to antimicrobial activity in the PMN phagosome *(1)* and defects in any component of the system compromises innate host defense. The profound morbidity and mortality associated with defects in H_2O_2 generation that define chronic granulomatous disease (CGD) have been described elsewhere in Chapter 31. In contrast, the phenotype of inherited MPO-deficiency is much less severe, with the most dramatic clinical syndrome being disseminated or visceral candidiasis in patients with concomitant diabetes mellitus *(2)*. The different morbidities associated with CGD and MPO deficiency may reflect the fact that H_2O_2 alone is microbicidal, although its activity is augmented approx 50-fold by MPO. That said, it is important to assess MPO activity in patients with unexplained infections with *Candida* species or with frequent infections consistent with PMN dysfunction (*see* Chapter 30).

Assessing MPO in PMN both enzymatically and immunochemically provides important complementary information. One can examine peroxidase activity as well as MPO-specific chlorinating activity, although the former is sufficient for the initial evaluation of an individual for MPO deficiency. The fundamental chemistry in all peroxidase (PX) assays is summarized in the following reactions:

$$PX + H_2O_2 \rightarrow \text{Compound I} \tag{1}$$

$$\text{Compound I} + AH \rightarrow A^+ + \text{Compound II} \tag{2}$$

where AH is a substrate that is chromogenic when oxidized. The substrate may be a histochemical dye (e.g., diaminobenzidine or benzidine dihydrochloride), thereby staining *in situ* the cellular compartment in which the peroxidase resides, or a soluble chromagen (e.g., *o*-dianisidine or guaiacol), thus providing a spectrophotometric means to quantitate activity. Immunoblotting of isolated leukocytes allows identification of the MPO protein, independent of its enzymatic activity, and thus makes it possible to identify missense mutations that compromise function but not the synthesis of the protein. Taken together, these three approaches provide a relatively comprehensive assessment of MPO deficiency, assessing the presence and subcellular location of MPO as well as its enzymatic activity.

2. Materials

2.1. Histochemical Staining of Peripheral Blood or Isolated Leukocyte Smears

1. Fixative: 10% formal-ethanol solution; prepare fresh each time. Mix 5 mL of 37% formaldehyde with 45 mL of absolute ethanol.
2. Staining solution: combine in the following order; 150 µg of benzidine dihydrochloride in 50 mL of 30% ethanol (*see* **Note 1**), 0.5 mL of 3.8% (w/v) $ZnSO_4$, 500 µg sodium acetate, 35 µL of 30% H_2O_2, 750 µL of 1.0 N NaOH, and 100 µg of safranin O (*see* **Note 2**). Filter sterilize, and store this solution at room temperature for up to 6 mo.
3. Counterstain: 1% aqueous cresyl violet acetate (500 µg/50 mL H_2O). Use if desired.

2.2. Spectrophotometric Quantitation of Peroxidase Activity in Lysed PMN Preparations

1. Lysates of isolated PMN: pellet the cells and resuspend in phosphate buffered saline containing 0.2% Triton X-100. Mix with frequent but careful pipetting to insure disruption of the azurophilic granules (*see* **Note 3**).
2. Solution A: 0.003% H_2O_2 in 0.1 M sodium phosphate buffer (pH 6.0). This must be made fresh daily and kept at room temperature.
3. Solution B: dissolve 5 mg o-dianisidine in 500 µL of methanol (*see* **Note 4**). Prepare solution B fresh daily and keep in a foil-covered tube on ice.

2.3. Immunoblotting for MPO-Related Proteins

1. Sodium dodecyl sulfate (SDS)-polyacrylamide gel electrophoresis (PAGE) sample buffer: 2.3% SDS in 62 mM Tris-HCl (pH 6.8) with 5% (v/v) β-mercaptoethanol and 5% glycerol.
2. Isolated cell pellets or cell lysates: solubilize in sample buffer for SDS-PAGE. Heat samples to 100°C for 3 min to denature and solubilze proteins. Cool before loading on the gel for electrophoresis.
3. Transfer buffer: 0.025 M Tris-base, 0.192 M glycine, 20 % methanol.

3. Methods

3.1. Histochemical Staining

1. Either a peripheral blood smear or a sample of isolated cells can be examined. In the latter case, dilute the PMN suspension to 5×10^5 PMN/mL in phosphate-buffered saline (PBS) or media, and centrifuge 100 µL at 500 rpm (Shandon Cytospin 2) to pellet PMN (5×10^4) onto the slide.
2. Flood the slide with fixative, and incubate at room temperature for 1 min.
3. Remove fixative by washing for 15–30 s with running tap water. Shake off excess water.
4. Immerse the slide in staining solution for 30 s.
5. Wash the slide in doubly-distilled water for 5–10 s.
6. Counterstain (optional) for 1 min and then rinse off with tap water.
7. Allow the slide to air-dry (*see* **Note 5**).

3.2. Spectrophotometric Quantitation of Peroxidase Activity in Lysed PMN Preparations

1. Set the spectrophotometer to read at 460 nm for 2 min.
2. Combine 2.965 mL of solution A with 25 µL of solution B in a 3 mL cuvette.
3. Add 10 µL of the sample and mix quickly.
4. Return cuvette to the sample compartment, and record absorbance at 460 nm (*see* **Note 6**).

3.3. Immunoblotting for MPO-Related Proteins

1. Separate proteins by SDS-PAGE, and immunoblot using standard conditions. Prepare transfer buffer and cool to 4°C before use. In the case of electroblotting for MPO-related proteins, transfer for 140–150 V-h (e.g., 70 V × 2 h).
2. Block and process using standard methods (*see* **Note 7**).

4. Notes

1. Benzidine dihydrochloride is a carcinogen and thus must be handled carefully.
2. A precipitate will form once the $ZnSO_4$ is added, but the subsequent addition of NaOH will clear the solution. The addition of $ZnSO_4$ stabilizes the oxidized benzidine dihydrochloride.
3. The peroxidase-containing granules must be disrupted either by solubilization of isolated cells in 0.2% Triton X-100 in Tris-buffered saline (TBS) (pH 7.5) or by sonication of the cells suspended in TBS or PBS.
4. Gentle warming and repeated vortexing are needed to dissolve *o*-dianisidine in methanol. Care in handling *o*-dianisidine is advised, as it is a potential carcinogen.
5. Peroxidases are normally packaged in intracellular granules in PMN, monocytes, and eosinophils. In PMN and monocytes, the MPO-containing azurophilic granules will appear blue, with faint blue staining of the surrounding cytoplasm. Although lymphocytes, basophils, erythrocytes, and platelets do not stain, the granules in eosinophils are intensely stained, reflecting the presence of eosinophil peroxidase (EPO). An important strength of including histochemical examination of peripheral blood or isolated leukocytes in assessment if an individual's MPO status rests on the ability to identify the cellular source of peroxidase activity detected by an enzymatic assay (e.g., *o*-dianisidine). In guinea pigs, eosinophils possess 10-fold more peroxidase than do the MPO-containing heterophils *(3)* and the specific peroxidase activity, expressed on the basis of cell number, is higher for eosinophils *(4)*. The same is true for humans, as human eosinophils contain 15 µg EPO/10^6 cells *(5)* (19.5 pmol/10^6 cells), whereas PMN possess approx 7.25 pmol/10^6 cells (based on spectroscopic determinations in our lab). Thus very little eosinophil contamination in a cell lysate can contribute significant peroxidase activity and obfuscate the detection of partial MPO deficiency as highlighted several years ago *(6)*. In addition, eosinophils have normal peroxidase activity in inherited MPO deficiency (**Fig. 2**).
6. One unit of activity in the *o*-dianisidine assay is the amount required to produce a change in absorbance at 460 nm of 0.001/ min at 25°C. International units are defined as the amount which utilizes 1 µmol substrate/min at 25°C. To convert *o*-dianisidine units to International Units, multiply the value by 2.655×10^{-4} (based on 1.13×10^4 $M^{-1}cm^{-1}$, the molar extinction coefficient of *o*-dianisidine).
7. Under reducing conditions, the 59-kDa and 13.5-kDa heavy and light subunits, respectively, are detected as separate immunoreactive bands. In addition, there may be immunoreactive proteins at approx 90 kDa and at 39 kDa. The former represents apoproMPO and proMPO, two precursors synthesized in the endoplasmic reticulum and destined for proteolytic processing into the subunits of mature MPO *(7)*. The 39-kDa species is product of autolytic cleavage of the heavy subunit *(8)*.

Normal PMN
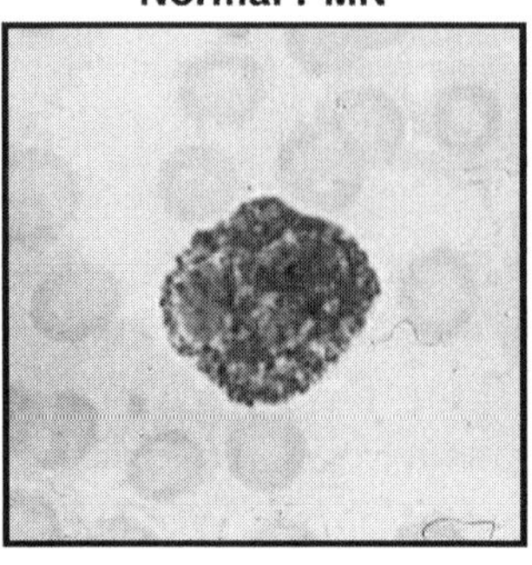
MPO-deficient PMN
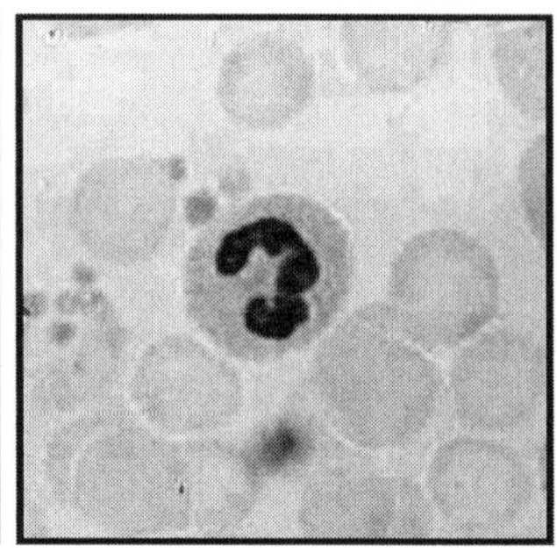
MPO-deficient eosinophil
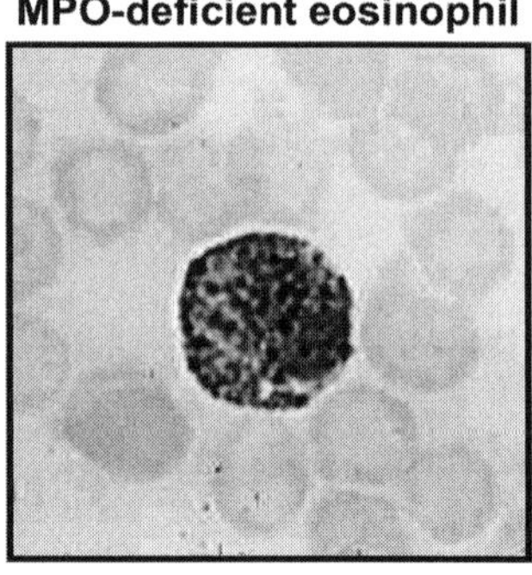

Fig. 2. Peroxidase activity in polymorphonuclear neutrophils (PMN) from a healthy subject (Normal PMN) or an individual with myeloperoxidase (MPO) deficiency (as indicated). Panel at far right indicates normal peroxidase activity in an eosinophil isolated from an individual with MPO deficiency. Histochemical staining was performed as described. (color insert will appear after page 300)

References

1. Klebanoff, S. J. (2005) Myeloperoxidase: friend and foe *J. Leukoc. Biol.* **77,** 598–625.
2. Dinauer, M. C., Nauseef, W. M., and Newburger, P. E. (2001) Inherited disorders in phagocyte killing, in *The Metabolic and Molecular Bases of Inherited Disease* (Scriver, C. R., Beaudet, A. L., Valle, D., et al. eds.), McGraw Hill Companies, pp. 4857–4837.
3. Mage, M. G., Evans, W. H., Himmelhoch, S. R., and McHugh, L. (1971) Immunochemical identification and quantification of guinea pig heterophil and eosinophil peroxidase. *J. Reticuloendothel. Soc.* **9,** 201–208.
4. Fabian, I. and Aronson, M. (1975) Deamination of histamine by peroxidase of neutrophils and eosinophils. *J. Reticuloendothel. Soc.* **17,** 141–145.
5. Carlson, M. G., Peterson, C. G., and Venge, P. (1985) Human eosinophil peroxidase purification and characterization. *J. Immunol.* **134,** 1875–1879.
6. Dri, P., Cramer, R., Soranzo, M. R., Comin, A., Miotti, V., and Patriarca, P. (1982) New approaches to the detection of myeloperoxidase deficiency. *Blood* **60,** 323–327.
7. Hansson, M., Olsson, I., and Nauseef, W. M. (2006) *Arch. Biochem. Biophys.* **445,** 214–224.
8. Taylor, K. L., Pohl, J., and Kinkade, J. M. Jr. (1992) Unique autolytic cleavage of human myeloperoxidase. Implications for the involvement of active site MET409. *J. Biol. Chem.* **267,** 25,282–25,288.

Index

Made in the United States of America